Mechanical Design and Manufacturing of Hydraulic Machinery

Hydraulic Machinery Book Series

HYDRAULIC MACHINERY BOOK SERIES

- **Hydraulic Machinery Systems**
  Editors: Prof D K Liu, Prof V Karelin
- **Hydraulic Design of Hydraulic Machinery**
  Editor: Prof H Radha Krishna
- **Mechanical Design and Manufacturing of Hydraulic Machinery**
  Editor: Prof Mei Z Y
- **Transient Phenomena of Hydraulic Machinery**
  Editors: Prof S Pejovic, Dr A P Boldy
- **Cavitation of Hydraulic Machinery**
  Editors: Prof Li S C, Prof H Murai
- **Erosion and Corrosion of Hydraulic Machinery**
  Editors: Prof Duan C G, Prof V Karelin
- **Vibration and Oscillation of Hydraulic Machinery**
  Editor: Prof H Ohashi
- **Testing of Hydraulic Machinery**
  Editor: Prof P Henry
- **Control of Hydraulic Machinery**
  Editor: Prof H Brekke

HYDRAULIC MACHINERY BOOK SERIES

Mechanical Design and Manufacturing of Hydraulic Machinery

Mei Zu-yan, Editor

LONDON AND NEW YORK

First published 1991 by Ashgate Publishing

Published 2016 by Routledge
2 Park Square, Milton Park, Abingdon, Oxon OX14 4RN
711 Third Avenue, New York, NY 10017, USA

Routledge is an imprint of the Taylor & Francis Group, an informa business

A CIP catalogue record for this book is available from the British Library and the US Library of Congress.

ISBN 13: 978-1-85628-820-0 (hbk)
ISBN 13: 978-1-138-26897-5 (pbk)

CONTENTS

Preface

The publishing of this book series on Hydraulic Machinery, organised and edited by the International Editorial Committee for Book Series on Hydraulic Machinery (IECBSHM), marks the results of successful cooperation between the Committee, the authors and the editors of each volume.

The Editorial Committee consists of 35 scholars from 20 countries. More than 100 academics and engineers from 23 countries have participated in the compilation of the book series. The volumes reflect the latest developments, gained from many countries, in concepts, techniques, and experiences related to specific areas of hydraulic machinery. This is a great joint exercise by so many experts on a world wide basis that will inevitably bring impetus to technical achievement and progress within the hydraulic machinery industry and promote understanding and cooperation among scholars and professional societies throughout the world.

The authors have devoted considerable time and energy to complete the manuscripts and even more time was occupied by the editors in revising and correlating the individual volumes. Dr A P Boldy, Secretary IECB-SHM, contributed greatly to the overall editing and preparation of the final manuscripts. Mr J G R Hindley, of Avebury Technical, offered many suggestions and remained helpful in every phase of the editorial and publishing work.

Great assistance in the preparation of the manuscripts from numerous persons in different parts of the world have been received, without which the publication of the book series would not have been possible. I would like to express appreciation both from myself and on behalf of the Editorial Committee to these people, the list of names being too lengthy to include here.

Duan Chang Guo
Chairman IECBSHM

Foreword of the Editor

This book is one of the nine volumes in the *Book Series on Hydraulic Machinery* organised by the International Editorial Committee. The subject for this volume is *Mechanical Design and Manufacturing of Hydraulic Machinery.* This is a topic that concerns the varied functions of design offices and workshops of all manufacturers. Experiences gained in these organisations are so numerous that they can hardly be outlined and compiled into one book. Besides, designs and manufacturing practices are so varied among firms that it is difficult to select any ones to be representative for description in this book. Therefore a large number of authors, fourteen from six countries, have been invited to contribute to the chapters of this volume, each contribution based on the author's extensive experience.

The question of reader level of this series of books was the major topic for discussion at the initial organising meeting. It was agreed that the book series would not primarily be used by college students as textbooks or by engineers as design handbooks. Rather it purports to convey to the profession the present status and latest developments in the hydraulic machinery field.

The structure of a hydraulic machine is evolved principally to satisfy the requirements of the fluid flow, therefore in a volume where the main interest is in mechanical design and manufacturing it is still necessary to constantly refer to the hydraulic characteristics of the machines. In many chapters of this volume, the authors very deftly illustrated the arrival of the best choice in mechanical construction from assessment of the hydraulic characteristics.

For the specific purpose of writing this book series, Prof. P. Henry of the Ecole polytechnique federal de Lausanne was entrusted to prepare a unified nomenclature of symbols for use in the various volumes. The nomenclature system proposed is based on the one that had been in use for many years within the Hydraulic Machinery and Fluid Dynamics Institute of the EPFL and by several Swiss manufacturers. Formulas and symbols in most parts of the manuscripts are expressed in conformity to the recommended system. but some drawings and curves were originally done in conventional symbols and are difficult to be converted.

The compilation of a volume on such a comprehensive subject within such a short time span and contributed by so many authors is altogether a new experience to all of us. perhaps more difficulties had been encountered by every author than if he is given more time, more space and also more freedom. For my part as editor of the volume, I shared with all the authors the labours of putting scattered information into one organised text and the expectation that these material be useful to our readers. I helped to revise and correlate many parts of the manuscripts and generally polished the English language. For all six chapters written by Chinese authors, I did the translation from Chinese into English. Through our joint efforts in nearly two year's time, this volume was finally completed in good form and in reasonably good time.

Contributions to this volume fall in two main categories, namely, hydraulic turbines (Chapter 1 through 11) and pumps (Chapters 11 through 16). Chapters 1 and 11 present general pictures on mechanical design of turbines and pumps respectively by giving specific examples of typical construction of each kind of machine.

Chapter 2 concerns the mechanical design of Francis turbines. The text on this most widely used type of turbine is arranged in a general form as the Francis turbine is the first major type of machine presented in this volume. The chapter covers general construction features of the turbine, determination of dimensions of major components and the sequence of assembly and dismantling of the machine.

Chapter 3 deals with axial-flow and diagonal-flow turbines, with particular interest in the Kaplan and Deriaz turbines. A major feature of these machines is the blade-adjusting machanism that allows runner blade angles to be changed during operation. Design considerations of the mechanism, and examples of stress calculation of its component parts are given.

Chapter 4 explains the design feature of tubular turbines that are widely used in low-head applications. Although the runner of a tubular turbine is similar to that of a Kaplan turbine, the overall arrangement of the flow passage is quite different such as the placing of the generator inside of a bulb or a pier, the guide vanes that are set on a conical surface and the inlet and draft tubes. Description is also made on the straight-flow turbine which places the generator rotor on the rim of the turbine runner allowing a more unobstructed passage through the machine.

Impulse turbines that operate on a different principle of energy exchange is the subject for Chapter 5. Essential relations in hydraulic layout of the Pelton turbine nozzle and bucket are explained. This is followed by a detailed description of the mechanical construction of Pelton turbines of both the horizontal and vertical types. Problems of scale effects and cavitation related to impulse turbines are also discussed.

Chapter 6 concerns the pump-turbines used in pumped-storage power stations with emphasis on the reversible pump-turbines. These machines, based on more than 30 years of development, are capable of offering satisfactory performance in both modes of operation. However, vibration induced by interaction of hydraulic forces between the runner and guide vanes is a special problem that must be carefully analyzed in the design stage. Merits of the reversible machine as against the multi-stage reversible and the tandem pump-turbines are compared.

Valves and gates used in conjunction with hydraulic turbines are described in Chapter 7. A main point of this chapter is on the sealing capability of valve seals since they are vital to the successful operation of the turbines.

Chapter 8 deals with the general topic of structural design of all types of hydraulic turbines. The question of growth rate of fatigue crack and allowable size of defects is examined in detail. Methods of stress and strain analysis are then explained and followed by description of numerical methods now widely used as computer–aided design tools.

Manufacturing methods of hydraulic turbine parts are the subject for Chapter 9. Production processes for the major components of various types of turbines such as the runners, shafts, guide apparatus and embedded parts are dealt with in detail. Emphasis is placed on the manufacturing of large turbines as it represents the major progress made in turbine production in recent years.

Small turbines which operate on the same principles as large ones do involve special problems in machanical design. Chapter 10 describes the essential features of small reaction turbines (tubular, axial–flow and Francis) and small impulse turbines (cross–flow and inclined jet) as well as the selection procedures of these machines.

Chapter 12 discusses various aspects of mechanical design of centrifugal pumps with respect to conditions dictated by their hydraulic characteristics. Detailed descriptions are given on the construction of suction and discharge chambers, pump casing (particularly for multi–stage pumps) and the bearing, seals and balancing devices.

Chapter 13 is concerned with three types of pumps used in the lower head ranges, that is the axial–flow, mixed–flow and tubular pumps. As the performance of low–head pumps are more dependent on the structure of the pumping station, questions of pump installation and design of intake structures are treated in detail. Examples are also given for low heads pumps with concrete volutes.

Various forms of corrosion and erosion which constitute the major damage to metallic material are analyzed in Chapter 14. Since most cases of pump material failure are from incorrect material selection and not from material defect. a thorough understanding of the corrosion and erosion mechanism is essential to proper selection of materials for pumps.

Chapter 15 explains the casting and machining requirements of pump parts and the balancing procedures for the rotating components. Importance of quality control is discussed.

Chapter 16 deals with the problem of selection of pump drives and compares the merits of the many types of pump drives in use today.

Mei Zu-yan,
Editor

Acknowledgements

Appreciation must be expressed to the manufacturers who have supplied drawings and pertinent information to various chapters of this volume. Their names are given in alphabetical order

Ateliers Constructions Electriques de Charleroi, Belgium
Alsthom–Jeumont, France
Burgmann Dichtungswerke GmbH, Germany
Byron Jackson, USA
DMW Corporation, Ltd., Japan
Dongfang Electrical Machinery Works, China
Dresser–Pleuger GmbH, Germany
EBARA Corporation, Japan
Fuji Electric Company, Japan
Harbin Electrical Machinery Works, China
Hayward Tyler, Ltd., UK
Hitachi, Ltd. Japan
Hydroart SpA. Italy
Jeumont–Schneider. France
Klein, Schanzlin und Becker, Germany
Kvaerner Brug A/S, Norway
Leningrad Metal Works, USS
Mather & Platt, Ltd., UK
Mitsubishi Heavy Industries, Ltd., Japan
Neyrpic Ltd., France
Ossberger–Turbinenfabrik GmbH, Germany
Pacific Pumps, Dresser Industries, Inc., USA
Shanghai Pump Works, China
Stal–Laval Turbin AB, Sweden
Sulzer Brothers, Ltd., Winterthur, Switzerland
Sulzer Escher Wyss, Ltd., Zurich, Switzerland
Toshiba Corporation, Ltd., Japan
Vevey Engineering Works, Ltd., Switzerland
Voith Hydro, Inc., USA
Voith, J.M. GmbH, Germany
Voith, J.M. AGUE, St. Polten, Austria
Weir Pumps, Ltd., UK

Appreciation is due to Dr. A P Boldy of the University of Warwick who kindly proof read the manuscript for language uniformity and of the IECBSHM secretariate for correlating the word processing systems used by various authors. We are indebted to Mr. J G R Hindley of Avebury Technical for his close cooperation since the first inception of the book Series.

Appreciation is also due to Miss Bao Yan-ying who expertly completed all drawings for the chapters whose authors were not able to supply originals of illustrations, and to Misses Xia Qing-zi, Zhang Xiao-ying and the clerical staff from the International Research Centre on Hydraulic Machinery (Beijing) for their efforts in the word processing work for the manuscript.

Mei Zu-yan, Editor

Contributing Authors

MEI, Zu-yan, Professor,
Department of Hydraulic Engineering, Tsinghua University, Beijing, China 100084.

Born 1924, China. B.S. Worcester Polytechnic Institute 1949 M.S. Illinois Institute of Technology 1951, USA. Pump and compressor design work with Worthington Corporation, USA, 1950 – 54. Teaching and research at Tsinghua University since 1954: Hydraulic design of hydraulic turbines and pumps; reversible hydraulic machinery; cavitation and silt erosion; transient problems of hydroelectric power stations. Mem. CMES, IAHR.

Turton, R K, Senior Lecturer,
Department of Mechanical Engineering, Loughborough Univeristy of Technology, Leicestershire LE11 3TU, U.K.

Visiting Fellow, School of Mechanical Engineering, Cranfield Institute of Technology. At Loughborough, responsible for honours undergraduate in fluid dynamics and turbomachinery and active in research on pump inducers and pump gas–liquid behaviour. Design consultant to a number of companies on problems in pumping and fluid dynamics. Author of *Principles of Turbomachinery* published by E & F N Spon Ltd. 1984 and *Principles of Centrifugal Pump Design* to be published by Cambridge Union Press Publishers. Mem. IAHR.

Vogt-Svendsen, C, Manager,
Quality and Training, Hydro Power Sector,
Kvaerner Brug A/S, N-0135 Oslo, Norway.
Born 1956, Norway.

Graduated from University of Oslo 1976, M.S. degree from Norwegian Institute of Technology 1980. Joined Kvaerner Brug A/S (N 0135 Oslo, Norway) in 1981. Served in various capacities as project engineer in Development and Accomplishment in Training on several hydroelectric power projects, and has edited four manuals on Hydroelectric Power.

Fang Qing–Jiang, Senior Enigineer (Retired)
Wuhan Steam Turbine and Generator Works, Wuchang, Wuhan, China 430074. Born 1927, China.

Graduated from Xiamen University in Mechanical Engineering 1949. Devoted whole professional life to designing of hydraulic turbines and associated equipment. Formerly connected with Harbin Electrical Machinery Works, Harbin Research Institute of Large Electrical Machinery and is now with Wuhan Steam Turbines and Generator Works as Senior Engineer. Participated in design and manufacture of equipment for over 40 domestic hydroelectric power projects ranging from large plants to small installations.

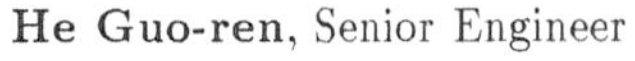

He Guo-ren, Senior Engineer
Tienjin Design and Research Institute of Electric Drive (TRIED), Hedong District, Tienjin, China 300180.

Graduated from the now Huazhong Univeristy of Science and Technology 1963 and has been working at Tienjin Design and Research Institute of Electric Drive (TRIED) on hydraulic turbine design and research since. In charge of development of bulb turbines for Jiangxia tidal power station, large bulb turbines for Baigou hyrdro power station and serial design of S–type tubular turbines.

Borciani, G, Senior Engineer (Retired),
Corso Garibaldi 32, 42100 Reggio Emilia, Italy. Born 1920 Italy, D. Eng. degree in Electrical Engineering from Politecnico di Milano, 1948.

Initially connected with Ufficio Impianti Idroeletrrici of Ansaldo and Franco Tosi doing research and design work on hydraulic equipment for various hydroelectric power projects. Served as Vice General Manager and Manager of Research and Development responsible for laboratories commissioning and acceptance tests at merger of technical staff of Riva Calzoni into Hydroart 1974. Very active in work with IEC (TC4 Hydraulic Turbines) and IAHR (Section on Hydraulic Machinery and Caviatation). Appointed professor at Universities of Bologna and of Florence.

Tanaka, H, Chief Engineer Hydraulic Machinery, Keihin Product Operations
Toshiba Corporation, Yokohama, Japan. Born 1933, Japan.

Graduated from University of Tokyo in Mechanical Engineering 1956, has been serving with Toshiba Corporation since. In first 20 years of service worked in research and development of both conventional turbines and reversible pump-turbines. Became Chief Engineer in charge of hydraulic machinery 1987 supervising all R & D, design and manufacturing of hydraulic turbines at Toshiba.

Zheng Ben–ying, Senior Engineer
Dongfang Electrical Machinery Works, Deyang City, Sichuan, China 618000. Born 1934, China.

Graduated from Shanghai Inst. of Naval Architecture 1954. First worked for Harbin Electrical Machinery Works and since 1966 for Dongfang Electrical Machinery Works. Specializes in welding technology and had supervised welding of turbines for Xinanjiang, Danjian Gezhouba and Longyangxia projects. Visiting engineer to

Westinghouse Corporation, USA in 1983, participated in inspection tours to large turbine factories and installations in USA., Canada, Brazil and Venezuela. Senior engineer 1983. Managing Director 1986. Has written widely in technical journals. Member of Executive Council, Welding Institution, CMES.

Lu Chu–xun, Senior Engineer
Tienjin Design and Research Institute of Electric Drive (TRIED). Hedong District. Tienjin, China 300180.

Born 1937, China. Graduated from Shanghai Institute of Power Machine Manufacture 1956 and started work at Tienjin Design and Research Institute of Electric Drive (TRIED). Engaged in design and research of medium/small turbines for many years and is now Deputy Director of Hydraulic Turbine Section of TRIED. Editor of *Atlas of Hydraulic Turbine Construction* 1978.

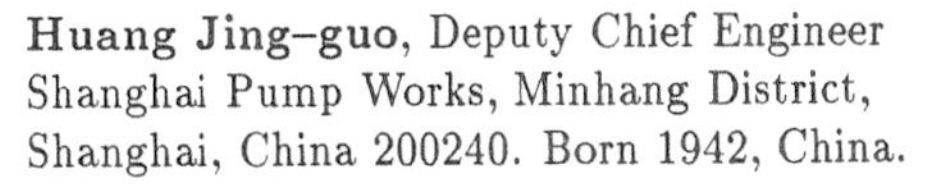

Huang Jing–guo, Deputy Chief Engineer
Shanghai Pump Works, Minhang District, Shanghai, China 200240. Born 1942, China.

Graduated from Huazhong University of Science and Technology, Department of Power 1964 and has since served with Shanghai Pump Works, first in research and development of various kind of pumps and from 1974 specialized in pumps for thermal power stations, and then nuclear power stations. Assumed a number of responsible positions, became Deputy Chief Engineer 1985 and holds title of Senior Engineer.

Matsumura, M, Senior Managing Director
DMW Corporation, Mishima City, Shizuoka 411, Japan.

Graduated from Kumamoto Univ. 1954 and has since devoted more than 35 years to design and development of pumps with Dengyosha Machine Works, Ltd. (now DMW Corp.). Acquired license of Consultant Engineer 1961 and became

Technical Managing Director at DMW 1980 and Senior Managing Director 1987. Author of a number of books and technical papers and holders of many patents on pump design.

Canavelis, R, Technical Director
and Managing Director
Bergeron–Rateau Company,
94132 Fontenay–sous–bois, Cedex, France.

Received engineering diploma 1962, worked as Research Engineer in Turbomachinery Department, of Electricite de France Directions des Etudes et Recherches. Doctorate degree 1967. Associated with Bergeron S.A. 1968 and worked on design and testing of pumps and other problems related to pump application. Head of Hydraulic Dept. and Vice Manager 1978. Became Technical Director and Managing Director of Bergeron–Rateau 1988.

Xu, Yi–hao, Deputy Chief Engineer
Shanghai Pump Works, Minhang District, Shanghai, China 299240.

Graduated from Zhenjiang Univ. in Mech. Eng. 1953, worked in design and development of centrifugal pumps at Shenyang Pump Works and Shenyang Pump Research Institute, becoming Director of Design Dept. and then of Standards Dept. Associated with Shanghai Pump Works since 1985.

Morita, H, Senior Chief Engineer
Systems and Equipments Group, Tsuchiura Works, Hitachi, Ltd., Japan.

Previously served as Manager of Turbomachinery Design Dept. of Tsuchiura Works of Hitachi, Ltd., (Chiyoda-ku, Tokyo 101, Japan) and is now Senior Chief Engineer of Systems and Equipments Group of the firm.

Salisbury, A G, Technical Director
Girdlestone Pumps, Woodbridge, Suffolk, IP12 1ER, U.K.

At present Technical Director of Girdlestone Pumps (Woodbridge, Suffolk IP12 1ER, England).

Chapter 1

General Concepts of Hydraulic Turbine Construction

R.K. Turton

1.1 Hydroelectric Generating Unit Layouts

Hydroelectric stations may be classified into high, medium and low head categories, so that it is possible to class the types of machines involved by referring to head and flow as shown in Figure 1.1. Figure 1.2 illustrates the types of runner used: from the impulse (Pelton) turbine for high head; the Francis turbine for medium/low head; and the axial-flow (Kaplan) turbine for low head installations.

High head and most medium head units are driven by water accumulated in dams through long tunnels and penstocks, and some medium and low head machines are placed in dam power houses or run-of-river stations which form part of barrages in rivers and estuaries. A further way of classification is by power output, with generating units of large outputs of $100MW$ or more, medium-size units of 25 to $75MW$, and small or micro hydroelectric stations of limited outputs.

1.2 Typical Construction of Turbines

1.2.1 Impulse turbines

These machines may be of either vertical or horizontal shaft design, with the pressure head converted into high velocity jets by specially designed nozzles. The jets impinge on a runner provided with buckets as shown in Figure 1.3. The head of a recent installation, for example, is $1200m$ and the generated

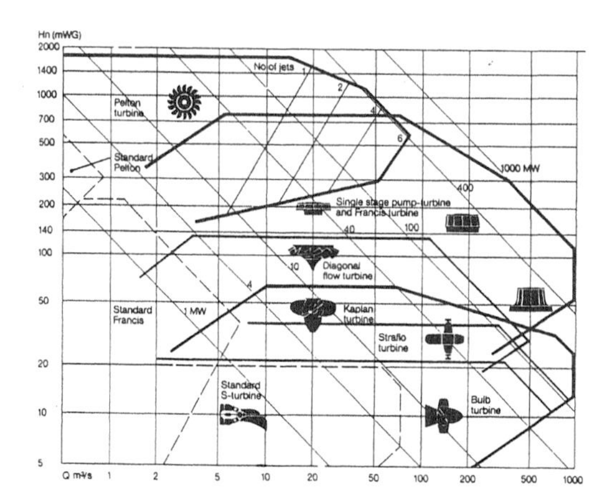

Figures 1.1 Typical ranges of application for turbines (Drawing courtesy Sulzer Escher Wyss)

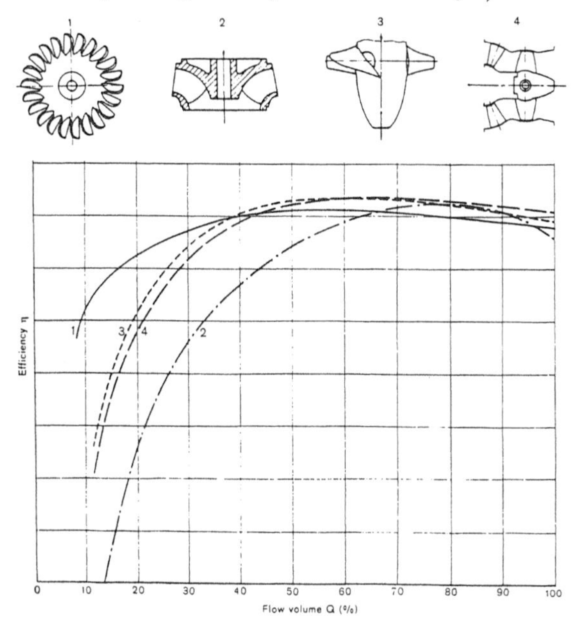

Figure 1.2 Comparison of efficiency characteristics of different types of turbines (Drawing courtesy Sulzer Escher Wyss)

output is $260MW$, which is typical of the latest high specific output schemes.

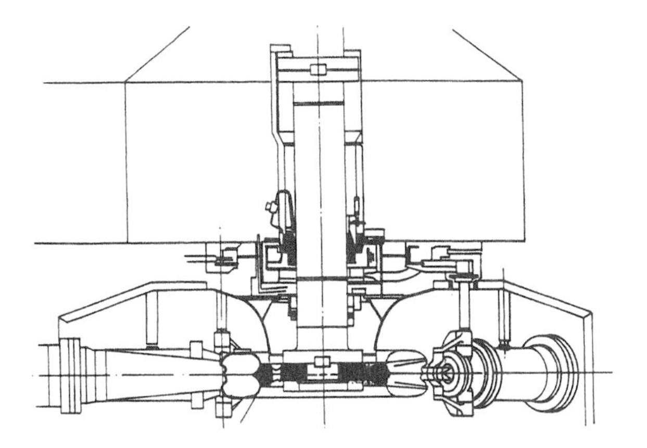

Figure 1.3 Typical vertical shaft multi-jet Pelton turbine (Drawing courtesy Sulzer Escher Wyss)

Small Pelton turbines may be installed where there is enough head, and units of $100kW$ or more are in operation. Many hydroelectric power stations use small Pelton turbines for station stand-by power.

1.2.2 Francis turbines

Figure 1.4 illustrates a large Francis turbine layout, the spiral case directs water through guide vanes (wicket gates) into the runner and out through the draft tube to the tail water. The turbine shown is a medium head machine that generates $715MW$ under $113m$ head. A Francis turbine is a mixed-flow machine, the characteristic shape of the runner varies with head and specific speed, as shown in Figure 1.5.

1.2.3 Axial-flow turbines

An axial-flow turbine may have fixed runner blades or variable-pitch blades, the former is commonly called a propeller turbine and the latter a Kaplan turbine. Figure 1.6 shows a typical Kaplan turbine layout. This machine with an output of $103MW$ under a head of $24.4m$ has a runner diameter of $8.4m$. This turbine is typical of low head axial-flow machines and has a

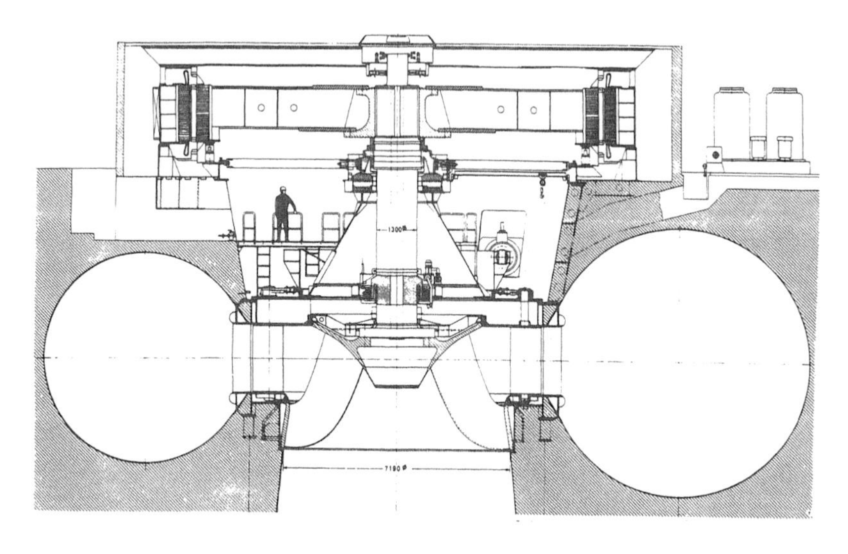

Figure 1.4 Large capacity medium head Francis turbine
(Drawing courtesy J.M. Voith)

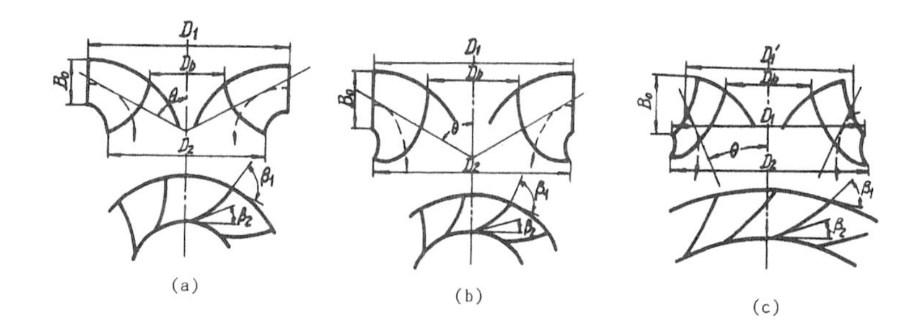

(a) Low specific speed (b) Medium specific speed (c) High specific speed

Figure 1.5 Characteristic profile of Francis runners

spiral case of near rectangular shape and stay vanes made up of separate columns.

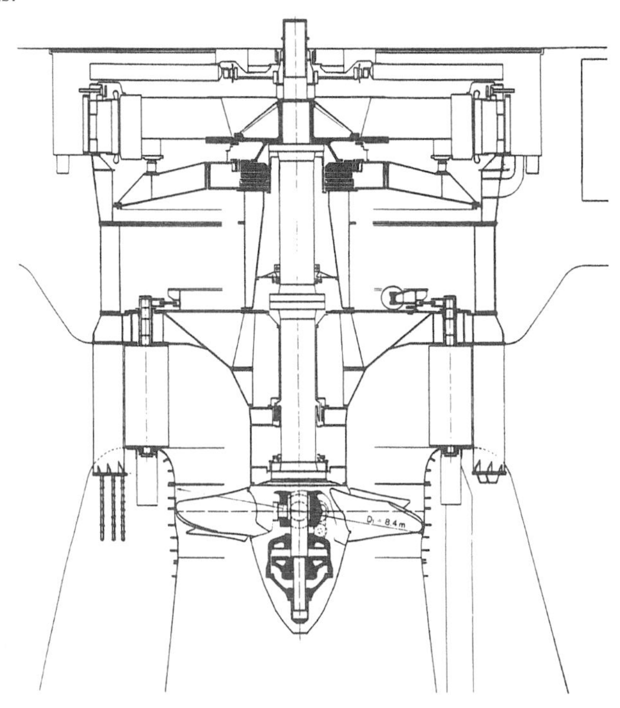

Figure 1.6 Typical Kaplan turbine with rectangular spiral case made of concrete (Drawing courtesy Escher Wyss)

A variation used in small, river barrage and estuary installations is the bulb turbine. Here water flows around the generator bulb, through the guide vanes which are set in a mixed-flow annulus section and into the axial-flow runner. The entire unit is effectively tubular in shape as there is no spiral case, hence it is also called a tubular turbine. The bulb turbine shown in Figure 1.7 has an output of $25MW$ under a $7m$ head, a runner diameter of $7.7m$ and a rotational speed of $62min^{-1}$.

1.2.4 Pump-turbines

Early pumped storage power stations used a *four machine* layout, with a turbine and generator as one set and a pump and motor as a second set.

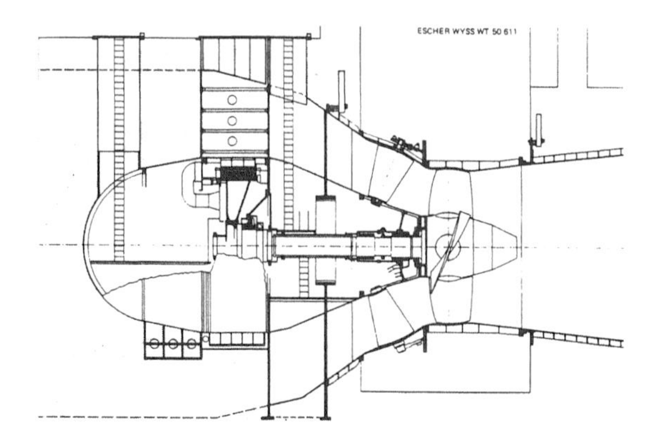

Figure 1.7 Cross-section of a large bulb turbine
(Drawing courtesy Sulzer Escher Wyss)

The turbine generates at high grid demand times, and the pump uses surplus power from the grid at non-peak demand times, to refill the supply reservoir for the turbine. Later the two electrical machines were combined into one to form a generator-motor with the turbine and pump connected one on each end of it. The aggregate is called a tandem set, or a *three machine* layout.

Present-day pumped storage power stations use a reversible pump-turbine which rotates in one direction as a pump and in the other direction as a turbine. A pump-turbine closely resembles a Francis turbine, as shown in Figure 1.8, but all components of the flow passage are designed to work in both directions of flow.

The most recent machines typically deliver over $100MW$ power with a head of 400-500m.

1.2.5 Micro hydroelectric units

Hydraulically the micro units may be Francis, axial-flow or bulb turbines, and are often laid out with the shaft either horizontal or inclined, the latter is used with the intent of saving civil engineering work. Typical arrangement of a micro hydroelectric unit is shown in Figure 1.9, where output may be 100-700kW with heads of 3-5m.

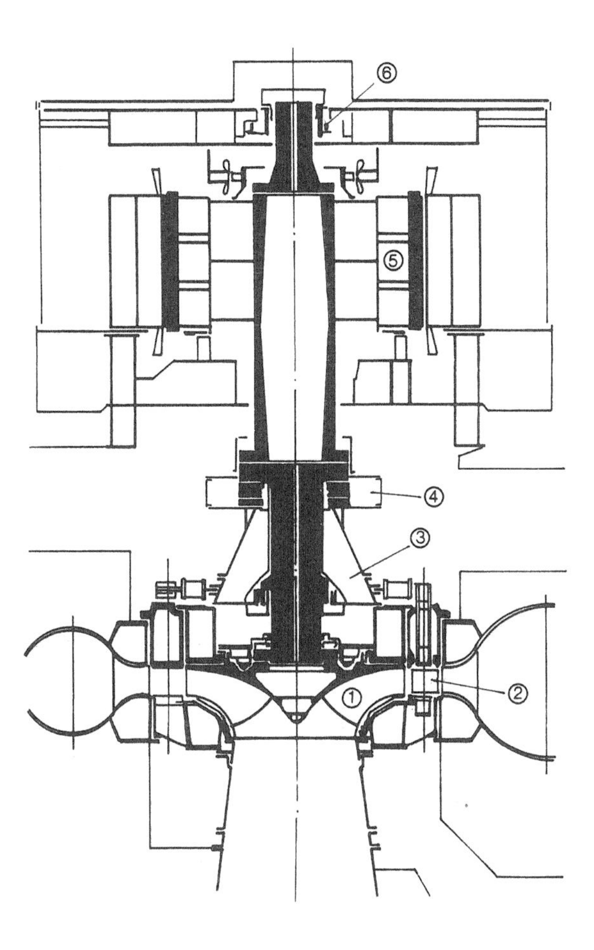

1: runner
2: guide vanes
3: lower guide bearing
4: thrust bearing
5: generator-motor
6: upper guide bearing

Figure 1.8 Typical Francis pump-turbine layout
(Drawing courtesy J.M. Voith)

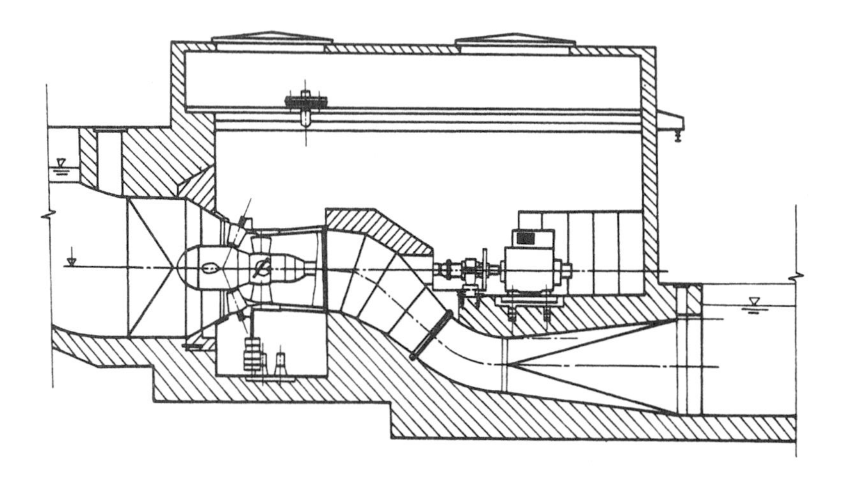

Figure 1.9 Layout of a low head micro hydro power station
(Drawing courtesy Sulzer Escher Wyss)

Chapter 2

Construction of Francis Turbines

C. Vogt-Svendsen

2.1 General

Hydraulic turbines may be classified into two main groups, namely, partial-admission turbines and full-admission turbines. A full-admission turbine is distinguished by the fact that water in the turbine is under pressure and fills the flow channels, including the runner, completely. The turbine power output is varied by changing the water flow through the runner, and is done by adjusting the exit angle of the guide vanes.

In a partial-admission turbine only part of the runner is in contact with water at any time. A typical example of the partial-admission turbine is a Pelton turbine where the water jet hits the runner buckets in turn. This type of turbine is also called a constant pressure turbine since the pressure energy is all converted into velocity in the nozzle and the driving of the bucket by the water jet is all done at atmospheric pressure.

The relative flow velocity through the runner of a partial-admission turbine is higher than that in a full-admission turbine but may vary in direction and magnitude with the change in power setting, which enables it to have higher efficiency with larger variation in output. Consequently, the full-admission turbine cannot be operated with varying power output with the same economy as the partial-admission turbine. The efficiency scale effect related to the size of the turbine is more pronounced for full-admission turbines than for partial-admission turbines. However, full-admission turbines have been in great demand in all phases of the development of hydraulic turbines. Various regulating methods have been devised for full-admission turbines. In 1849 the American engineer James Francis designed the type of full-admission turbine that bore his name, although the regulating system with adjustable guide vanes that always go with a Francis turbine was not developed until 30 years later.

An example of this regulating system is shown in Figure 2.1. The flow area at the exit of the guide vanes is varied by rotating the guide vanes which are controlled by a common regulating ring.

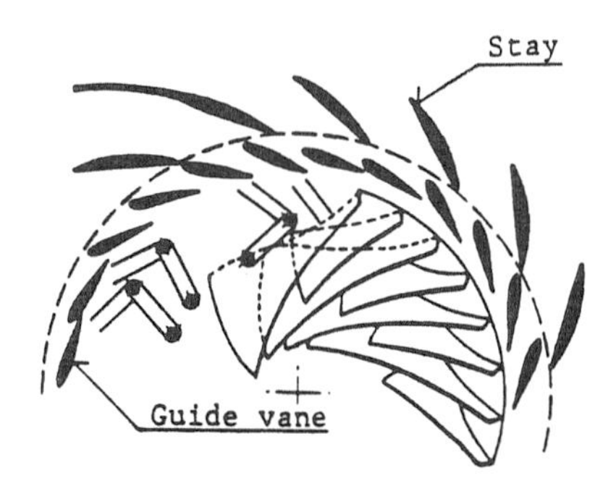

Figure 2.1 Regulating system for Francis turbine

Francis turbines may be built with horizontal or vertical shaft arrangements according to the demand of the power house and/or the practice of the manufacturer. Turbines with small dimensions are almost always horizontal while the largest turbines are almost all vertical. Turbines in the intermediate range are designed from various structural and operating considerations and no definite criterion exists. Horizontal and vertical turbines are the same for most parts but certain details may be significantly different.

2.2 Description of Francis Turbine Construction

Figure 2.2 shows a high head vertical Francis turbine in isometric projection along with designations of the various parts of the turbine. Figure 2.3 shows the details of a medium-head Francis turbine in a cut out drawing.

The mode of operation may be described as follows: water flows from the conduit into the spiral case and is distributed around the complete periphery. Water is then guided by the stay vanes and the guide vanes in the proper direction for entering the runner. The adjustable guide vanes are operated by the governor servomotor and are set in accordance with the load output requirement. After transferring its energy to the runner, water proceeds to the draft tube and departs through the outlet channels. Power developed in the runner is transmitted to the generator which is directly connected to the turbine shaft.

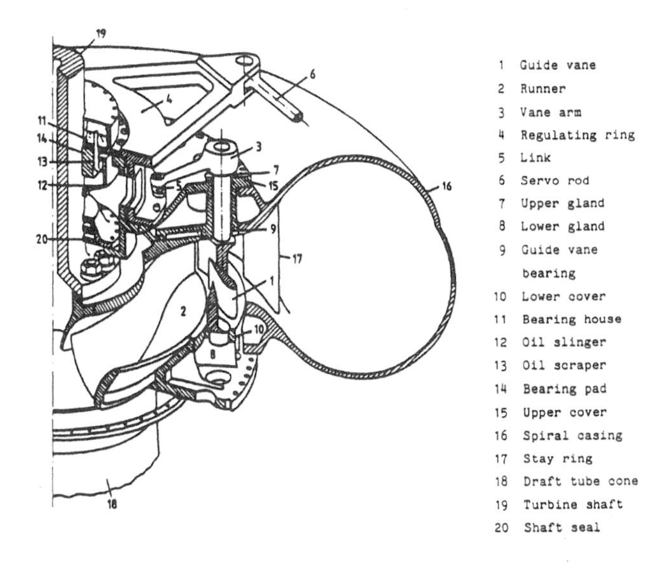

Figure 2.2 Construction of high head Francis turbine

Figure 2.3 Construction of medium head Francis turbine

The Francis turbine runner design will vary with the operating head the same way as other turbine types for respective head ranges.

Typical runner passage shapes for high head, medium head and low head runners have been given in Figure 1.5, overall views of a high head runner and a low head runner are shown here in Figures 2.4 and 2.5 respectively.

Figure 2.4 High head runner

Figure 2.5 Low head runner

2.3 Development of Francis Turbines

In countries where the terrain is mountainous, most hydroelectric power plants have machinery installed for working under high heads, that is within the range of 200 to 1200m. In this head range both the high head Francis turbines and Pelton turbines may be employed, although a practical upper limit for the Francis turbine is about 700m.

In other countries the hydroelectric power development started with low head projects where relatively low head Francis turbines are used. The general operating requirements and mechanical design conceptions for low head turbines may be quite different from high head turbines.

Construction features of Francis turbines are quite varied from different manufacture designs, such as runner blading based on different theoretical concepts; spiral cases calculated from different velocity distribution criteria; arrangements of stay vanes and guide vanes from different experiences; constructions of the stay ring (speed ring) with flared flanges or parallel plates; various characteristic shapes of draft tubes as well as the different ways of dismantling the main parts of the turbine. Efforts have been made all along to make turbines as compact as possible. By increasing the rotational speed and the water velocity it is possible to reduce the dimension of the turbine.

However, in view of the danger of cavitation, this necessitates a deeper submersion of the turbines which in turn brings about the increasing use of underground power stations. The submergence of a turbine is defined as the vertical distance from the tailwater level to the turbine centre (or more exactly, to the highest point where cavitation is likely to occur). The present trend in turbine development is to go to more compact turbine designs at the expense of greater submergence for better overall economy. Following the rapid advances in hydraulic development, mechanical construction of turbines has also been greatly improved in recent years. With the introduction of materials that allow higher stresses and offer stronger protection against cavitation and fatigue in the rotating components and other parts exposed to high water velocities, a reduction in turbine weight per kilowatt for the same head is realized.

In the early period cast iron was the most commonly used material in water turbine manufacture. In the 1920 to 30s cast iron was replaced by cast steel and riveted steel plate designs. Since the 1940s these constructions have been replaced to a large extent by welded structures following the development of fine grain steel and the improved weldability of plate materials. Table 2.1 shows the development of welded designs for Francis turbines in the 400 to 600m head range over the period of 1958 to 1976 for three examples of spiral cases. Spiral case III would have weighed 250 tonnes by 1958 technology. It should however be noted that in the pursuit of greater weight reduction with all-welded spiral cases, splitting of the stay ring should be avoided for structural considerations.

Table 2.1 Development in reduction of spiral case weight

	I	II	III
Design year	1958	1965	1976
Turbine output, MW	110	165	315
Weight, tonnes	100	100	100
Weight factor	1.0	1.46	2.58

Figure 2.6 shows a section through a high-head Francis turbine manufactured around 1960. All the major parts with the exception of the runner are made of cast steel. The runner has hot-pressed blades welded to the cast crown and band.

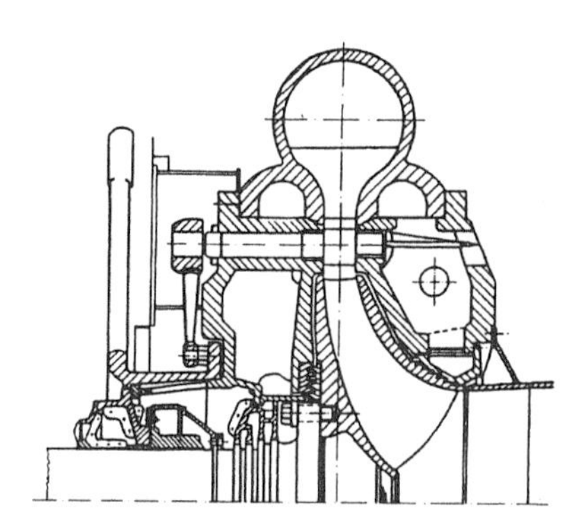

Figure 2.6 High head Francis turbine of cast steel construction

Parts made from castings in comparatively small quantities require more labour time in their manufacture such as the making of patterns, forming, and the removal of trim allowance. With the improved welding technology plus good weldability of fine grain steels large savings of labour can be attained by use of all-welded designs.

In recent years the advances in hydro power development has also led to an increase in the size of turbines. This has proved profitable because of the increased efficiency attainable with larger units, and also for the savings in civil works by building plants with fewer and larger units. There are of course practical limits to the size of the units such as by the capacity of machine tools and transportation limitations.

Steels with low carbon content and good weldability are now used in fabricated parts, which enable more welding to be done at site without the need for subsequent stress relieving.

Figure 2.7 shows the transporting of a large spiral case for a reversible pump-turbine by trailer. A part of the spiral case is cut off and will be rejoined by welding at site while the stay ring is retained as one piece to save the high cost of flange connections necessary for a split construction.

Figure 2.7 Transportation of spiral case/stay ring

The trend of development is now to apply Francis turbines to even higher heads and at present there are turbines operating in excess of $700m$ head. If the necessary suction head is provided, Francis turbines can be used for up to $800m$ without cavitation or wear problems if the water is clean.

When water contains quartz sand and feldspar, erosion may occur at heads exceeding $250m$, and for heads over $400m$ the erosion damage will be considerable and will lead to expensive and time consuming repairs. Special design solutions are required for employing machinery in these operating conditions, if dismantling time and maintenance work are to be kept within reasonable limits.

The pressure head limit for Francis turbines will be determined by the clearance between the guide vanes and the covers. Large clearances between the guide vanes and the covers will lead to large losses and decreased efficiency. In order to keep these clearances small, the covers must be designed with sufficient rigidity to keep deformation small which is true also for the guide vanes and stay vanes. This will however result in larger dimensions for the turbine parts and higher material costs. The highest profitable head for Francis turbines will probably be about 800 m under clean water conditions.

2.4 Main Components of Francis Turbines

The Francis turbine has the following main components:

Rotating parts
Turbine bearing
Shaft seal
Regulating mechanism
Spiral case
Draft tube

2.4.1 Rotating parts

The rotating parts of a turbine consist of the runner, turbine shaft and turbine bearing oil slinger, which are shown in Figure 2.8. The runner is the most critical part of the turbine since the pressure energy of the water is converted into mechanical energy in the runner. The number of blades of a runner depends on the operating head, runners of higher head will require a larger number of blades. The purpose of using more blades is first from strength consideration, and then for the purpose of lessening the pressure loading on the blades which will help to avoid cavitation, and also to prevent separation in the blade channels at the runner inlet under low loads. This however will increase the wetted surface of blades with subsequent higher frictional losses.

In certain designs, blades with reduced length which are called splitter blades are introduced between the full blades. These blades will effectively reduce separation at the inlet while creating no serious turbulence in the blades passages since the flow is converging there. This runner design has proved to be favourable with respect to efficiency, especially at part load conditions. In low head turbines, strength and rigidity must be taken into consideration when the number of blades is determined. For obtaining high efficiency the runner blades should be as thin as possible, but sufficient thickness must always be maintained to provide the required strength, rigidity and operational flexibility. The shape of the flow passage through the runner determines the hydraulic efficiency to a high degree so the utmost

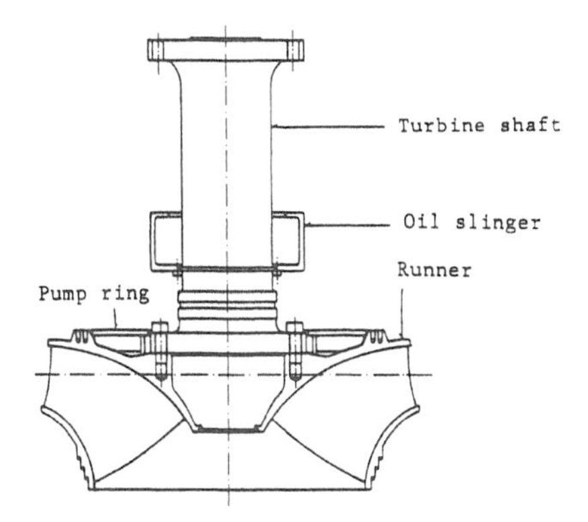

Figure 2.8 Rotating parts of a Francis turbine

importance should be attached to its design. Hydrodynamic analysis and laboratory tests combined with experience from site measurements are used to finalize the design.

The runner of a Francis turbine may be either cast in one piece or welded from hot-pressed plate or cast blades to the crown and band. For medium and high head applications the runner is manufactured completely of stainless steel.

The runner is provided with labyrinth seals working in close clearance against the upper (head) cover and lower cover (bottom ring). The water flowing through the labyrinth seals is not utilized by the runner and becomes a source of loss that is dependent on the condition of the seals. A new turbine has small seal clearances, but during operation the seals wear and the efficiency of the turbine will decrease with a growing rate of leakage. Higher silt content in the water will cause the seals to wear faster. Regular measurement of the leakage flow will give a good indication of the seal wear. Together with other factors, it is possible to assess the condition of a turbine and to consider possible repair work.

On high and medium head turbines the leakage water through the labyrinth seal may be utilized as cooling water, and will contribute to improvement of the overall plant efficiency. The upper side of the runner is sometimes fitted with a horizontal pump ring, welded on by means of radial ribs. In this way the shaft seal and eventual air inlet through the shaft flange are kept dry during operation.

Figure 2.9 shows a section through the turbine during operation and at standstill. The pressure just outside the pump ring will be equal to the head on the labyrinth outlet, where the leakage water will flow through a hole

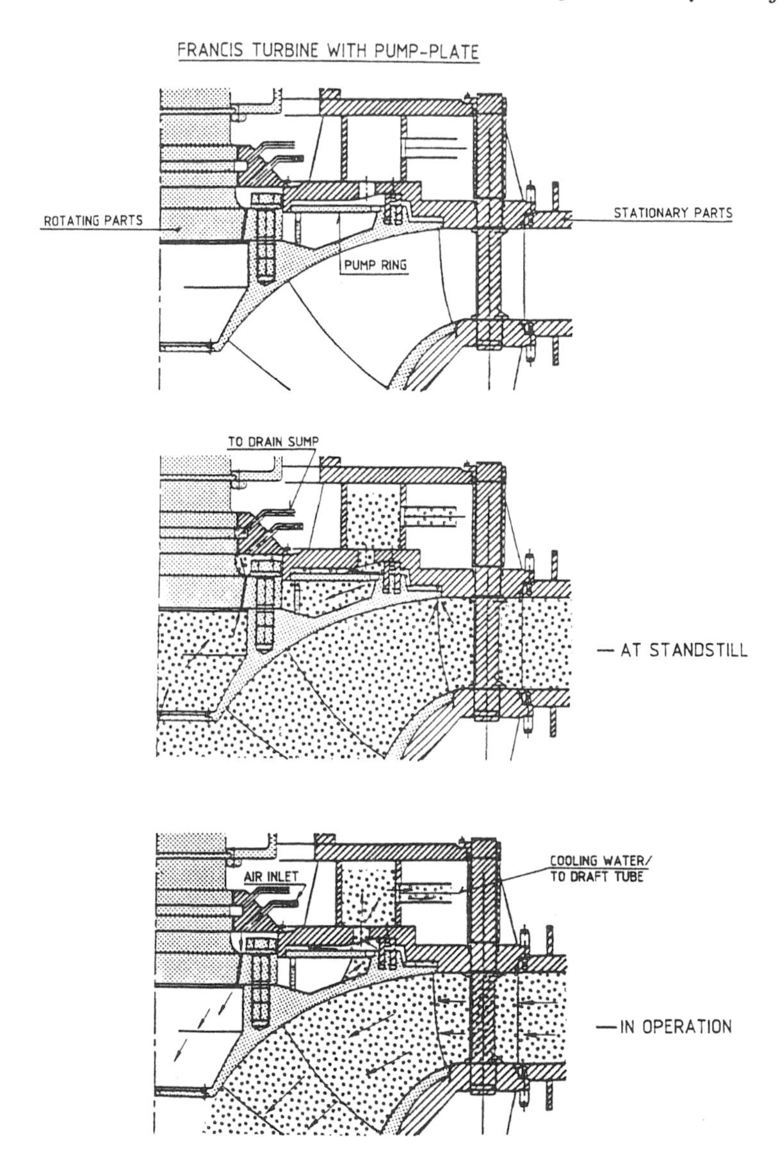

Figure 2.9 Function of the pump ring of a Francis turbine

in the upper cover and then by way of a pipe to the cooling water basin. Delivery head of the pump ring is given by:

$$H = \frac{1}{2g}\left(U_y^2 - U_i^2\right) \tag{2.1}$$

where U_y and U_i are peripheral velocities respectively at the outer and inner diameters.

The leakage water flow is found from experience to be 0.25 to 0.3 % of the full power discharge when the turbine is new. Low head turbines cannot be provided with a similar pumping device because in order to attain sufficient cooling water pressure the pump ring would have to be too large for the design. Here the sealing water must flow through balancing holes in the runner crown and directly to the draft tube.

The runner torque is transmitted to the turbine shaft through a bolted friction joint or a combined friction/shear joint. For large dimensions the bolts in this joint are prestressed by means of heating. The bolts are manufactured from high tensile strength steel and provided with a centre bore for measuring the elongation during prestressing.

The turbine shaft is made from SM-steel and has forged flanges in both ends. The turbine and generator shafts are connected by a flanged joint. This joint may be a ream-bolted or friction coupling where the torque is transmitted by shear or friction.

2.4.2 Turbine bearings

A simple and commonly used bearing design is shown in Figure 2.10. The bearing is of robust design and has a simple working principle that requires minimal maintenance. The bearing housing *1* is split into two halves and mounted to the upper flange of the upper cover. The bearing pad *3* consists of two segments bolted together and mounted to the underside of the housing. The pad has four oil pockets and four white metal bearing surfaces with correctly shaped leading ramps to ensure centering of the turbine shaft. The upper part of the housing has a split cover *5*, with inspection openings. A sleeve from the cover skirts around the shaft to keep the oil in the housing from rotating with the shaft. The bearing pad is surrounded by the oil slinger *2* which is split into two halves and bolted to the shaft and rotates with it. A riser *4* scoops the drained oil to the oil cooler.

Gravity force will exert on the shaft of a horizontal turbine as a bending load which is not present on the shaft of a vertical turbine. This must be taken into account in the design phase. The bearings on horizontal turbines will have an asymmetrical load due to gravity, that is all bearings on a horizontal turbine must be designed for asymmetrical radial forces. A startup

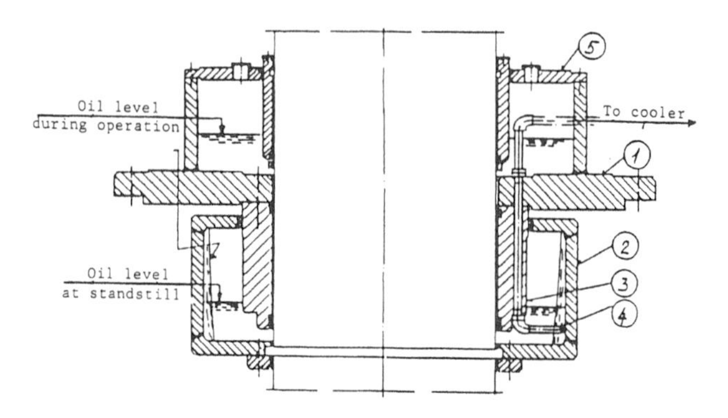

Figure 2.10 Turbine bearing

lubricating device is required in order to ensure lubrication at startup of the turbine.

2.4.3 Shaft seals

The shaft seal casing is usually split in halves and mounted to the upper cover as shown in Figure 2.11. The seal surfaces are made of white metal and have a small radial clearance of 0.2 to $0.4mm$ against the seal surface of the shaft that is often a stainless steel sleeve mounted on the shaft.

When the turbine is at standstill with the draft tube gate open, water from downstream will seep through the upper labyrinths and into the seal box. This leakage water must be removed by an overflow pipe and brought to the pump sump.

At certain turbine loads, unstable flow may develop downstream of the runner which in some cases can be stabilized by air injection. This air supply may come from the overflow pipe or through a separate air supply connected to the shaft seal box. In order to allow repair work on the shaft seal without having to empty the draft tube, an additional seal A may be installed as in Figure 2.11. This seal is a hollow rubber tubing inflated by compressed air or pressurized water to provide sealing against the shaft and is actuated only at standstill.

2.4.4 Regulating mechanism

The regulating mechanism consists of the guide vanes, upper and lower cover surfaces, vane arms and links as shown in Figures 2.2, 2.12 and 2.13.

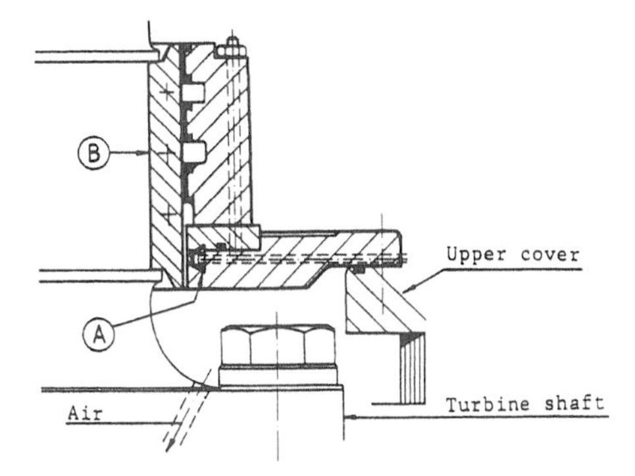

Figure 2.11 Shaft seal

The guide vanes are shaped in accordance with hydraulic design, and the surfaces are machined to a high degree of smoothness. The vanes, together with the trunnions on both ends, may be cast in steel or die-forged in one piece. The material used for guide vanes is mostly stainless steel. On low head turbines, however, the vanes may be welded from fine-grain carbon steel.

Guide vane bearings are fitted in the upper and lower covers. In turbines of recent design, plain bearings made of Teflon are used thereby eliminating the need of external lubrication. In some cases the bearings are mounted inside sleeves which are able to accommodate some misalignment in order to avoid excessive edge pressure on the bearings. The guide vane arms are fastened to the upper (longer) trunnion by means of a wedge or shear pins.

Signals from the turbine governor are transmitted to the servomotor and through a set of control rods to the regulating ring which then transfers the movement to the guide vanes through an arm/link construction. The vane arms are connected to the regulating ring through link bars that are fitted with non-lubricated spherical bearings on stainless steel trunnions both on the regulating ring and the vane arms. The guide vane trunnions are positioned in such a manner that the hydraulic forces acting on the guide vanes are balanced, in order to minimize the regulating force required.

Safety measures in mechanical design are always provided in the guide mechanism in case a guide vane is blocked by debris or other foreign objects. Some member in the linkage system is designed to break or move so the guide vane(s) will not be damaged by the blockage.

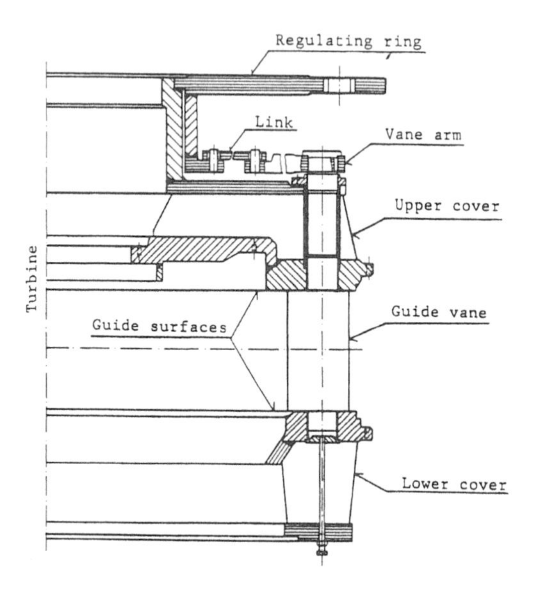

Figure 2.12 Regulating mechanism

Figure 2.13 Erection of guide vanes in workshop

Commonly used measures are:

1. Breaking links: Every second link is made with a slimmer cross section or a machined neck, see Figure 2.12, that will break under the increased tension caused by blockage.

2. Bending arms: The guide vane arms are the weakest points of the regulating mechanism and will be bent by abnormal forces acting either in the opening or in the closing direction due to blockage, see Figure 2.14. Normally every second arm is dimensioned for lower bending stress.

3. Friction coupling between the arms and guide vanes: The connection between the two parts is formed by a friction coupling which normally withstands the maximum operating force, but will slip when excessive force occurs, as shown in Figure 2.15. This solution is the best design from an operating point of view, but it is also the more expensive.

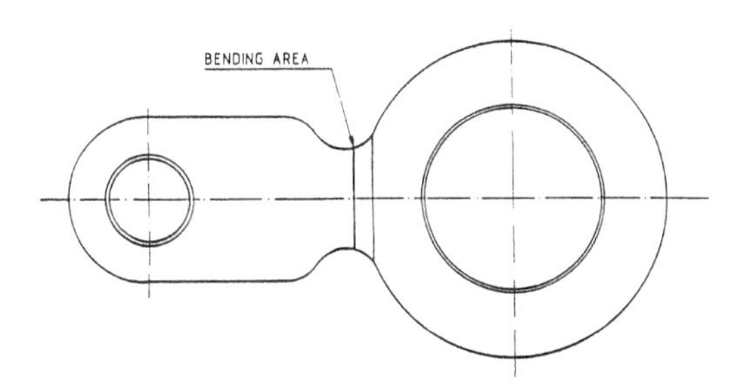

Figure 2.14 Bending arm of guide vanes

The upper and lower cover surfaces adjacent to the guide vanes are normally protected with high strength stainless steel which may be fastened to the covers by bolts or by clad welding.

The upper and lower covers are normally made of welded structure using fine-grain steel plates and are usually bolted to the stay ring. They are designed with sufficient rigidity so that the deformation caused by water pressure is kept to a minimum. This is necessary to guarantee the effective operation of the regulating mechanism since the guide vane trunnions are supported by the covers as bearings. In addition, the upper cover supports the regulating ring bearing, labyrinth ring, turbine bearing and shaft seal box, and alternatively an inner cover. The lower cover supports the lower labyrinth seal and the draft tube cone.

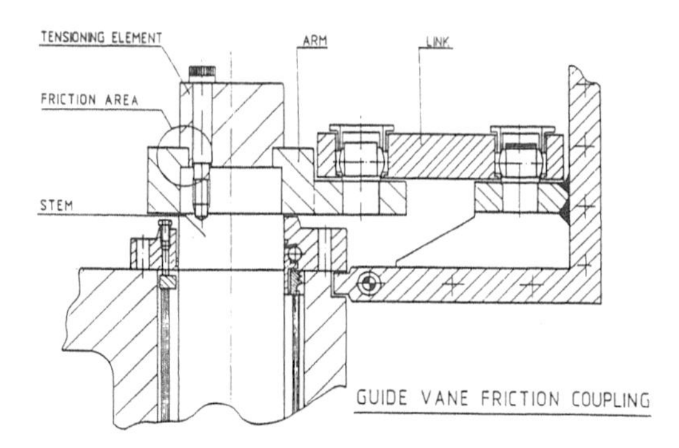

Figure 2.15 Friction coupling between guide vane and arm

2.4.5 Spiral case and stay ring

The spiral case forms the water conduit between the penstock and the regulating mechanism, its design must provide steady and uniform flow of water around the periphery of the turbine to the best practical extent.

The stay ring is the fundamental member of the turbine structure that supports the various covers on the inside and the spiral case on the outside. It is made up of an upper and a lower ring and a number of stay vanes. The vanes are designed to have good hydraulic profile for directing the water flow to the movable guide vanes with minimized losses.

The area where the spiral case joins the stay ring presents critical problems both in hydraulic and mechanical designs. The best construction that combines good hydraulic efficiency and sound structural concept is one main subject concerned by all designers.

Today, the spiral case and stay ring are normally all of fully welded construction from-fine grain steel plates. Vertical turbines have the spiral casing embedded in concrete for firm support and reduction of vibration. The spiral case is provide with flanged connections for pressure and index flow measurements, drain, vent and manholes.

2.4.6 Draft tube

The draft tube of a Francis turbine consists of a draft tube cone, a bend or elbow and a diffusing section in the case of surface power stations as shown in Figure 2.16.

The primary function of the draft tube is to recover the dynamic energy from the water flow at the runner exit. This is achieved by shaping the draft tube with increasing cross sectional area in the direction of flow, but in the

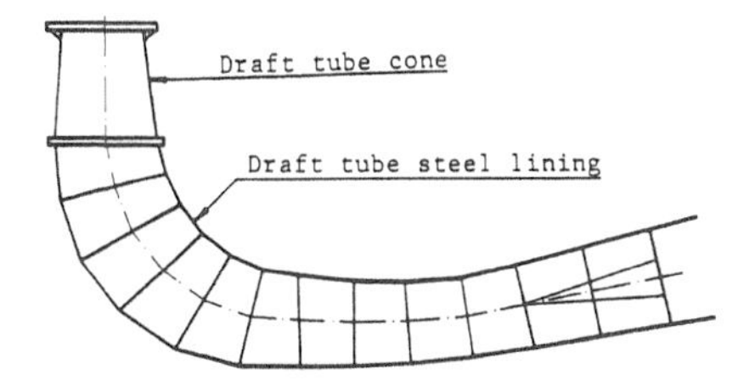

Figure 2.16 Outline of the draft tube

portion around the bend the cross section is reduced somewhat to form a converging flow in order to reduce losses from flow separation.

The draft tube cone is of welded construction, and consists of two sections when the unit is designed for removal of the runner from the underside. The upper cone may be lined with stainless steel at its inlet and is normally provided with two manholes for inspection of the runner from below. The lower cone is made removable for dismantling of the runner.

2.4.7 Turbine assembly

Figure 2.17 shows the complete turbine assembled from the various component parts described in the preceding sections. The upper and lower labyrinth rings A and B are mounted onto the upper (head) cover and lower cover (bottom ring) respectively. In order to reduce the leakage flow from the runner to a minimum, the labyrinth clearances are kept as small as possible. The runner is normally made of stainless steel and the stationary rings are made of bronze to reduce the risk of seizing in the labyrinths.

A balance of the axial hydraulic forces acting on the runner is obtained to some extent by making the upper and lower labyrinth rings of approximately the same diameter. Periodic measurements of the leakage flow through the labyrinths are conducted for indication of wear development in the labyrinths. It is normally possible to measure the physical clearance of the lower labyrinth through holes in the draft tube cone flange.

2.4.8 Drainage and filling facilities

A normal drainage and filling arrangement is shown in Figure 2.18. The penstock and the spiral case can be drained to a level corresponding to the tailwater level by gravity through the draft tube. The remaining water has to be pumped out by special purpose pumps with the draft tube gate or valve closed.

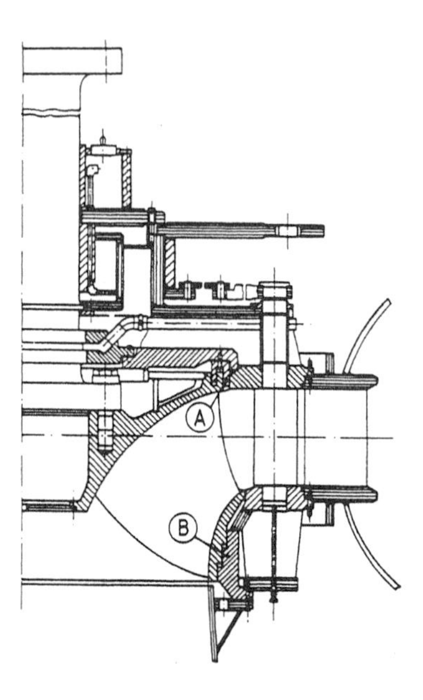

Figure 2.17 Turbine assembly

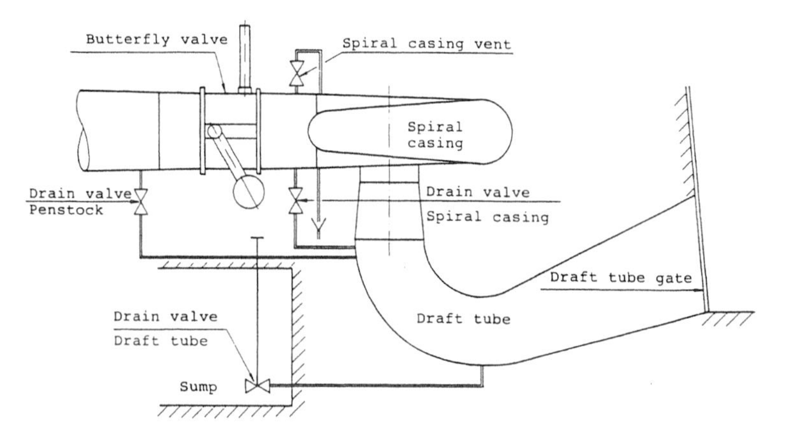

Figure 2.18 Drainage and filling facilities

The draft tube is filled through a pipe leading from downstream of the draft tube gate, these gates are usually designed to operate only under condition of equalized pressure on both sides. The spiral case is filled by the drain pipe from the draft tube. If the turbine setting is not below the tailwater, the draft tube and spiral case are both filled from the penstock.

2.5 Establishing of Main Dimensions

2.5.1 Determination of runner parameters

The turbine output P, the water flow quantity Q and effective head H_e, must be given in order to determine the runner parameters. Sometimes the preferred speed n and the optimal flow Q^* or guide vane position α are also given.

The main dimensions of the runner are determined from the basic equation

$$\eta = \frac{1}{gH_e}(U_1C_{1U} - U_2C_{2U}) \tag{2.2}$$

The hydraulic efficiency η may be assumed to be 0.96 and the turbine is first assumed to have no rotation in the exit flow, that is $C_{2U} = 0$. The basic equation is then simplified to

$$U_1C_{1U} = 0.48 \cdot 2gH_e$$

The runner exit may be the place to start the dimensioning. Assuming the velocity is uniform, the exit flow velocity is

$$C_2 = \frac{4Q^*}{\pi D_2^2} \tag{2.3}$$

The peripheral velocity at the runner band is

$$U_2 = \frac{\pi D_2 \cdot n}{60} \tag{2.4}$$

The exit flow angle β_2 is defined by

$$\tan\beta_2 = \frac{C_2}{U_2} = \frac{240Q^*}{\pi^2 n D_2^3} \quad \text{or} \quad D_2 = \left[\frac{240Q^*}{\pi^2 n \tan\beta_2}\right]^{1/3} \tag{2.5}$$

From this it appears that a large diameter D_2 will give a small exit velocity C_2 which is beneficial for reducing friction losses but the exit angle determined this way tends to be too small for access for welding and grinding in the vane passages so $\beta_2 = 16^0$ is assumed as a practical value, and the following is obtained

$$D_2 = 4.4 \left[\frac{Q^*}{n} \right]^{1/3} \quad (2.6)$$

The peripheral velocity at runner exit, U_2, is in this connection often used as a parameter for the turbine blade loading. Some 25 years ago, a U_2 value of $30ms^{-1}$ was a normal upper limit and in 1985 the general limit has risen to $40ms^{-1}$. For heads between 400 and $700m$ it may be tempting to use U_2 up to $50ms^{-1}$ in order to arrive at a good specific speed for best possible efficiency. Such a velocity will put a high demand on the blade surface finish. The runner inlet area is often made 10% larger than the exit area since an

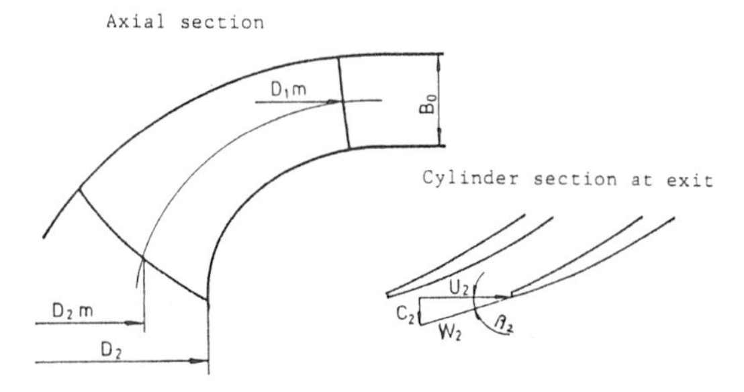

Figure 2.19 Axial section of a low specific speed runner

accelerated flow will have less losses, then

$$\pi B_0 D_1 = 1.1 \frac{\pi D_2^2}{4} \quad \text{or} \quad B_0 = 0.275 \frac{D_2^2}{D_1} \quad (2.7)$$

It appears that by choosing a small U_1 for given speed n a small inlet diameter D_1 may be attained and will result in a low-cost turbine. With a smaller U_1, it is seen from the basic equation, C_{1u} and thereby C_1 will become larger. From the energy equation the highest attainable velocity is

$$C_{1max} = \sqrt{2g\,(H - H_s + H_b - H_{va})} \quad (2.8)$$

here, for high head turbines, the suction head H_s, atmospheric pressure H_b and the vapour pressure H_{va} are small compared to H so that $C_{1max} \approx \sqrt{2gH}$. As C_{1u} is slightly smaller than C_1, see Figure 2.20, the lowest peripheral velocity is in accordance with the basic equation

$$U_1 = 0.48\sqrt{2gH} \quad (2.9)$$

Under this condition, all energy in the water is converted to velocity energy before the runner inlet, as in the case of a free jet turbine. Such a turbine will have large friction losses in the guide vanes and runner blades and local cavitation may occur at the guide vanes and runner inlet. Also at flows other than the designed flow rate large impact losses will occur at the runner inlet. The turbine runner thus designed will have seriously curved blades that are difficult and expensive to make, as shown in Figure 2.20. Apparently this is a poor choice in design.

Francis turbines are usually designed with an inlet velocity C_1 in the same range as U_1, the inlet diagram for which is shown in Figure 2.21. The blade inlet angle β varies in the range of 65° for high head turbines to nearly radial for low head turbines.

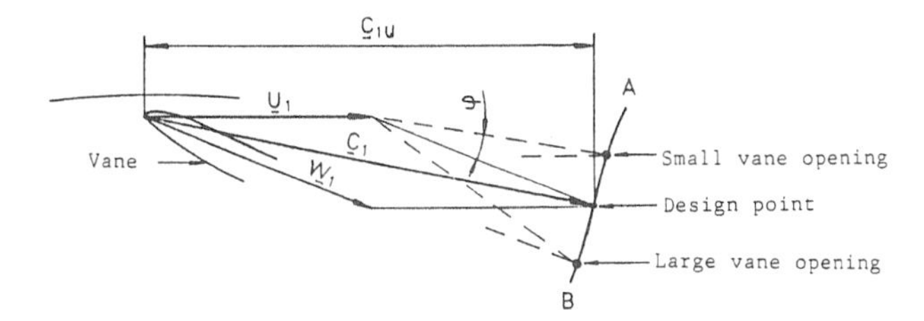

Figure 2.20 Inlet velocity diagram (Case A)

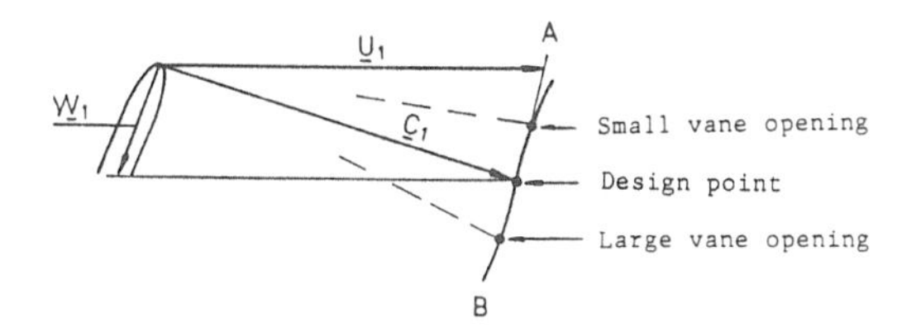

Figure 2.21 Inlet velocity diagram (Case B)

For low head turbines, the inlet diameter D_1 tends to be smaller than the exit diameter D_2. This will present no problem for the flow channels close to the crown, but at the runner band D_1 must be made approximately equal to D_2 in order to attain continuity in the blade surface.

For high specific speed turbines, there is a deviation in the velocity C and circulation UC_u along the inlet width due to the influence of the conduit and the distorted exit flow from the guide vanes. This implies that a vertical exit flow will not exist at the runner outlet and the efficiency will therefore be lower as compared with low specific speed turbines.

For very low specific speed turbines, the friction losses external to the runner will become larger so the efficiency is also lower. Hence, there is a best efficiency range which lies between $\Omega = 0.3$ and 0.55 (n_q between 16 and 30). For the lowest possible price of the turbine and generator, the highest possible specific speed is always favourable. But in view of the submergence requirement, a smaller specific speed is then preferred. As is well known, the turbine design is always a compromise between the aforementioned factors, selection of parameters may differ somewhat depending on the power station requirements.

2.5.2 Determination of suction height

Modern turbines of medium to large size are rarely installed above the tailwater level, the practical meaning of the suction height is really its negative value, that is, the submergence of the turbine. In order to calculate the necessary submergence of the turbine, it is common to estimate the required net positive suction head $NPSH_r$. The condition is that the pressure of the fluid in the runner should be higher than that for occurrence of cavitation

$$H_b - H_s - NPSH > H_{va} \quad \text{or} \quad H_s < H_b - H_{va} - NPSH \tag{2.10}$$

here, H_s is the vertical distance from the tailwater level to the critical point for cavitation in the turbine. For most turbines, H_s is normally negative, that is the turbine is in submersion. The required $NPSH$ may be estimated from

$$NPSH_r = \lambda_1 \frac{C_{2m}^2}{2g} + \lambda_2 \frac{U_2^2}{2g} \tag{2.11}$$

where C_{2m} is the mean meridian velocity at runner exit, $C_{2m} = 4Q/D_2^2$, and λ_1 and λ_2 are coefficients determined from experience.

The values of λ_1 and λ_2 depend on the number of runner blades and their shape. For Kvaerner runner designs with 14 to 16 blades it is usual to set λ_1=1.12 and λ_2=0.055 for $\Omega < 0.55$ and λ_2=0.1Ω for $\Omega > 0.55$.

To indicate the influence the speed has on the suction head, suitable values are substituted in the above expressions to get the following relations

$$H_s = 10 - 0.00046 Q^{2/3} \cdot n^{4/3} \tag{2.12}$$

In the final determination of the turbine setting, the speed giving the best efficiency, the civil works costs and the machine costs must be considered together. In certain topography a plant may have a submergence of 50 to $200m$. In such cases, as is true with all high head plants, care should be taken not to choose a speed that gives velocities close to the cavitation limit. Small irregularities in the blade surfaces may cause local cavitation and tests have shown that once cavitation sets in, the erosion damage will increase with the flow velocity to the power of five or six. When the runner dimensions are established from these considerations, the runner blades and the crown and band must be designed to give the best hydraulic performance.

2.5.3 Dimensioning of other flow passage components

In establishing dimensions for the spiral case, stay (speed) ring and guide vanes, the situation requires a compromise between manufacturing costs and hydraulic performance. As it is difficult to calculate all components involved in such a compromise, it is normal to set the spiral case inlet diameter in a certain relation to the runner exit diameter, and then design the spiral case sections and the guide and stay vanes so that the water flows as uniformly as possible to the runner. In recent years, the spiral case inlet diameter has been defined as

$$D_{sp} = (0.833 + 0.333\Omega)D_2$$

Exit flow angle α from the guide vanes is determined by manipulating the turbine basic equation and the continuity equation as follows

$$\tan\alpha = \frac{nQ}{30B_0\phi}\left[\eta 2gH_e + U_2^2\left(1 - \frac{n^*}{n}\cdot\frac{Q}{Q^*}\right)\right] \tag{2.13}$$

here ϕ is the vane contraction in the guide vane exit and may be set to be 0.9.

The guide vane mechanism diameter is chosen so that the guide vane exit diameter is at least 4% larger than the runner inlet diameter. If the guide vanes are too close to the runner, each time a runner vane passes the wake of a guide vane (shadow) a shock will occur and create a high pitched noise. Even if the distance between the guide vanes and the runner is generous, unfavourable combinations of a number of guide vanes and runner blades may still cause hydraulic interference, resulting in strong vibration of the turbine.

The number of guide vanes may generally be determined by considering that fewer and larger vanes will need larger diameters for the upper and lower covers and thereby greater weights, while a larger number of smaller vanes will demand higher manufacturing costs. Accordingly, large turbines are designed with more vanes and small turbines with fewer vanes. In recent years, 16 to 32 guide vanes have been used, with the smaller number for smaller turbines. The vane profiles are designed to cause as little loss as possible along with small operating torque and preferably with self-closing trend.

The number of stay vanes is also a compromise between performance and cost. A large number of thin vanes will give a more favourable stress distribution in the stay ring and smaller welding cross sections but a larger wetted surface and hence greater loss. When a large number of smaller stay vanes are used, the requirement for compliance between the flow angle and the vane angle is higher than with fewer and thicker vanes. For small and low head turbines, 12 stay vanes are often chosen while for large machines with head in the range of 500 to $700m$, up to 24 vanes may be used.

The draft tube is normally composed of a straight conical section with a length of (1 to $2)D_2$ and a cone angle of $7 - 8^o$ and then a bend with flattened cross sections. Downstream of the bend the cross section increases to the equivalent of a cone angle of 10°. The draft tube is steel lined when the flow velocity is above 4 to 5 ms^{-1}.

Most manufacturers of turbines have developed a series of models. The task is to find a particular serial turbine which suits the chosen specific speed for the project at hand and scale it to the size in question.

The serial turbines may be models or operating units whose characteristics such as efficiency, cavitation number, runaway speed and guide vane torque are known.

Table 2.2 presents examples of turbines with different parameters and the subsequent calculated dimensions. Examples 1 and 2 are variants for a high head turbine ($H = 500m$) while examples 3, 4 and 5 are those for a low head turbine ($H = 100m$).

Turbine 1 has low blade loading but is difficult and costly to manufacture and has poor efficiency. Turbine 2 has comparatively high blade loading and requires a high runner surface finish because of the high flow velocities through the turbine.

Of the low head group, Turbine 3 has the largest dimension but has good efficiency over a large range of output. Turbine 4 has smaller dimensions but still with good efficiency. Turbine 5 represents the level of blade loading used in 1985.

2.6 Structural Design

2.6.1 Stay ring/spiral case

(i) Load condition

The major load on the stay ring and spiral case is the internal pressure whose distribution varies with the turbine flow and whose magnitude varies with the head. For structural dimensioning it is natural to choose the condition with maximum pressure and a uniform distribution on the stay ring. Under normal running condition there is a negative pressure gradient from spiral case to runner.

Table 2.2 Examples of turbine designs

Example No.	1	2	3	4	5	Dim.
Given Q	10	10	10	10	10	m^3/s
Given H	500	500	100	100	100	m
Given h_s	0	−12	2	0	−5	m
Given ∇u. v.	100	100	100	100	100	m.a.s.l
Given max. water temperature	15	15	15	15	15	℃
Derived from equ. (2−2): n	500	1000	428.7	500	750	min^{-1}
$\underline{\omega}=\frac{\pi}{30}\frac{n}{\sqrt{2gH}}$	0.529	1.057	1.014	1.182	1.773	m^{-1}
Assumed Q^*	7.5	8.0	9	9	9.5	m^3/s
Assumed $\underline{Q}^*=\frac{Q^*}{\sqrt{2gH}}$	0.0757	0.0808	0.203	0.203	0.214	m^2
Assumed $\Omega^*=\underline{\omega}\sqrt{Asu.Q^*}$	0.145	0.300	0.457	0.533	0.821	–
$k=1.3444-0.2222\cdot\Omega^*$	1.31	1.28	1.24	1.22	1.16	–
$Q^*=\frac{Q}{k}$	7.6	7.8	8.1	8.2	8.6	m^3/s
$\underline{Q}^*=\frac{Q^*}{\sqrt{2gH}}$	0.077	0.079	0.183	0.185	0.194	m^2
$\Omega^*=\underline{\omega}\sqrt{\underline{Q}^*}$	0.147	0.297	0.435	0.508	0.780	–
Chosen $\underline{u}_1$	0.72	0.72	0.72	0.72	0.72	m/s
Assumed η	0.96	0.96	0.96	0.96	0.96	–
$D_1=\frac{u_1\cdot 60}{\pi\cdot n}$	2.724	1.362	1.420	1.218	0.812	m
$D_2=4.4\sqrt[3]{\frac{Q^*}{n}}$	1.090	0.872	1.172	1.118	0.992	m
$B_0=0.275\frac{D_2^2}{D_1}$	0.120	0.154	0.266	0.282	0.333	m
$u_2=\frac{D_2\cdot\pi\cdot n}{60}$	28.6	45.8	26.4	29.3	33.8	m/s
$c_{2m}=\frac{4\cdot Q}{\pi\cdot D_2^2}$	10.7	16.7	9.3	10.2	12.9	m/s
$NPSH=1.12\frac{c_{2m}^2}{2g}+0.055\frac{u_2^2}{2g}$	8.9	21.7	7.0	8.2	14.3	m
Max. $h_s=h_b-h_{va}-NPSH$	1.1	−11.7	3	1.8	−4.3	m
$\alpha=\text{arc tg}\frac{nQ}{30B_0\varphi(\eta 2gHe+u_2^2(1-k))}$	9.6°	15.2°	19.2°	21.2°	26.2°	
$D_{sp}=D_2(0.833+0.333\Omega^*)$	0.96	0.82	1.15	1.12	1.08	m
Assumed η_T^*	90	92.8	94.1	93.9	93.2	%
Runaway speed $n_{run}=n\ (1.3+0.8\ \Omega^*)$	710	1540	710	850	1450	min^{-1}
Chosen runner blade number	15+15	15+15	16	16	16	–
Chosen guide vane number	24	24	24	24	24	–
Chosen stay vane number	24	16	12	12	12	–

(ii) Structural model

There are two different approaches to the analysis with different degrees of simplification and accuracy. The more accurate solution is obtained by using the finite element analysis (see Chapter 8) which is used to control the conventional calculation methods described here.

(iii) General conditions of the structure

The spiral case centre line is in a plane of symmetry.

The deformation of the stay ring and stay vanes consists of a rotation and a translation. The angle of rotation in the joint between the ring and vanes is assumed constant in the axial section. The rotation will be about the centre of gravity of the stay ring cross section which will not be changed due to the rotation (neglecting deformation from shear and bending stress in the axial section). The translation of the joint between vane and ring will be constant Δy_0 in the y-direction.

The actual vane is replaced by an equivalent radial vane with the same cross

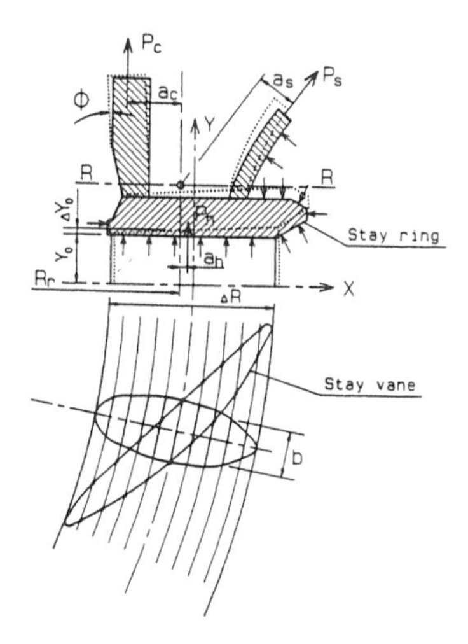

Figure 2.22 Forces acting on the stay ring

sectional area. The neutral axis of the vane is tangential to a circle through the centre of gravity of the vane.

The spiral case is treated as a circular torus.

(iv) Loading on the stay ring

Tensile force from the spiral case shell P_s (bending moment neglected).

Lifting force from upper (head) cover P_c.

Resultant force from internal pressure on stay ring P_h.

(v) Limitations

The analysis holds for medium and high-head turbines where $2 \cdot y_o$ is of the same order as ΔR (see Figure 2.22).

(vi) Analysis

The stress in the stay vane expressed by rate of strain is

$$\sigma = E\left(\frac{\Delta y_0}{y} + \phi\frac{x}{y}\right) \tag{2.14}$$

The moment taken up by the stay vane can be found by numerical or graphical integration

$$M_s = \int_{-x_1}^{x_2} x\sigma b \cdot \mathrm{d}x \tag{2.15}$$

The moment taken up by the stay ring is

$$M_R = \phi\frac{2\pi E I_R}{Z R_t} \tag{2.16}$$

where Z is the number of stay vanes. The total applied twisting moment about the centre of gravity of the stay ring must equal the sum of the two internal moments:

$$M_v = M_s + M_R \quad \text{or} \quad M_v = P_c a_c - P_h a_h - P_s a_s$$

The magnitude of M_v varies with the arm a_s, and a change in a_s will result in a change in ϕ which will then influence the equilibrium of the stay ring. This is important for the stress distribution in the stay vanes and must be taken into account in the design.

The stresses in the spiral case are computed by the torus formula at the point of juncture with the stay ring in the different cross sections as shown in Figure 2.23:

$$\sigma_s = p r_s \frac{R + R_t}{2Rt} \tag{2.17}$$

where t is thickness of the spiral case shell.

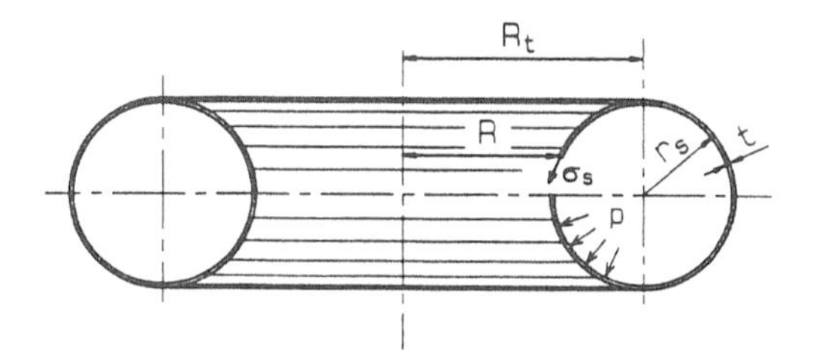

Figure 2.23 Schematic of loading on a spiral case

2.6.2 Upper and lower covers

(i) Load condition

Unlike the stay ring, the design criterion for the upper and lower covers is deformation and not stress. In order to keep the leakage flow between guide vanes and the covers to a minimum, the covers must be made stronger than necessary from strength consideration, in order to keep the deflection low, particularly for the case of high head turbines. The major loads on the covers are the internal pressure in the turbine and the bolt forces that hold the covers in place. Because the structure is designed for resistance to deflection the important case is at normal operating condition, where the internal water pressure is distributed in a decreasing trend from the outside toward the centre, as shown in Figure 2.24.

(ii) Structural model

The following analysis is simplified as the deformation of the ribs in the cover is neglected. The actual deflection measured as the angle of rotation ϕ may be 1.5 to 2 times higher than that calculated with this simplified model. Figure 2.25 shows the deflection of an upper cover (exaggerated) computed with finite element analysis taking deformation of ribs and the welding geometry into account.

(iii) Analysis

The tangential stress in the ring sections of the cover is

$$\sigma_t = \frac{\phi y E}{r} \tag{2.18}$$

An integration of the tangential stress multiplied with the distance from the neutral axis over the cross section and circumference gives the moment M_y

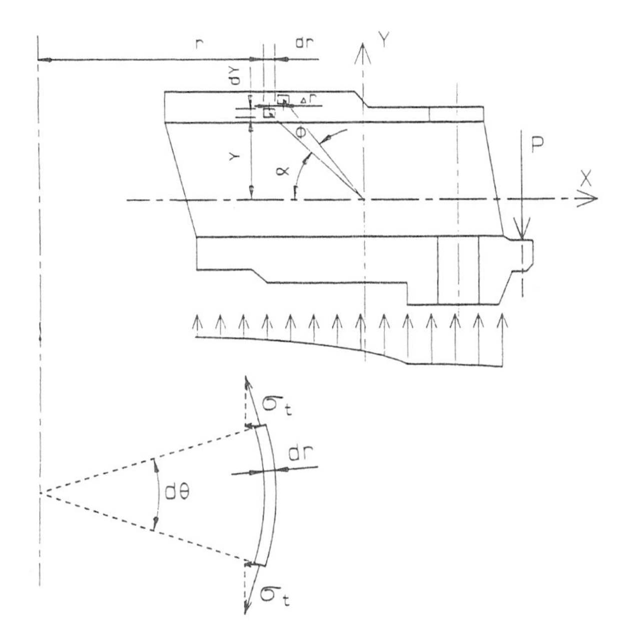

Figure 2.24 Loading on the upper cover

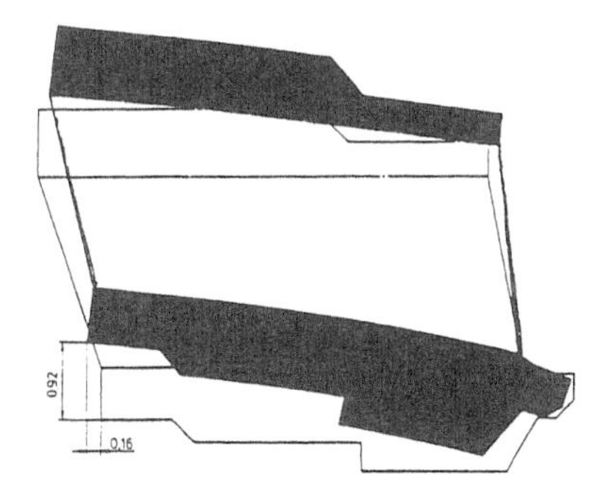

Figure 2.25 Upper cover deflection by finite element analysis

that balances the moment caused by hydraulic load

$$M_y = 2\pi E\phi \int_{r_1}^{r_2} \int_{-Y}^{+Y} \frac{Y^2}{r} \, \mathrm{d}r \, \mathrm{d}y \,, \tag{2.19}$$

Under the condition that the cover is not restricted in the radial direction, the position of the neutral axis is determined from

$$\int_A \sigma_t \, \mathrm{d}A \approx 0 \tag{2.20}$$

which means that the sum of forces in the tangential direction is zero. By substituting the value of σ_t it is possible to find the position of the neutral axis ($y = 0$) in the cross section that satisfies the above equation.

The hydraulic load on the cover is found by integrating the internal pressure over the cover area. The pressure distribution is found from the following equation as a function of the radius

$$p_1 = p_0 - \int_0^1 \frac{\rho C_x^2}{r} \, \mathrm{d}r \tag{2.21}$$

where C_x is the absolute tangential velocity of the flow in the area between the cover and the runner. The hydraulic moment found from the above calculation must equal the internal moment in the cover, that is $M_h = M_y$. This gives the deflection expressed by

$$\phi = \frac{M_h}{2\pi E} \int_{r_1}^{r_2} \int_{-Y}^{+Y} \frac{Y^2}{r} \, \mathrm{d}r \, \mathrm{d}y \tag{2.22}$$

And the stresses may be found in different parts of the cover from

$$\sigma_t = \frac{\phi y E}{r}$$

2.6.3 Guide bearing

The guide bearing for a vertical turbine takes no gravitational load, but is designed to withstand unbalanced hydraulic or other dynamic forces acting in the radial direction. The bearing will usually be of hydrodynamic type with mineral oil as bearing fluid. A typical design is shown in Figure 2.26.

The oil is forced to flow through the converging channel formed by the tapered white metal pad and the shaft by the rotation of the shaft. This leads to a pressure rise and a force acting perpendicular to the pad which is given by the following equation

$$F = \frac{6W\eta U \ell^2 b}{h_0^2} \tag{2.23}$$

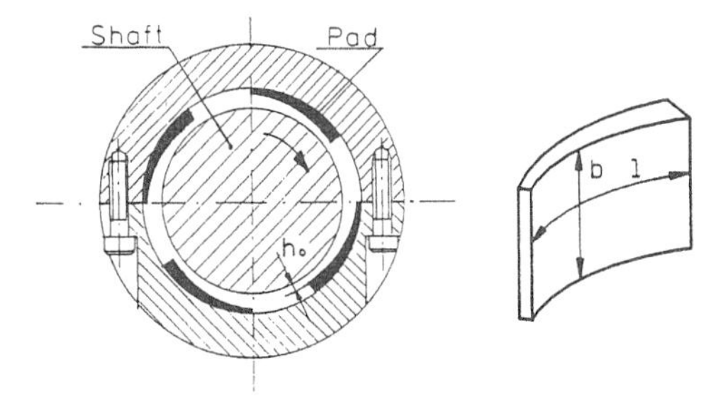

Figure 2.26 Guide bearing

The design shown has four pads equally spaced. The forces are equal when the shaft rotates in the centre of the bearing as the clearance h_o is the same around the bearing. A movement of the shaft in one direction caused by an external force will lead to an increase in pad force in the opposite direction due to the reduction of h_o and will tend to restore the shaft to its original position. The following equation gives the relation between the external force F_e and the displacement Δr

$$F_e = \sum F = \frac{6W\eta U\ell^2 b}{(h_0 - \Delta r)^2} - \frac{6W\eta U\ell^2 b}{(h_0 + \Delta r)^2} \tag{2.24}$$

when considering $\Delta r << h_o$, (2-24) becomes

$$F_e = 24\frac{W\eta U\ell^2 b}{h_0^3}\Delta r \tag{2.25}$$

The dimensions of the bearing is decided by assuming the maximum external load and the permissible displacement. In addition to the force acting perpendicular to the pad, there is another force acting in the opposite direction of rotation. This is due to the internal friction in the fluid which will result in power loss for one pad as

$$P = 1 \cdot 10^{-5}\frac{E\eta U^2 \ell b}{h_0} \quad kW \tag{2.26}$$

The equation shows that a bearing designed with increased pad force will cause higher power loss. The need for cooling is mainly dependent on the shaft surface speed and has to do with the heat energy generated by friction.

2.6.4 Regulating ring

The function of the regulating ring is to distribute the actuating force from the servomotors to the guide-vane links. To avoid mechanical distortion in the regulating mechanism the deflections of the regulating ring must be small and accordingly the stress level must be low.

The regulating ring is moved by the servomotors as shown in Figure 2.27, after overcoming the hydraulic force and friction force on the guide vanes. The servomotor force F_s is used as the maximum load on the ring for design calculations.

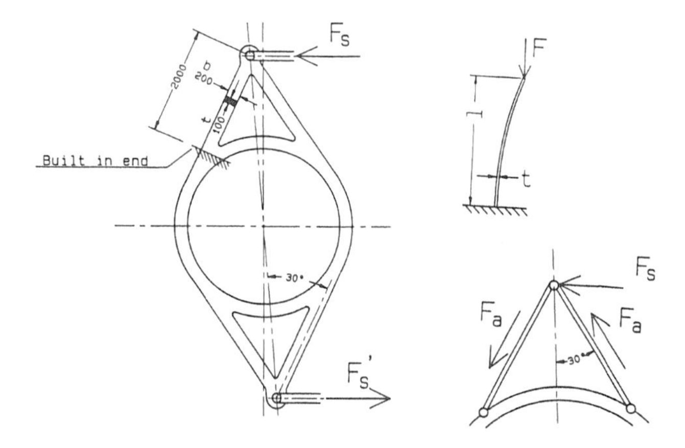

Figure 2.27 Loading of regulating ring

With the design shown, the arms of the regulating ring have to be dimensioned against buckling in the stress calculation. The critical stress is

$$\sigma_{cr} \approx \pi^2 E \left(\frac{k}{2\ell}\right)^2 = \frac{\pi^2 E}{\lambda^2} \tag{2.27}$$

where k is the smaller radius of gyration

$$k = \sqrt{\frac{I}{A}} = \sqrt{\frac{\frac{1}{12}t^3 b}{tb}} = 0.29t \tag{2.28}$$

With slenderness ratio $\lambda = 2\ell/K = 6.9(\ell/t)$, equation 2.28 becomes

$$\sigma_{cr} = \frac{\pi^2 E}{(6.9\ell/t)^2} = \frac{t^2}{\ell^2} \cdot 0.207E$$

For a typical case, $t = 100mm$, $l = 2000mm$, then $\sigma_{cr} = 109MPa$. With a reasonable safety factor against buckling about 3, the stress level will be very moderate. To calculate the allowable servomotor force, F_s is determined by:

$$F_s = 2F_a \sin\alpha \quad \text{with} \quad F_a \ll (\sigma_{cr}/3)tb = 730 \ \ kN$$

2.6.5 Flange connections

To allow for easy dismantling, bolt connections are used extensively in a hydraulic turbine. Some of the connected parts are designed for stiffness such as the bearing support structure, so the total required bolt cross section is small compared with the structure. The support may then be designed without flanges and the bolts integrated in the body. On the other hand, some flanges are necessary to provide space for bolts as on the spiral case inlet and shaft coupling because of their high loading.

The dimensioning of pipe flanges involves calculations similar to those of turbine cover and stay ring, the external moments acting on the flanges are balanced by internal stress resulting in rotational deformation. The joint between the flange and the pipe is vulnerable, and care must be taken to avoid overloading of this part. The dimension of the flange connection is dependent on the bolt dimensions and bolt stress level. By using high strength bolts the flange can be made smaller and cheaper, see Figure 2.28.

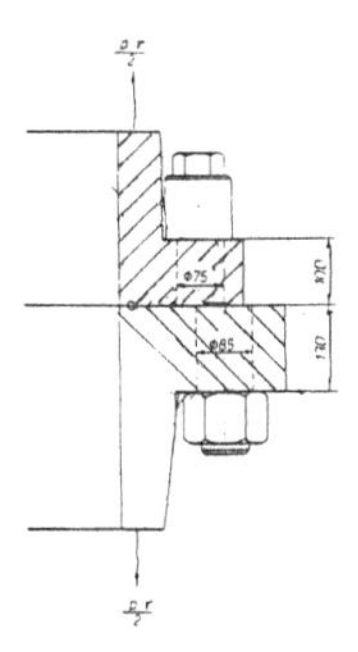

Figure 2.28 Comparison of flange connection with high strength and ordinary bolts

Prestressing of bolts in a flange connection is important especially with varying loads. Because of the difference in deformation of bolt and flange during prestressing, the bolt is elongated far more than the flange is compressed, the stress variation in the bolt will be small compared to the variation in external load. This is especially important when considering fatigue resistance of the bolts. The shaft coupling can be a pure friction connection if the bolt prestressing is sufficiently high and the friction factor of the surfaces is in a suitable range.

This coupling has great advantages over the reamed bolt couplings in machining and assembling costs. The prestressing may be carried out by hydraulic tensioning and/or heating of the bolt (with the nut turned on without torque). The prestressing level is controlled by measuring the elongation of the bolts.

2.7 Assembly and Dismantling of Francis Turbines

It is of great importance that the turbines are made for easy assembly and dismantling. Depending on the water quality and the operating head, wear will in time occur in the guide vane mechanism and in the upper and lower runner labyrinths that will require repairs. Medium and high-head Francis turbines generally have free access to the draft tube cone so that vital parts of the turbine can be dismantled and reassembled readily. This has great importance for efficient maintenance and will contribute to bringing the unit quickly back to service after repair or alterations.

A widely used method is to remove the draft tube cone and dismantle the main parts downward and then lift the parts up to the machine hall with the main crane. The parts requiring repair are usually the runner, lower cover and the labyrinth rings. The turbine bearing should also be easily removable. Dismantling the upper cover is normally not required. Possible damage to the surface adjacent to the guide vanes is easily repaired at site.

Dismantling or reassembling from below requires a comparatively large opening below the spiral case which may be difficult to achieve with low head turbines with large dimensions, as the load carrying concrete around the spiral case takes up so much space in the area where the access way must be located. These turbines, on the other hand, are not so exposed to wear in the labyrinths and other parts so less repair is needed. When the draft tube cone is totally embedded in concrete, the turbine must be dismantled from above after removal of the generator rotor.

2.7.1 Assembly of turbine

Figure 2.29 shows the sequence of assembly of embedded parts for a vertical high-head Francis turbine. The draft tube is first installed and embedded in concrete. At the same time foundations for the spiral case are embedded. Lower cover *1* is temporally placed on the draft tube flange. The spiral case *2* is lowered down on to the foundations, and the lower cover is lifted and mounted to the stay ring. The draft tube cone *3* is placed temporarily on the draft tube flange. The turbine may now be prepared for pressure testing of the spiral case which will be temporally sealed by bulkhead *4*.

Figure 2.30 shows the assembly of the rotating parts and guide apparatus. The guide vanes *6* complete with seals, the pressure test bulkhead *7* and the upper cover *5* are installed for pressure testing of the spiral case. After

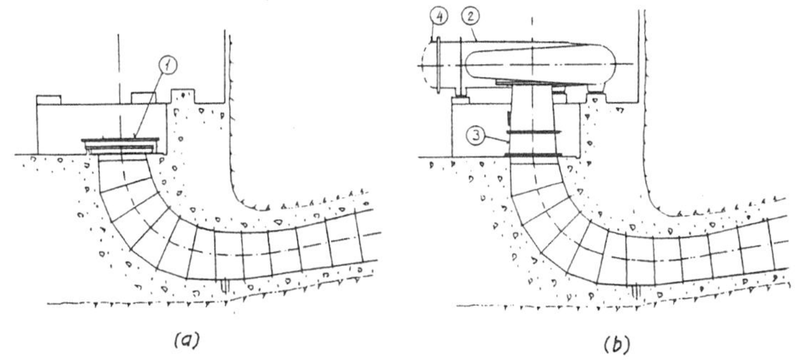

Figure 2.29 Sequence of assembly of embedded parts

this is completed, the spiral case is accurately aligned and welded to the foundations. The pit liner *8* is then welded in preparation for the spiral case to be embedded. After embedding and curing of the concrete, the upper cover is dismantled and the pressure test bulkheads *4* and *7* are removed. The runner *10* and turbine shaft *9* which are usually bolted together in the shop or on the machine hall floor can now be lowered and positioned on top of the draft tube cone and properly aligned. The turbine is now ready for generator erection. At the same time, links *12* guide vane arms *13* and the regulating ring *11* are installed. The turbine and generator shafts are temporarily joined with bolts. The draft tube cone which so far has served as support for the runner, can now be removed. The shaft assembly is correctly aligned and fixed in the correct position. The lower labyrinth ring *14* is centred in relation to the runner and is installed in the lower cover. The locking of the shaft assembly can now be removed, and the turbine bearing *15* is centred and installed. Shaft seal box *16* floor grid plates *17* and drain pipe *18* are then installed. Photographic records of installation of the Kvilldal turbine runner are shown in Figure 2.31.

2.7.2 Dismantling of runner

The draft tube cone *3* shown in Figure 2.29 is made up of two parts. The lower part is connected to the draft tube flange by a telescopic joint and can be removed sideways. The upper part is then loosened and lowered. At the same time the turbine bearing and shaft seal box are dismantled. The lower labyrinth ring *14* (Figure 2.30), is lowered after which the lower cover *19* is dismantled from the stay ring, lowered and removed sideways.

The flange connection between the turbine and generator is dismantled, and runner *10* and shaft *9* are lowered by means of hydraulic jacks. The

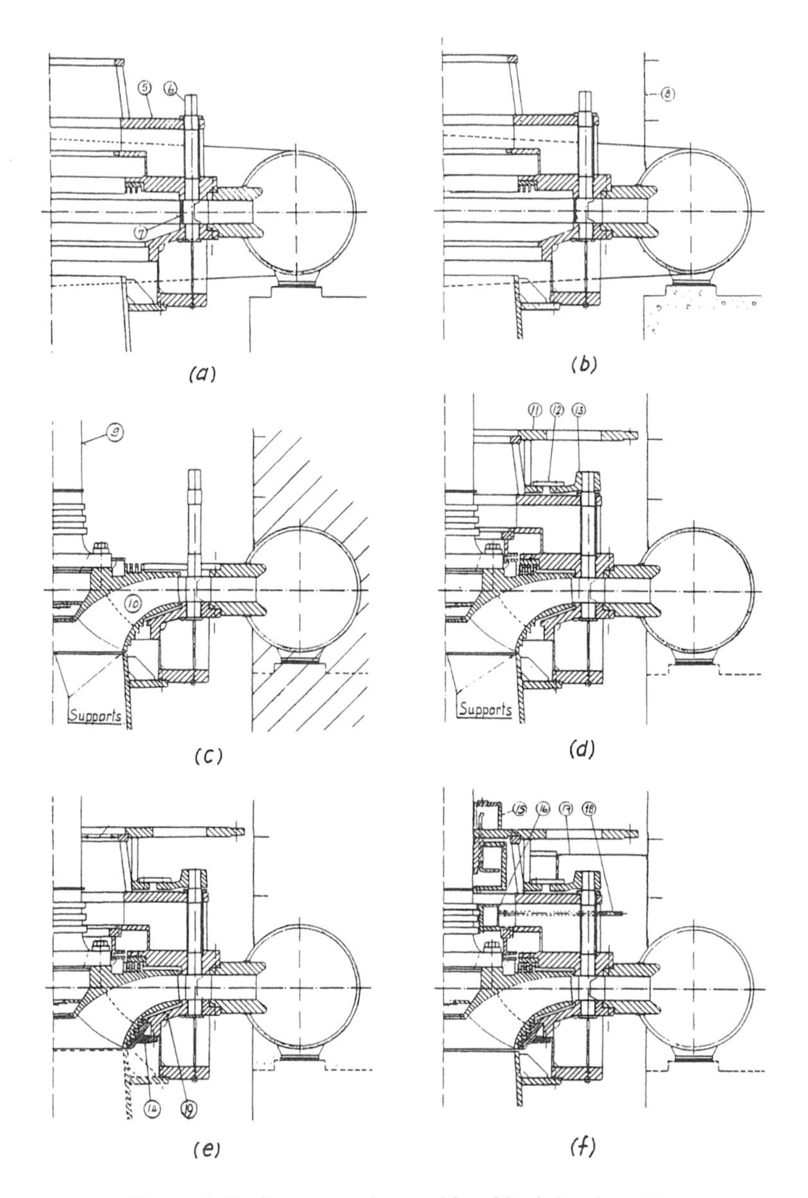

Figure 2.30 Sequence of assembly of high head turbine

Figure 2.31 Photographic records of installation of Kvilldal turbine runner (manufactured by Kvaerner Brug A/S)

runner is dismantled from the shaft which now can be lifted. The runner is placed aside while the guide vanes *6* are lowered one by,one from the upper cover. The upper labyrinth ring is now dismantled from the upper cover. Dismantling of the upper parts is done upwards either through the generator stator or sideways over the spiral casing if the power house permits.

Chapter 3

Construction of Axial–Flow and Diagonal–Flow Turbines

Fang Qing–jiang

3.1 General Features of Axial–flow and Diagonal–flow Turbines

Axial–flow and diagonal–flow turbines are designed for low and medium head service. The axial–flow turbine has a longer history than the diagonal–flow turbine, but both are comparatively recent in development compared with the mixed–flow (Francis) turbine.

Figure 3.1 shows a schematic diagram of the three distinctive types of low and medium head turbines: (a) is the axial–flow turbine that emerged first and is so named because the flow passing through the runner is essentially in the direction of the rotating axis; (b) shows an evolution from the axial–flow turbine where the runner blades are arranged in a conical surface and the flow is in an diagonal direction to the rotating axis, hence the name diagonal–flow turbine; (c) is the other evolution from the axial–flow turbine toward the lower head range where the water flows through the turbine in a nearly straight–through passage and is axial in direction to make it a truly axial–flow turbine, but is now known as tubular turbine because of its passage shape and is also called a bulb turbine since the generator is often enclosed in a bulb shape casing submerged in the water flow. The tubular turbine is the subject of Chapter 4.

An axial–flow turbine with fixed runner blades is called a propeller turbine whilst the one with adjustable blades (changeable in angle while in operation) is known as a Kaplan turbine after the inventor of this design.

The major advantages of the axial–flow turbine over the Francis turbine is that in the low head range for the same runner diameter it has a higher flow capacity and rotates at higher speed, that is, it has a higher specific

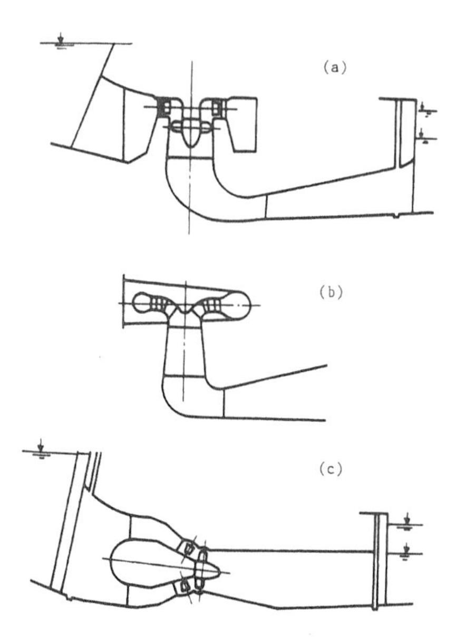

Figure 3.1 Layout of axial–flow, diagonal–flow and tubular turbines [3.1]

speed than the Francis turbine. But the runner exit kinetic energy of an axial–flow turbine is higher than a Francis turbine, therefore it has poorer cavitation characteristics and requires a deeper setting. In terms of mechanical construction, as the runner blades of Kaplan turbines (and some propeller turbines) are detachable, runner diameters of axial–flow turbines maybe made larger than Francis turbines. The diagonal–flow turbine is a late addition to the turbine world and has been developed mainly for medium head application where it competes with the Francis turbine. Evolution of the diagonal–flow turbine stems essentially from the following facts

- Francis turbines are not efficient in coping with large variations in head, especially in the lower head range.
- High head Kaplan turbines require a larger number of runner blades which tend to reduce the flow capacity and make the blade–adjusting mechanism rather cumbersome.
- The large amount of excavation due to the low setting required by high–head Kaplan turbines poses difficulty in power station layout and construction.

3.2 Axial-flow Turbines

3.2.1 Hydraulic characteristics

Statistical relations of axial-flow turbine performance parameters have been studied by [3.2, 3.3, 3.4, 3.5]. The charts shown in Figure 3.2 give the specific speed n_q as function of working head H in (a), and the speed coefficient K_u, energy coefficient ψ and flow coefficient φ all as functions of n_q in (b), (c) and (d) respectively. From these coefficients, the runner diameter and rotating speed of the turbine may be calculated and in turn the through flow and power output determined.

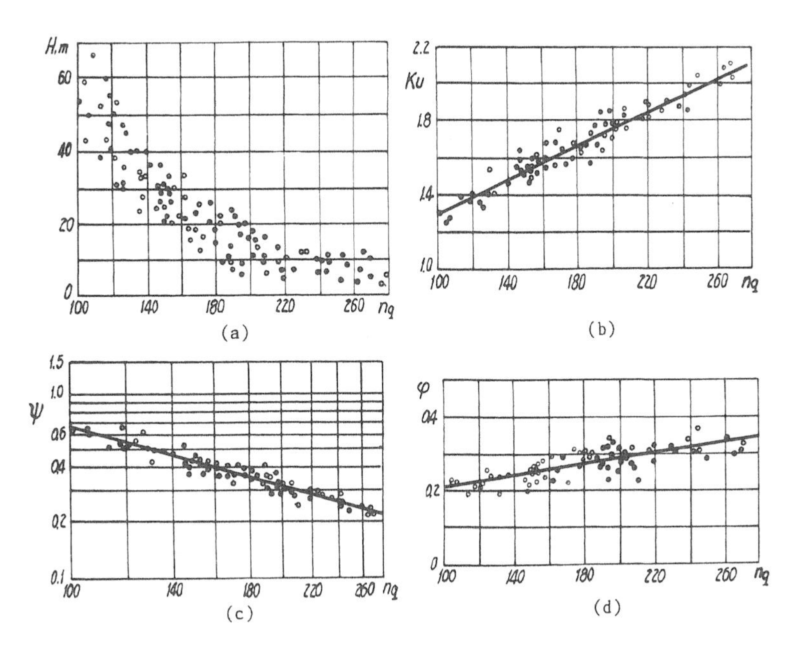

Figure 3.2 Design coefficients for axial-flow turbines [3.4]

The coefficients quoted here are defined as:

$$\text{specific speed} \quad n_q = \frac{n\sqrt{Q}}{H^{3/4}} \tag{3.1}$$

$$\text{speed coefficient} \quad K_u = \frac{\pi D_1 n}{60\sqrt{2gH}} \tag{3.2}$$

$$\text{energy coefficient} \quad \psi = \frac{gH}{K_\psi n^2 D_1^2} \tag{3.3}$$

$$\text{flow coefficient} \quad \varphi = \frac{Q}{K_\varphi n D_1^3} \tag{3.4}$$

where H is in m, Q in $m^3 s^{-1}$, D_1 in m and n in min^{-1}; and the conversion constants $K_\psi = 0.00137$, and $K_\varphi = 0.0411$.

The coefficients ψ and φ are related to the unit speed n_{11} and unit flow Q_{11} as follows

$$n_{11} = \frac{84.7}{\sqrt{\psi}}, \quad \text{and} \quad Q_{11} = \frac{3.48\varphi}{\sqrt{\psi}}$$

in consistent units.

3.2.2 Construction of Kaplan turbines

The general proportions of an axial–flow turbine is shown in Figure 3.3. All dimensions are given in fractions of runner diameter D_1, and the factors are given in order of increasing turbine head. Turbines of $30 to 40 m$ head and higher generally have steel spiral cases of circular cross sections and those below usually have concrete spiral cases with cross sections in the shape of polygons.

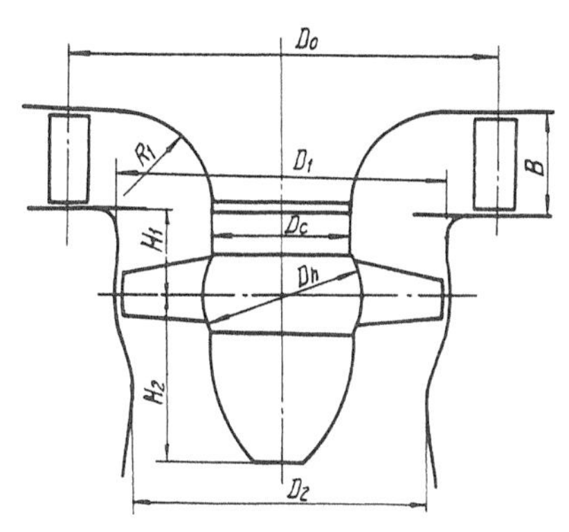

$$\begin{aligned} D_0 &= (1.16 - 1.20)D_1 & H_1 &= (0.21 - 0.24)D_1 \\ D_2 &= 0.975D_1 & H_2 &= (1.1 - 1.5)D_1 \\ D_h &= (0.6 - 0.33)D_1 & R_1 &= (0.23 - 0.27)D_1 \\ D_c &= D_h - 0.05D_1 & B &= (0.35 - 0.5)D_1 \end{aligned}$$

Figure 3.3 Profile of axial–flow turbine

A low head Kaplan turbine, currently the largest, in size, in the world with runner diameter of $11.3m$ is shown in Figure 3.4. This turbine operates under a rated head of $18.3m$ and develops over $170MW$ and is characterized by its concrete spiral case of polygonal cross section, rather slender stay vanes and long guide vanes ($0.4D_1$). The runner has four blades and is designed for a maximum head of $27m$. Most structural and pressure sustaining parts of the turbine are fabricated, such as the stay vanes, the head cover, the top cover, and the thrust bearing support, whilst the thrust chamber (made up of three sections) may be cast of welded construction.

Figure 3.5 shows a medium head Kaplan turbine with steel spiral case. The runner has six blades made of stainless steel whilst the guide vanes are actuated by individual servomotors. With exception of the rotating components, the turbine is nearly all of welded construction. The thrust bearing is mounted on the top cover and supports both the weight of the runner (plus its hydraulic thrust) and the rotor of the generator. With other large capacity low head units, a more compact layout is realised by making the turbine and the generator share one common main shaft. This arrangement not only reduces the overall height of the machine but also contributes to improvement in the stiffness of the rotating system.

The largest Kaplan turbines on record, based on capacity, are the Tabka turbines in Syria manufactured by LMZ of the USSR (1970). The turbine runners, with eight blades, have diameters of $6.0m$ with unit output of $242MW$ under $60m$ head and a speed of 150 min^{-1} [3.6].

A very high head Kaplan turbines was manufactured by CKD in the sixties at Orlik in Czechoslovakia. The runners have diameters of $4.6m$ and ten blades, output being $94MW$ under maximum head of $71m$.

3.2.3 Runner geometry

Axial-flow turbines may be designed for heads over a rather wide range, from 3–$5m$ to 70–$80m$, and the runner geometry will change greatly according to the flow conditions at the different heads.

The number of runner blades may vary from three to ten. The variation of blade angles is between 20° and 35°. The blade angle β_b is generally designated 0° for the optimum performance condition, opens to plus values and closes to minus values. Another convention is to set $\beta_b = 0^o$ for the fully closed position and all openings have plus angles.

The early axial-flow turbines have runner hubs of cylindrical shape. When the blades are made adjustable, a minimum clearance of 2–$5mm$ must be provided at the blade root to avoid interference. This clearance will increase many times at larger blade angles and will become a source of leakage that is estimated to result in loss of efficiency of around 1%. Propeller run-

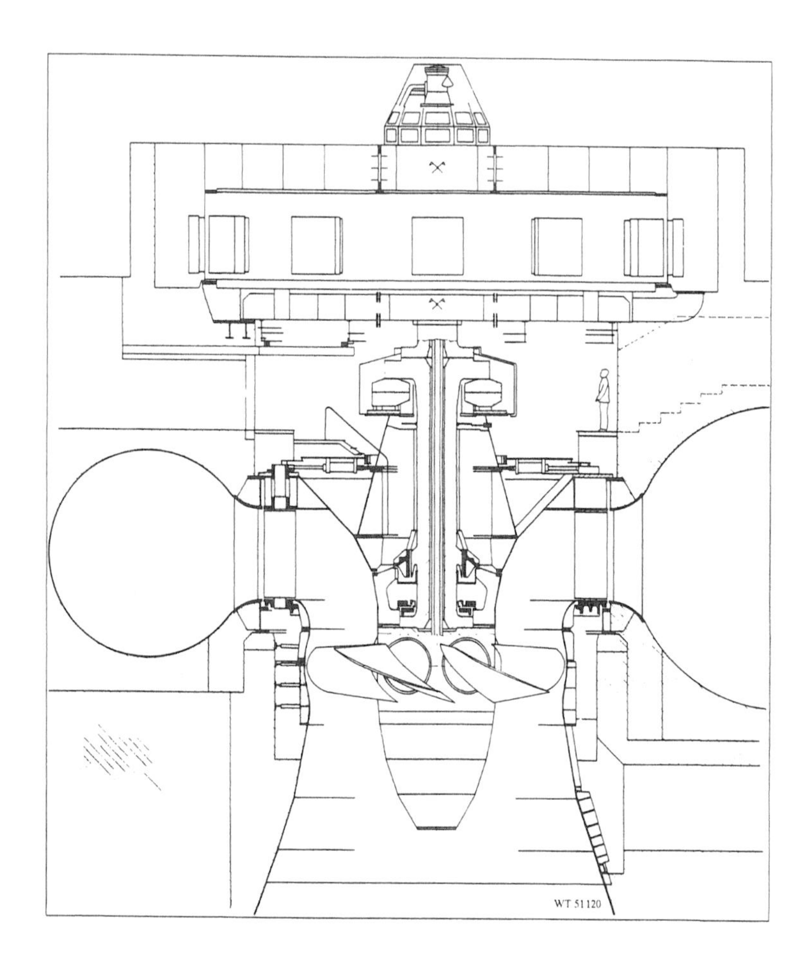

$H = 18.6m$, $Q = 1130m^3s^{-1}$, $P = 170MW$,
$n = 54.6min^{-1}$, $D_1 = 11300mm$.
Gezhouba Hydro Power Station, China Units 1 and 2, 1981

Figure 3.4 Low head Kaplan turbine
(Drawing courtesy Dongfang Electrical Machinery Works)

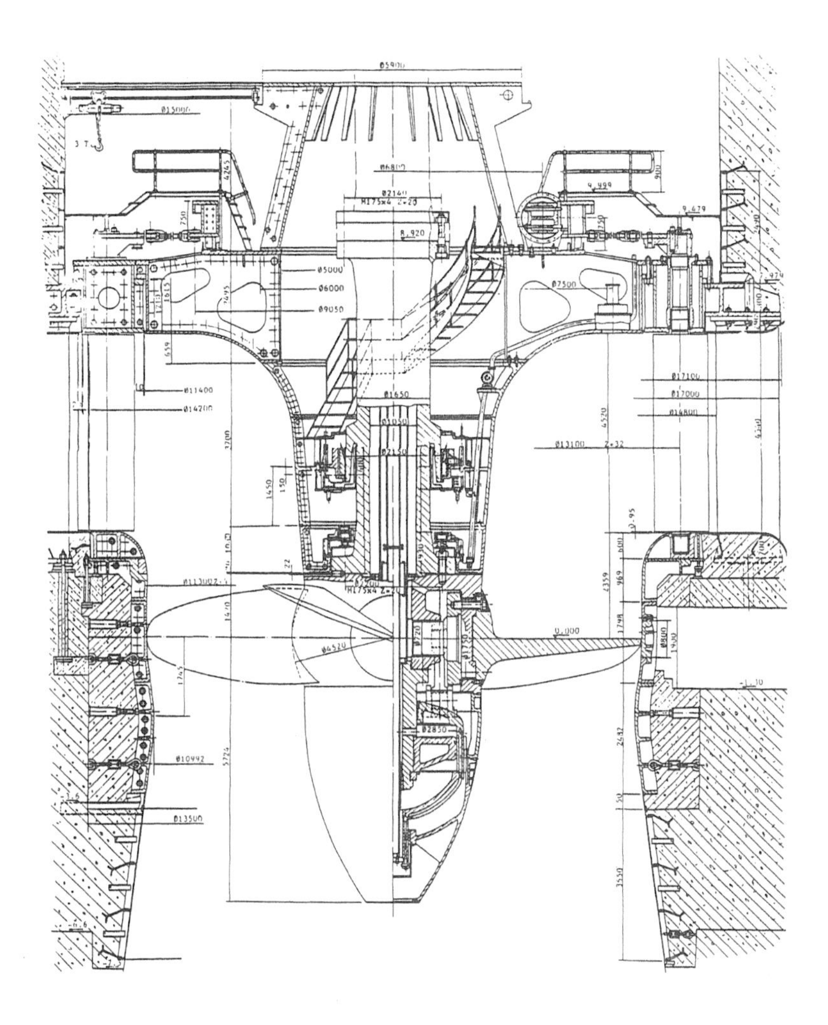

$H = 43m$, $P = 110MW$, $n = 93.8min^{-1}$, $D_1 = 7600mm$
Cedillo Hydro Power Station, Spain, 1986

Figure 3.5 Medium head kaplan turbine
(Drawing courtesy Sulzer Escher Wyss)

ners without such clearance gain an advantage in this respect.

Modern Kaplan turbine runners have a spherical segment in the region covered by the blade root and is able to minimize the clearances at all blade angles. But this spherical *bulge* reduces the through flow area somewhat and causes higher local velocities that have a detrimental effect on the cavitation characteristics of the turbine.

The runner hub must increase in diameter following the increase in the number of blades and the larger space required by the blade–adjusting mechanism with increasing head. These factors contribute to a shorter but stronger blade to withstand the higher hydraulic load. The range of variation of the runner hub is roughly as shown in Table 3.1.

Table 3.1 Runner hub geometry of Kaplan turbines

	Head range (m)	Number of blades	Hub ratio (D_h/D_1)
Very low head	5	3	0.33
Low head	5–25	4	0.35–0.4
Medium head	25–45	5–6	0.4–0.45
High head	50–80	7–8,10	0.5–0.6

The obvious drawback of having too large a hub is the resulting restriction of water passage so that for any given output and head, the size of a Kaplan turbine at high heads must be larger than that of a Francis turbine, and the advantage gained by higher specific speed is somewhat reduced. At 20–30m head the guide vane pitch circle diameter for a Kaplan turbine is about the same for a Francis turbine. For lower heads than this the Kaplan is smaller [3.7]. But a larger number of runner blades for the Francis turbine will increase the total blade surface area and lower the pressure difference across the blade per given area. This in effect increases the pressure level on the low pressure side of the blade and improves the cavitation characteristics.

3.2.4 Runner blades

The blades of an axial–flow runner has to withstand the bending moment produced by the pressure difference across the blade, the torsional moment caused by the hydraulic action and the tensile stress from the centrifugal force. The highest stressed region is at the blade root, so the blade is formed thin at the tip and thick at the root which then requires a fairly large flange to ensure a longer joint at the root.

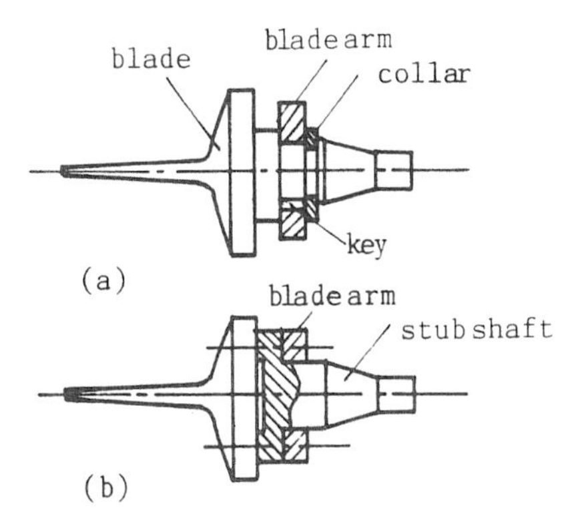

(a) integral stem (b) with stub shaft

Figure 3.6 Construction of blade stem

A blade stem at the inner end of the blade flange is necessary to support the blade in the hub and for connection with the blade–adjusting mechanism. This stem may be made integral with the blade and flange or as a separate stub shaft bolted to the flange, as shown in Figure 3.6. The choice between an integral and a separate blade stem is based on the following considerations:

(a) A two–piece design will allow the individual parts to be smaller and easier to manufacture whilst at the same time make possible the use of different materials for the blade and the stub shaft, for example stainless steel for the blade and carbon steel for the stem.

(b) The runner blade may be removed for repair without having to dismantle the whole runner with the two-piece arrangement, and replacement of flange seals is also made easier.

(c) However, it should be noted a two–piece blade stem will require a larger flange and will reduce the strength of the hub where the stem bores are already very close to each other.

Many attempts have been made over the years to apply the adjustable–blade axial–flow turbine to even higher head ranges. One solution realised in the USSR was to fix two runner blades on the same flange thus reducing the number of linkages of the blade–adjusting mechanism by two and allowing the use of a smaller hub. A schematic diagram of this twin-blade design is shown in Figure 3.7. For a turbine head of $50m$, the conventional Kaplan turbine would need eight blades and a hub ratio of 0.525 while the twin-blade turbine requires ($4\ x\ 2$) blades and a hub ratio of only 0.45 and has better cavitation characteristics. The Soviet hydroelectric station Serebiansk employed twin–bladed Kaplan turbines (manufactured by LMZ in 1981) working between heads of $82.7m$ and $71m$ with rated output of $68MW$ at $250min^{-1}$ [3.6]. The drawback of this design is obviously the uneven spacing of the runner blades at angles other than the optimum. This would naturally limit the use of the turbine to small head ranges or small changes in blade angles.

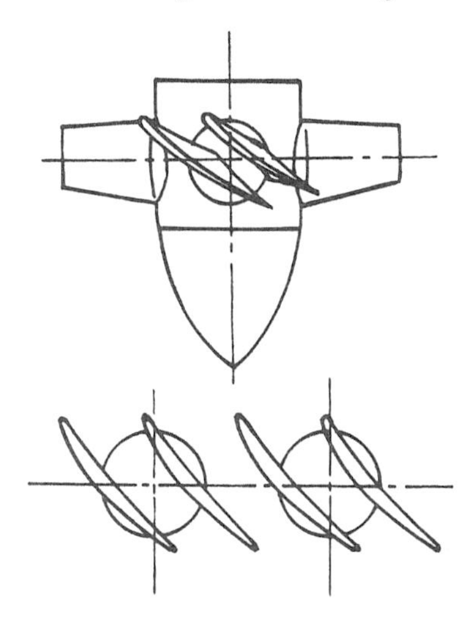

Figure 3.7 Schematic of twin–blade design

When the blades of Kaplan turbines are installed at initial assembly, the runner is placed on a boring mill with blades in the fully closed position and the outside diameter turned to the required dimension. A clearance between the runner tip and the runner chamber of 0.1% D_1 is normally allowed but more recent machines have much smaller clearances.

Carbon steel and low–alloy steels may be used to make runner blades for small turbines or installations with low cavitation tendencies. Such materials are low in cost, easy to cast and has good weldability. The common practice for repairing runner blades from cavitation erosion is to fill the pits with stainless steel weld rods and grind smooth to the original contour. Some machines have stainless steel overlays 3–5mm thick welded on beforehand at positions where cavitation is likely to occur. Welding will generally cause deformation due to heating and would require extensive correction. Overlaying with stainless steel strips has not been successful because of difficulties in the technique and results in strips becoming loose and fall off.

In applications where the cost is justified stainless steel blades are always used. Cast steel with 13Cr content was initially used but the casting, heat treatment and the repair of defects all proved to be difficult. Much improvement was later found in CrNi steels, a common grade is the OCr13Ni5MoCu cast steel which has good mechanical properties, weldability and cavitation resistance.

Figure 3.8 shows three typical designs of seals for runner blade flanges. Design (a) is the well known λ type seal that employs a single sealing ring to retain the oil on the inside and the water on the outside. There are many variations of this type of seal from different manufacturers. Designs (b) and (c) both use separate sealing rings to seal the oil and water.

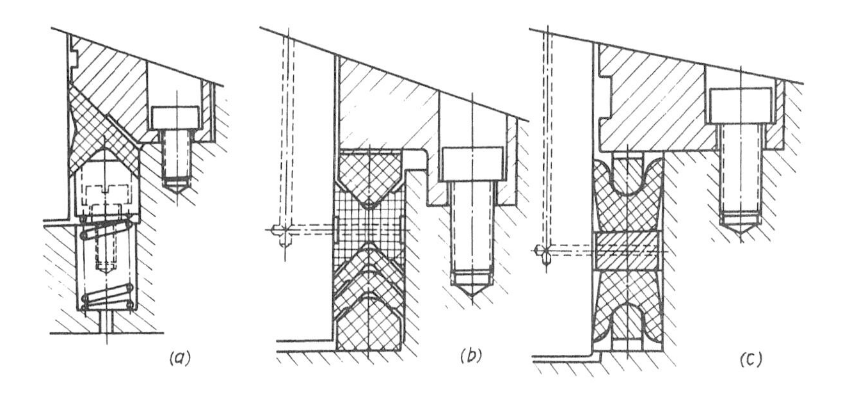

Figure 3.8 Details of blade flange seals

3.2.5 Blade–adjusting mechanism

Oil pressure is introduced to the servomotor cylinder by action of the selector valve and pushes the piston to the desired position. The servomotor may be placed in a number of locations in the turbine, such as in the runner hub below the blade–adjusting mechanism as in Figure 3.4 and above the blades as in Figure 3.9. In other designs the servomotor is placed in an enlarged section between the turbine and generator couplings, whilst others have it inside the space above the hollow hub of the generator rotor.

Figure 3.9 shows the details of construction of a five-bladed Kaplan runner. The vertical movement of the servomotor piston *1* is transmitted to the blade stem *3* via a five–prong crosshead *7* and lever arm *6*. An upward motion of the crosshead would cause the blade to turn in the clockwise direction and reduce the blade angle. The rotational force is transmitted through round keys *5*/ and the centrifugal force is retained by the collar *4* and bronze hushing *2*.A different type of mechanism does away with the crosshead and the lever arms connect directly to the servomotor piston via a swivel joint fitted inside the piston, as in Figure 3.10.

The blade–adjusting mechanism with crosshead is used in Kaplan turbines of all sizes and in all head ranges. It permits a greater degree of adjustment and requires comparatively less accuracy in manufacture. The disadvantage of this mechanism is that it has more parts and greater weight and requires overturning the runner hub during assembly. The mechanism with no crosshead on the other hand has less parts and is lighter in weight and can be assembled with the runner hub upright. However, the mechanism needs a higher degree of accuracy in manufacture and occupies greater space for each linkage, so it is used only in large turbines with smaller numbers of blades (six and less).

The problem constantly facing the turbine designer is how to optimise the function of the blade–adjusting mechanism whilst keeping the hub diameter to a minimum. One solution is to set the linkage arm at an angle from the vertical, as shown in Figure 3.11, so a longer crank than in the conventional design can be used to reduce the force in the axial direction. This mechanism can be used in runners with more than six blades. However, because of the inclination of the link, the lateral reaction at the pinion is higher and results in greater friction.

Another solution is to *stretch* the linkage system downward thus reducing the crosswise dimension such as the compound linkage system shown in Figure 3.12. This is a three–link system employing a bell crank between two connecting bars. The disadvantage of this design is also the increased friction with more pivot joints, but with the three–link arrangement it is possible to obtain a better combination of displacement and rotation of the mechanism so that the actuating force can be better matched with the blade hydraulic torque at different blade angles.

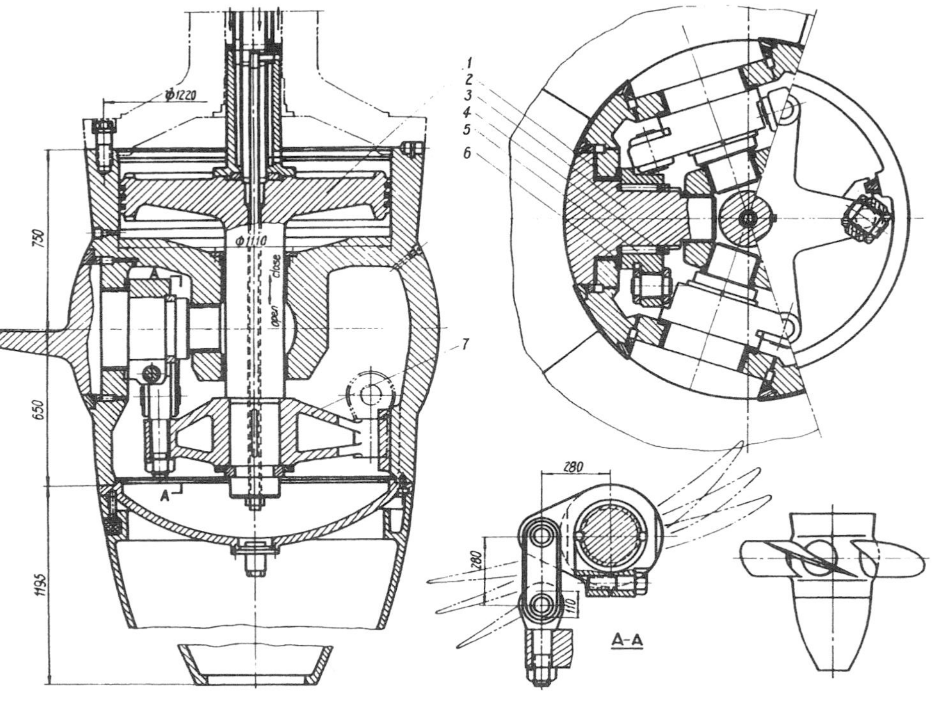

Figure 3.9 Construction of five–bladed Kaplan runner
(Drawing courtesy Harbin Electrical Machinery Works)

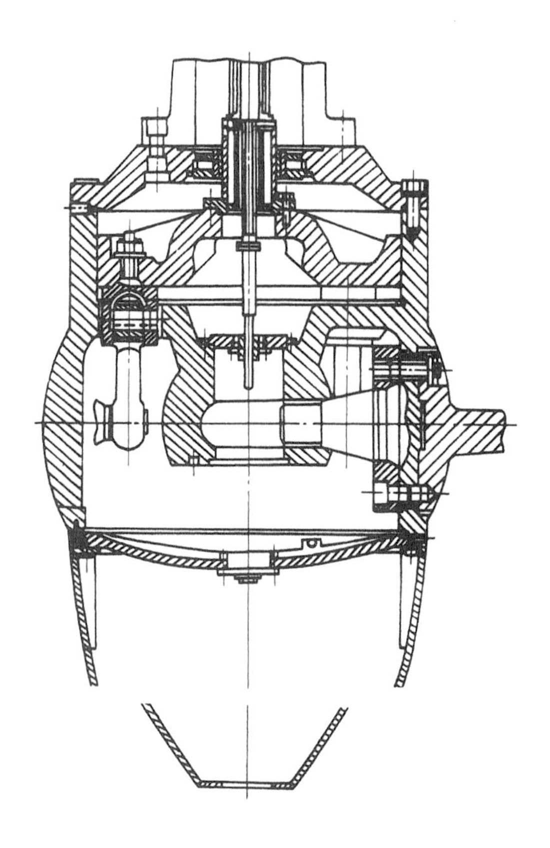

Figure 3.10 Construction of blade–adjusting mechanism with no crosshead [3.6]

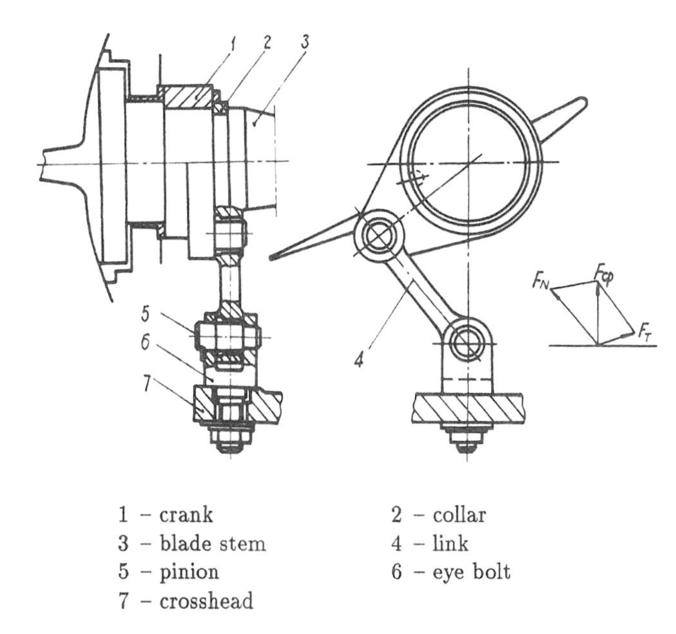

1 – crank
2 – collar
3 – blade stem
4 – link
5 – pinion
6 – eye bolt
7 – crosshead

Figure 3.11 Blade–adjusting mechanism with inclined linkage

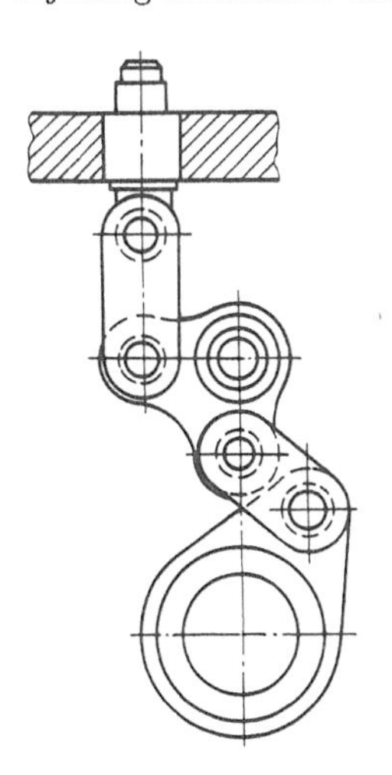

Figure 3.12 Compound linkage system

3.2.6 Oil pressure distributor

The oil pressure distributor has different forms of construction depending on the location of the servomotor. Figure 3.13 shows the distributor system for a turbine with the servomotor fitted inside the runner hub. It is a system of concentric pipes passing inside the hollow shafts of the turbine and the generator.

The oil pressure in the inner pipe *c* and the outer space *b* may alternate as controlled by a sliding valve at the top to determine whether the servomotor piston is to move upward or downward. The concentric pipe system is itself mounted on the piston so the motion of the piston is transmitted to the governor as a feedback signal via steel cables or a linkage system. As the top part of the pressure distributor is situated on the generator stator, good insulation from the generator must be ensured in order to prevent turbine parts from erosion by shaft current.

3.3 Diagonal–flow Turbines

3.3.1 Construction of diagonal–flow turbines

The diagonal–flow turbine retains many features of the Kaplan turbine in both hydraulic and mechanical design. It also has adjustable blades and is better known as the Deriaz turbine after its inventor.

Typical proportions of a diagonal–flow turbine is shown in Figure 3.14. The runner blades are mostly inclined from the vertical by 45° but depend on model investigation and conceptual design [3.9], diagonal–flow turbines may have the blade axes inclined from 30°–60°, depending on the operating head range roughly as: $\theta = 60^o$ for $H = 40$–$80m$; $\theta = 45^o$ for $H = 60$–$130m$ and $\theta = 30^o$ for $H = 120$–$200m$. The nominal diameter of the runner D_1 is measured from the points where the runner blade axes intersect the runner chamber. Because of its peculiar geometry, it has a large guide vane pitch circle D_0 and consequently a larger spiral case than a Kaplan turbine of the same runner diameter but a smaller draft tube throat diameter.

An example of the mechanical construction of a Deriaz turbine is shown in Figure 3.15. Some design features of the turbines in general are

1. As the runner blades are adjustable, diagonal-flow turbines are capable of coping with rather wide ranges of head variation. A ratio of H_{max} to H_{min} of 2 is easily attainable while in some applications it may be as high as 2.8. This puts the diagonal–flow turbine in a very competitive position compared with the Francis turbine.

2. The runner blades are distributed on a sphere of fairly large diameter, hence more blades can be accommodated than in a Kaplan turbine.

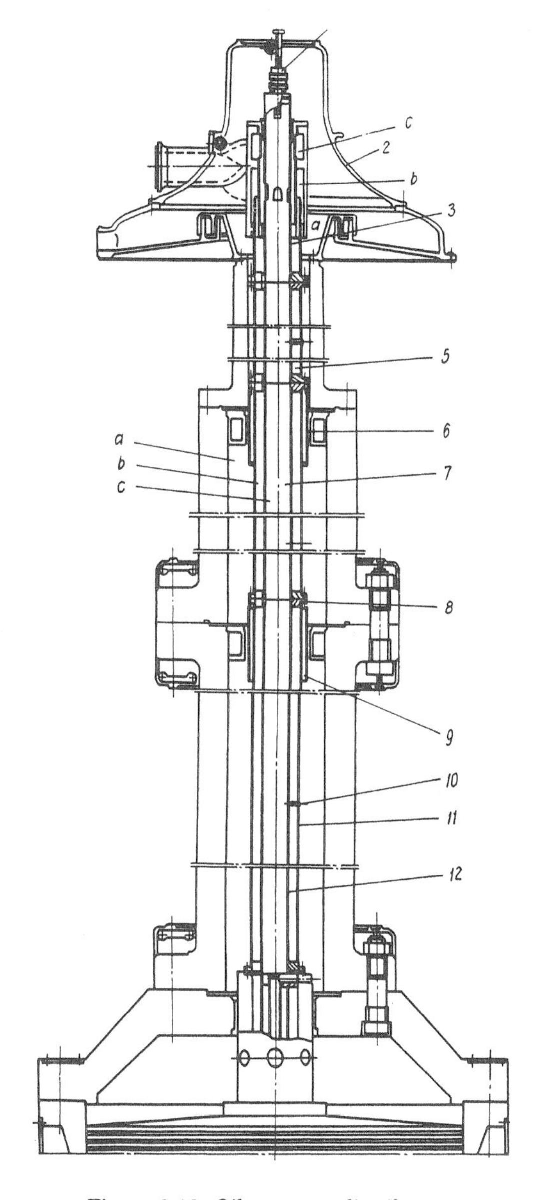

Figure 3.13 Oil pressure distributor

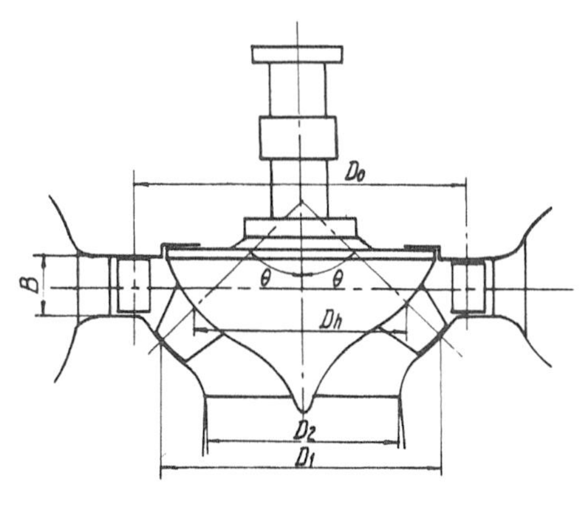

$D_0 = (1.28\text{-}1.44)D_1$ $\quad D_2 = (0.8\text{-}0.68)D_1$

$D_h = (0.55\text{-}0.65)D_1$ $\quad B = (0.30\text{-}0.18)D_1$

$\theta = 30^o\text{-}60^o$

Figure 3.14 Profile of diagonal–flow turbine

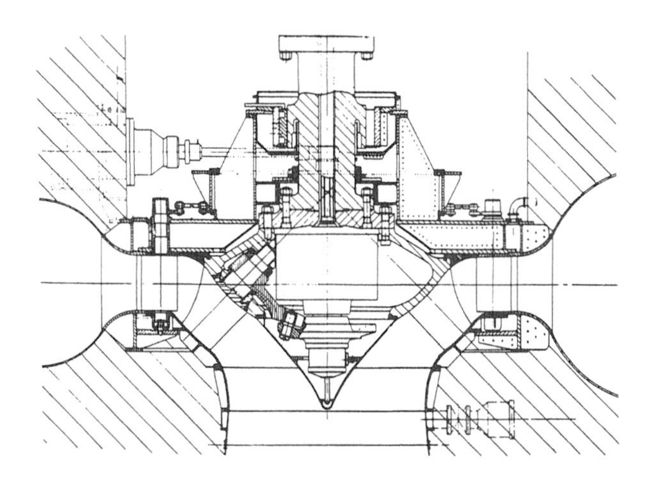

Turbine mode: $H = 74.2m$, $Q = 128m^3s^{-1}$, $P = 80.6MW$, $N = 150min^{-1}$, $D_1 = 5380mm$.

Figure 3.15 Typical Deriaz reversible pump–turbine
Valdecanas pumped storage power station, Spain. (Drawing courtesy IMH, EPEL)

Diagonal–flow turbines commonly have eight to ten blades while some experimental models go up to 12 and 14 blades. The highest head on record for a diagonal flow–turbine is $125m$ and the same for a diagonal flow pump–turbine is $136m$. Heads of 50–$150m$ is generally considered a logical range of application for diagonal–flow turbines.

3. Since the blades are installed at an angle to the axis, the centrifugal force generated in rotation will partly counteract the bending force due to hydraulic pressure and reduce the stresses in the blades. For reasons of this geometry it also has lower axial thrust and lower runaway speed than the Kaplan turbine.

4. The inclined blades, for the same span, have less difference in diameter between the blade tip and root, thus resulting in less curvature for the blade design. With a flatter profile the blades work more effectively in the pumping direction. This explains the very wide use of Deriaz type reversible pump–turbines in pumped storage projects like the example shown in Figure 3.15.

Although the diagonal–flow turbine has the above advantages over the Kaplan turbine, it is admittedly larger in size and heavier in weight and more complicated in construction than a comparable Kaplan turbine.

3.3.2 Runners of diagonal–flow turbines

Runners of diagonal flow turbines also have blade–adjusting mechanism and retain most features of Kaplan runners except for the following

1. All bores for the blade stems are on inclined planes, mostly at 45° angle, and require special fixtures for machining.

2. As the blades are inclined, the blade–adjusting mechanism requires a more elaborate articulation system whether actuated by a piston type or a sliding-vane type servomotor.

3. The axial–flow runner is allowed a small amount of adjustment in its vertical position, however, the diagonal flow runner must be accurately aligned to give the specified clearance between the runner blades and the runner chamber. Monitoring devices must be installed to check the runner position at all times. The turbine will be automatically stopped in case the clearance drops below an allowable limit.

Blade–adjusting mechanism for diagonal–flow turbines fall in two categories, the piston type and the sliding vane type. The former moves up and down while the latter rotates. Figure 3.16 depicts the principle of the piston type mechanism. The vertical motion of the servomotor piston *1* is transmitted via bell crank *3* into revolving motion of the blade arm *6* with pinion *4* as a fixed pivot. Slots are machined at the open ends of both bell crank *3* and arm *6* so that they can slide over pinions *2* and *5* respectively in action. A downward motion of piston *1* will rotate the blade to open.

Figure 3.17 shows the scheme of a sliding vane type mechanism. A number of vanes (four shown in this diagram) are attached to the actuator spindle *1* and are able to revolve in cylinder *4* where the same number of fixed vanes *3* divide the cylinder into individual compartments. When oil pressure is introduced into chamber A and drained from chamber B the sliding vane *2* will push spindle *1* and pinion *5* in the clockwise direction and cause the blade *7* to open through action of lever arm *6*.

The piston type mechanism employs the compound linkage system like that shown in Figure 3.12 but is laid out on an inclined surface, which will conceivably involve more difficulty in aligning the component parts. The sliding vane mechanism has a force transmitting system that appears to be more direct but the mechanical design must assure good sealing along the peripheries of the sliding vanes by providing ample rigidity of the housing and cover to prevent deformation of these parts when under oil pressure.

3.4 Embedded Parts of Axial–flow and Diagonal–flow Turbines

3.4.1 Spiral case

The axial–flow turbine works in the low and medium head ranges where the turbine flow may become quite large so it is more advantageous to use a concrete spiral case of polygonal cross-section. The diagonal–flow turbine works mainly in the medium head range and uses only steel spiral case of circular cross–section like the Francis turbine.

Most concrete spiral cases are called partial spiral cases that envelope only part of the turbine periphery. Partial spiral cases are widely used in low head applications with the purpose of reducing the crosswise dimension of the turbine and thus lessening the distance between machine centres which is vital for power stations with large numbers of units. However, the performance of the turbine will suffer slightly from this arrangement.

Figure 3.18 shows the outlines of two typical spiral cases, (a) with a wrap angle of 270° and Γ-shape cross sections and (b) with a wrap angle of 180° and Γ-shape sections. A general rule for selection is to use wrap angles of 225–270° for 30–40m head, 180–225° for less than 30m head and 135° for very low heads. Steel spiral cases of circular cross sections must be used for turbines over 40m head, like the examples shown in Figures 3.5 and 3.15.

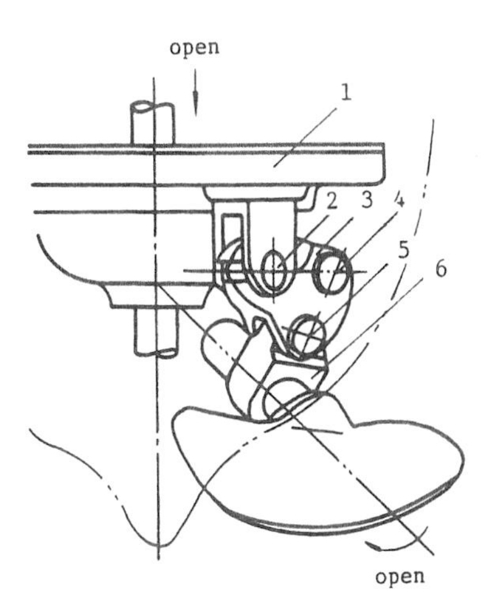

Figure 3.16 Piston type blade–adjusting mechanism

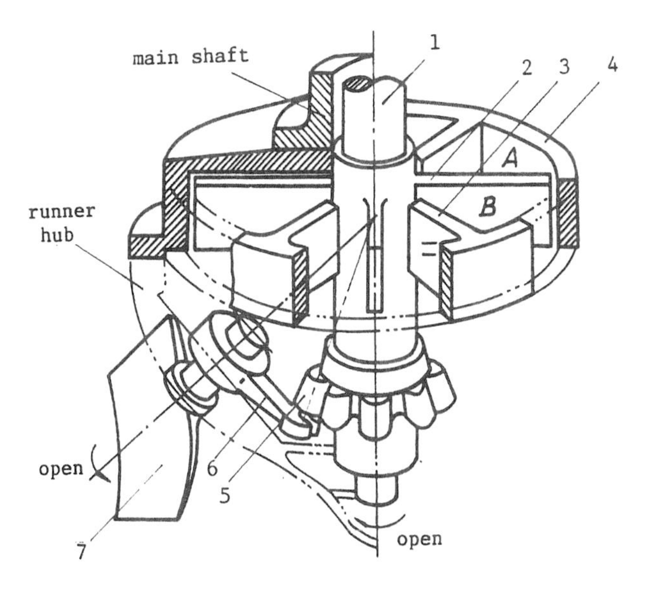

Figure 3.17 Sliding vane type blade–adjusting mechanism

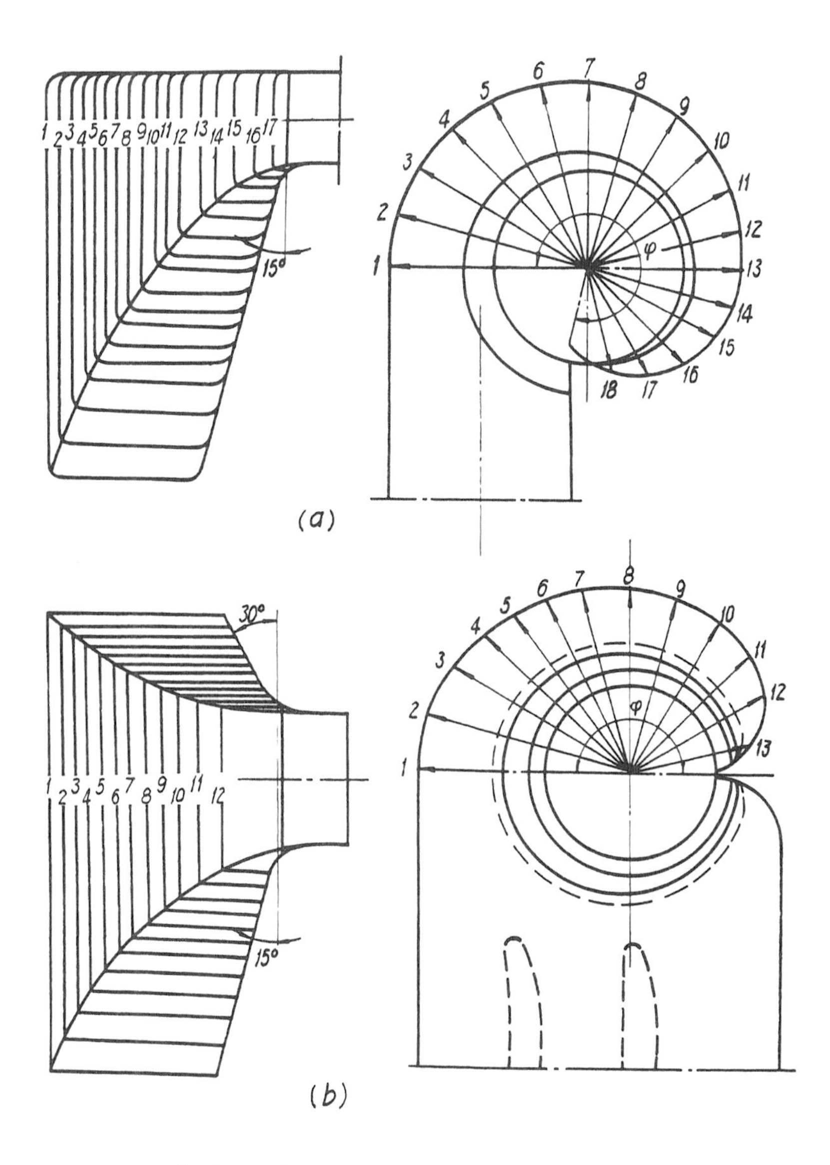

Figure 3.18 Outlines of partial spiral cases constructed with concrete [3.5]

In concrete spiral cases, when the tendency of seepage through the concrete at higher heads becomes more serious, a sheet metal liner is often used to prevent leakage and serves also to protect the concrete form being eroded.

3.4.2 Stay ring

The stay ring (stator or speed ring) of axial–flow turbines with concrete spiral cases is primarily a structural member to support the weight of the generator form above, but it also has to take up the loading of the spiral case since the reinforcement bars for the spiral case are normally welded to the stay ring. The stay ring is usually of fairly large dimension, so problems in manufacture and shipment must be carefully considered in the design stage.

The cross section and spacing of the vanes in stay rings vary from designs. For full spiral cases, the vanes are of the same profile and spaced equally around the periphery of the turbine. For partial spiral cases, at the upstream part of the case water flows nearly straight into the turbine, stay vanes in this region must be made curved to create the necessary amount of circulation and are therefore spaced closer together. At the farther end of the spiral case where circulation is already formed by the curved passage the stay vanes are much flatter and more sparsely spaced, as shown in Figure 3.19. The designer must bear in mind the asymmetry in layout when deciding on division of the stay ring into segments for transportation.

Figure 3.20 shows typical structures of stay rings for axial-flow turbines. In (a) is a conventional stay ring complete with upper and lower rings and stay vanes like those for Francis turbines. The components are usually of all welded construction and shipped to the hydroelectric station site in segments. Scheme (b) employs an upper ring and a number of columns that are individually embedded in the foundation concrete and joined only at the top with the upper ring. Manufacture and transportation are made easier this way but the stay ring has less rigidity than the previous design. This arrangement is used in large turbines but only in cases where the thrust bearing is not to be mounted on the head cover.

Scheme (c) does away with the upper ring also and has only individual columns that are embedded in the concrete separately. Integrity of the stay ring is relied solely on the head cover (serving as the upper ring) which is also anchored to the concrete. Special provisions must be made for assembly and removal of the guide vanes with head cover made fixed in this manner and also for that of scheme (b) (see 9.5.2).

3.4.3 Runner chamber

The axial–flow and diagonal–flow turbines have no bands on the blade tips, so accurately machined runner chambers are necessary to ensure the required

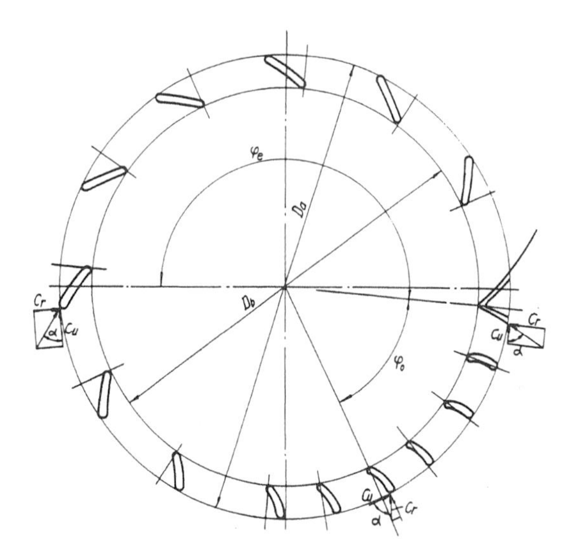

Figure 3.19 Arrangement of stay vanes for a partial spiral case

running clearance. For axial–flow turbines the runner chamber is usually cylindrical in the upper part and has a spherical section in the middle to match the blade tip contours at all blade angles. The lower part of the chamber contracts somewhat to form the throat of the flow passage that is slightly smaller than the runner diameter (see Figures 3.3 and 3.4). The runner chamber for diagonal–flow turbines is conical in shape and likewise has a spherical section in the area facing the runner blades (see Figure 3.15).

The runner chamber is actually subjected to very strong alternating loads in operation and should be provided with sufficient stiffness on the outside for strength and securely fastened to the anchors embedded in the foundation concrete.

Axial–flow turbine runners damaged by cavitation to an extent that cannot be weld repaired in place have to be brought to the workshop for repair entailing the removal of the generator rotor and the turbine guide apparatus. In some designs provisions are made for removal of runner blades only by making a section of the runner chamber detachable (to slide outward) so that a blade at a time may be swung out sideways and retracted through an enlarged inspection door in the draft tube liner, as is the case of the turbine shown in Figure 3.5.

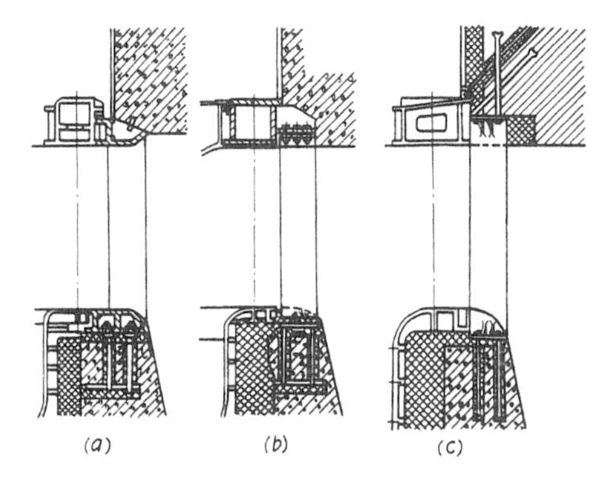

Figure 3.20 Construction of stay rings for axial-flow turbines [3.6]

3.5 Dimensioning of Turbine Components

Strength calculations for three typical types of axial-flow turbine parts, namely, turbine blade and component, blade-adjusting mechanism and stay ring for concrete spiral case, are cited as examples of design considerations.

3.5.1 Runner blade and component

(i) Runner blade

The runner blade of an axial-flow turbine is subject to complicated stress conditions due to its irregular shape and unevenly distributed hydraulic load. There is no simple and accurate method to calculate the strength of the blade. The following are design considerations derived from experience. The runner blade is assumed to be a cantilever beam fixed at the flange and strength is checked mainly for stresses at the blade root in addition to certain weak points in the blade outlet region.

The blade root is subject to three loads: (a) bending moment due to hydraulic pressure and weight of blade; (b) centrifugal force of the blade both at normal speed and runaway; (c) hydraulic moment about the blade axis. The bending moment exerted on a unit length of the blade root is

$$M_u = \frac{1}{D_F}\left[F_h(R_S - R_B) + G(R_g - R_B)\right] \tag{3.5}$$

where D_F – diameter of blade flange
F_h – hydraulic force on the blade
R_s – radius of point of hydraulic action

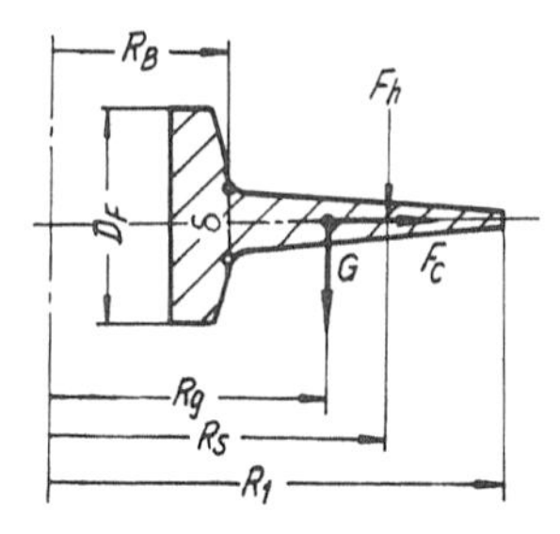

Figure 3.21 Loading on runner blade

R_B – radius of runner hub
R_g – radius of blade centre of gravity
G – weight of blade

The bending stress at blade root is

$$\sigma_W = \frac{7.8M_u}{\delta^2} \tag{3.6}$$

where δ is the thickness of blade root at the hub periphery.

The centrifugal force exerted at the blade root is

$$F_c = \frac{G}{g}R_g\omega^2$$

where ω is the angular velocity at normal/runaway speed.

The corresponding tensile strength is

$$\sigma_c = \frac{F_c}{D_F\delta} \tag{3.7}$$

Torsional moment at the blade root is

$$M_k = M_{11}D_1^3H$$

and the corresponding stress is

$$\tau_k = \frac{M_k}{K_2D_F\delta^2} \tag{3.8}$$

where K_2 is an empirical factor depending on ratio of D_F/δ, see Ref. [3.5].

The resultant stress at the blade root is

$$\sigma_{re} = \sqrt{(\sigma_W + \sigma_c)^2 + 4\tau_k^2} \tag{3.9}$$

Several sections along the outlet edge of the runner blade must be checked for strength mainly due to bending.

(ii) Blade arm

The blade arm as shown in Figure 3.22 transmits the control force F_n from the blade–adjusting mechanism to the blade and sustains the centrifugal pull of the blade F_c through a collar or a number of bolts.

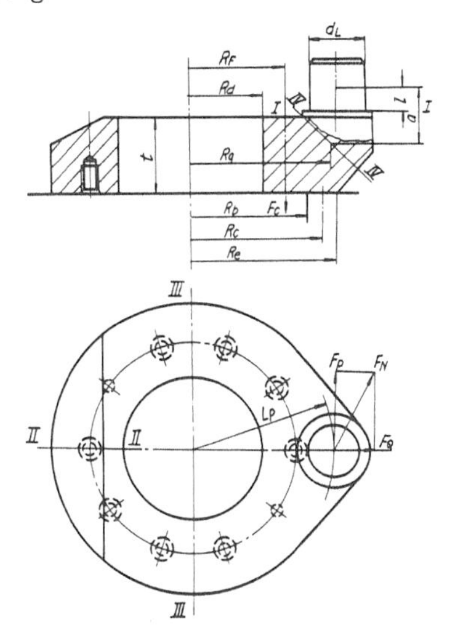

Figure 3.22 Loading on blade arm

The combined stress at section I–I is

$$\sigma_I = \sqrt{\left(\frac{F_N \ell}{0.1 d_L^3}\right)^2 + 4\left(\frac{4F_N}{\pi d_L^2}\right)^2} \tag{3.10}$$

At section II–II:

$$\begin{aligned} \text{bending moment } M_t &= F_c(R_c - R_f)/2\pi \\ \text{tangential stress } \sigma_{t1} &= \frac{6M_t}{t^2 \cdot r \cdot log_e(R_e/R_d)} \end{aligned} \tag{3.11}$$

where R_c – average radius of the thrust bearing surface
r – radius of calculated point

At section III–III:

$$\text{bending moment } \sigma_b = \frac{3F_p L_p r}{2t(r_e^3 - r_d^3)}$$
$$\text{total stress } \sigma_{III} = \sigma_b + \sigma_{t1} \tag{3.12}$$

At section IV–IV:

$$\text{bending stress } \sigma_{b1} = \frac{3F_p a}{(R_e - R_d)t^2}$$
$$\text{tangential stress } \sigma_{t2} = \frac{F_p}{rA_{IV}} \cdot \frac{L_p}{R_a} \tag{3.13}$$

where A_{IV} – area of cross section IV–IV

The resultant stress is given by

$$\text{total stress} \quad \sigma_{IV} = \sigma_{t1} + \sigma_{t2} + \sigma_b \tag{3.14}$$

3.5.2 Blade–adjusting mechanism

The pitch of runner blades of a Kaplan turbine is controlled by the oil pressure actuated servomotor via a linkage system known as the blade–adjusting mechanism. Figure 3.23 shows a control mechanism employing a single link arm in the near–vertical position (as in Figure 3.9). An example is given here on the determination of the forces and torques transmitted through members of the mechanism. The strength of each member can be calculated by conventional methods.

The control torque of one blade is

$$M_c = \pm M_s + M_t \tag{3.15}$$

Tangential force exerted on the arm pin is

$$F_p = \frac{\pm M_s + M_t}{L_p} \tag{3.16}$$

where M_s is the hydraulic torque which has to be obtained from model test data, the minus sign is used when M_s is in the same direction as M_c, otherwise the plus sign is used; M_t is the friction torque that can be calculated for a given design.

The friction torque of the blade stem is

$$M_t = \mu R_a \frac{d_a}{2} + \mu R_b \frac{d_b}{2} + \mu R_c \frac{d_c}{2}, \tag{3.17}$$

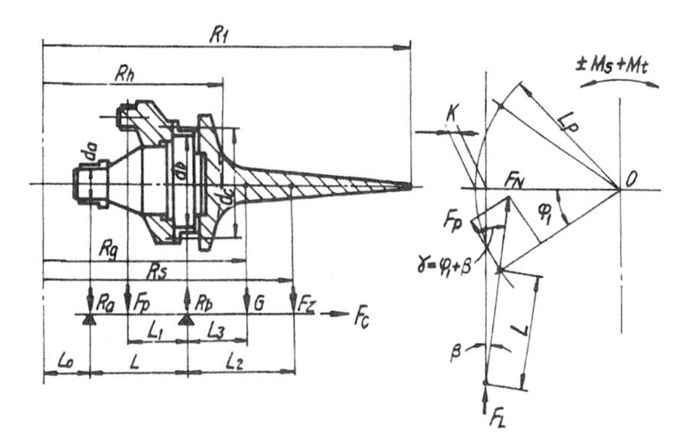

Figure 3.23 Forces acting on blade–adjusting mechanism

where R_a, d_a – reaction and diameter of stem at pivot a, respectively,
R_b, d_b – same at pivot b
R_c, d_c – reaction and average diameter at axial thrust bearing surface
μ – coefficient of friction, normally 0.15-0.20

The reaction at pivots a and b are

$$R_a = \mp\frac{L_1}{L}F_p + \frac{L_2}{L}F_z + \frac{L_3}{L}G,$$
$$R_b = \pm\left(1 - \frac{L_1}{L}\right)F_p + \left(1 + \frac{L_2}{L}\right)F_z + \left(1 + \frac{L_3}{L}\right)G \tag{3.18}$$

where G is the weight of blade and component parts, and F_z is the axial hydraulic pressure on one blade at radius.

$$R_s = \frac{4}{3}\left(R_{av} - \frac{R_1 R_h}{4R_{av}}\right)\frac{\sin\varphi}{\varphi} \tag{3.19}$$

where $R_{av} = (R_1 + R_h)/2$, and φ is one half of the wrap angle of the blade at radius R_{av}.

The axial reaction is

$$F_c = F_{c1} + F_{01} \tag{3.20}$$

where F_{c1} is thecentrifugal force of blade and component, and F_{01} is the thrust from oil pressure inside the runner hub.

Substituting the various reaction forces into equation (3-17) and rearranging

$$M_t = \mp\,\mu A_1 F_p + \mu A_2 F_z + \mu A_3 G + \mu A_4 F_c \tag{3.21}$$

where constants A_1, A_2, A_3 and A_4 are relations of geometrical details. When the direction of F_p is the same as F_z, it takes the minus sign in equations 3.18 and 3.21, otherwise it takes the plus sign.

By substituting 3.21 into 3.16, force F_p becomes

$$F_p = \frac{\pm M_s + \mu A_2 F_z + \mu A_3 G + \mu A_4 F_c}{L_p \pm A_1} \tag{3.22}$$

The vertical actuating force of each blade (Figure 3.23) is

$$F_{av} = \frac{F_p \cos\beta}{\cos\gamma} \tag{3.23}$$

Since β is rather small, then $\cos\beta \approx 1$. By considering additionally the friction at linkage joints and seals, the above becomes

$$F_{av} = 1.2\left[\frac{\pm M_s + \mu A_2 F_z + \mu A_3 G + \mu A_4 F_c}{(L \pm A_1)\cos(\phi_1 + \beta)}\right] \tag{3.24}$$

The relation between the link swing angle $\beta$and blade angle φ_1 from Figure 3.23 is

$$\sin\beta_1 = \frac{L_p(1 - \cos\varphi_1) - K}{L} \tag{3.25}$$

The minimum oil pressure of the servomotor is then

$$P_{min} = \frac{Z_1 F_{av}}{A} \tag{3.26}$$

where Z_1 is the number of blades, and A is the effective area of servomotor piston.

3.5.3 Stay ring for concrete spiral case

The stay ring for a concrete spiral case is loaded in a rather irregular way and is difficult to calculate accurately. Approximate methods are commonly used to decide on the major dimensions and subsequent numerical analysis applied when necessary.

The upper and lower rings of the stay ring are assumed to have very little stiffness so they are treated as uniformly loaded horizontal beams with fixed ends supported by individual vertical columns. Loading is considered under the following three operating conditions:

(i) Spiral case dewatered

The stay ring is to support the weights of the generating unit G_u and the concrete mass above it G_c. The latter is composed of the weight of concrete directly exerted on the stay ring G_{c1} and the weight of the spiral case roof shared by each column G_{c2}. The load G_{c1} is determined from:

$$G_{c1} = \frac{\pi}{4}\left(D_{ou}^2 - D_p^2\right) h\rho_c \tag{3.27}$$

where D_{ou} is the maximum diameter of stay ring, D_p the diameter of turbine pit, h the thickness of concrete above stay ring, and ρc the specific gravity of concrete.

The load G_{c2} is normally determined by drawing fan-shaped segments of the spiral case each containing one stay vane as shown in Figure 3.24 and calculating the weight of concrete in each segment. The total load on one particular column (stay vane) is then:

$$G_{Ii} = \frac{\alpha_i}{360}(G_u + G_{c1}) + G_{c2} \tag{3.28}$$

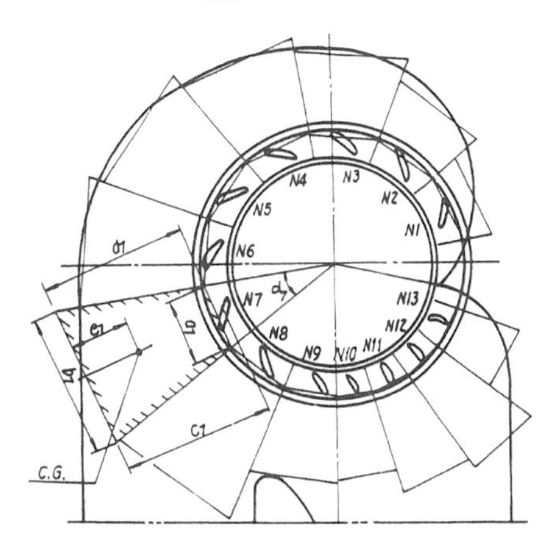

Figure 3.24 Load distribution on concrete spiral case

(ii) Turbine under normal operation

In normal operation, the stay ring must sustain the additional loads such as the axial thrust of runner F_z, the hydraulic pressure within the outer diameter of the stay ring, and the hydraulic pressure in the spiral case shared by the stay vane F_{szi}. F_{s1} is given by

$$F_{s1} = p_1\frac{\pi}{4}(D_{ou}^2 - D_s^2), \tag{3.29}$$

where:

$$p_1 = g(H_1 - C_1^2/2g),$$

with H_1 being the elevation difference between normal pool level and the centre line of the guide vane, C_1 the flow velocity under the head cover, normally $C_1 = 1.5Q/(\pi D_0 B)$, and D_s the shaft diameter.

The load shared by each stay vane is

$$F_{szi} = p_2 \frac{a_i + b_i}{2} \cdot d_i e_i \cdot \frac{1}{c_i} \tag{3.30}$$

where:

$$p_2 = \rho g(H_2 - C_2^2/2g)$$

with H_2 the elevation difference between maximum pool level and the upper ring of the stay ring, C_2 the average velocity in the spiral case, and a_i, b_i, c_i, d_i, e_i, as shown in Figure 3.24.

The load on one stay vane under normal operation is then

$$G_{IIi} = G_{Ii} - F_{szi} + \frac{\alpha_1}{360}(F_z - F_{s1}) \tag{3.31}$$

(iii) Under turbine load rejection

During load rejection the following forces are exerted on the stay ring besides G_{Ii}: hydraulic pressure between outside diameter of stay ring and guide vane pitch circle F_{s3}; vacuum force in region between guide vane pitch circle and shaft F_{s4}; hydraulic pressure in spiral case F_{s5i}. They are determined from

$$F_{s3} = p_3 \frac{\pi}{4}(D_{ou}^2 - D_o^2) \tag{3.32}$$

where p_3 is the hydraulic pressure in the spiral case;

$$F_{s4} = p_4 \frac{\pi}{4}(D_o^2 - D_s^2) \tag{3.33}$$

where p_4 is the vacuum within the guide vane circle, normally $0.1MPa$

$$F_{s5i} = p_3 \frac{a_i + b_i}{2} \cdot d_i e_i \cdot \frac{1}{c_i} \tag{3.34}$$

Therefore the total force on one stay vane at load rejection is

$$G_{IIIi} = G_{Ii} + \frac{\alpha_i}{360}(F_{s4} - F_{s3}) - F_{s5i} \tag{3.35}$$

References

3.1 'Kaplan and tubular Turbines', Escher Wyss News, 1964, 1/2.

3.2 Graeser, A.-E. 'Abaque pour turbines hydrauliques' IMH, EPFL, Lausanne, 1974.

3.3 di Siervo, F., de Leva, F. 'Modern Trends in Selecting and Designing Kaplan Turbines', *Water Power and Dam Construction*, Dec. 1977/Jan. 1978.

3.4 Schweiger, F., Gregori, J. 'Developments in the Design of Kaplan Turbines', *Water Power and Dam Construction*, Nov. 1987.

3.5 Harbin Large Electrical Machinery Institute *Hydraulic Turbine Design Manual*, Mechanical Industry Press, Beijing, 1976 (in Chinese).

3.6 Kovalev, N.N. (editor) 'Hydraulic Turbine Handbook', Mashinostroenia, Leningrad, 1984 (in Russian).

3.7 Brown, J.G. 'Hydroelectric Engineering Practice', Vol. 2, Blackie and Son, Ltd., 2nd edition, 1970.

3.8 Institut de machines hydrauliques 'Feuilles de cours illustrees', B 2e edition, EPFL, 1975.

3.9 Krivchenko, G.I. *Hydraulic Machines - Turbines and Pumps*, MIR Publishers, Moscow, 1986.

Chapter 4

Construction of Tubular Turbines

He Guo-ren

4.1 General Features of Tubular Turbines

4.1.1 Application ranges of tubular turbines

Tubular turbines which have, in general, excellent hydraulic characteristics such as higher specific speed, larger discharge capacity and higher efficiency are best suited for development of low head hydroelectric potential in the plain regions and of tidal energy. The general working head range is 2–25m, to as high as 40m with straight–flow turbines. Based on features of their mechanical construction and arrangement of installation, tubular turbines may be classified into:

- Tubular turbine
    - Straight–flow turbine
    - Semi-straight–flow turbine
        - Bulb turbine
        - S–type tubular turbine
        - Pit–type turbine

Each type of turbine has a different power capacity and head range to which it is best suited. Various manufacturers offer their recommended series of tubular turbines for selection by prospective users. Figures 4.1 through 4.3 are typical selection diagrams.

4.1.2 Arrangement and construction features of tubular turbines

Tubular turbines are sometimes known by the following types from their construction features, especially from the arrangement of the generators.

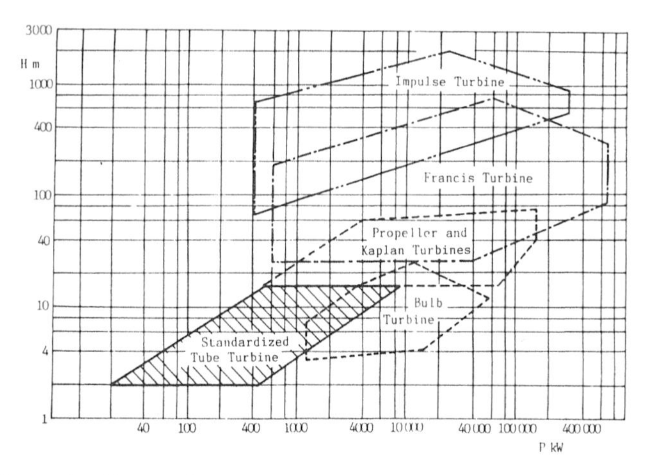

Figure 4.1 Range of application of tubular turbines in relation to other types of turbines (Diagram courtesy Voith Hydro Inc.)

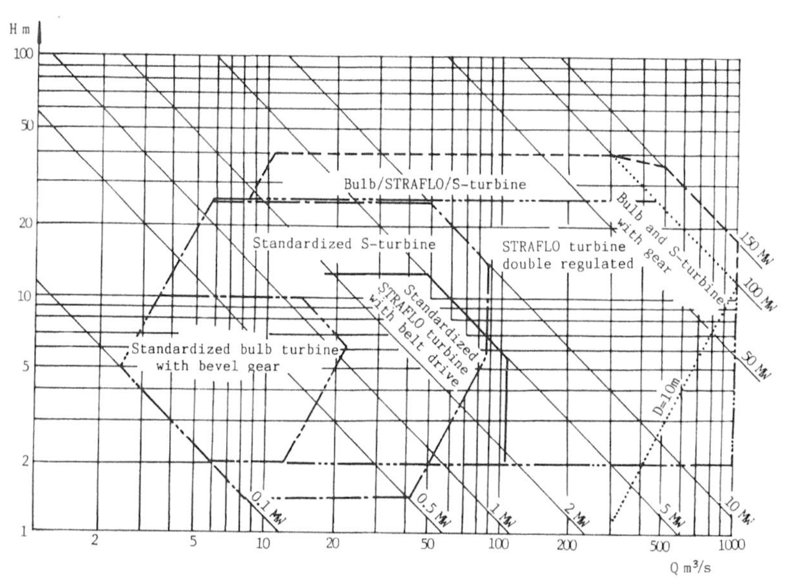

Figure 4.2 Range of application of bulb, S-type and STRAFLO turbines (Diagram courtesy Sulzer Escher Wyss)

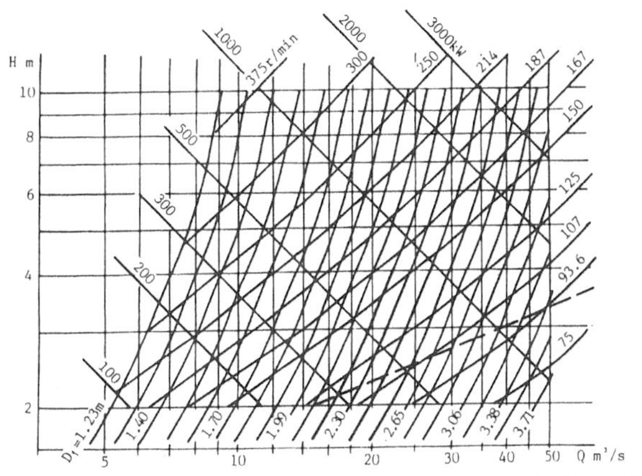

Figure 4.3 Monograph for small capacity pit–type turbular turbines (Diagram courtesy Voith St. Polten)

(i) Bulb turbines

In a bulb turbine, the generator is encased in a metallic shell commonly called a bulb and placed either upstream or downstream of the turbine. Figure 4.4 shows a typical bulb turbine with the bulb casing upstream of the turbine runner which is currently the most commonly used arrangement. Water entering from the straight conical inlet through the conical guide apparatus into the runner and leaves by way of the straight conical draft tube.

Some installations have the bulb casing placed downstream of the turbine, in which the water flows around the bulb casing in an annulus passage at the outlet of the turbine, as shown in Figure 4.5. Comparative model tests have been made on turbines with upstream and downstream bulbs and showed advantage for the former. In the head range of 3–9.5m, the upstream bulb unit has an efficiency of 89–86% while the downstream bulb unit has only 88–80%which is about equal to vertical axial–flow turbines [4.2]. The bulb turbine, due to its use of a straight–through flow passage, eliminates the need for a spiral case as in vertical shaft Kaplan turbines and allows the shortening of the distance between units, as shown in Figure 4.6. It is also apparent from the diagram that civil engineering work is greatly reduced for the installation with bulb turbines.

Most bulb turbines are supported by two bearings placed in the bulb casing. One arrangement is to have the turbine runner and generator rotor mounted on one common shaft and overhung on both sides of the bearings, as shown in Figure 4.4. Another arrangement uses separate shafts for the

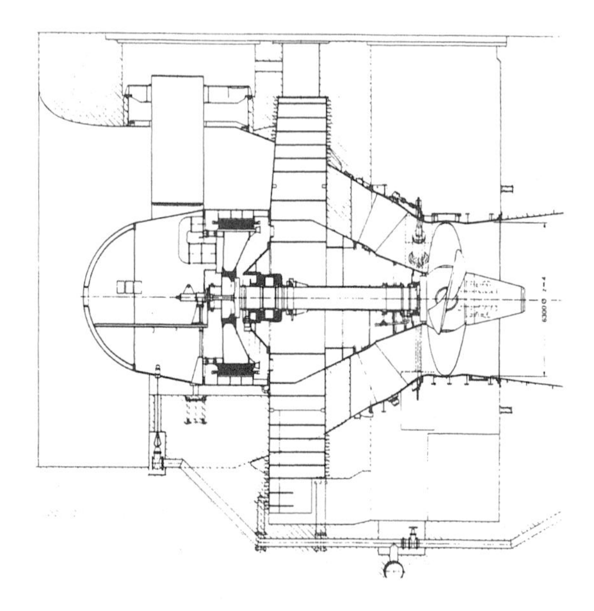

$H_r = 6.6m,\ n = 75min^{-1},\ P = 18.0MW,\ D_1 = 6300mm$

Figure 4.4 Section of bulb turbine, Majitang, China
(Drawing courtesy J.M. Voith)

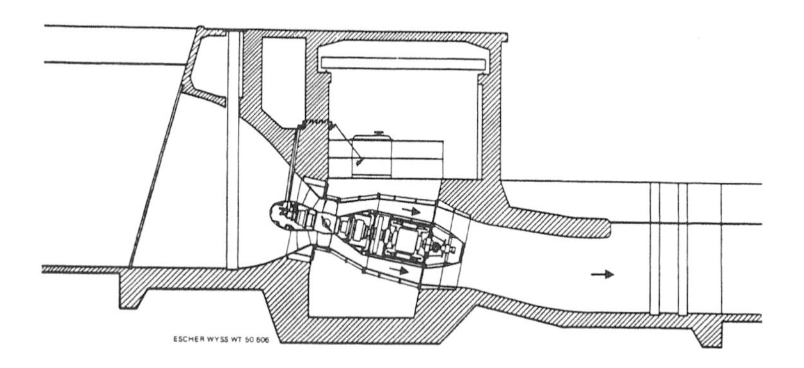

Figure 4.5 Tubular turbine with downstream bulb casing
(Drawing courtesy Sulzer Escher Wyss)

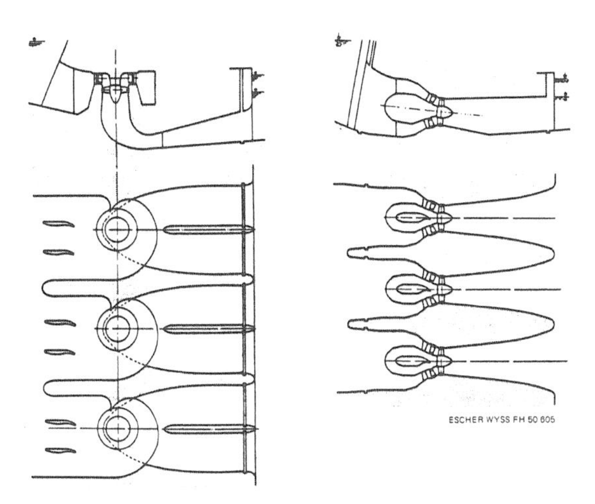

Figure 4.6 Comparison of installations of bulb turbine with Kaplan turbine of identical parameters (Drawing courtesy Sulzer Escher Wyss)

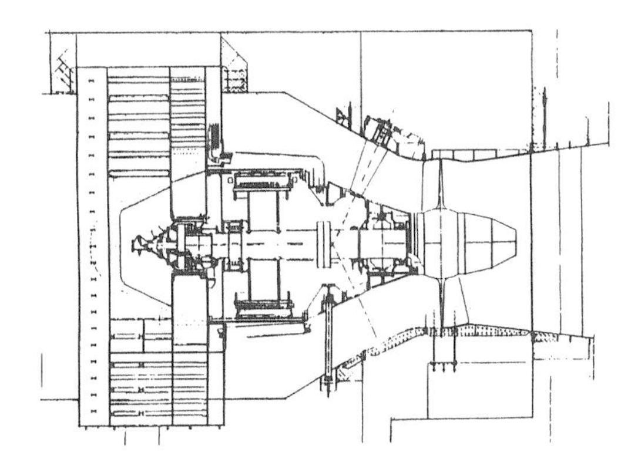

$H_r = 14.7m,\ P = 45MW,\ D_1 = 6.4m,\ n = 107min^{-1}$

Figure 4.7 Bulb turbine with generator rotor between two bearings Belley, France (Drawing courtesy Neyrpic)

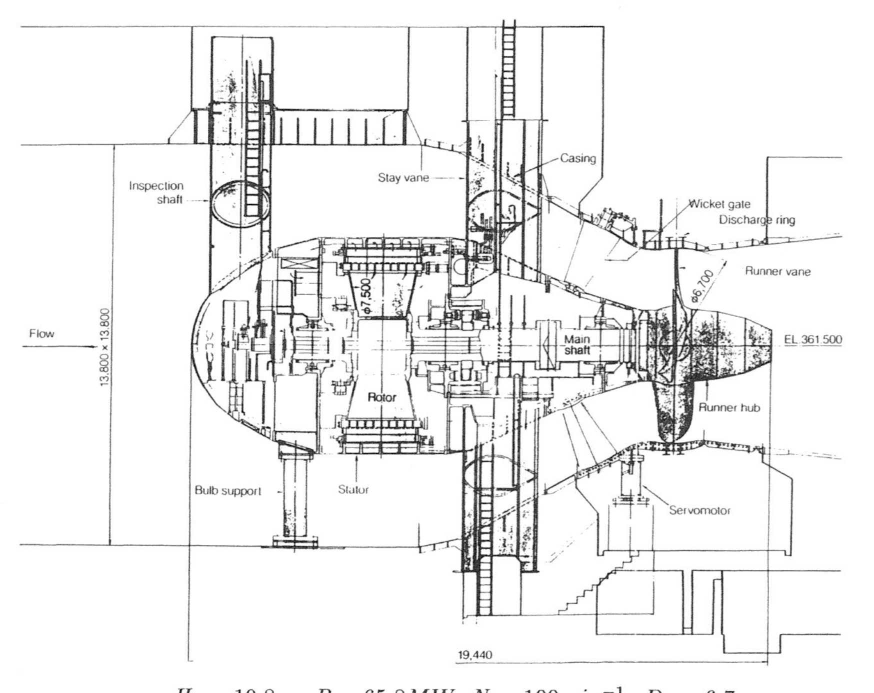

$H_r = 19.8m,\ P = 65.8MW,\ N = 100min^{-1},\ D_1 = 6.7m$

Figure 4.8 Bulb unit with three bearings, Tadami, Japan (Drawing courtesy Hitachi)

turbine and generator with the generator rotor situated between the turbine guide bearing and the upstream combined thrust guide bearing like the unit shown in Figure 4.7. The construction employing one main shaft with overhung runner and rotor has been used widely in large capacity bulb units of recent manufacture. One large bulb unit uses a three bearing arrangement by placing an additional upstream guide bearing beyond the generator, as in Figure 4.8, since the overhung would be too long due to the large capacity of the generator.

(ii) Pit–type tubular turbines

The generator is placed in an open vertical concrete pit upstream of the turbine. The generator and other machine parts may be lifted by way of the pit to the upper floor for repairs. In low head applications, when a planetary gear increaser is incorporated in the drive shaft, the generator may then be of standard design, like the one in Figure 4.9.

A main guide bearing is located near the turbine runner while the thrust bearing is placed before the increaser gear box, separate bearings are provided at both ends of the generator rotor. In other units where a parallel–shaft increaser is used, the turbine and generator shafts will not be in line. In Figure 4.9, the machine axis is inclined from the horizontal by about 10° so that some civil engineering work may be saved at the upstream section, especially for installations with a long intake structure [4.3].

(iii) S–type tubular turbine

There are two main configurations of S–type tubular turbines, the one with main shaft passing through the S-shape draft tube and connecting with the downstream generator, the other with main shaft passing through the inlet conduit and connecting with the upstream generator. S–turbines may be installed horizontal or inclined, the drive may also be direct connection or through increaser gearing. Figure 4.10 shows an example with the extension shaft in the upstream and Figure 4.11 shows one in the downstream.

4.2 Hydraulic Characteristics and Flow Passage Design of Tubular Turbines

4.2.1 Hydraulic characteristics

The tubular turbine utilizes a straight converging inlet passage, a set of conical guide vanes and a straight conical draft tube to replace the spiral case, radial guide vanes and bent draft tube in a vertical axial–flow turbine. The water flow in the tubular turbine is basically in the axial direction with the least amount of curvature and obstructions that gives it the advantage

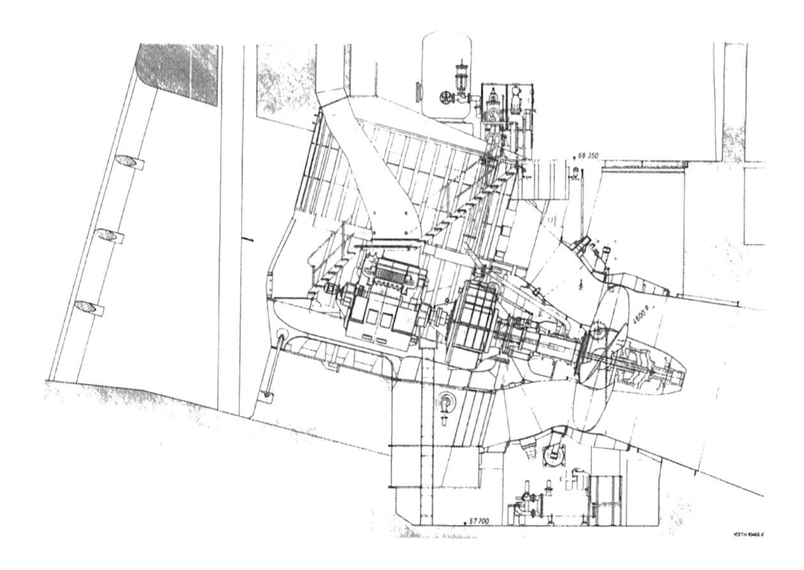

$H_r = 5.3m,\ P = 4.64MW,\ D_1 = 4.6m,\ n = 85/750min^{-1}$

Figure 4.9 Pit–type tubular turbine, Lehmen, F.R.G.
(Drawing courtesy J.M.Voith)

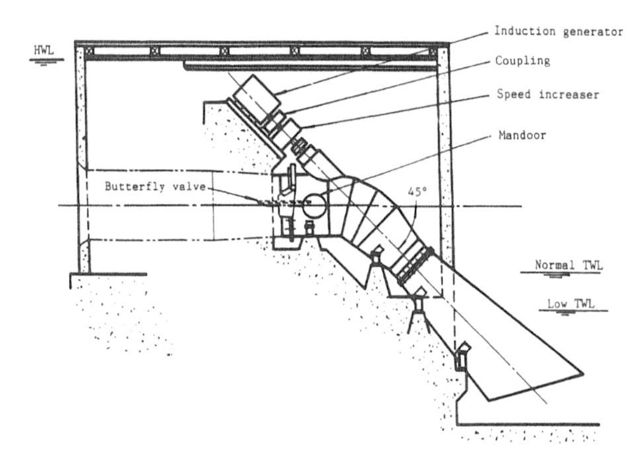

Figure 4.10 S–type tubular turbine with upstream generator
(Drawing courtesy Voith Hydro, Inc.)

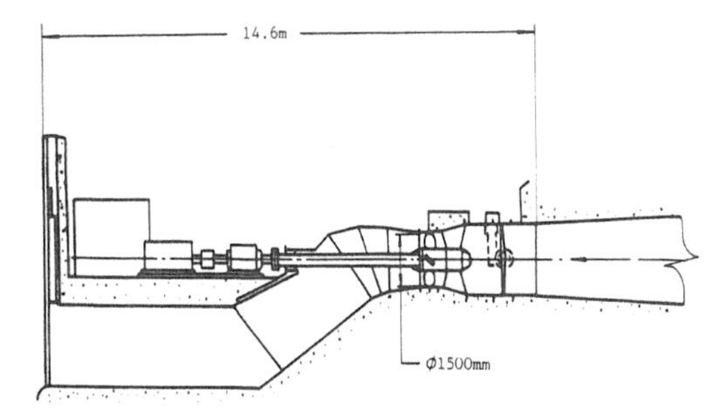

Figure 4.11 S–type tubular turbine with downstream generator [4.1]

of high discharge capacity and hydraulic efficiency.

The major advantages of a tubular turbine are:

1. With the absence of the spiral case, the flow is better distributed and with less loss. Greater simplification in civil engineering design is possible;

2. As compared with vertical units, the elimination of bends in flow passage from the guide vanes to the runner will bring about a reduction in flow losses;

3. As is well known, the straight conical draft tube is more efficient than a bent draft tube;

4. Due to its larger discharge capacity and higher efficiency, the tubular turbine will have a higher power output than an axial–flow turbine of the same runner diameter, or a smaller diameter for the same output.

A comparison between the parameters of a tubular turbine and a Kaplan turbine is given in Table 4.1.

4.2.2 Bulb casing, inlet passage and supports of tubular turbines

The proper design of the flow passage components of a tubular turbine greatly influences its hydraulic performance. Aside from the conical guide apparatus and the straight conical draft tube that have decided effect on the discharge capacity and efficiency of the unit, the design of the bulb casing, the inlet passage and the support structure are also important factors.

Table 4.1 Comparison of parameters of bulb turbine with Kaplan turbine

Machine type	Bulb tubular turbine	Vertical Kaplan turbine
Model	GZ003-WP-550	ZZA30-LH-650
Runner diameter (m)	5.5	6.5
Design head (m)	6.2	6.2
Maximum head (m)	15.0	15.0
Output power (kW)	10,420	10,420
Speed (min^{-1})	78.9	54.7
Weight of turbine (t)	420	530
Centreline spacing (m)	13	21

(Data supplied by the Tienjin Generating Equipment Works based on design studies for the Baigou Hydropower Station in Guangdong)

(i) Bulb casing

The size of the bulb casing which houses the generator depends mainly on the overall construction of the unit. A small bulb with straighter flow passage is obviously desirable for decreasing hydraulic losses. However, under allowable hydraulic dimensions, a larger bulb will facilitate the installation and operation of the generator. For medium/small capacity units, speed increasers with planetary or parallel shaft gearing are used to reduce the size of the generator and consequently the diameter of the bulb (but at the expense of some added length) as shown in Figure 4.12.

A rough rule for estimating the bulb size for application with increaser gearing is

$$\frac{d_b}{D_1} = 0.80$$

where d_b is the diameter of the bulb and D_1 the diameter of runner.

For large capacity units, the tubular turbine is mostly directly connected to the generator which then requires a larger bulb [4.6]

$$\frac{d_b}{D_1} = 1.12 \text{ to } 1.20$$

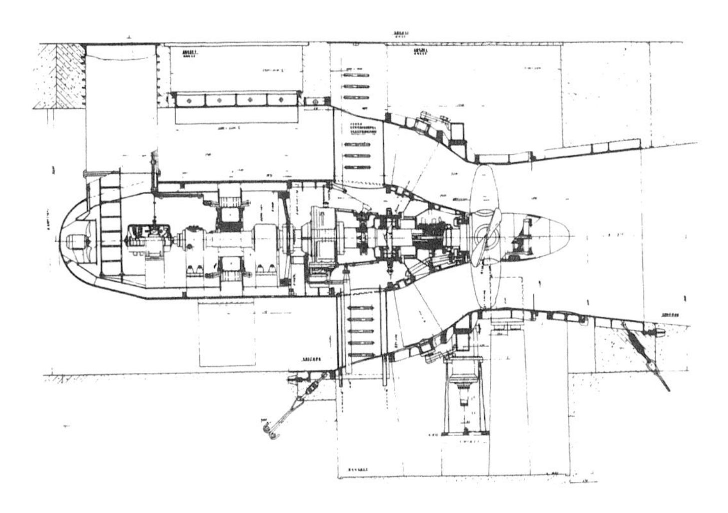

$H_r = 25m,\ P = 600kW,\ D_1 = 2.5m$

Figure 4.12 Bulb turbine with planetary gear increaser Jiangxia tidal power station (Drawing courtesy TRIED)

(ii) Inlet passage

In order to reduce the inlet hydraulic losses, the inlet passage is usually made fairly large compared with the runner. Well-shaped passage contours will help to reduce loss and induce good flow distribution. But from engineering considerations, replacement of curved surfaces with a series of straight surfaces and reduction of the inlet length will greatly reduce construction costs; the reduction of inlet dimensions will also result in savings of equipment in gates, trash racks and auxiliary facilities.

The size of the inlet passage is further decided by the bulb casing size required by the particular type of turbine. For the case with directly driven generators, the inlet passage diameter may be

$$\frac{d_i}{D_1} = 1.95 \text{ to } 2.25$$

here d_i may be reduced appreciably when generators driven through speed increasers are used. Figure 4.13 shows the overall dimensions of a typical bulb turbine flow passage and Figure 4.14 shows the detailed proportions of a GZ003 bulb turbine.

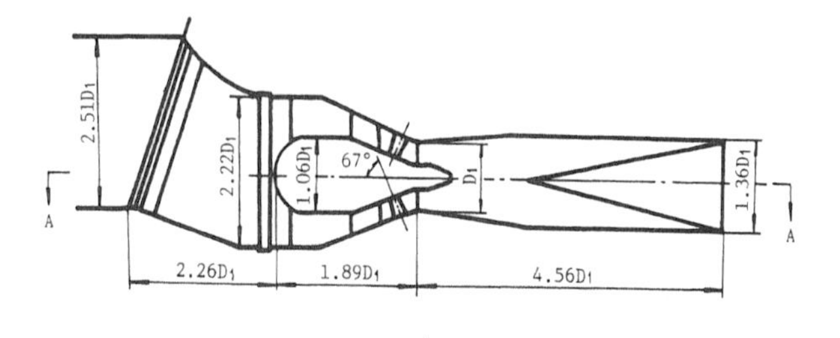

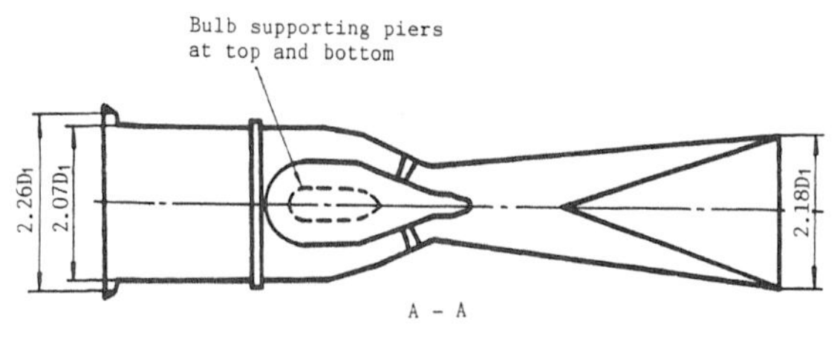

Figure 4.13 Typical flow passage proportions of a bulb turbine [4.5]

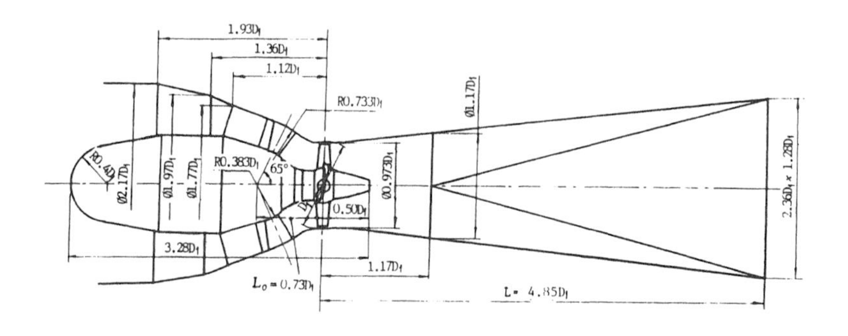

Figure 4.14 Flow passage dimensions of GZ003 bulb turbine
(Drawing courtesy TRIED)

(iii) Support structures

The bulb turbine is usually supported by two cylindrical vertical columns rigidly connected to the inner part of the bulb casing. All forces and torques exerted on the bulb casing are transmitted through these columns to the surrounding concrete structure. The support columns must be designed with sufficient strength and rigidity to accommodate the dynamic loads. They are covered on the outside with fairings of good streamline shape.

4.2.3 Runner chamber, guide apparatus and draft tube

(i) Runner chamber

Because bulb turbines are mostly used in the low head range, the working head varies over a wide range relative to the design head, so runners for medium/large tubular turbines are made of the adjustable–blade type. Small tubular turbines sometimes have fixed runner blades. For fixed–blade machines, cylindrical runner chambers are usually used to obtain the maximum throat area for greater discharge capacity and to reduce the clearance between the chamber and the runner perimeter. For adjustable–blade machines, semi–spherical runner chambers are used as in Kaplan turbines. The throat diameter has great influence on the discharge capacity of the runner and the hydraulic efficiency. Investigations have been conducted at the Tienjin Design and Research Institute of Electric Drive (TRIED) to find the optimal throat size for semi–spherical runner chambers [4.6]. Of the three throat sizes tested, it is found that the one with $d_m = 0.973D_1$ has the highest efficiency in the optimum region, the one with $d_m = 0.984D_1$ has best performance in the large flow region while the one with $d_m = 0.943D_1$ has best performance in the small flow region.

(ii) Conical guide apparatus

The conical guide apparatus must be so designed from hydraulic consideration to satisfy the circulation distribution at the runner inlet and to close tightly to prevent water leakage at standstill of the machine. The first objective is obtained by proper selection of geometrical parameters of the guide vanes according to the distribution of circulation along the radial height of the vanes. Vanes designed according to potential flow assumptions, that is C_z = constant and C_uR = constant will provide irrotational flow entering the runner accompanied by low hydraulic losses in this region, but will result in a spatially twisted vane shape.

Some designs use planar guide vanes, similar to those for axial–flow turbines, for easy solution of tightness at closed conditions but this will destroy the potential flow assumption and produce greater hydraulic loss. Guide vanes so designed have poorer hydraulic performance than the twisted vanes. For the latter, tightness at closing may still be accomplished by modifying the central camber and thickness distribution of the vane while retaining the potential flow assumption.

The relative height of guide vanes $\overline{B_0}$ varies with the specific speed of the machine and its maximum operating head. Normally, the $\overline{B_0}$ for tubular turbines is in the range $(0.32–0.42)D_1$ with larger values for low head and high specific speed machines; smaller values for high head and low specific speed machines.

The selected $\overline{B_0}$ value must satisfy the discharge capacity of the turbine runner as the guide vane characteristics directly effects the overall performance of the unit. Too small a height will increase the flow velocity in the guide vane space and create more losses and even diffusing of flow from guide vane to runner. With a larger height, the flow condition in the guide vane region will improve due to the lower flow velocity but the machine size will increase accordingly and become more expensive. The selection of $\overline{B_0}$ therefore must be made from comprehensive consideration of hydraulic performance and compactness in structure. and not the least, strength requirements.

The number of guide vanes may vary from 12, 16, 18, to 24, with 16 mostly used for large/medium machines, and 8 or even less for small machines. Aside from properly designed vane shapes, too few vanes may cause uneven flow distribution and consequently lowering of efficiency. The inclination angle of guide vanes may be $60^o, 65^o$ or 70° from the horizontal with 60° and 65° most commonly used. The distance between the intersection of the guide vane axis and the machine axis, to the centre line of the runner blades is designated by L_o (Figure 4.14) and its value lies in the range of $(0.65–0.80)D_1$. A large L_o value will afford better flow conditions but will increase the length of the machine. Consequently, larger values of L_o are used for smaller machines and smaller values for larger machines.

(iii) Dimensioning of draft tubes

The kinetic energy at runner exit is quite high for low head, large capacity machines like the tubular turbine so the machine performance depends to a large extent on the efficiency of the draft tube. Great importance must be attached to the proper selection of the draft tube size and development of its hydraulic contours. Tubular turbines are horizontal machines for which the highly efficient straight conical draft tube is best suited. The important factors in the design are the relative length, ratio of inlet/outlet areas and the conical angle. The draft tube length L is defined as the distance from the runner blade axis to the outlet as shown in Figure 4.14. The relative value of L varies in the range of $(4.5–5.0)D_1$ with $5.0D_1$ more commonly used. A length greater than $5.0D_1$ will cause more friction loss and result in no appreciable improvement in recovery of energy.

Conical angles of $\theta = 10^o$–13° are mostly used for the conical section of straight draft tubes. The outlet losses depend on the change of section along the axial length, that is the shape of the diffusing section as well as the outlet area of the draft tube. The inlet/outlet area ratios for high specific speed tubular turbines are usually taken as $A_i/A_o = 0.2$–0.25.

Two main types of conical draft tubes are now used, the first is a completely conical diffuser from inlet to outlet which is mainly used for medium and small machines. In order to reduce the distance between machine centres and from considerations of submergence and excavation, the straight conical diffuser is sometimes cut flat on four sides as shown in Figure 4.15 (a). The other type is a combination of a straight conical diffuser at the inlet and a transition from round to rectangular cross section, mainly for the purpose of reduction of excavation, as shown in Figure 4.15 (b).

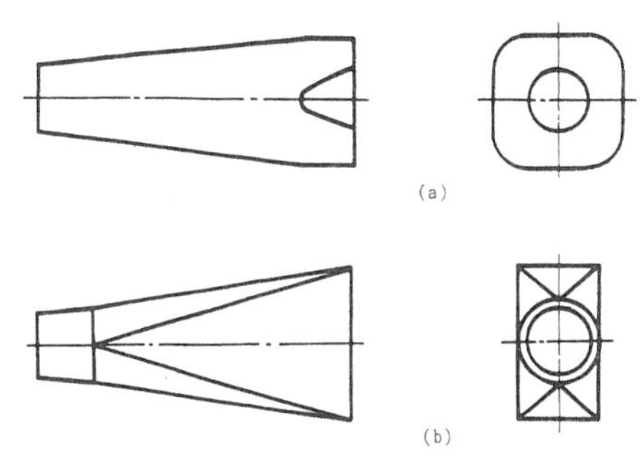

Figure 4.15 Schematic of draft tubes for tubular turbines

The round conical diffuser has better hydraulic performance which is demonstrated by comparative tests made at the TRIED on model turbine runner GZ003 with draft tube length of $5.0D_1$. The results show the draft tube with round conical diffuser has a higher efficiency over the one with rectangular outlet to the extent [4.6]

$$\text{At blade angle} \quad \phi = 15^o, \quad \Delta\eta = +0.8\%$$
$$\phi = 13^o, \quad \Delta\eta = +1.6\%$$

4.2.4 Flow passages of S–type tubular turbines

The S–type tubular turbine is an effective design for developing low head energy mainly in the head range of 3–15m and power output up to 5000kW with runner diameters up to 3m.

The outstanding feature of this type of machine is the location of the generator and the speed increaser outside the flow passage. This solves the great difficulty of placing the generator in a small bulb casing for small turbines and allows the use of a standard–line generator that is much less costly. Other features are the simplification of machine construction, reduction of power house space and civil engineering work involved. From these considerations it is seen that the S–type tubular turbine has clear advantages over conventional vertical turbines so they are used as replacements for vertical units in many countries. When comparing the merits of the S–type tubular turbine with the bulb turbine, the following difference should be noted:

- A much smaller bulb casing is required to house only the bearing assembly; the draft tube is made in an S–shape to allow the main shaft to connect directly, or through increaser gearing, with the generator.

- Both the conical and the radial guide apparatus may be used; the size of the conical assembly is smaller than in a bulb turbine.

- The main shaft passing through the inlet passage or the draft tube will create greater hydraulic losses in the high flow region.

The S–shape draft tube is usually made up of three parts, namely, the inlet conduit, the bend and the horizontal outlet diffuser. The major losses are diffusing loss, outlet loss and the head loss due to the rotation shaft. A good proportion for the conical inlet is about $L_1/D_1 = 1$ with a conical angle of $\theta = 16.5^o$, while a good proportion for the inlet and outlet areas is $A_o/A_i = 1.18$ [4.7]. The cross sectional area should increase according to a certain rule from the inlet to the outlet, as shown by the example in Figure 4.16.

The horizontal diffuser may have a horizontal top surface with a bottom surface slanting downward (16.5^o) and two parallel side walls with all cross sections rectangular.

An example of the general proportions of an S–type tubular turbine is shown in Figure 4.17.

4.3 Construction of Main Components of Tubular Turbines

4.3.1 Main shaft, bearings and blade-adjusting mechanism

The main shaft is made of forged steel and has flanges on both ends for connection to the turbine runner and the generator rotor. A centre hole in the shaft allows the passage of oil pressure from the servomotor and

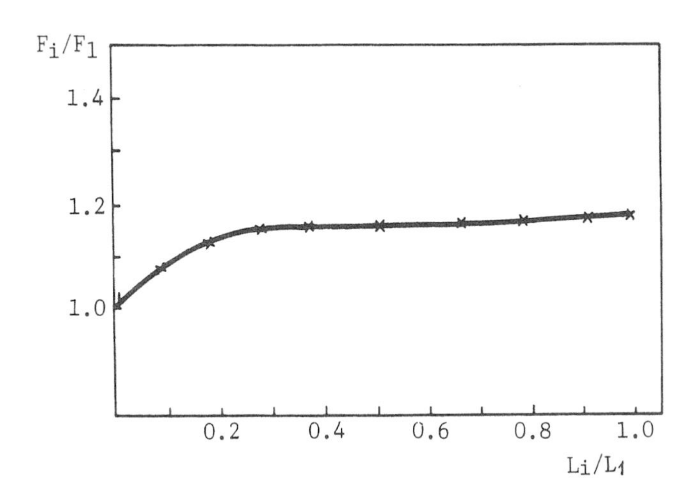

Figure 4.16 Variation of cross sectional area of draft tube

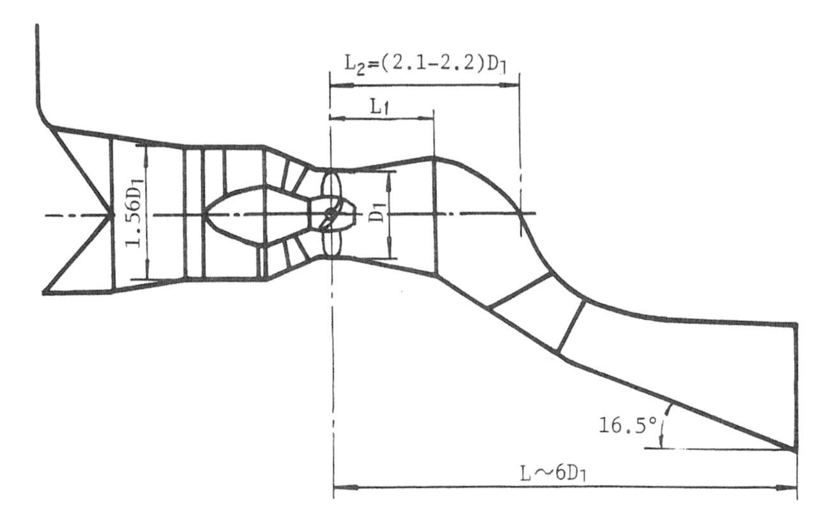

Figure 4.17 Typical proportion of an S-type tubular turbine

installation of the feedback mechanism. Stainless steel is usually overlaid on the shaft at the place where it passes through the seals. Waterproof covers are provided on main shaft bolts and transition parts of the shaft.

Main bearings for tubular turbines are normally of the sleeve type with pads made of babbit metal working under forced lubrication. Lubricating oil may come from a gravity oil reservoir providing oil pressure at 0.1–0.15MPa or from a pump supply system at 0.2–0.5MPa pressure. The oil is cooled either by a cooling pump system with water pressure of 0.2–0.5MPa or by an external cooling system.

For protection against wear of the bearings in the process of starting and stopping and operation at 40 to 60% rated speed, a separate high pressure oil system under 14–16MPa is used to provide the necessary oil film in the bearings. Length/diameter ratio of the bearings are usually $L/d = 0.7$–0.8.

Most tubular turbines have a blade–adjusting mechanism similar to those of Kaplan turbines. A more recent design is to place the mechanism inside the runner discharge cone with the servomotor piston fixed and the cylinder movable to drive the runner blades through links and cranks. The servomotor works under 4–6MPa oil pressure but the runner hub is filled with only low pressure leakage oil and is maintained at around 0.1MPa from the oil reservoir to prevent water from leaking in. Sealing requirements are very easily met with this construction, which is shown in Figure 4.18.

Another arrangement is to have the servomotor placed at the flange between the turbine and the generator, also a Kaplan turbine practice, with the piston driving the runner blades through a cross head in the runner hub.

4.3.2 Speed increaser

In the selection of hydroelectric power units in the low/medium head range, a major problem is the matching of the optimum speed of the turbine and the generator. Turbines working in low heads, especially under 5m head will have very low speeds. The generators to match these speeds will be large in diameter and high in cost. But, by using speed increasers of proper design, the speed for the turbines and the generators may be selected independently and allow the use of standard line generators with eight, six or even four poles of much smaller size and lower cost.

For units with capacities larger than 10MW and turbine speeds lower than 150min^{-1}, the benefits from using speed increasers far exceed their added initial costs and operating losses. This advantage is especially apparent with bulb turbines since the reduction in bulb size will appreciably reduce the civil engineering costs and construction time [4.9]. There are two main types of speed increasers used in tubular turbines. The first is the conventional parallel shaft one–stage gear unit commonly used for pit-type and S–type tubular turbines, as in Figures 4.10 and 4.11.

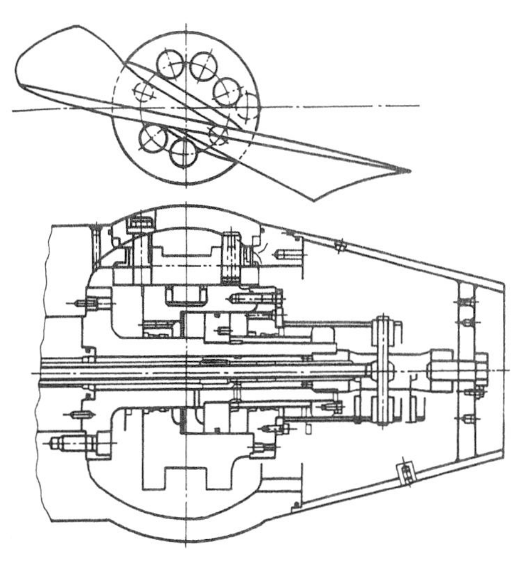

Figure 4.18 Runner blade-adjusting mechanism with servomotor in runner hub (Drawing courtesy Sulzer Escher Wyss)

For bulb turbines, the planetary gear speed increaser is more often used, such as the example in Figure 4.9. The efficiency of the increaser depends on the speed ratio and the speed of the driver. Normally, a loss in overall efficiency of 0.6–1.2% is unavoidable, but this may be reduced to 0.2–0.5% as compensated by the use of a more efficient high speed generator. Besides, a few tenths of a percent of efficiency may be recovered by the gain in selecting a speed that is closer to the optimum speed of the turbine.

The speed ratio of planetary speed increasers may be from three to eight for one–stage arrangement. Up to nine planet gears may be used but not more than three for speed ratios over six. Figure 4.19 shows a planetary gear speed increaser of CPG design [4.9]. The planet gears are mounted on the rack which is a part of the low–speed shaft connecting to the turbines. When this shaft rotates, the planet gears turn with it and revolves about their own axes at the same time. The sun gear which is connected to the generator shaft is turned by the planet gears and rotates at a much higher speed.

4.3.3 Components of guide apparatus

The conical guide apparatus used in tubular turbines comprises the inner and outer guide casings, movable guide vanes (16, 18 or 24), bushings, linkages,

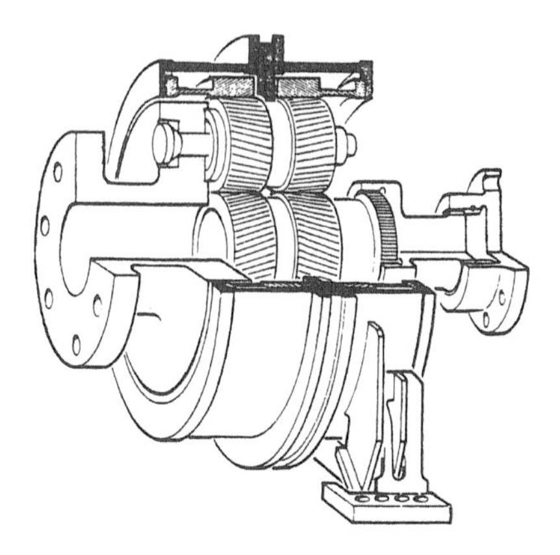

Figure 4.19 Planetary gear speed increaser
(Drawing courtesy Stal–Laval Turbin AB)

control ring and servomotor. The guide apparatus works in much the same manner as in other types of turbines.

The guide vanes are designed to have a closing tendency throughout the operating range, and will be closed by their own hydraulic torque plus the action of the counter weight in case of loss of oil pressure. The guide vanes are also required to close to a high degree of tightness to prevent leakage of water at standstill.

The control ring serves the purpose of synchronizing the movement of all the guide vanes through their respective linkages. It slides on the outer guide casing over single row or double row steel rollers fitted in machined grooves. With a low sliding friction, the control ring can be made of less rigidity such as a band cut from thick plating. The control ring may also be of box construction and move on the outer guide casing over sliding blocks. Figure 4.20 shows an example of control ring construction with rollers.

Guide vane linkages are made in different designs according to the experience of various manufacturers. One traditional construction is the use of rigid linkage with shear pins, others make use of stiff linkage with a machined neck or a flexible link. Their differences lie in the manner of protection in case of vane blockage – by shearing, by fracture or by bending. In all three cases, when blockage occurs, the machine must be shut down and the linkage in question replaced.

Fluid pressure controlled self–restoring linkages are used in bulb turbines by three Austrian firms: Voith employs a combination of a rigid linkage with shear pins and a fluid–actuated linkage, Voest–Alpine and Andritz use

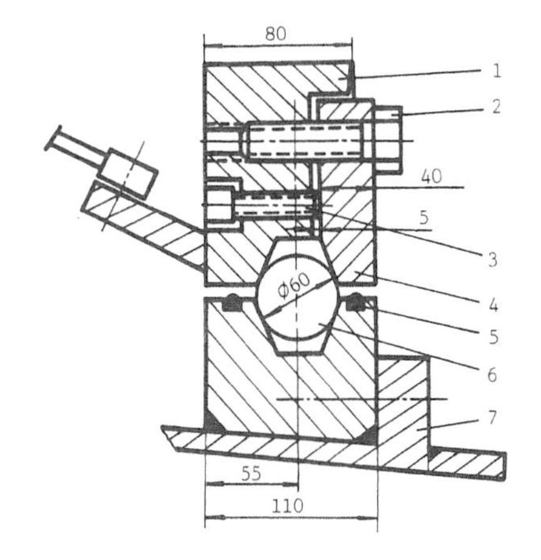

1 – control ring 2 – hold-down screw 3 – adjusting screw
4 – cover ring 5 – rubber seal 6 – roller
7 – runner chamber

Figure 4.20 Construction of guide apparatus control ring

complete fluid pressure controlled linkages.

The fluid pressure linkage as shown in Figure 4.21 works through a high pressure cylinder that is filled with oil and helium under a pressure of $13.5MPa$. In case of guide vane blockage the axial load on the link surpasses the normal value and pushes a small piston which in turn injects a small amount of oil into the high pressure cylinder and compresses the fluid–pressure linkage. As the linkage is retracted to a certain extent, a signal is set off to open the guide vane to release the debris and returns the linkage to its normal length. This action will be repeated if the blockage is not removed the first time. Should the debris still remain it will have to be removed manually after draining the flow passage.

The fluid pressure linkage saves the trouble of replacing shear pins, broken or bent linkages that require outage of the machine, but it is more complicated in construction and expensive to manufacture.

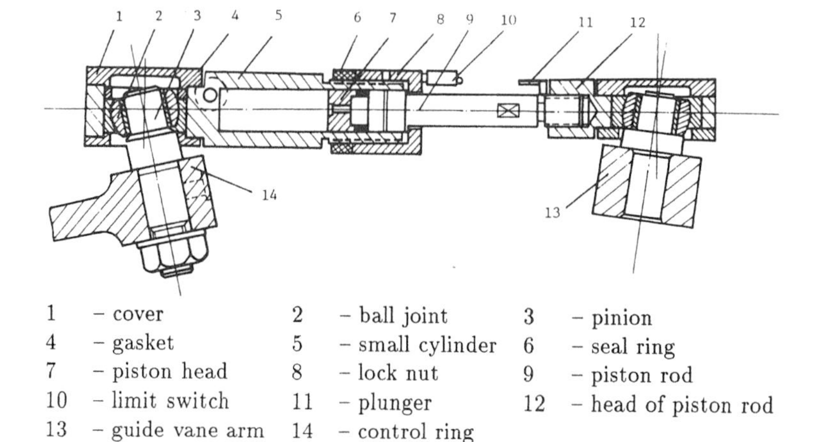

1	– cover	2	– ball joint	3	– pinion
4	– gasket	5	– small cylinder	6	– seal ring
7	– piston head	8	– lock nut	9	– piston rod
10	– limit switch	11	– plunger	12	– head of piston rod
13	– guide vane arm	14	– control ring		

Figure 4.21 Construction of fluid pressure linkage for guide vanes

4.4 Construction of straight–flow Turbines

4.4.1 Features of construction and arrangement

The straight–flow turbine is a machine designed with the maximum degree of unobstructed water flow from the inlet to the outlet. The annulus–shaped generator rotor (rim generator) is mounted on the outer perimeter of the turbine runner blades, leaving only a fairly small bulb in the flow that supports the runner bearings. Whilst the S–type tubular turbine endeavours to improve the effectiveness of the machine by employing high speed generator exterior to the flow passage, the straight–flow turbine takes advantage of the large diameter of the turbine rim to mount the low speed generator that effectively matches the turbine characteristics. Straight–flow turbines have been manufactured by Sulzer Escher Wyss under the patented name of STRAFLO and poses a number of ingenious features in construction that make them highly competitive in applications throughout the world.

In a STRAFLO turbine, specially designed seals are fitted to either side of the generator rim to keep water from leaking into the generator while keeping the friction power loss low as the dimensions of the seals are quite large. The successful solution of the sealing problem is one of the key factors that make these machines reliable in operation. Figure 4.22 shows a schematic of construction of medium/small size STRAFLO turbine made by Sulzer Escher Wyss and ACEC.

The stay ring *5* that is embedded in concrete, supports the bulb casing

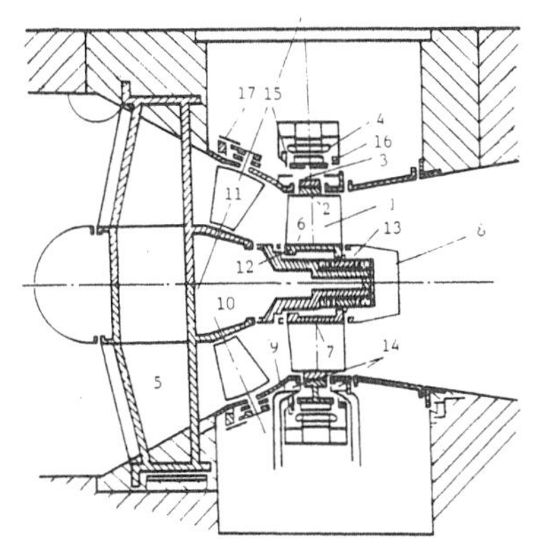

1	– runner blade	2	– external runner ring
3	– generator rotor rim	4	– generator stator
5	– upstream stay ring	6	– shaft
7	– runner hub	8	– runner downstream cover
9	– outside guide vane casing	10	– inner guide vane casing
11	– wicket gates	12	– guide bearing (upstream side)
13	– downstream guide bearing and combined thrust bearing	14	– sealing boxes (water boxes)
15	– brakes	16	– collector ring
17	– regulating ring		

Figure 4.22 Schematic of STRAFLO turbine construction (Drawing courtesy ACEC)

10 and the outer guide casing *9* which in turn support the runner *1* and the guide vanes *11* respectively. The runner rotates on the upstream guide bearing *12* and downstream guide/thrust bearing *13*. The outer rim of the runner blades are fixed to the external ring *2* and holds the generator rotor rim *3*. Sealing boxes *14* keep the water leakage to a minimum and collects all seepage at standstill into the station drain sump.

A large capacity STRAFLO turbine of recent manufacture which has quite different arrangement features from the previously described schemes is shown in Figure 4.23.

Because of its large size, the inlet duct and draft tube of the turbine are divided by central piers. The inlet pier replaces the steel bulb casing and supports the upstream turbine bearings. This arrangement brings about a saving in material as well as better stability in operation. The downstream turbine bearing is supported by steel struts near the draft tube inlet. The

$H_r = 5.5m,\ H_{max} = 7.1m,\ P_r = 17.8MW,\ P_{max} = 19.9MW,$
$Q_r = 378m^3/s,\ n = 50min^{-1},\ D_1 = 7.6m,\ z_1 = 4,\ z_0 = 18$

Figure 4.23 Perspective view of large STRAFLO turbine Annapolis Royal Canada (Drawing courtesy Sulzer Escher Wyss)

turbine and the generator are placed in the centre of the machine pit with the stator fixed on the side wall of the pit. This will allow the stator to be moved axially out of the rotor plane to facilitate assembly and repair.

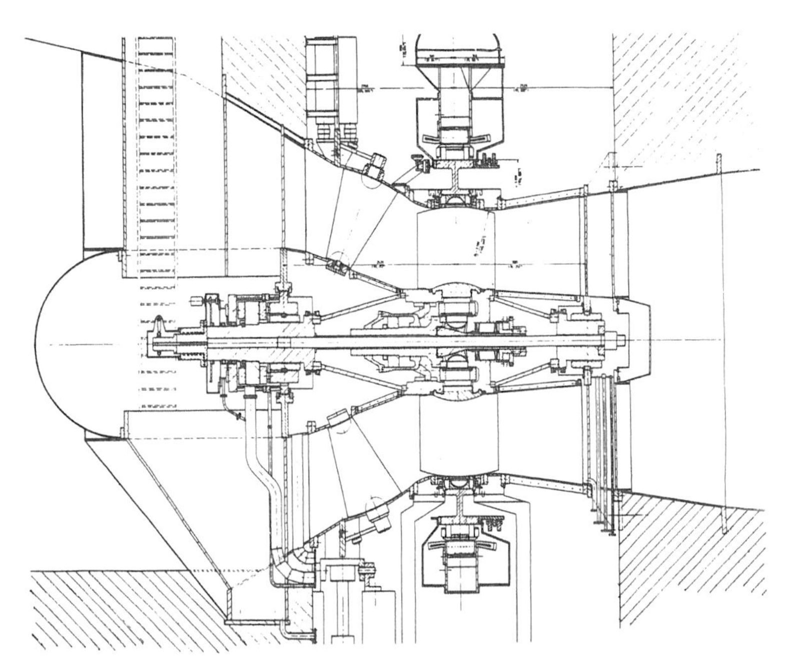

$H = 10.3m,\ P = 2 \cdot 8.35MW,\ D = 3.7m$

Figure 4.24 Double–controlled straight–flow turbine
Weinzodl, Austria, 1982
(Drawing courtesy of MFA, Sulzer Escher Wyss and ELIN)

The runner assembly comprises the runner blades, the hub and outer ring which are welded into one part and assembled to the upstream, and downstream sections of the large diameter hollow shafts. The assembly rotates on fluid static bearings. Water leakage is prevented by fluid static seals mounted on both sides of the generator rim. STRAFLO turbines may be controlled in two ways. The commonly used and technically well developed is the design with fixed runner blades and adjustable guide vanes. Attempts have been made to have the runner blades also adjustable which will greatly improve the hydraulic performance of the machine. But mechanical difficulties are encountered in adjusting the runner blades while driving the generator at the same time. There have been double–controlled straight–flow turbines in operation today although they are still in the trial stage. Figure 4.24 shows an example of this construction.

4.4.2 Support elements and rotor seals of STRAFLO turbines

(i) Hydrostatic support elements

For low head STRAFLO units, the weight of the turbine runner and the generator rotor are taken up by the traditional guide and thrust bearings located in the runner hub assembly. But for high head large capacity units, the generator poles and the outer rim may weigh several times that of the turbine runner and poses new problems in bearing support. For this purpose, a new scheme is devised using individual support elements working on fluid static principles that acts directly on the rotor rim.

These bearing elements function as shown in Figure 4.25. The support piston is held in a fixed housing by a seal of special polymer material and is free to move in all directions. The interior of the housing is supplied with suitable fluid from a pressure vessel, and the piston is forced against the rotor. The side of the piston facing the rotor has four pockets which are connected via capillaries to the pressure chamber. A pressure cushion is thus built up between the support piston and the rotor, whose resultant force is in equilibrium with the force acting on the other side of the piston. A constant, precisely defined gap is established between the rotor and the support piston. The piston itself is retained as a floating element between the two pressure cushions, that is it follows any radial displacement of the rotor precisely.

If the piston becomes tilted relative to the rotor surface, the gap between the piston and rotor becomes smaller on one side and larger on the other, and, due to the throttling effect of the separate feed capillaries, pressure is increased on one side and reduced on the other. The resultant force of the pressure cushion thus becomes eccentric, producing a restoring moment until the support piston is again parallel to the rotor surface [4.2].

In contrast to hydrodynamic bearings, hydrostatic bearings must have an external supply system. If this fails, a fully functional protective system must immediately take over to prevent damage to the bearings and the rest of the machine. The reliability of this system is to a certain extent just as important as the governor system.

(ii) Seals for rotor rims

Good sealing at the outer rim of the rotor is of paramount importance to the successful operation of a STRAFLO turbine [4.2]. For medium/small units,

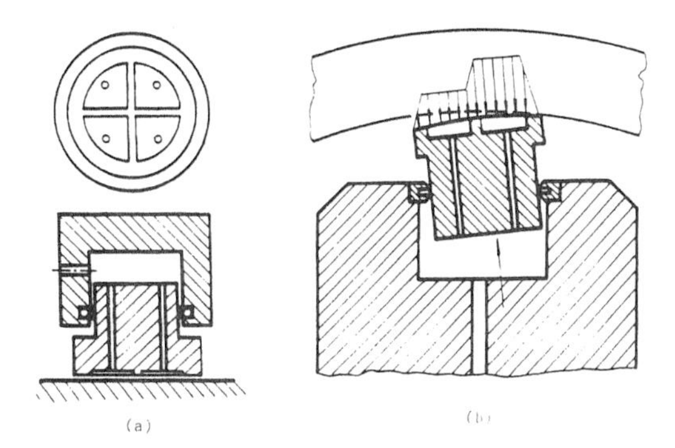

Figure 4.25 Working principle of hydrostatic support elements (Drawing courtesy Sulzer Escher Wyss)

the lip–type seal is commonly used where an elastic rubber lip fastened to the rotor works in contact with a stainless steel surface on the turbine casing, as shown in Figure 4.26. The sealing element can be removed from inside the machine for replacement.

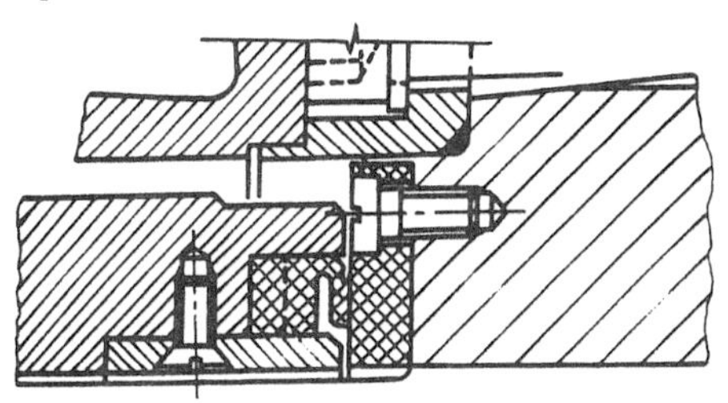

Figure 4.26 Lip–type contact seal [4.1]

The lip–type seal consumes too much power and is not reliable enough for large machines, instead a series of individual sealing elements working on hydrostatic principles are used. The elements are pressed against the surface to be sealed but are not in contact with it.

The sealing element is a curved fan–shaped strip with isolated pockets made by special composite material and works free in a recess in the turbine casing and is pressed against it by a hose in the back of it, as shown in Figure 4.27. The individual pockets on the working face of the element is filled with filtered water at a pressure higher than the turbine inlet pressure through capillary action so that they are in a balanced condition. A minimum clearance is obtained between the sealing element and the rotor and allows the least amount of water to leak through. The clearance will automatically readjust if it deviates from the set value.

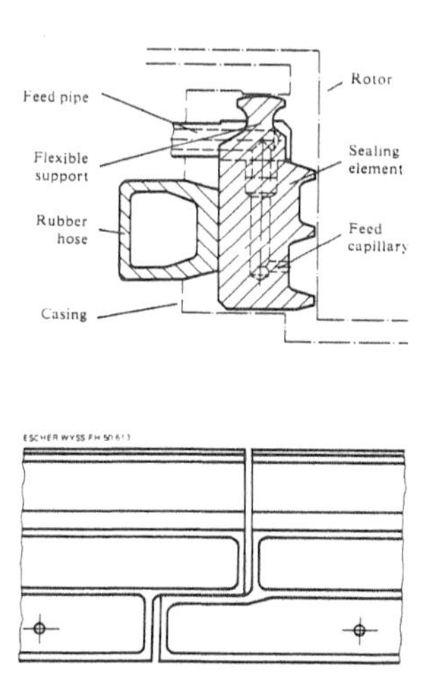

Figure 4.27 Hydrostatic sealing elements for rotor rim (Drawing courtesy Sulzer Escher Wyss)

The use of filtered water (removal of particles larger than $20\mu m$) will guarantee the least amount of wear on the sealing element. One third of the sealing water will flow across the clearance into the draft tube and two thirds will be recirculated. The amount of outflow will effectively prevent foreign matter from entering the seal [4.10]. Though provided with a standby filter that can be put into service automatically, the sealing element has proved to be able to run dry for a considerable time. The same seal will function as the maintenance seal when the machine is at standstill.

A later development of the sealing arrangement from the above described scheme has two pressurizing hoses pressing against the sealing element and an additional hose on the inner periphery especially for sealing at standstill, as shown by Figure 4.28.

4.4.3 Development prospects of straight–flow turbines

The straight–flow turbine is a new type of machine suitable for development of low head, large flow energy that can be used in river hydro stations as well as in tidal power stations. With the exception of slightly lower efficiency, it

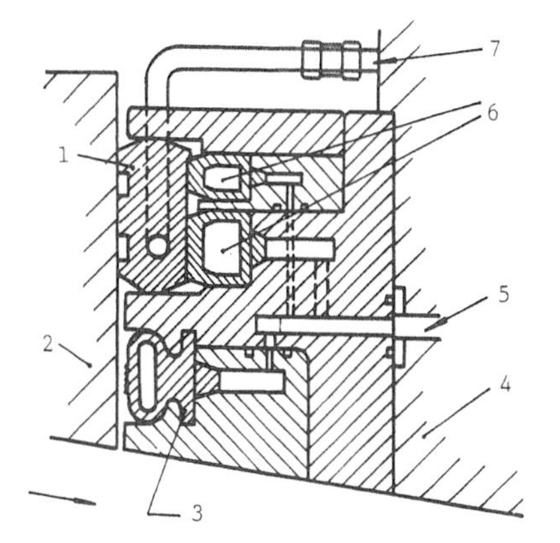

1	– sealing segment seal	2	– turbine rim
3	– standstill seal	4	– generator casing
5	– compressed air supply	6	– pressurizing hoses
7	– sealing water supply		

Figure 4.28 Improved hydrostatic sealing element for rotor rim (Drawing courtesy Sulzer Escher Wyss)

has the following advantages over the bulb turbine:

1. The generator rotor is fitted on the outer rim of the turbine runner so that only one machine pit is needed instead of two, with the resulting reduction in power house dimensions and civil engineering costs.

2. The generator stator may be moved sideways along its supports so that the rotor poles and stator windings may be maintained without too much dismantling.

3. With the large diameter of the generator, cooling by air is very effective, auxiliary blowers remove the warm air from the stator to the outside.

4. With the generator rotor mounted on the outer periphery of the turbine runner the rotating inertia of the revolving parts is quite sufficient for stable operation.

5. With the resulting smaller power house and a more compact machine, construction time is appreciably reduced.

The straight–flow turbine is the modern development of the axial–flow turbine with generator on its rim first patented by Leroy Harza in 1919. Sulzer Escher Wyss Limited manufactured a low head turbogenerator unit under the patented name of STRAFLO in 1974 which was based on the experiences gained from the operation of 73 straight–flow turbines built between 1938 and 1950. These machines are rated at 1.5-1.9MW in the head range of 7.3-9.0m, with runner diameter of about 2.0m. The four STRAFLO turbines installed in run–of–river plants in Austria, Belgium and Sweden have runner diameters up to 3m. The very well publicised large STRAFLO unit is the tidal power unit built by Sulzer Escher Wyss for Annapolis Royal in Canada in 1983 [4.11].

References

4.1 Gladwell, J. S., Warnick, C. C. 'Low Head Hydro - An Examination of an Alternative Energy Source', Idaho Water Resources Research Institute.

4.2 Holler, K., Miller, H. 'Bulb and STRAFLO Turbines for Low-head Power Stations', *Escher Wyss News*, 1977,No.2.

4.3 'Electrical and Mechanical Equipment for the Ampsin-Neuville Power Station', ACEC Presentation 1981.

4.4 He Guo-ren. 'Bulb Turbine Using Tidal Energy for Generating Electric Power from Both Directions', ASME Small Hydro Power Fluid Machinery Symposium, 1984, 13-20.

4.5 Pugh, C. A. 'Flow Passage Design for Bulb Turbine Intakes', ASME Small Hydro Power Fluid Machinery Symposium, 1982, 99-106.

4.6 He Guo-ren. 'Flow Passages for Bulb Tubular Turbine and Studies on Hydraulic Characteristics of GZ003 Runner', *Hydraulic Equipment*, 1980, No.3, 1-13 (in Chinese).

4.7 He Guo-ren. 'The Research and Testing of the Hydraulic Performance of S-type Tubular Turbines', 4th International Symposium on Hydro Power Fluid Machinery, ASME, 1986, 11-17.

4.8 Bosc, J., Megnint, L. 'Present Design of Tidal Bulb Units Based on the Experience in the Rance Tidal and in River Bulb Units', International Symposium on the 20th Anniversary of the Rance Tidal Power Plant, Saint Malo, 1986.

4.9 Cederberg, B. 'Hydro Turbine Geared for Higher Powers', *Water Power and Dam Construction*, 1981, No.9.

4.10 DeLory, R. P. 'Prototype Tidal Power Plant Achieves 99% Availability', *Sulzer Technical Review*, 1987/1.

4.11 Duoma, A., Stewart, G. D. Meier, W. 'Straflo Turbine at Annapolis Royal - First Tidal Power Plant in the Bay of Fundy', *Escher Wyss News*, 1981, No.1/1982, No.1.

Chapter 5

Construction of Impulse Turbines

G. Borciani

5.1 General

The impulse turbine has a basic feature, in that all the available energy in the water is fully converted in the distributor system into kinetic energy which is transferred to the runner according to the impulse principle. This is the first difference between impulse and reaction turbines, the second is that the impulse (or action) turbine has a partial water admission, that is water impinges on the runner at one or several points on the periphery only (see Volume *Hydraulic Design of Hydraulic Machinery* of this Book Series.

Small turbines, like the inclined jet turbine and the cross–flow turbines, are usually classified as impulse turbines although they have some degree of reaction. These turbines will be described in detail in Chapter 10.

The most important impulse turbine is the Pelton turbine, named after L. A. Pelton who in 1880 invented the central ridge (splitter) for the buckets. Around 1900, A. Doble used a double–elliptic shape bucket and a needle for the control of discharge for the first time.

Most Pelton turbines built up to 1960 are of the horizontal type. Only after the wider application of the vertical design did the power of Pelton turbines exceed the $100 MW$ mark, see Figure 5.1. The two units at the Sy–sima plant (Norway, 1981) each with five jets are rated at $315 MW$ at $n = 5 s^{-1}$ (300rpm) under a specific hydraulic energy [1] E of $8690 J \cdot kg^{-1}$($H = 885 m$). The three Pelton turbines at Reisseck (Austria, 1970) operate under the highest value of E in the world of 17318 $J \cdot kg^{-1}$ ($H = 1765 m$).

The operating scheme of a Pelton turbine is fairly simple, as shown by the horizontal two–jet machine in Figure 5.2. Water from the penstock *1*

[1] see Note 1 at the end of this chapter

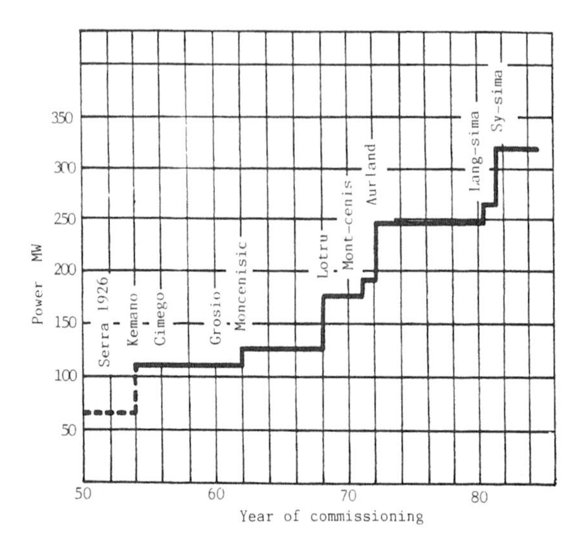

Figure 5.1 The most powerful Pelton turbines in the world from 1950 [5.10]

passes through the spherical valve *2* which has a two–fold task: to permit inspection and maintenance of the turbine, and stopping the machine in an emergency, for example the failure of the normal closing system. The feeding system (distributor) consists of the two branches *4* connecting the spherical valve to the two nozzles *5* through which the energy of the water is transformed into kinetic energy in the form of a free jet. The jet impinges on a series of buckets *6* set at the periphery of the runner and causes it to rotate. In the diagram, D is the jet circle diameter and d_0 the jet diameter.

The theoretical absolute velocity of the water flow at atmospheric pressure under a specific energy E is given by

$$C_{th} = \sqrt{2E} = \sqrt{2gH}$$

and the actual velocity is

$$C_1 = \varphi\sqrt{2E}$$

where φ is the coefficient of discharge that depends on the efficiency of the feed system and the nozzle with an approximate value of 0.98.

The jet diameter d_0 is determined from the equation

$$Q_j = \frac{\pi d_0^2}{4} \cdot \varphi\sqrt{2E} \tag{5.1}$$

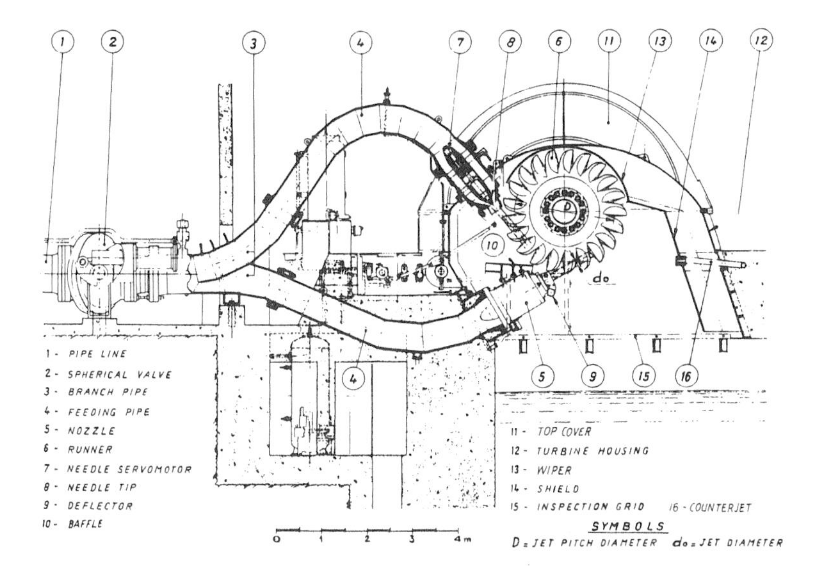

Figure 5.2 Cross section of a horizontal two–jet Pelton turbine (Drawing courtesy Hydroart)

where Q_j is the discharge of a single jet.

As the runner rotates in an atmospheric setting except in particular cases (see 5.3.8), the Pelton turbine is known as a constant pressure turbine. It is sometimes called a tangential turbine because of the direction action of the jet on the runner. The basic characteristic of the Pelton turbine, as mentioned previously, is given by the complete transformation of the water energy into kinetic energy at the runner inlet.

Consequently, the machine may be considered as made up of two parts placed on either ends of the space where the jet develops so it is possible to change the geometry of one part (for example the stator size and number of jets) without changing the other part (runner). This feature of the two parts not being strictly interrelated constitutes a major difference in hydraulics from reaction turbines.

5.1.1 Range of application

Pelton turbines operate almost exclusively under high specific hydraulic energy (over $8,000 J \cdot kg^{-1}$) and with relatively small discharges (less than $100 m^3 s^{-1}$). Considerations from the hydraulic and mechanical aspects are:

(i) Hydraulic consideration

Figure 5.3 shows the relation between the peripheral velocity factor $u = U/\sqrt{2E}$ and the specific speed, [2] n_{qj}, defined by equation 5.2, relevant to one jet [5.1].

$$n_{qj} = \frac{n\sqrt{Q_j}}{H^{0.75}} \tag{5.2}$$

The velocity factor, u, for Pelton turbines is very low, ranging between 0.45 and 0.49, which is about 50% of that for Francis turbines. This fact requires the Pelton turbines to operate under high specific hydraulic energies in order to reach sufficiently high peripheral velocity and rotational speed to become economically efficient.

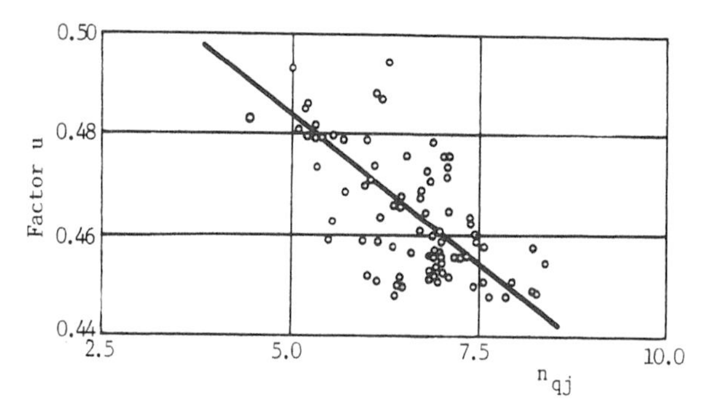

Figure 5.3 Peripheral velocity factor versus specific speed [5.1, Fig. 5]

(ii) Mechanical consideration

Due to its very simple construction the Pelton turbine is able to withstand high pressures, whereas reaction (Francis) turbines cannot operate over $10,000 J \cdot kg^{-1} (H \approx 1,000 m)$ because of mechanical limitations of the distributor. The range of 2,000 to 8,000 $J \cdot kg^{-1}$ ($H \approx$ 200 to 800m) is where the Pelton and the Francis turbine can both be applied.

[2] see Note 2 at the end of this chapter

The following points show the main features of the Pelton turbine [5.1, p.47] [5.2]:

- Hydraulic performance:
  - lower peak efficiency and higher partial load efficiency as shown in Figure 5.4;
  - more sensitive to specific hydraulic energy variations;
  - lower rotational speed, see Figure 5.5;
  - runner generally installed above the maximum tailwater level;
  - greater loss of energy in case of variation in tailwater level;
  - possibility of installation in cases where Francis turbines cannot be used due to lack of sufficient counter-pressure.
- Operation:
  - higher reliability (simpler design);
  - smoother hydraulic behaviour;
  - in general no cavitation;
  - higher erosion rate due to high velocities, but easier accessibility for repair and maintenance.
- Costs:
  - higher cost for the generator;
  - requires wider and longer power house.

5.1.2 Basic hydraulic and structural principles

With the design specific hydraulic energy E and the design discharge Q given, a predetermined value of n_{qj} allows the defining of the peripheral velocity factor u from Figure 5.3 and the peripheral velocity $U = u/\sqrt{2E}$ can be calculated.

The cost of the turbine and in particular that of the generator will be reduced if the rotational speed is increased, therefore the trend is to choose a rotational speed as high as possible. This aim is achieved by increasing the number of jets z_j as given by

$$n_{qj} = \frac{n(Q/z_j)^{0.5}}{H^{0.75}}$$

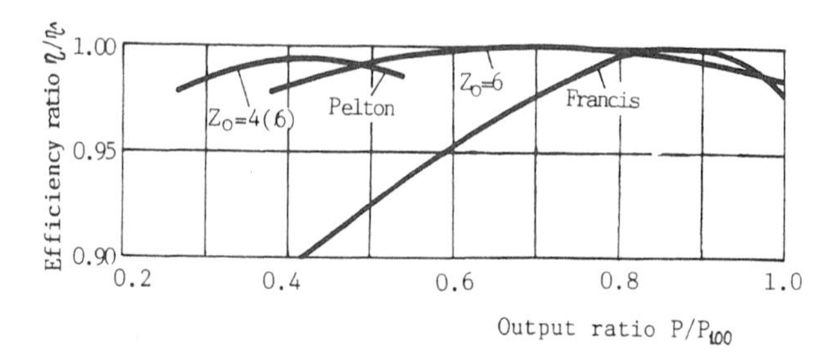

Figure 5.4 Comparison between a six-jet Pelton turbine and a Francis turbine of the same power ($100MW$) and under the same E ($3600\ J \cdot Kg^{-1}$) [5.3, Fig. 10]

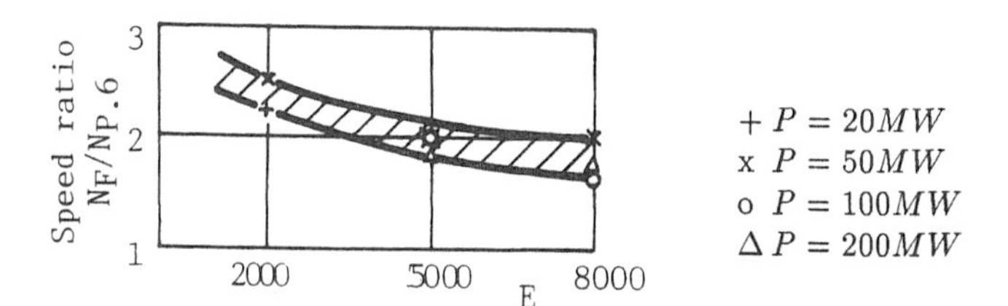

Figure 5.5 Speed ratio between Francis turbine and six-jet Pelton turbine [5.3, Fig. 9]

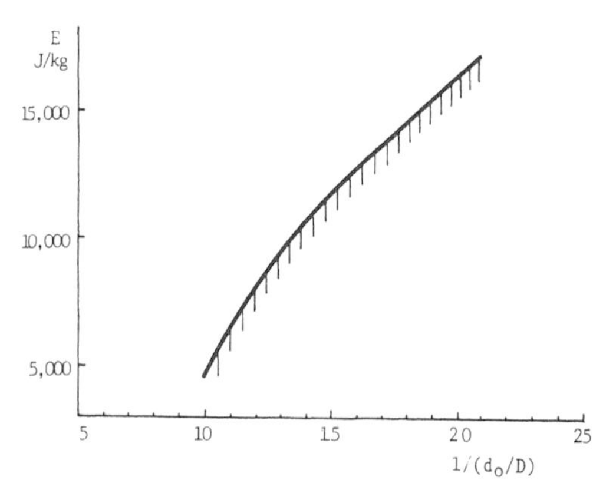

Figure 5.6 Limit curve of E versus $1/(d_0/D)$

and the ratio d_0/D as

$$n_{qj} = k \cdot \frac{d_0}{D}$$

The ratio d_0/D is basically decided when taking into account the mechanical limits represented by the curve of Figure 5.6. The choice of the number of jets is made on the basis of the maximum values allowed by the arrangement of the unit as

(a) horizontal shaft: four jets (two runners with two jets each), or
(b) vertical shaft: six jets.

The six-jet vertical Pelton turbine have good performance only over a narrow range of specific hydraulic energy (see Section 5.3.1.). In the case of powerful turbines operating under high specific hydraulic energy, the problem of the stresses and the presence of sediments may lead to a selection of a lower number of jets than six. If the required power of the unit is very high, construction reasons may make it impossible to have a generator of high enough rotational speed. In this case compromise between economical and technical requirements must be found.

Having chosen the ratio d_0/D and the number of jets, n_{qj} is checked and the following parameters are determined

(a) the single jet discharge $Q_j = Q/z_j$
(b) the jet diameter d_0 as determined from equation 5.1
(c) the jet circle diameter D
(d) the rotational speed $n = U/\pi D$.

All the above parameters are linked by the curves shown in Figures 5.7 and 5.8.

5.1.3 Layout

As mentioned before, the ratio d_0/D for a given specific hydraulic energy cannot be increased beyond certain limits, therefore the discharge per jet is limited and the power of a Pelton turbine can be raised only by increasing the number of jets. For hydraulic reasons related with the feeding system, the maximum number of jets for horizontal Pelton turbines is two for one runner or four in the case of two runners. For vertical Pelton turbines the number of jets may be up to a maximum of six. The possible hydraulic interference between the jets do not allow an even higher number. The increased number of jets acting on the runner has brought about the increase of power and speed and consequently the development of the vertical Pelton turbines which have replaced the horizontal units in the range of power of over $100 MW$.

The horizontal Pelton turbine and the vertical Pelton turbine are described in Sections 5.2 and 5.3 respectively which illustrate the close links

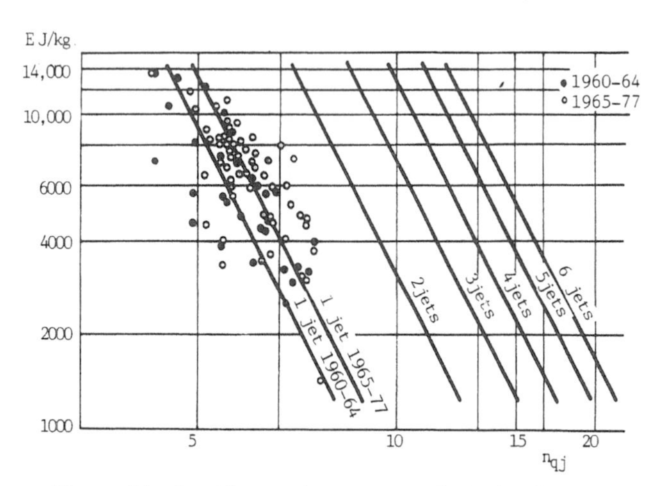

Figure 5.7 Specific speed versus specific hydraulic energy for a different number of jets [5.1, Fig. 1]

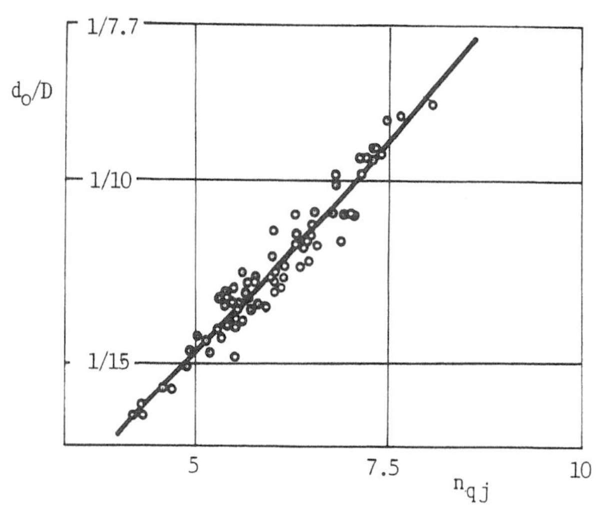

Figure 5.8 Ratio d_0/D versus specific speed [5.1, Fig. 6]

between hydraulic and mechanical design considerations. Runners for the two types of turbines are described in 5.2.1 and 5.3.1 with the former dealing mainly with the basic hydraulic and structural design and the latter devoting more specifically to problems of hydraulic performance, fatigue, cavitation and erosion in multi-jet vertical Pelton turbine runners.

5.2 Horizontal Shaft Pelton Turbine

With horizontal units the generator may be driven by a single runner or twin runners, Figure 5.9 (a), and each runner may be fed by one or two nozzles, Figure 5.9 (b). The runner is usually mounted overhung on the generator shaft. In cases where the jet dimension is large, the overhung may become too great so a second bearing is provided to limit the deflection of the shaft.

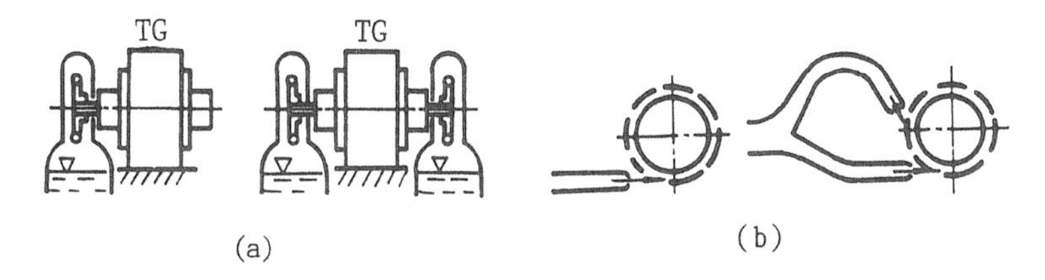

Figure 5.9 Layout of horizontal shaft Pelton turbines [5.4, Fig.3.18]

Twin runners are normally mounted one on each side of the generator. If the Pelton turbine is part of a ternary pumped storage unit which includes a storage pump, the two hydraulic machines will be placed at either ends of the electrical machine. If the Pelton turbine has twin runners, a second bearing is required to support the added length of the shaft. The runner can be easily dismantled and reassembled by removing the top cover.

Component parts of a Pelton turbine fall into three major groups

(a) Rotating parts – runner (5.2.1), shaft and bearing (5.2.2)
(b) Stationary parts – housing (5.2.3), nozzle (5.2.4) and feeding system (5.2.5)
(c) Control mechanism – (5.2.6)

These parts will be described in detail in the following sections.

5.2.1 Runner

Figure 5.10 shows a Pelton runner where the main components (discs and buckets) are all cast in one piece. Each bucket has two bowls that are divided by a splitter. The shape at the outer contour of the bucket (the outcut) is so designed as to avoid an abrupt impact when the bucket edge enters into the jet. The buckets are reinforced by two ribs on the back side.

(i) Basic hydraulic design

(A) Determination of the number of buckets

The analysis is based on the principle that water particles should deliver the maximum amount of their energy to the runner. The interaction between the jet and the bucket is illustrated by Figure 5.11 showing the section of the full jet in the plane perpendicular to the axis of the runner. The design rotational speed and the design specific hydraulic energy are the basic parameters.

The relative paths of the jet referred to the runner are drawn as follows: the water particle A_1 impinges on the tip of the bucket with the relative velocity W_1. The jet covers the segment $\overline{A_1a_1}$ in Δt before leaving the bucket, where $\overline{A_1a_1} = C_1\Delta t$. The runner covers the arc $\overline{A_1a}$ in the same time, where $\overline{A_1a} = U\Delta t$ with δ the corresponding angle of travel. To obtain the relative path of the jet, the runner must be *stopped* and the radial line $O - a_1$ moved back by an angle δ to establish the terminal point A_2 of the relative path according to the relation [Ref. 5.5, p126]

$$\frac{\overline{A_2a_1}}{\overline{A_1a_1}} = \frac{U}{C_1}$$

The relative path $A_1 - A_2$ (dark line) cuts the jet circle at two points M and N determined by the same procedure as for A_2, with

$$\frac{\overline{Mm}}{\overline{A_1m}} = \frac{\overline{Nn}}{\overline{A_1n}} = \frac{U}{C_1}$$

$A_1 - M - P - N - A_2$ is then the relative path of a water particle on the upper contour of the jet and $B_1 - B_2$ is the relative path of a water particle on the lower contour of the jet. All the relative paths of the jet are covered between the paths $A_1 - A_2$ and $B_1 - B_2$.

The ideal bucket pitch would be equal to the arc $\overline{B_1B_2}$ corresponding to the relative path of the water particle on the outer contour of the jet. Energy of some particles will not be utilized if the pitch is greater than this value, or the next bucket will interfere with the jet if it is smaller. To take into account of other factors related to the jet, a value of about $0.8 \cdot \overline{B_1B_2}$ is usually chosen for the pitch. The resulting number of buckets is rounded off to the next

Figure 5.10 Pelton turbine runner for New Colgate Power Station, USA (Drawing courtesy J.M. Voith)

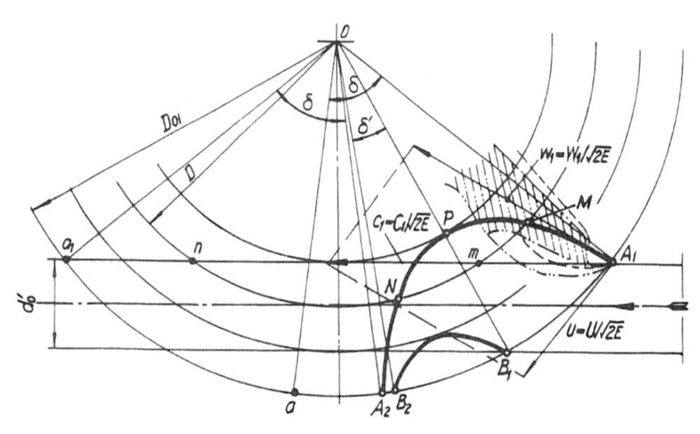

Figure 5.11 Relative paths of the jet

higher whole number [Ref. 5.5, p392]. In case of Pelton runners having a low ratio of d_0/D, the above procedure gives a lower value of the angle δ' corresponding to the arc $\overline{B_1 B_2}$, therefore the number of buckets increases as d_0/D (and n_{qj}) decreases. For $d_0/D = 1/7$ the number of buckets may be between 17 and 22, for $d_0/D = 1/20$ the number of buckets is usually higher than 24.

(B) Position of the buckets on the runner

The jet is deflected in both directions of the generatrices of the bucket ($a - a$, $b - b$, $c - c$, $d - d$ in Figure 5.12 (a)) only when it impinges perpendicularly on the splitter. This condition cannot be always satisfied because the angle between the relative path and the splitter changes continuously with the rotation of the runner. Therefore the inclination of the splitter is chosen so that the above condition of best incidence is attained when the jet is acting fully on the bucket.

(C) Geometry of the bucket

The profile of the splitter outer edge is such that the relative jet impinges on it perpendicularly at the beginning of its action, avoiding an abrupt impact that would impair the performance of the runner. At the outer extremity, a portion of the splitter and the bucket is cut according to the width of the jet and is called the outcut as shown in Figure 5.12. This is the only way to solve the problem of the impact of the jet at the entrance into the bucket that is caused by the direction of the relative velocity W_1 (point A_1 in Figure 5.11).

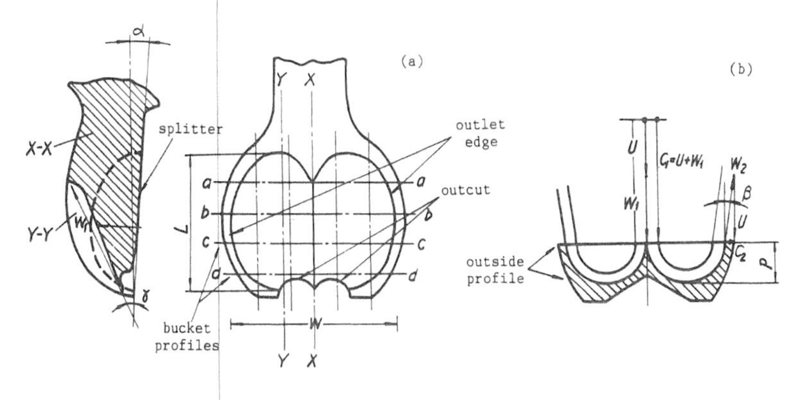

Figure 5.12 Basic geometrical data of the bucket

The angle γ of the back of the bucket (Figure 5.12 (a)) is designed to assure that the angle between the relative velocity W_1 and the back has a correct value to avoid any danger of cavitation.

The discharge angle β, shown in Figure 5.12 (b), must be small to reduce the outlet energy losses. Assuming the distribution of the relative velocity is uniform and that $W_2 = 0.98W_1$ because of friction losses, it is easy to demonstrate that for a value of $\beta = 9^o$ the outlet energy $C_2^2/2$ is less than 0.5% of the specific hydraulic energy. The outlet angle must not be too small or else the discharging water will impinge on the back of the following bucket.

The main dimensions of the bucket depend on the specific speed and are approximately as follows (Figure 5.12)

length, $L = (2.3\text{–}2.8)\ d_o$
width, $W = (2.6\text{–}3.8)\ d_o$
depth, $P = (0.8\text{–}1.0)\ d_o$

Small buckets have their maximum efficiency at lower discharge and large buckets at higher discharge, but the value of the maximum efficiency is scarcely influenced by their size.

All the considerations developed above are based on a number of simplifications including

(a) The velocity distribution of the jet is uniform.
(b) The outlet relative velocity distribution is uniform.
(c) The cylindrical jet is reduced to a plane perpendicular to the axis of the runner.

These conditions are close to the actual only for runners having a very low d_o/D ratio, where the jet diameter may be considered negligible compared with the jet circle diameter.

In recent years the shape of the Pelton bucket has been determined by three–dimensional analysis of relative paths of the jet as is shown in Figure 5.13. The analysis is based on the fact that the calculated acceleration acting on a particle on the water surface is perpendicular to that surface. This condition must be checked step–by–step for all particles flowing across the bucket. The theoretical analysis is checked by a stroboscopic investigation on the model runner [5.7, p. 8].

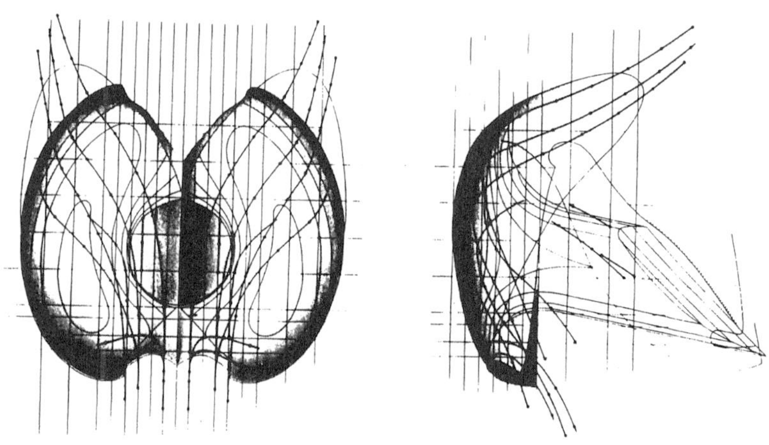

Figure 5.13 Theoretical analysis of the relative paths across the Pelton bucket [5.7, Fig. 24]

(ii) Efficiency curve

The runner is the component of the Pelton turbine which determines the efficiency of the machine. The same simplifications for the basic hydraulic design of the runner are applied for the inlet and outlet triangles shown in Figure 5.12 (b). The tangential force acting on the runner when operating at full discharge Q is

$$F_u = \rho Q\,(C_{1u} - C_{2u}) = \rho Q\,(C_1 - U)\left(1 + \frac{W_2}{W_1}\cos\beta\right) \tag{5.3}$$

The relevant hydraulic power converted to mechanical power is

$$P_t = F_u \cdot U = \rho Q\,(C_1 - U)\left(1 + \frac{W_2}{W_1}\cos\beta\right) \cdot U$$

The hydraulic power entering the runner is

$$P_h = \rho Q \frac{C_1^2}{2}$$

The hydraulic efficiency of the runner is then

$$\eta_t = \frac{P_t}{P_h} = \frac{2(1 + (W_2/W_1)\cos\beta)}{C_1^2}(C_1 - U) \cdot U.$$

where $C_1 = \varphi\sqrt{2E}$, φ reflects the efficiency of the distributor and the nozzle. Introducing the peripheral velocity factor term $u = U/\sqrt{2E}$ and assuming $\varphi = 0.98$, the hydraulic efficiency of the runner will become

$\eta_t = 0$	when $U = 0$,	$u = 0$	turbine at standstill
$\eta_t = max$	when $U = C_1/2$	$u = 0.49$	turbine at optimum speed
$\eta_t = 0$	when $U = C_1$	$u = 0.98$	turbine at runaway

The curve $\eta_t(u)$ is shown in Figure 5.14. If the above simplifications are not assumed and the following factors taken into account

- the full jet impinges on the bucket in the same direction with the peripheral velocity in one moment, hence the inlet velocity vectors can not be considered aligned as in Figure 5.12 (b)
- the windage losses due to the presence of water droplets, increase very rapidly in the range of the peripheral velocity factors higher than the best efficiency value
- the outlet relative velocity decreases because of the increased losses in the bucket, consequently the condition of minimum outlet losses will shift towards the lower peripheral velocity factors

The runaway speed will decrease and the best efficiency peripheral velocity factor will also shift towards the values shown in Figure 5.3.

The peripheral velocity factor is lower for high specific speed runners which have high d_o/D values and therefore are farther from the simplified scheme. The actual efficiency becomes η_t' as shown in Figure 5.14.

The basic influence of the peripheral velocity factor on the runner losses can be shown by analysing again the relative path of the jet. If the peripheral velocity factor u increases beyond the best efficiency value, the two paths $A_1 - A_2$ and $B_1 - B_2$ of Figure 5.11 will become $A_1 - A_2'$ and $B_1 - B_2'$ as shown in Figure 5.15. This is due to the fact the pitch of the buckets is about 0.8 times the arc $\overline{B_1B_2}$, a part of the water passes by without doing work on the bucket. Also the changed direction of W_1' is such that the water impinges on the back of the next bucket and slows down the runner.

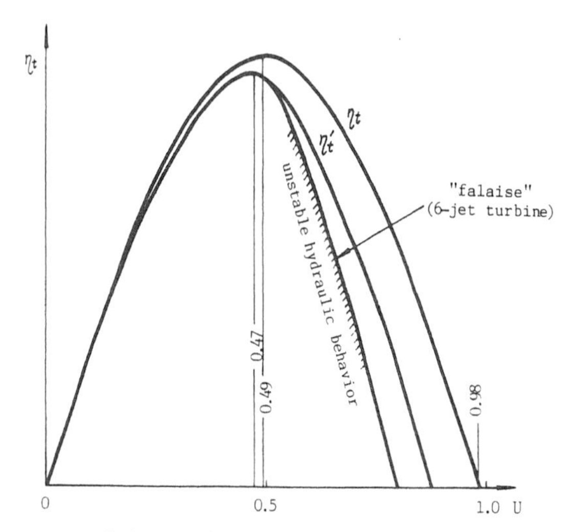

Figure 5.14 Pelton turbine runner hydraulic efficiency curve

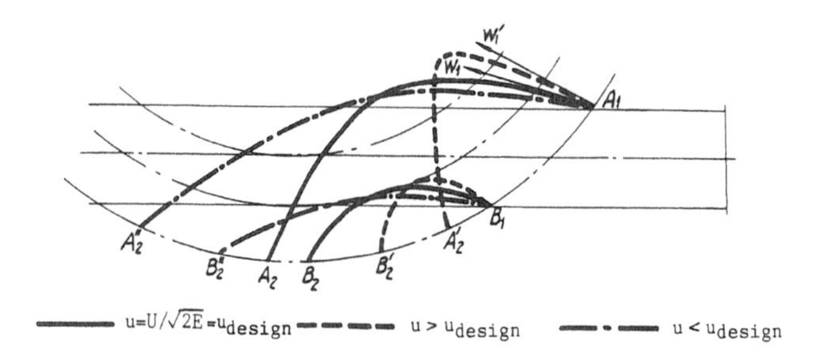

Figure 5.15 Influence of U on the relative paths [5.6, Fig. 3.13]

This negative action is increased by the higher resistance of the environment in which the runner rotates.

On the contrary, if u decreases, the paths will become $A_1 - A_2''$ and $B_1 - B_2''$ and the quantity of water acting on each bucket is greater with the risk of flooding and the relative velocities becoming greater, result in an increase of friction losses.

All these considerations confirm the pattern of the curve $\eta_t'(u)$ of Figure 5.14 and explains why the Pelton turbine is so sensitive to variations in the specific hydraulic energy (from expression $u = U\sqrt{2E}$).

(iii) Basic structural design of the runner

The structural analysis of the bucket starts from the determination of the two forces acting on it, the centrifugal and jet forces. The centrifugal force acts radially and is given by

$$F_c = 4\left(u^2 \frac{E}{D}\right) M$$

M being the mass of the bucket and D the jet circle diameter where the centre of gravity of the bucket is approximately located.

If the turbine is at standstill, $F_c = 0$. If it is operating at synchronous speed under the design specific hydraulic energy E, $F_c = 0.88(M \cdot E)/D$ assuming $u = 0.47$. If it is rotating at runaway speed under the design specific hydraulic energy E, then $F_c = F_{c_{max}} = 3.03(M \cdot E)/D$ assuming $u = 0.87$.

The centrifugal force can be considered static when the turbine is running at steady state conditions.

The jet force acts tangentially to the runner and is given by

$$F_j = 2\rho Q_B(C_1 - U)$$

which is a simplified form of equation (5.3) by assuming $\beta = 0$ and $W_2 = W_1$. Here, Q_B is the varying discharge acting on one bucket and will reach a maximum when the jet discharge is fully acting on the bucket, that is $Q_{Bmax} = Q_j$. If the turbine is at standstill under the design specific hydraulic energy E

$$F_j = F_{jmax} = 2\rho Q_j C_1$$

If it is operating at synchronous speed under the design specific hydraulic energy, E

$$F_j \approx \frac{F_{jmax}}{4} \quad \text{since } U \approx \frac{C_1}{2} \text{ and } Q_{Bmax} \approx \frac{Q_j}{2}$$

If it is at runaway speed, then

$$F_j = 0 \quad \text{since } C_1 \approx U$$

The jet force is alternating and changes rapidly when the turbine is running under load. Section 5.3.1 will refer to this problem again with regard to fatigue.

The stress calculation is usually carried out for the three operating points mentioned above and under condition of the highest specific hydraulic energy and for two positions of the bucket, that is action of the full jet at the first moment and the last moment. Experience shows that the maximum stresses will occur at the transition part connecting the bucket to the disk (root of the bucket) and at the splitter root whose stress versus time curve is shown in Figure 5.16. In addition to the stresses derived from centrifugal and jet forces, there is a further stress due to the vibration of the buckets. Determination of this stress is complex as it depends on the dynamic response of the buckets to the jet pulses and on the comparison of the Fourier spectrum of the jet force with the natural frequencies of the bucket [5.8, p.3].

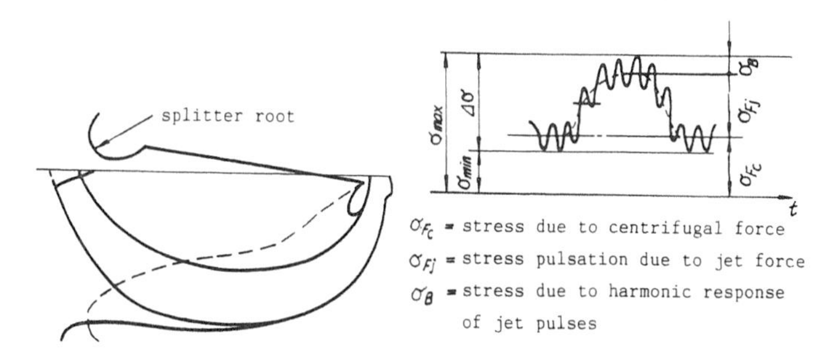

Figure 5.16 Typical stresses in the splitter zone

A more accurate calculation of the stresses due to centrifugal and jet forces is made by development of three dimensional analysis of the relative paths of the jet as shown in Figure 5.13 and by determining the pressure distribution on the active surface of the bucket by means of the calculated accelerations and of the thickness of the water body in the bucket [5.7, p. 9]. The calculated stresses are checked by strain gauge measurements carried out on models and prototype machines.

The statistical curve of Figure 5.6 shows the minimum values of the ratio $1/(d_o/D)$ necessary to act against the mechanical stresses: the ratio increases with the specific hydraulic energy. Stresses in the disk and its coupling to the shaft are usually lower than those in the bucket and may be calculated by classical methods. The thickness of the disk is chosen by taking into account of possible vibrations.

In the past, runners were made in many pieces where the buckets were cast separately and fastened to the disk of the runner by fitted bolts with conical shanks. Nowadays, almost all the Pelton runners are integrally cast. Material chosen is usually stainless steel for its good resistance to fatigue and cavitation pitting (see 5.3.1). Great care must be exercised in the manufacture of the buckets whose active surfaces must be ground according to the designed shape and checked by templates. A smooth finish is essential, waviness and roughness must be reduced to within the limits set by International Standards [5.9] in order to guarantee resistance to fatigue and cavitation, and for increased efficiency. The surface roughness R_a (centre line average) recommended by IEC Code is 0.8 to 1.6μm.

The runner is fastened on the shaft by key(s) or by prestressed bolts.

5.2.2 Shaft

Depending on the arrangement of the turbine and of the unit, the shaft is supported by one or more bearings. The shaft transmits the torque generated by the runner and withstands the bending moment due to the radial forces induced by the jets and by weight of the runner. The radial forces must be duly considered also in the design of the bearings in order to prevent overheating because of asymmetry in the layout. Axial thrust on the shaft may be considered to be zero because of the symmetrical action of the jet on each bucket, therefore simple collars are sufficient to keep the runner(s) centred with reference to the jet(s).

5.2.3 Housing

The housing of a Pelton turbine is shown as *12* in Figure 5.2. Its main dimensions are decided by the radius of the top cover *11* and the width of it, experience has shown that they must be carefully calculated to reduce the windage losses to a minimum. The wiper *13* is built in such a way to have a very small clearance between it and the outer contour of the buckets to prevent air and water from entering the highest part of the top cover and causing additional losses. The baffle *10* has the purpose to protect the lower nozzle from the water discharging from the buckets or in case of jet deflection. In case of accidental runaway, the baffle *10* deviates the lower jet towards the tailwater where the shield *14* and the grid *15* located below the runner together help to destroy the energy of the jets.

The housing is split at the runner axis with the top cover removable so that the runner is readily accessible for inspection. The runner is braked by a counterjet *16* which acts on the back of the buckets. It is normally only turned on when the runner is at a reduced speed so as to avoid excessive stresses in the buckets.

5.2.4 Nozzle

The turbine shown in Figure 5.2 has two nozzles *5* each containing a needle which controls the discharge through its linear stroke.

(i) Nozzle design

Figure 5.17 shows the construction of a straight nozzle in detail. The outlet part of the needle and the nozzle mouthpiece are conical, a good geometry is obtained for the jet by properly selecting the angles 2α and 2β, for example their values are chosen to be in the ranges 40° to 55° and 60° to 90° respectively. The conical shape gives a rapid convergence to the jet that is compact and without rotational components. Guide fins (three to eight) help to eliminate swirls in the flow and to strengthen the nozzle frame.

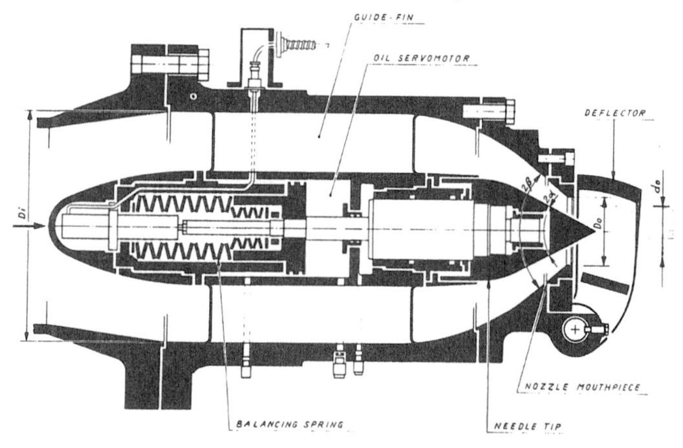

Figure 5.17 Straight nozzle for a Pelton turbine (Drawing courtesy Hydroart)

Upon leaving the nozzle orifice, the diameter of the jet first decreases to the contracted section and then gradually increases due to the reduction of velocity in the outer layers from exchange of energy with the surrounding air. The kinetic energy of the jet will decrease with the distance from the nozzle, therefore the nozzle should be placed as near as possible to the runner. When the servomotor is built inside the nozzle, the flow through the nozzle is straightforward and the jet attains a higher efficiency. The nozzle is usually designed to have a water velocity in the central part not higher than $0.1\sqrt{2E}$ in order to keep the losses low.

At full opening the jet discharge is

$$Q_j = \frac{\pi d_0^2}{4} \cdot \varphi\sqrt{2E}$$

referring to the section where the mean velocity is given by $\varphi\sqrt{2E}$. The discharge through the nozzle depends on the stroke of the needle and it is given by the coefficient

$$k_Q = \frac{Q}{\pi(D_0^2/4)\sqrt{2E}}$$

where D_o is the diameter of the nozzle orifice (see Figure 5.17). The $k_Q - (s/s_{max})$ relation shown in Figure 5.18 relates to a particular nozzle since it is affected by the size and shape of each nozzle.

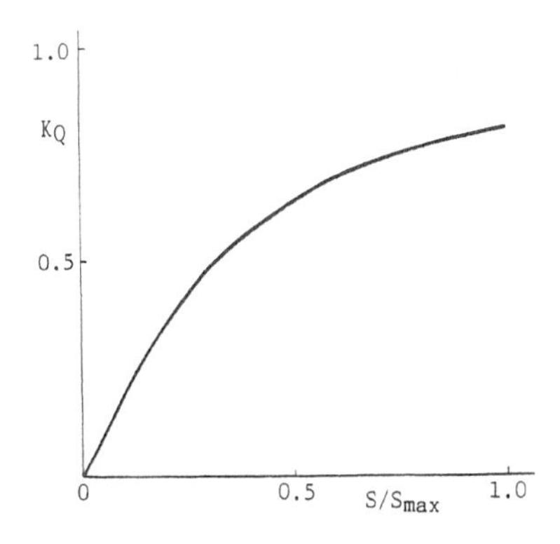

Figure 5.18 Trend of nozzle discharge coefficient curve

(ii) Forces acting on the needle

Reference is made to the simplified nozzle shown in Figure 5.19. When the needle is closed, the hydraulic force acting on it is given by

$$F_h = \frac{\pi}{4}(D_0^2 - D_2^2)\rho \cdot E$$

which acts in the closing direction. When the needle opens, this force at first rises slightly and then drops almost linearly because of the counteracting force arising on the downstream side of the needle. In order to reduce the work for controlling the needle, it is convenient to counterbalance the hydraulic force as much as possible by the action of a set of springs and a relief piston as shown in Figure 5.19.

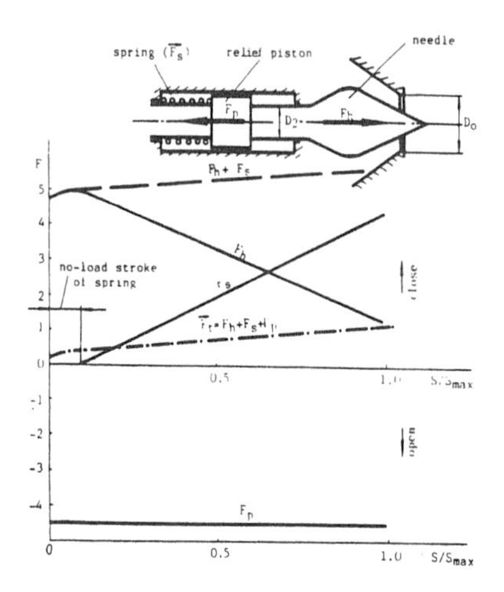

Figure 5.19 Forces acting on the Pelton needle

The actual work required to control the needle by the oil servo motor is actually quite small even taking into account the friction losses not indicated in the diagram. As the hydraulic force acting on the needle depends on the geometry of the needle, and of the nozzle, the above example represents only one of many solutions devised by turbine manufacturers. In this figure, the resultant force assures a closing trend as in the usual case, however, the trend can be modified by varying the characteristics of the different components.

Sand erosion can seriously damage both the nozzle mouthpiece and the needle tip, as shown in see Figure 5.20, that will impair the function of the jet and consequently cause erosion of the buckets and loss in efficiency. Special resistant materials such as chromium–cobalt alloys are used to make the needle for machines working in high sediment content waters.

5.2.5 Feeding system

The feeding system is composed of one or two branch pipes connecting the penstock to the nozzle through a main valve (usually of the spherical type) which can be operated as an emergency shutdown device in case of failure of the governing system. The design of the feed pipes for a turbine with two nozzles is very important for the performance of the machine. Small deflection angles of the Y–branch and small curvature of the pipes at its ends will help to reduce or eliminate flow disturbances and swirls at admission to the nozzles, thus improving the efficiency and avoiding cavitation.

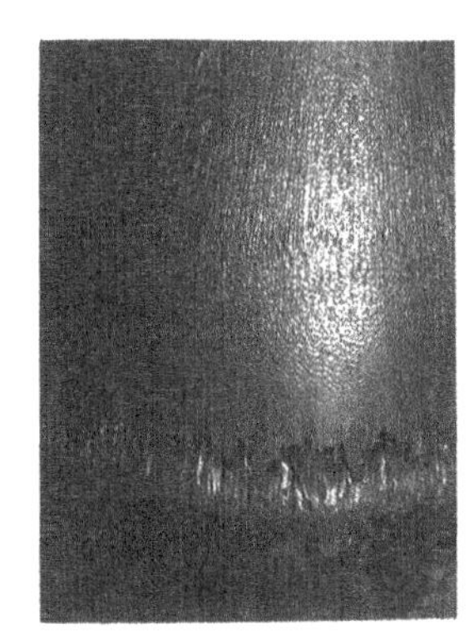

Figure 5.20 Needle tip when new and worn by glacier–type erosion (Photographs courtesy Sulzer Escher Wyss)

5.2.6 Control mechanism

The control mechanism is composed of the needle, the deflector and their servomotors. Water discharge and power are controlled by the stroke of the needle. Given the usual length of pipeline, the needle closure time must be sufficiently long to avoid too high momentary pressure rise in the penstock. To prevent too high transient speed rises, it is necessary to have a device cutting the jet flow to the runner in a very short time but without causing undue waterhammer. The deflector shown in Figures 5.2 and 5.17 serves this purpose by cutting the jet rapidly (in two to three seconds) and deviating part of it into the tailwater, and thus affects a sudden drop of power without altering the discharge in the penstock. The needle then closes slowly in 20 to 40 seconds. The control system, therefore, has a dual function and acts similarly to the runner blade/guide vane control system for adjustable–blade reaction turbines. The position of the needle is correlated with the position of the deflector through a cam which is replaced by electric devices in more modern designs. The deflector is normally poised ready at a distance of a few millimetres from the edge of the jet.

The control systems for all types of turbines are analysed in detail in the Volume *Control of Hydraulic Machinery* of this Book Series to which the reader is referred for more information.

(i) Deflector

There are two possible ways for the action of the deflector, the arrangement shown in Figure 5.17 moves downwards to cut off the upper part of the jet while the one shown in Figure 5.21 moves upwards to cut off the lower part of the jet. A partial penetration into the jet is sufficient to destroy its useful energy. The forces acting on the deflector are shown in Figure 5.22. Neglecting the tangential force T and the friction force F, the force acting on the deflector is

$$N = \rho Q_d C_1 \sin \alpha$$

where Q_d is the deflected flow. The resulting moment referred to the point O is $M = N \cdot d$. Moment M and force N are taken into account in the mechanical design of the deflector and of the control system mechanism and also in the determination of the governor capacity.

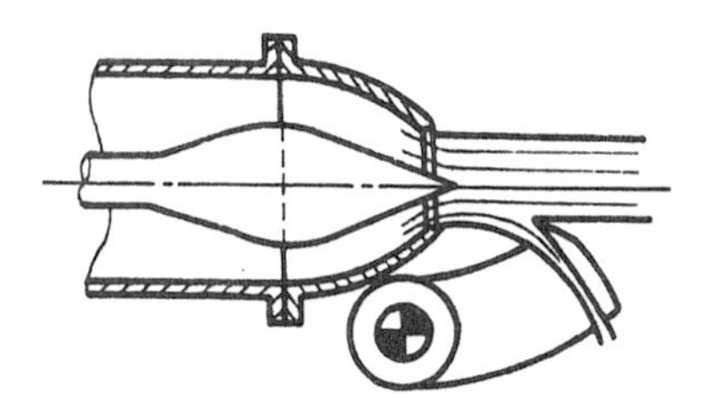

Figure 5.21 Deflector acting on the lower part of the jet

(ii) Needle

The needle is the main component of the nozzle. As mentioned above, it regulates the discharge and determines the necessary control action.

(iii) Deflector servomotor

The deflector and the needle are operated by two servomotors having different operating times. The deflector is usually actuated by an oil servomotor where the counterforce is in some cases supplied by water from the penstock or by a spring. In case of oil pressure failure, the deflector is actuated by water pressure or by the spring. Normally there is only one servomotor to serve all the deflectors for a multi-jet Pelton turbine, the movement of the deflectors is synchronized mechanically or electrically.

(iv) Needle servomotor

Older designs of Pelton turbines often have the servomotor fitted outside the

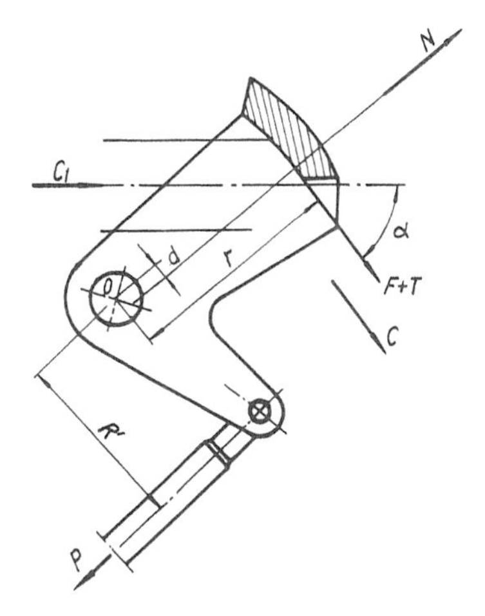

Figure 5.22 Forces acting on the deflector

nozzle pipe which, of necessity, is made curved as shown by the example in Figure 5.23. More recent designs have the servomotor built inside the nozzle frame and are called straight nozzles whose construction is shown in Figure 5.17. The straight nozzle affords a more compact design that eliminates the long control rod which has to pass through the inlet pipe bend.

In multi–jet Pelton turbines the movement of the needles is synchronized by hydraulic or electric restoring devices that control the servomotors. In case of failure of oil pressure, the needle is usually closed automatically under effect of the forces acting on it (Figure 5.19).

5.3 Vertical Shaft Pelton Turbine

In comparison with the horizontal shaft arrangement, the vertical shaft turbine is able to produce greater power and also to increase the rotational speed of the unit with economical advantages (see 5.1.3). The higher speed is achieved by increasing the number of jets, up to an optimum number of six. Laboratory research has shown that it is possible to concentrate six jets on a vertical axis runner without having any hydraulic interference. Many elements that have particular influence on the vertical Pelton turbines with high power, such as, large size, high frequency of the alternating forces on the buckets and operation under high specific hydraulic energy, must be taken into account in their design. The basic features developed for horizon-

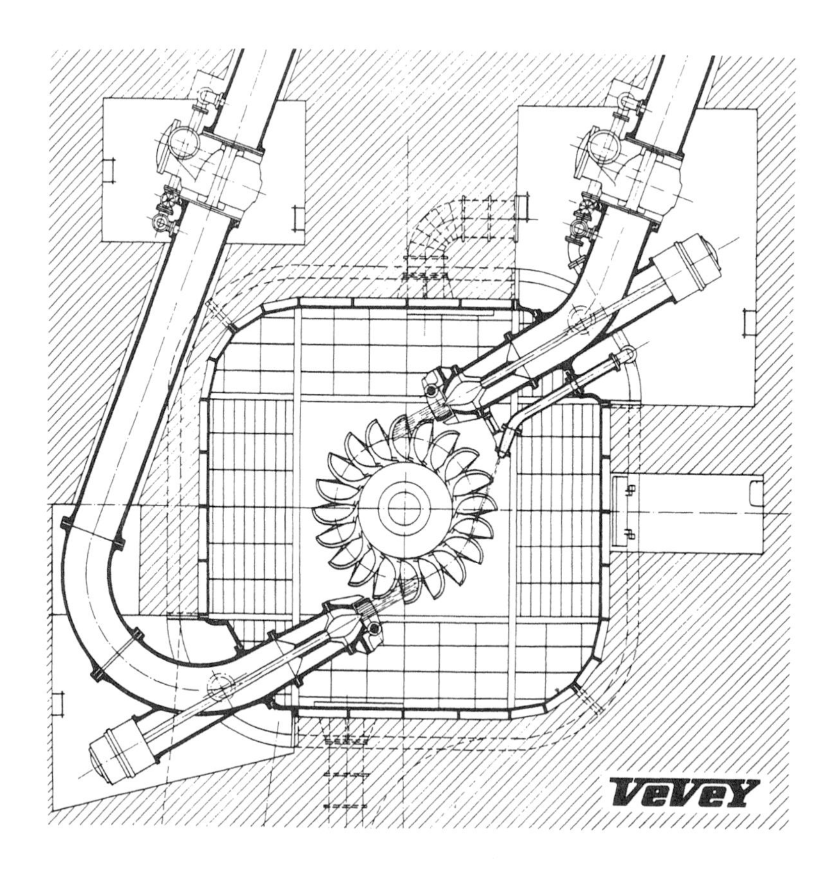

Figure 5.23 Pelton turbine nozzles with external servomotors
(Drawing courtesy VeVey)

tal Pelton units must be integrated by further consideration on the specific problems of the vertical multi–jet turbines and their runners.

5.3.1 Runner

(i) Hydraulic performance

In 5.2.1 it has been demonstrated that the Pelton turbine is sensitive to variations in specific hydraulic energy. In the case of a six–jet turbine, the efficiency hill diagram shows a phenomenon which is typical of this type of machine. When the specific hydraulic energy becomes lower than the design value the efficiency drops suddenly. This symptom, known as *falaise*, is attributed to the interference of the jets when the peripheral velocity factor u increases, causing instability of the machine and reducing the runaway speed as shown in Figure 5.14.

For multi–jet turbines, in particular the six–jet units, the efficiency can be increased by bending the bucket entrance upwards and increasing the inlet diameter so as to reduce the water coming out through the bucket entrance [5.10, p. 13]. This change has the drawback of increasing the cavitation on the back of the bucket (occurring when the angle between the back of the bucket and the relative velocity is greater than 7° on the sides and greater than 2° in the splitter of the inlet) and also increasing the alternating stress amplitude. A correct compromise must be found between the hydraulic performance and the service life of the runner. Large variations in the specific hydraulic energy will make the solution of the problem even more difficult.

(ii) Fatigue

Fatigue problems of the runner are particularly serious for multi–jet Pelton turbines. Figure 5.16 shows the typical stress cycle in the splitter root which amounts to about 10^{11} times during the life of a six–jet Pelton turbine.

The maximum stress calculated according to 5.2.1 must be compared with the resistance curve. Figure 5.24 shows that the tested material can assure a life of more than 10^{11} cycles if the peak-to-peak amplitude is less than 45 MPa and if the surface flaw size is less or equal to $2\ x\ 1\ mm$. Research has further indicated the submerged cracks must be smaller than $2\ x\ 2\ mm$ [5.10, p.14].

Material selection is extremely important in the design of Pelton turbines [5.11]. The fatigue phenomena dictate the choice for the right material (results of fatigue life time for 16/5 and 13/4 Cr/Ni stainless steels are practically equivalent) and the predetermination of the acceptable defects in the stressed areas together with the calculation of the maximum stresses that must be compatible with detectable defects.

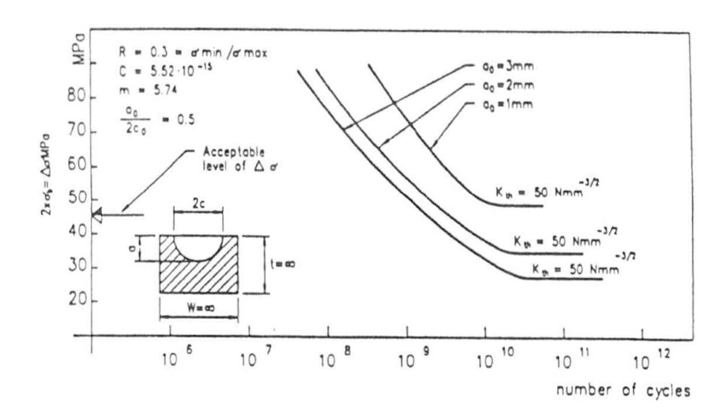

Figure 5.24 Life of Pelton runners versus peak–to–peak stress amplitude [5.10, Fig. 3]

Only qualified foundries should be chosen for casting Pelton runners because the size of defects smaller than 2 x 2 mm can only be achieved by very strict quality control.

(iii) Cavitation

The curve shown in Figure 5.25 defines the maximum allowable specific speed per jet as function of the specific hydraulic energy, in order to avoid any cavitation pitting when the Pelton turbine operates in clean water. The same figure shows the limit curve for six–jet turbines due to a drop in efficiency at full load. Cavitation pitting which determines the limit curve of Figure 5.25 is also caused by the water droplets thickly populating the surrounding of the runner in multi-jet Pelton turbines operating under high specific hydraulic energy. These water droplets produce an impact on the buckets similar to rain erosion as shown in Figure 5.26.

(iv) Sand erosion

This problem cannot be solved only by a suitable design and is most serious with the multi–jet Pelton under high specific hydraulic energy. Sand erosion has also an effect on the cavitation damage since the latter may be started by wear due to sand erosion. Sand erosion occurs where there are high accelerations, the theoretical analysis and stroboscopic laboratory observations have shown that acceleration may reach values up to $100,000 m \cdot s^{-2}$ locally in the bucket [5.11, p.12].

The erosion rate is very high and the buckets may be damaged in a very short time. As acceleration is inversely proportional to the size of bucket,

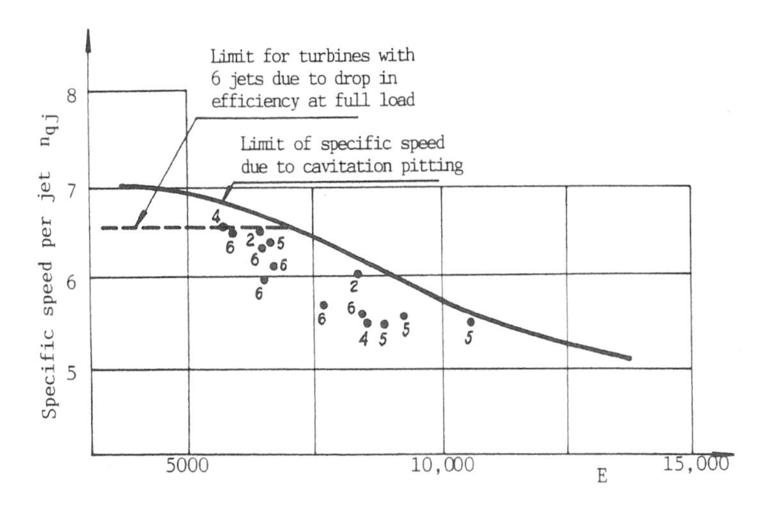

Figure 5.25 Specific speed per jet for Pelton turbines with cavitation pitting-free operation in clean water [5.7, Fig. 23]

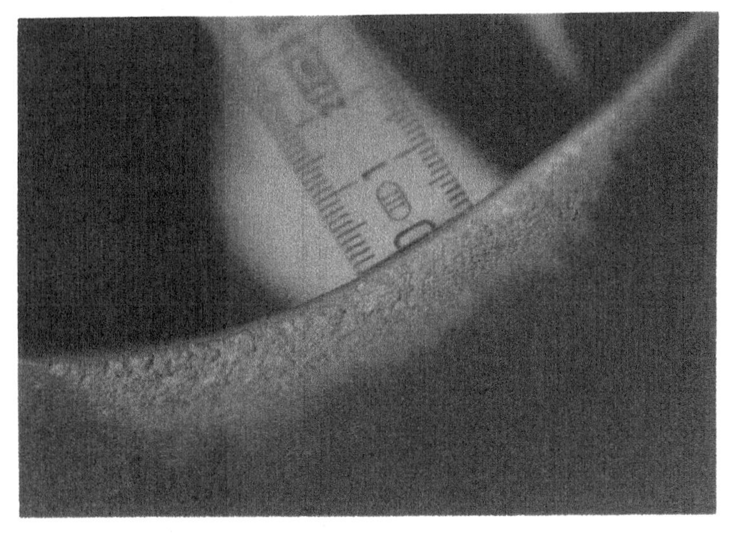

Figure 5.26 Cavitation pitting at the rear of bucket leading edge [5.8]

the buckets should be made as large as possible if sand erosion is expected.

So the erosion problem too suggests choosing not too high a specific speed per jet when designing multi-jet Pelton turbines operating under high specific hydraulic energy. As mentioned in 5.2.1, profile of the buckets must be carefully studied and smoothly ground to assure good hydraulic performance and high resistance to fatigue, cavitation and sand erosion.

5.3.2 Shaft

Layout of vertical Pelton turbines is similar to vertical reaction turbines, but there is no hydraulic axial thrust, the thrust bearing is loaded only with the weight of the rotating parts (turbine and generator). The guide bearings are acted on by the usually negligible radial loads due to small hydraulic and electrical asymmetries. When the jets acting on the runner are not symmetrical, the guide bearings must be designed to withstand the resulting hydraulic radial thrust. Normally there are two guide bearings for each unit, but for turbines of high specific hydraulic energy and power, three guide bearings are usually required because of the lower critical speed due to the increased height of the high speed generator, and to the difficulty of getting a sufficiently rigid support for the upper guide bearing of the generator.

5.3.3 Housing

The housing for vertical Pelton turbines is usually prismatic with few exceptions of circular design. Dimensions of the housing must allow a good outflow of the water discharging from the runner into the tailrace channel.

The depth F, and tailrace channel dimensions H and I, as shown in Figure 5.27, are strictly related to the particular layout of the plant. Long tailrace channels or wide variations of the tailrace level require a deeper setting of the housing foundation and/or a greater channel width to give the necessary distance between the runner centreline and the maximum water level in the housing (about $1.2D$ as recommended by [5.12, p. 33]), taking into account the rise of water level during transient conditions. Figure 5.28 shows a part of the housing and the central structure supporting the turbine guide bearing.

Usually the housing is built from steel plate elements which are welded on site. It is not subject to mechanical forces and is mainly used to line the concrete which surrounds it. A grid is located below the runner to break the flow discharging from the runner and is used also as a support for the inspection platform. An access door is provided in the housing.

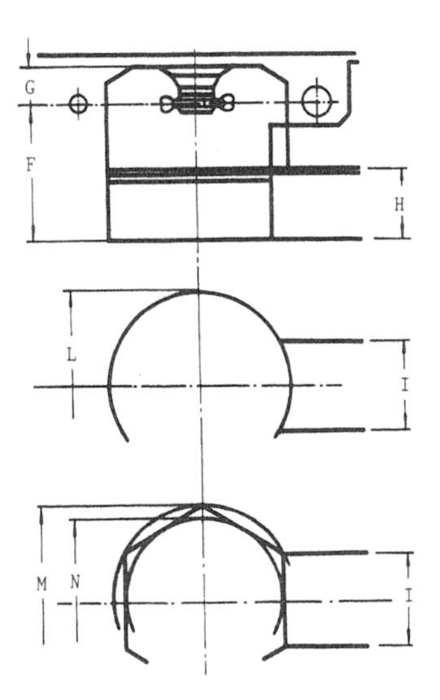

Figure 5.27 Pelton housing dimensions [5.1]

5.3.4 Nozzle

Figure 5.28 shows details of three straight nozzles of a multi–jet Pelton turbine together with their deflectors and control mechanism.

5.3.5 Feeding system

As mentioned in 5.2.3, the design of the feeding system is very important in obtaining good hydraulic performance of the machine.

Figure 5.29 shows two arrangements for a six-jet Pelton turbine, in (a) all nozzles are supplied by one pipeline while in (b) the feeding system is divided into two pipelines before entrance into the power house and thus allowing reduction of diameter and thickness of the distributor pipes. The latter scheme is shown in actual construction in Figure 5.30. When the hydraulic design of the bifurcations and the bends are carefully considered, the losses of the feeding system including the nozzles are very low compared with the global losses of the turbine.

The present trend is to increase the water velocity to over $13m \cdot s^{-1}$ for specific hydraulic energy, E higher than $10,000J \cdot kg^{-1}$, which would allow the distributor inlet diameter to be reduced without impairing the hydraulic performance of the machine. Together with the runner, the distributor is

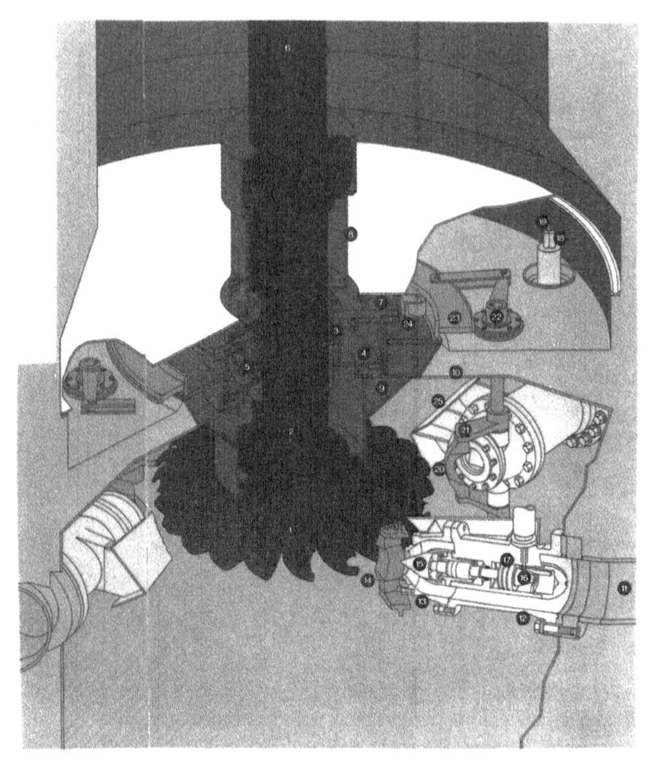

1 – runner
2 – turbine shaft
3 – bearing
4 – rotating oil sump
5 – lubricator
6 – generator shaft
7 – bearing support
8 – shaft casing
9 – interior housing
10 – pit liner
11 – distributing pipeline
12 – nozzle pipe
13 – nozzle body
14 – seat ring
15 – nozzle tip
16 – needle control servomotor
17 – piston
18 – pressure oil
19 – restoring device
20 – deflector
21 – control lever
22 – deflector drive
23 – adjusting ring
24 – adjusting ring guide
25 – protective cover

Figure 5.28 Design of a multi-jet vertical Pelton turbine showing three nozzles[5.13]

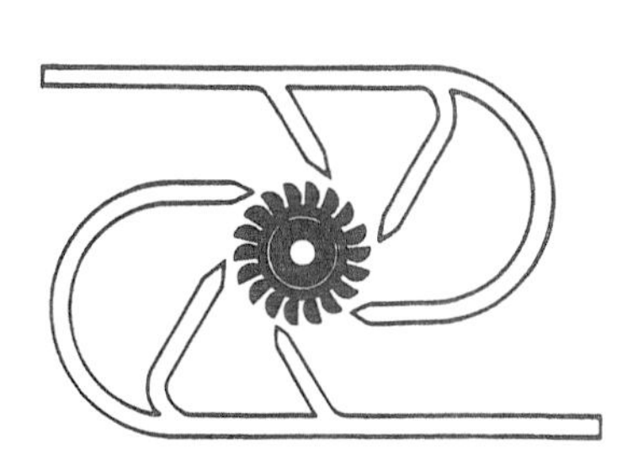

Figure 5.29 Different schemes for the feeding system of a six–jet Pelton turbine

Figure 5.30 Housing and two distribution pipes of a six–jet Pelton turbine prior to concreting (Photo courtesy Sulzer Escher Wyss)

one of the most important structural components. In a cavern type power house, the embedded penstock is assured of support by the surrounding rock, but at the entrance to the power house the pipes are exposed and must withstand the pressure alone. Basically the distributor is statically loaded, but a turbine in typical peaking operation may be loaded and unloaded three times a day. Therefore fatigue problems must be carefully considered. The stresses, including stress concentration and residual stresses, must be limited to such values that the acceptable defects will not reach a size which may lead to unstable fracture (rupture) within 20 to 50,000 loading cycles.

Nowadays the Pelton distributor is fully fabricated from low carbon micro-alloy steel. The joints at the bends between the bifurcations on the inner part are the critical points where the maximum working stresses should be limited to $200MPa$ [5.11, p.8].

5.3.6 Control mechanism

The multi–jet vertical shaft Pelton turbine has the same components in the control mechanism as for horizontal units, such as the needles, deflectors and their servomotors (see 5.2.6). These components are shown in detail in Figure 5.28.

For optimum performance not all jets of multi-jet Pelton turbines, are required to be operative at reduced load. Figure 5.31 shows the advantages derived from reduction of the number of operative jets in the case of a six–jet Pelton turbine.

The Sima power plant Pelton turbines ($2\ x\ 315MW + 2\ x\ 260MW$) are utilized for peak load operation controlled by a central computer system. The turbine governor is equipped with a control system which can automatically select, at any moment, the needle opening so as to obtain the best global efficiency from the machines in operation.

5.3.7 Pumped storage plants

Pumped storage plants were originally equipped with ternary units employing one electrical machine (motor–generator) and two hydraulic machines (turbine and pump).

Pelton turbines are normally used in such plants operating under high head. When a vertical unit is used, like the one shown in Figure 5.32, the Pelton turbine is installed above the pump because the pump usually requires a greater submergence from cavitation considerations. Consequently the middle section of the shaft has to be made very long and the unit requires a large number of guide bearings in addition to a very good balancing of the rotating parts.

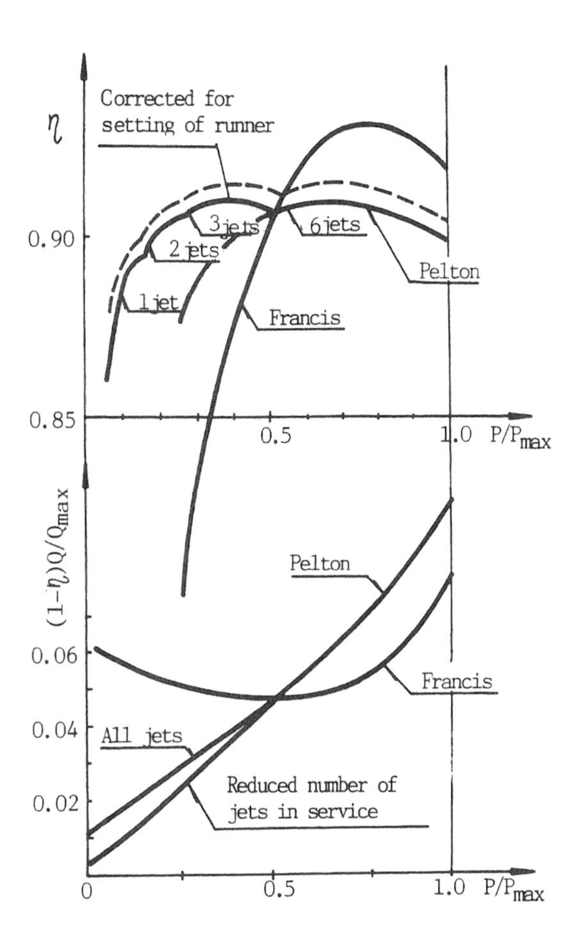

Figure 5.31 Comparison of efficiency and losses for Pelton and Francis turbines [5.7, Fig. 10]

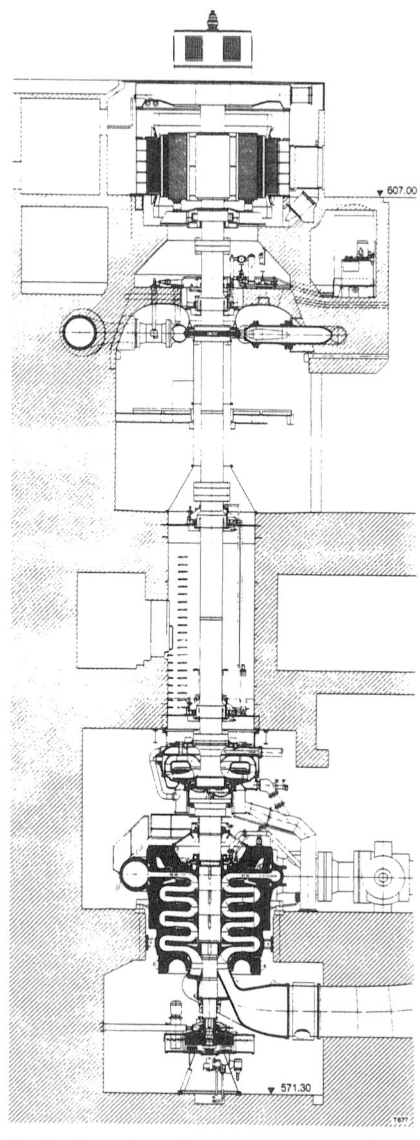

Figure 5.32 Vertical section of a ternary pumped storage unit (Drawing courtesy J. M. Voith)

The Pelton turbine in a ternary unit may be used to assist starting of the machine in pumping mode, and is shut off when the motor-generator is synchronized to the grid. In the case of binary units (reversible pump–turbines, see Chapter 6), a small Pelton turbine is sometimes used as an auxiliary machine to start the main unit in the pumping mode with the impeller immersed in water.

5.3.8 Operation under back pressure

The setting of the Pelton turbine is defined with reference to the highest tailwater level. If the range of variation of this level is great, it may be convenient to avoid the loss of specific hydraulic energy and power by setting the turbine deeper and by lowering, when necessary, the water level in the housing by depression with compressed air [5.12].

A baffle must be installed in the tailrace channel to prevent the air from escaping with the discharging water. Figure 5.33 shows the baffles S placed in three different positions on the model during laboratory tests. The arrows in the upper part of the channel show the air released by the foaming water flowing back to the turbine housing through the tailwater channel and the recirculation channels C. The correct choice of the distance L and the depth H of the baffle helps to reduce the air injected by recovering the air released from the water in the tailrace channel. The minimum aeration rate required is found by throttling the air supply to the housing until the efficiency starts to decrease because of contact of the runner with the foam layer.

Special shaft seals are used to prevent air from escaping into the machine hall when the unit is operating under back pressure. The seals will be retracted and the compressor stopped when the machine operates at a low tailrace level [5.13, p15].

5.4 Research

Research on various types of hydraulic turbines is extensively dealt with in the Volume *Testing of Hydraulic Machinery* of this Book Series.

Only two research problems specific to Pelton turbines are discussed here, namely, scale effects and cavitation.

5.4.1 Scale effects

The IEC Publication 995 scale effects states in Clause 8 that *Different manufacturers have shown scale effects on impulse turbines to be influenced mainly by Froude, Reynolds and Weber numbers. However, since these effects are not yet sufficiently analyzed and no theoretical approach exists, it is not possible to indicate a proven procedure of calculation.*

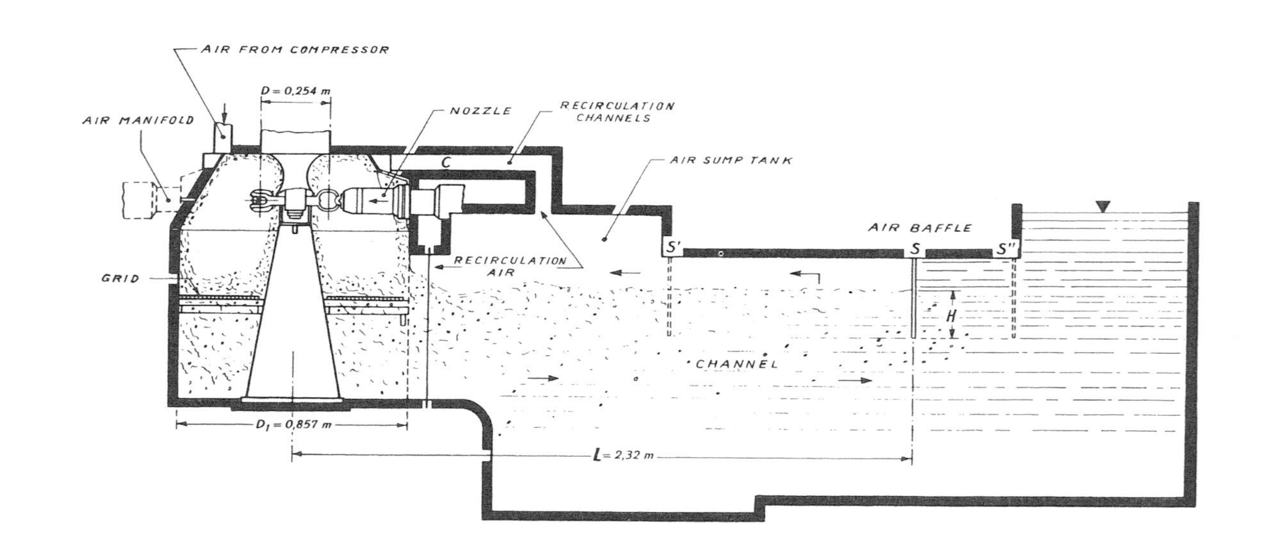

Figure 5.33 Model Pelton turbine for research on operation under back pressure [5.14, Fig. 9]

Most probably this is due to the fact that the flow phenomena associated with the Pelton turbines (pipe flow, free jet flow, unsteady flow with free surface in the bucket, two–phase flow in the housing) are so varied that it is extremely difficult to find a scale effect formula which is able to take into account of all of these factors.

One way to solve the problem is to determine the efficiency step–up relations separately as functions of

(a) Reynolds number that correlates the inertia and viscosity forces
(b) Froude number that correlates the inertia and gravity forces
(c) Weber number that correlates the inertia forces and surface tension.

By measuring the efficiency of the model with different specific hydraulic energy and discharge parameters, (with nozzles of different cross–sections), and with housing of different sizes, the model efficiency is compared with that of the prototype [5.15]. The general formula

$$\eta_P = \eta_M + \Delta\eta_{R_e} + \Delta\eta_{F_r} + \Delta\eta_{W_e}$$

allows to calculate the step-up efficiency for various discharge parameters.

Another approach to the problem is based on the analysis of the power losses, starting from the well–known formula [5.16, p. 1]

$$\Delta P = a + bP + cP^3$$

where a is the coefficient for runner windage loss in dry air, frictional loss in buckets and a portion of water by–pass loss; b is the coefficient for bucket outlet loss, a portion of water by–pass loss and a portion of runner windage loss in air with water droplets; c is the coefficient for distributor and nozzle frictional loss, plus a portion of runner windage loss in air with water droplets; and ΔP is the power loss of the machine. Figure 5.34 shows the loss coefficients at design speed and design specific hydraulic energy measured on the model and on the five–jet Pelton turbines of Grytten and Sima power plants by the thermodynamic method.

Coefficient a decreases and coefficients b and c increase with the dimension of the machine. As the bucket outlet, distributor and nozzle frictional losses, are small and in any case do not increase with the size of the turbine, it can be concluded that the windage losses, due to the high density of the water droplets, strongly influence the trend of the coefficients b and c. This phenomenon is particularly important to the six–jet Pelton turbines operating under high specific hydraulic energy in view of the large amount of water droplets circulating around the runner. On the contrary, no trace of water droplets is observed on the model, since specific hydraulic energy, and discharge in the test, are low so that actual prototype windage losses are not reproduced in the model. The negative scale effect at full load or

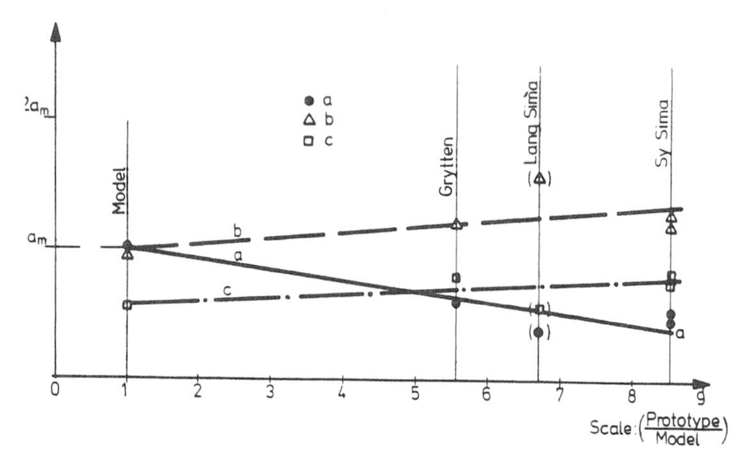

Figure 5.34 Loss coefficients for estimating scale effect of Pelton turbine [5.16, Fig.1]

at overload may thus be explained. A complete survey on the experience by all manufacturers will contribute to formulate a step-up formula that takes into account all the different parameters influencing the Pelton scale effect including the different designs and profiles.

5.4.2 Cavitation

The IEC Code states that the Pelton turbines *usually are not subject to cavitation pitting, but if operating conditions are so extreme that pitting is expected, they would have to be treated as special cases* [5.17]. Actually there are some cases of cavitation in Pelton runners, part of these is due to poor quality of the jet and others are due to the incorrect profiles of the buckets or wrong choice of specific speed. Some cavitation phenomena occur in Pelton turbines when they are operating out of the range of design conditions.

Cavitation tests can be performed on a model in the laboratory where cavitation is reproduced under different operating conditions. The plant cavitation factor must be reproduced on the model. Considering the high specific hydraulic energy of the prototype, the pressure in the model housing has to be scaled down to a very low value. An example is for a prototype of $E_P = 10,000 J \cdot kg^{-1}$, and the model $E_M = 1,000 J \cdot kg^{-1}$, then plant cavitation factor, $\sigma_P = NPSE/E \approx 0.01$, assuming the pressure inside the prototype housing is atmospheric. The net positive suction energy $NPSE$ in the model housing must be $10 J \cdot kg^{-1}$, corresponding to $NPSH = 1m$.

The model tests performed in accordance to these conditions produced good results.

5.5 Future Programmes

The present trend is to develop multi–jet large vertical Pelton turbines [5.16] with advantages in

(a) greater resistance against cavitation and sand erosion due to larger hydraulic radius and smaller acceleration
(b) higher efficiency which compensates for the higher setting of the runner
(c) easier access for inspection and less maintenance per installed kilowatt
(d) automatic selection of number of operative nozzles to increase the efficiency at part loads.

However, there are also some disadvantages

(a) problems of transport of generators and transformers
(b) larger space required in power house
(c) lower critical speed of the unit, mainly due to height of the high speed generator.

Based on existing technology, a design of a Pelton turbine for $860MW$ output under an E of $20,000J \cdot kg^{-1}$ has been projected with the following features (Figure 5.35)

(a) large size for reduced acceleration and stresses in the buckets
(b) power house of the cavern type to render rock support to the embedded penstock
(c) distributor divided into two branches for reduction of diameter and thickness of pipes and their stresses
(d) lowest possible specific speed to reduce stress amplitude at bucket roots and risk of cavitation
(e) working in water relatively free of hard grain sand.

Future trends in Pelton turbine design include not only the above projects of large machines, but also the development of simple, reliable, multi–jet and high speed turbines to replace the traditional horizontal one– or two–jet types on small–scale developments.

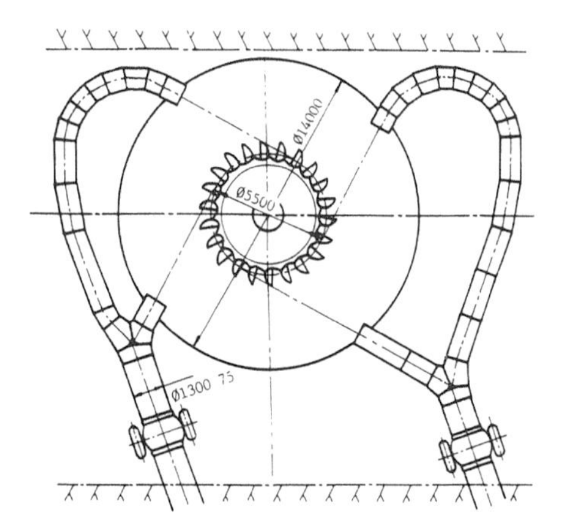

Figure 5.35 Conception of an 860 MW Pelton turbine for $E = 20,000 J \cdot kg^{-1}$ (2,000m head) [5.10, Fig. 6]

Acknowledgements

The author would like to thank Mr. Giacomo Manfredi and Mr Vittorio Zanetti for their precious collaboration, and Professor Hermod Brekke and all the companies which have helped him with information and figures.

Notes

1. The general definition of the specific hydraulic energy of a turbine is given in the Volume *Testing of Hydraulic Machinery*, which gives also the simplified formulae for horizontal and vertical Pelton turbines.

2. In this formula, H is exceptionally used instead of E to maintain specific speed values not too far from the worldwide known range. It is

$$n_{qi} = 60n\frac{(Q/z_j)^{0.5}}{H^{0.75}} \approx 60n\frac{(P/z_j)^{0.5}}{H^{1.25}} \cdot \frac{1}{\sqrt{8.83}} \approx 0.336 n_{sj}$$

   where

$$\frac{Q}{z_j} = \frac{P/z_j}{9.81 H \eta} = \frac{P/z_j}{8.83 H}$$

   where η is assumed to be 0.9, therefore the range of n_{qj} is 4 to 9, corresponding to the range 12 to 27 of n_{sj} for Pelton turbines. The equation

$$n_{qi} = 60n \frac{Q_j^{0.5}}{H^{0.75}}$$

can easily be transformed into

$$n_{qi} = k \frac{d_0}{D} .$$

References

5.1 de Siervo, F., Lugaresi, A. 'Modern Trends in Selecting and Designing Pelton Turbines', *Water Power and Dam Construction*, December 1978, p. 40 to 48.

5.2 Raabe, J. 'Hydro Power - The Design, Use and Function of Hydromechanical, Hydraulic and Electrical Equipment' (book), VDI-Verlag GmbH, 1985.

5.3 Grein, H., Hauser, H. 'Francis or Pelton Turbines in the Head Range between 200 and $800m$', *Escher Wyss News*, 1979/2.

5.4 Krivchenko, G. I. 'Hydraulic Machine Turbines and Pumps', Mir Publishers, Moscow, 1986.

5.5 Nechleba, M. 'Hydraulic Turbines Their Design and Equipment', Artia-Prague/Constable & Co., 1957.

5.6 Vivier, L. 'Turbines hydrauliques et leur regulation', Edition Albin Michel, Paris, 1966.

5.7 Brekke, H. 'High Head Hydro Power Plants-Choice between Pelton and Francis Turbines with a Description of the Design and Experiences from the Operation of the Turbines', General Doctoral Lecture, Norwegian Institute of Technology, Trondheim, 1984.

5.8 Grein, H. et al., 'Inspection Periods of Pelton Runners', 12th IAHR Symposium, Stirling, 1984, Paper 4.4

5.9 'International Code for Model Acceptance Tests of Hydraulic Turbines', Amendment No. 1, IEC Publication 193, 1977.

5.10 Brekke, H. 'Recent Trend in the Design and Layout of Pelton Turbines', *Water Power and Dam Construction*, Nov. 1987, p. 13-16.

5.11 Brekke, H., 'Design and Material Quality for High Head Turbines', 13th IAHR Symposium, Montreal, 1986, Paper 12.

5.12 Grein, H., Holler, H. K. 'Operation of Pelton Turbines under Back Pressure Conditions', *Escher Wyss News*, 1981/1-1982/1, p.32-36.

5.13 'Pelton Turbines', *Sulzer Escher Wyss* Publication 21.16.30 e.

5.14 Ceravola, O. 'Alcuni problemi relativi alle turbine Pelton', *L'Energia Elettrica*, Vol. XLVII, 1970.

5.15 Grein, H., Meier, J., Klicov, D. 'Efficiency Scale Effects in Pelton Turbines', 13th IAHR Symposium, Montreal, 1986, Paper. 76.

5.16 Brekke, H. 'Experience from Large Pelton Turbines in Operation', 11th IAHR Symposium, Amsterdam, 1982, Paper 55.

5.17 'Cavitation Pitting Evaluation in tydraulic Turbines, Storage Pumps and Pump Turbines', IEC Publication 609, 1978.

Chapter 6

Construction of Pump–Turbines

Hiroshi Tanaka

6.1 General

Pumped storage power plants are generally divided into two categories: *mixed pumped storage* which utilizes natural inflow to the upper reservoir for generation besides pumping water by pump–turbines to the reservoir for storage, and *pure pumped storage* which uses only the pumped up water for generation.

Potential sites for mixed pumped storage are limited by geographical and hydrological conditions to utilize natural inflow effectively and they are, in many cases, constructed in combination with multi–purpose projects for flood control, irrigation, water diversion and so forth. The operating heads for these plants are therefore not very high.

On the other hand, in case of pure pumped storage power plants, hydrological conditions are of less importance and in order to achieve greater economy, the operating head is selected as high as possible to minimize the size of reservoir for the required amount of energy to be stored.

In pumped storage plants, the following types of pump–turbines are used:

(i) Reversible pump-turbines

The runner/impeller rotates in one direction as a pump and in the other direction as a turbine and may be subdivided as

- Francis pump–turbines: for application to a wide range of operation from low head to high head, mostly with single–stage runners except for very high head plants where multi–stage machines are used.

- Diagonal–flow (Deriaz) pump–turbines: for application in medium and low head ranges.

- Axial–flow (Kaplan, tubular) pump–turbines: for application in low and very low head ranges.

(ii) Tandem pump–turbines

The pump and turbine are separate units connected to one common electric machine and rotates in the same direction; the pumps may be of single– or double–suction, single– or multi–stage designs; the turbines are predominantly Francis–type but Pelton turbines are used in very high head sites.

6.1.1 Francis reversible pump–turbines

An example of high head Francis reversible pump–turbines is shown in Figure 6.1 General construction features of Francis reversible pump–turbines are similar to conventional Francis turbines except for the following particulars.

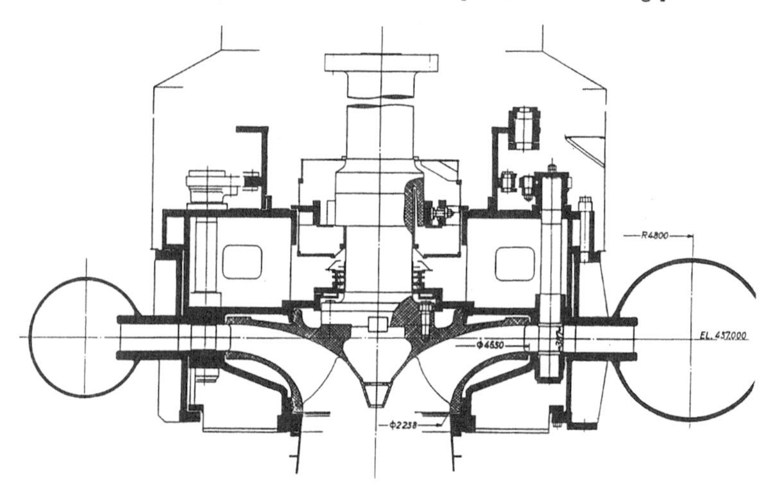

Figure 6.1 High head Francis type reversible pump–turbine

(i) Runner

Runner blades are smaller in number and longer in length. The runner design appears quite similar to that of a centrifugal pump impeller rather than of a conventional Francis turbine runner (Figure 6.2).

(ii) Guide vanes

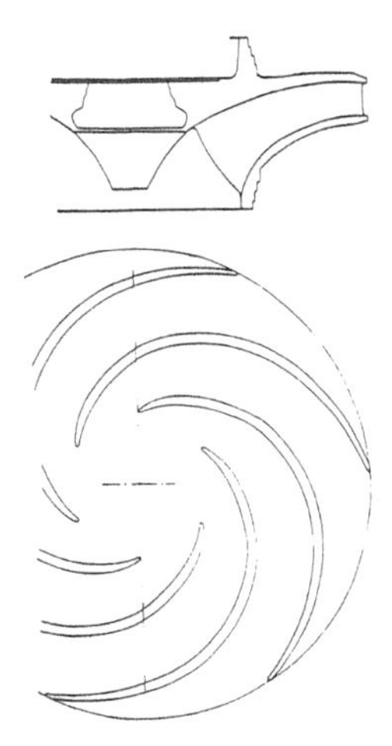

Figure 6.2 Runner of Francis reversible pump–turbine

The inner edges of the guide vanes (inlet edges in pump operation and outlet edges in turbine operation) are shaped rather blunt and the stems of guide vanes are made stronger than for conventional turbines to withstand the greater vibration likely to occur in pump operation.

(iii) Guide bearing and thrust bearing (as part of pump-turbines)

These bearings are designed to operate in both directions of rotation for generation and pumping.

(iv) Auxiliary equipment for pump starting

When starting the unit as a pump by dewatering the runner chamber with compressed air, various control pipes are necessary for admission and release of air and for drainage of leakage water through the guide vane clearances. A special compressed air system is required for this operation.

6.1.2 Deriaz reversible pump–turbines

Deriaz–type reversible pump–turbines are in general similar to conventional Deriaz turbines but they also have different features from conventional turbines similar to the case for Francis pump–turbines, as can be seen in Figures 3.7 and 6.3

Runner blades of conventional Deriaz turbines have their axes located near the centre of hydraulic moment, which is usually at 30–40% of the chord

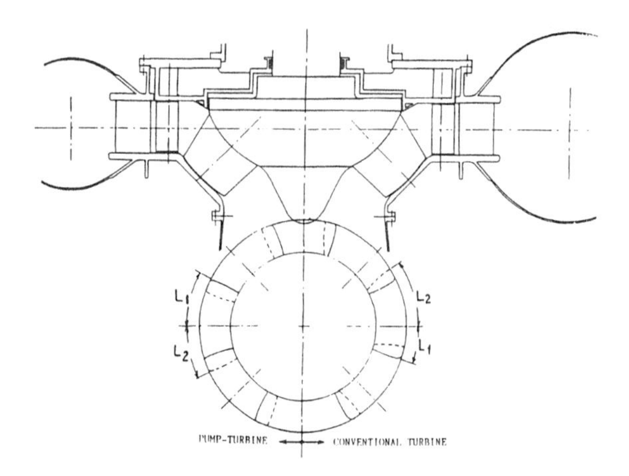

Figure 6.3 Comparison of Deriaz pump–turbine and conventional turbine

length from the leading edge in order to minimize the operating capacity of the runner blade servomotor. But, as shown in Figure 6.3, the runner blades of reversible pump–turbines have their axes placed near the middle of their chord in order to balance the operating forces required in both generating and pumping. However, the required operating force for the runner blades of Deriaz pump–turbines is larger than for conventional turbines by several folds and consequently servomotors of much larger capacity are required.

6.1.3 Axial–flow reversible pump-turbines

Most axial–flow type reversible pump–turbines have very similar construction to those of conventional bulb turbines, tubular turbines or conventional axial–flow pumps. Their runner blades are, however, designed to meet the specific duties for both generating and pumping. The axes of the adjustable runner blades are also positioned at the middle of the runner chord as in the case of Deriaz pump–turbines.

Bulb–type reversible pump–turbines are often used in tidal power stations where they are required to operate both as a turbine, and as a pump, in either directions for efficient utilization of low head tidal energy. An example of this type of machine is shown in Figure 4.12.

6.2 Francis Reversible Pump–Turbines

Detailed designs of the principal components of Francis–type reversible pump–turbines are described below.

6.2.1 Runner

Figure 6.2 shows an example of runner blade design for Francis reversible pump–turbines. It resembles to a large degree the impeller blading for conventional centrifugal pumps.

To improve the overall economy of pure pumped storage power plants, many high head reversible pump–turbines with operating heads of $500m$ or more are being employed. The runners for such high head machines have a peripheral speed of $100m \cdot s^{-1}$ or more and are subjected to both high static stresses due to pressure and centrifugal force, and very large dynamic stress due to pressure pulsation.

Figure 6.4 shows the distribution of the static stress calculated by the finite element method for the case of steady state operation of a $570m$ head pump–turbine. Most of the static stress is developed by centrifugal force rather than by hydraulic force (pressure). The stress due to hydraulic force has a maximum value at the inlet of the runner blades just after the machine is started as a turbine.

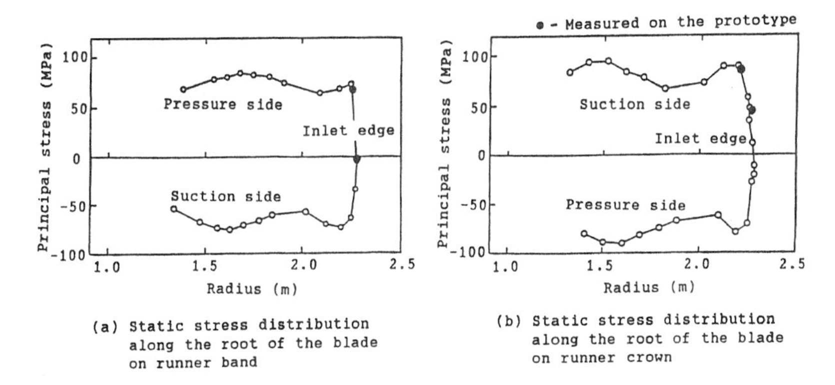

Figure 6.4 Static stress distribution of a reversible pump–turbine runner calculated by the finite element method

Figure 6.5 shows an oscillogram record of the stress measurement taken during turbine start of a $600m$ head reversible pump–turbine.

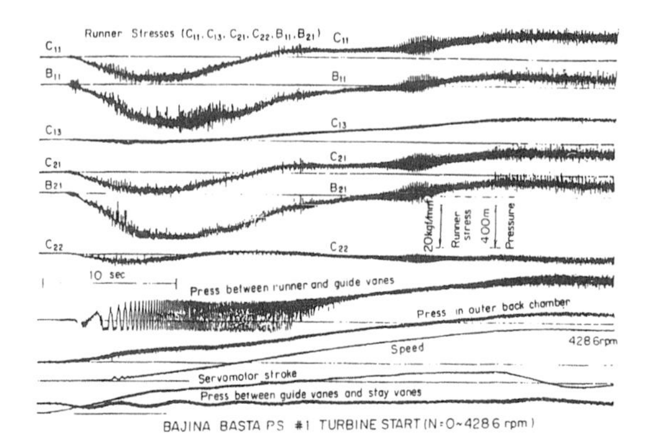

Figure 6.5 Runner stress measured during starting as a turbine

The dynamic stress of a pump–turbine runner consists mostly of regular alternating stress caused by pressure pulsations due to the interference between runner blades and guide vanes. As shown in Figure 6.6, this alternating stress is induced at the inlet part of the runner blade when it passes across the wake of each guide vane. In case of high head pump–turbines, the wake at the exit of the thick guide vanes is very strong and will result in very high amplitudes of alternating stress in the blade under conditions of high flow velocity.

The hydraulic force caused by this interference appears in the manner of a cyclic excitation force on the runner and induces vibration in the runner structure. It has a principal frequency of $Z_g \cdot (n/60)\ Hz$ with higher harmonics $j \cdot Z_g \cdot (n/60)\ Hz$, where Z_g is the number of guide vanes, n is the speed of rotation (min^{-1}) and j is an arbitrary integer.

The interference between runner blades and guide vanes occurs at a certain phase shift and time lag around the periphery of the runner, determined by the combination of Z_g and number of runner blades Z_r. Figure 6.7 shows an example of such a phase shift in the case of interference for $Z_g = 20$ and $Z_r = 6$. As indicated in Figure 6.7, runner blades *1* and *4* are first excited and then blades *2* and *5* as in (a). In this case, runner blades *1* and *4*, or *2* and *5* are excited in phase and induces a vibration having a mode with two diametral nodes. In addition, the nodes of this vibration mode move around the runner in opposite direction to the rotation of the runner, as in (b). In this way, the excitation force caused by interference between runner blades and guide vanes induces certain modes of vibration having a specific number of diametral nodes which rotate around the runner.

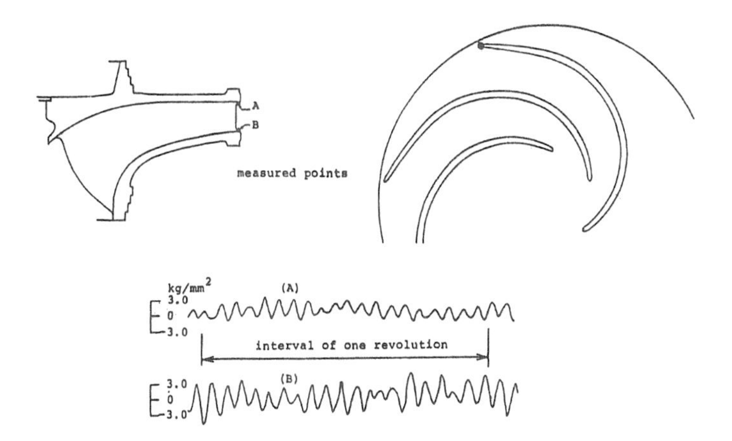

Figure 6.6 Alternating stress at inlet of runner blades

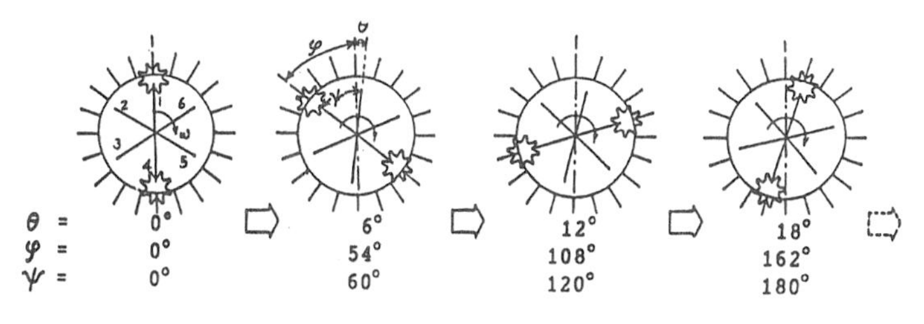

a) time and phase lag of hydraulic impacts

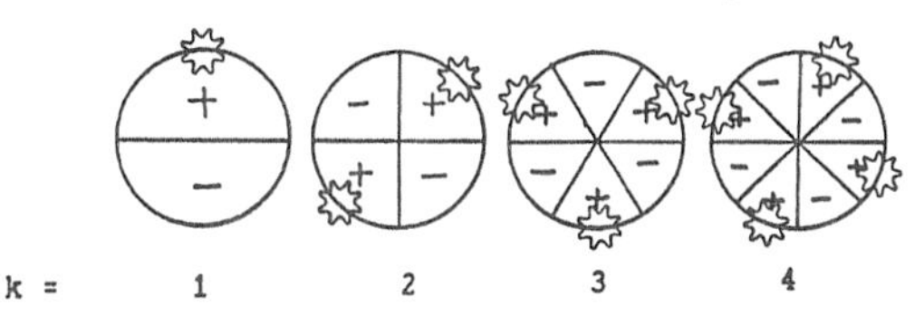

b) vibration modes with k diametral nodes

θ : angle of runner rotation ω : angular velocity of runner
φ : traveling angle of hydraulic impacts on stationary coordinates
ψ : " " on rotating coordinates

✵ : hydraulic impacts

Figure 6.7 Time and phase lag of hydraulic impacts caused by interference between runner blades and guide vanes

The number of diametral nodes k is determined by the combination of Z_g and Z_r as given by the following formula [6.1, 6.2].

$$j \cdot Z_g \pm k = m \cdot Z_r \tag{6.1}$$

where j and m are arbitrary integers.

In the above expression, the first term on the left–hand side indicates the ratio of the frequency of the excitation imposed on the runner over the runner rotation frequency. In other words, the frequency of the hydraulic excitation imposed on the runner is

$$f = j \cdot Z_g \cdot (n/60) \tag{6.2}$$

The sign of the second term of equation (6.1) represents the direction of rotation of the spinning mode. A plus sign means the spinning mode rotates in the same direction as the runner rotation and a minus sign the opposite direction. The right hand term of equation (6.1) corresponds to the ratio of the frequency of the hydraulic excitation on the runner to the frequency of the runner rotation, when observed from stationary coordinates. This means the runner itself undergoes the hydraulic excitation with a frequency of $j \cdot Z_g \cdot (n/60)$, but as the runner is rotating and the mode of the hydraulic excitation is spinning around the runner axis, the frequency of the hydraulic excitation observed from the stationary coordinates becomes $m \cdot Z_r \cdot (n/60)$. This frequency also represents the dominant frequency of the vibration induced at the stationary parts of the pump–turbine.

In general, the vibration modes of a runner that correspond to $k \gg 1$ have high eigenvalues or high natural frequencies. They usually far exceed the excitation frequency $j \cdot Z_g \cdot (n/60)$ and would hardly cause a resonance problem. Therefore, the vibration normally observed in actual cases is of the mode corresponding to some value between $k = 1$ to 4.

A typical example of the frequency spectrum of the pressure fluctuation measured between runner and guide vanes is shown in Figure 6.8. It consists of the fundamental harmonic with a frequency of $Z_r \cdot (n/60)$ and its higher harmonics, which are all induced by the hydraulic interference of the runner blades due to downstream wakes of the guide vanes. These harmonics have their own specific modes around the runner which have k diametral nodes as given by equation (6.1). If one or more of these frequencies are close to the natural frequency of the stationary part with corresponding mode of k diametral nodes, considerable vibration will be developed in the stationary part. Similar to the vibration of the runner, the vibration induced in the stationary part in this way is mostly of the modes corresponding to $k = 1$ to 4. Hence the frequencies of the vibration observed at the stationary part are, in most cases, equal to $m \cdot Z_r \cdot (n/60) Hz$ corresponding to $k = 1$ to 4.

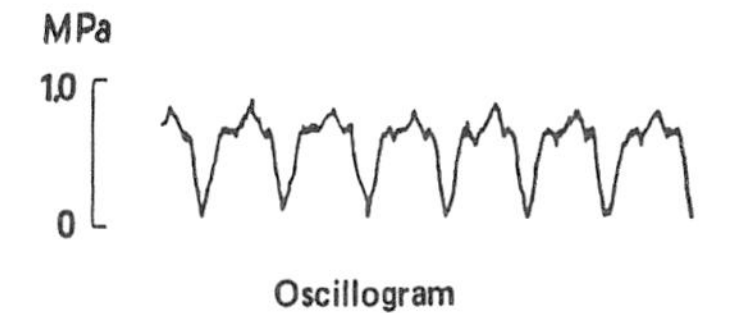

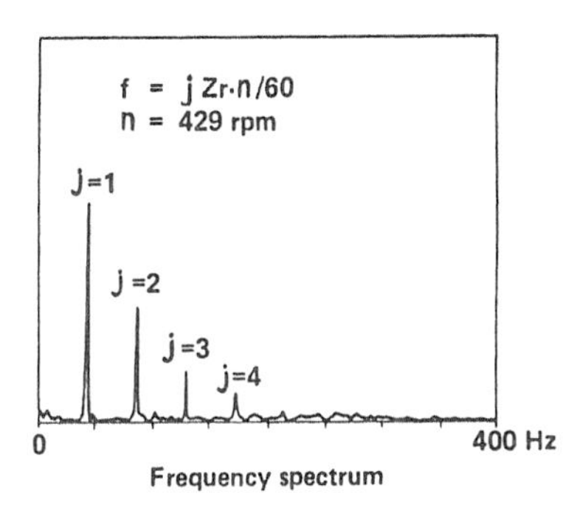

Pressure Fluctuation from Field Tests of the 600 m Head Prototype Pump-turbine

Figure 6.8 Frequency spectrum of pressure fluctuation between runner and guide vanes

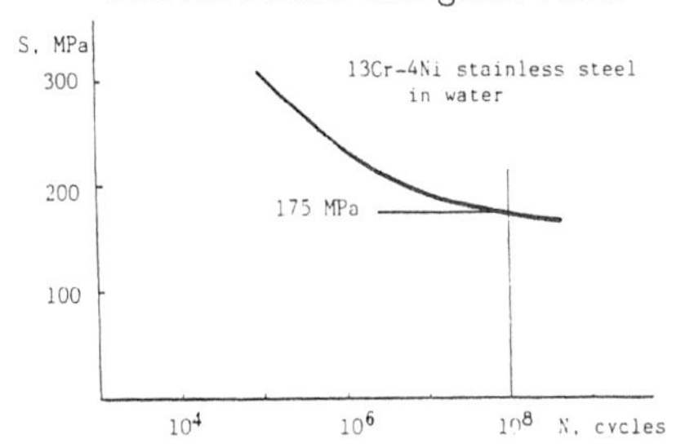

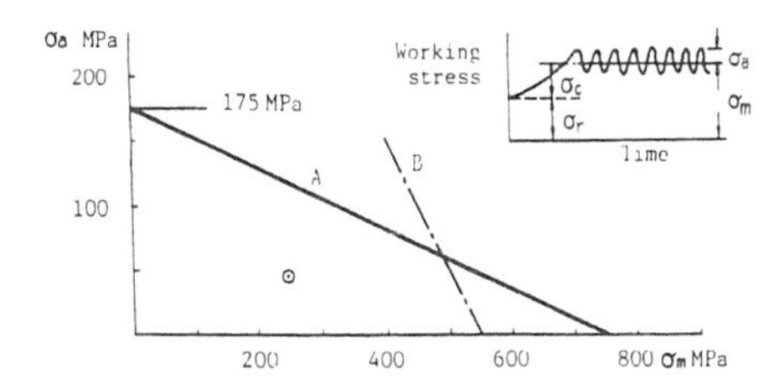

Figure 6.9 S–N curve of 13Cr4Ni stainless steel and modified Goodman's diagram

In Figure 6.5, the amplitude of the stress pulsation of the runner shows a slight rise during acceleration of the unit. This is due to the resonance of one of the natural frequencies of the runner with a harmonic of the hydraulic excitation having the same mode. Such resonance of the vibration of the runner is seldom observed in low head pump–turbines, or if it takes place, the intensity of the resonance is very weak and hardly recognizable. On the other hand, in high head pump–turbines of over $500m$ heads, the resonance becomes of significant intensity and the dynamic stress developed by such resonance may reach appreciable magnitude to cause fatigue failure of the runner.

The vibration of the runner will cause high concentration of alternating stress at the roots of the inlet edges of runner blades. In order to prevent fatigue failure by such high stress pulsation, design studies by the modified Goodman's diagram as shown in Figure 6.9 are recommended.

In general, allowable amplitude of alternating stress σ_a decreases with increasing magnitude of mean stress σ_m as indicated by line A in Figure 6.9.The value of alternating stress at zero mean stress represents the fatigue strength in terms of completely reversed alternating stress and the value of mean stress at zero alternating stress represents the static ultimate strength of the material. Line B represents the stress condition where the sum of alternating stress and mean stress reaches the yield strength of the material at which the material may undergo plastic deformation. If the point plotted at $\sigma_m - \sigma_a$ in the modified Goodman's diagram is located in the area below both lines A and B, the stress condition is regarded as safe with respect to fatigue failure.

For examination of the stress condition of the runner, the following stresses are taken into account

(i) Mean stress ($\sigma_m = \sigma_c + \sigma_r$):

1. Stress due to centrifugal force (σ_c)
   This is estimated (by calculation) by the finite element or other methods.

2. Residual stress (σ_r)
   Since the cooling rate of each part of the runner structure, after heat treatment is not uniform, some residual stress is left in the material. The magnitude of the residual stress varies with the creep strength of the material under high temperature and other thermal properties of the material. Usually the following values are used:

Carbon steel	$\sigma_r = 100\text{MPa}$
13% C_r stainless steel	$\sigma_r = 150\text{MPa}$

(ii) Alternating stress (σ_a):

1. High cycle alternating stress
   This is the alternating stress caused by vibration of the runner described previously. In case of high head pump–turbines, this alternating stress governs the fatigue life of the runner. Since the frequency of this stress is as high as 100 to 200Hz, the expected total cycles of stress pulsation during the whole life of the runner may reach around 10^{11}. A procedure to assess theoretically the magnitude of this alternating stress has not been established. Therefore, it must be determined from either empirical data measured on prototype machines in operation, or from actual head test data of model turbines conducted under dynamic similitude of hydro–elasticity [6.2, 6.3].

2. Low cycle alternating stress
   Alternating stress of this category includes the stress fluctuation in turbine start, turbine load rejection, pump input power failure and so forth. The expected total cycles of these stress pulsations for the whole life of a runner are relatively low, being around 10^5 to 10^6 .

For assessment of the combination of high cycle and low cycle alternating stresses, Miner's rule (cumulative damage rule) as described in Section 8.3 can be applied. Since the value of fatigue strength for a higher number of cycles beyond 10^8 is hardly available, line A of Figure 6.9 is usually drawn based on the fatigue strength for 10^8 cycles. The fatigue strength which is obtained with notched specimens tested in water is intentionally used for the assessment of runner design.

Notched specimen data is needed since cast steels usually used for runners contain minor casting defects such as micro shrinkage which might trigger fatigue failure as a notch. The fatigue strength of steel in water will decline continually beyond 10^7 or 10^8 cycles as different from that in air. Consequently, when the modified Goodman's diagram based on 10^8 cycle fatigue strength is used for the assessment of a runner design, it is necessary to keep a safety factor of 2 or more against line A by considering the deterioration of fatigue strength in the range of 10^8 to 10^{11} cycles.

In addition, the criterion for non–destructive examination of the runner material must be established before manufacturing. Such criterion for allowable size of defects can be established by calculation according to the theory of linear elastic fracture mechanics for the design stress condition. Detailed procedures for such calculation are given in Reference [6.4].

In case of large size pump–turbines, the runners may be split into two or three parts to meet the limitation of transportation. A typical example of split runner construction is shown in Figure 6.10. When the operating head of the runner is high, the stress in the bolts holding the split pieces together becomes very high due to centrifugal forces acting on them. At present, split runners have been manufactured for diameter up to 5.56m and operating under 406m head.

After the connecting bolts are properly fastened, the flanges are protected with cover plates to eliminate churning loss while in rotation. The cover plates are then fixed to the runner by weldment or bolting.

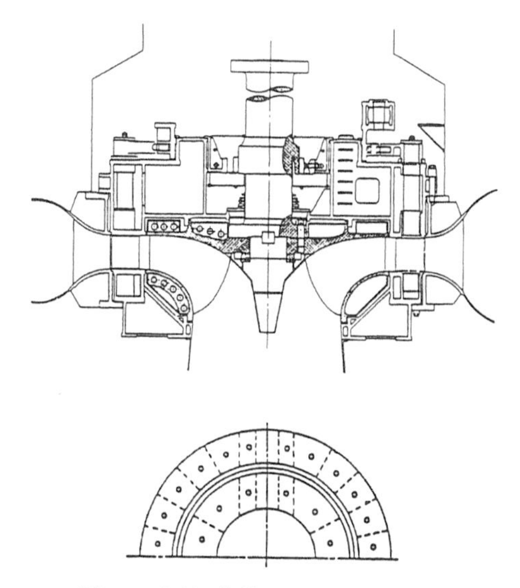

Figure 6.10 Split runner construction

6.2.2 Guide Vanes

For pump–turbines, the inlet edge of the guide vanes in pump operation are designed to have a blunt nose as shown in Figure 6.11, to avoid vibration by separation of the flow. The hydraulic moment acting on a guide vane in the operating ranges *turbine* and *reverse pump* is shown in Figure 6.12, where n_{11} is unit speed of the model under one metre head and a_0 is guide vane opening.

As seen from Figure 6.12, the hydraulic moment becomes very large in the operating range of reverse pump. A pump–turbine transient condition after turbine load rejection will go through the reverse pump range in overspeed condition with flow reversal toward the pumping direction, as shown by points *3* to *5* in Figure 6.13.

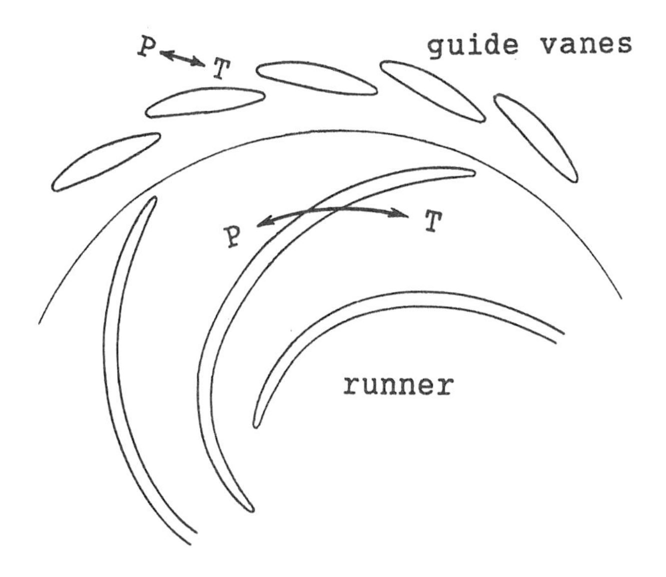

Figure 6.11 Guide vane profile of reversible pump–turbines

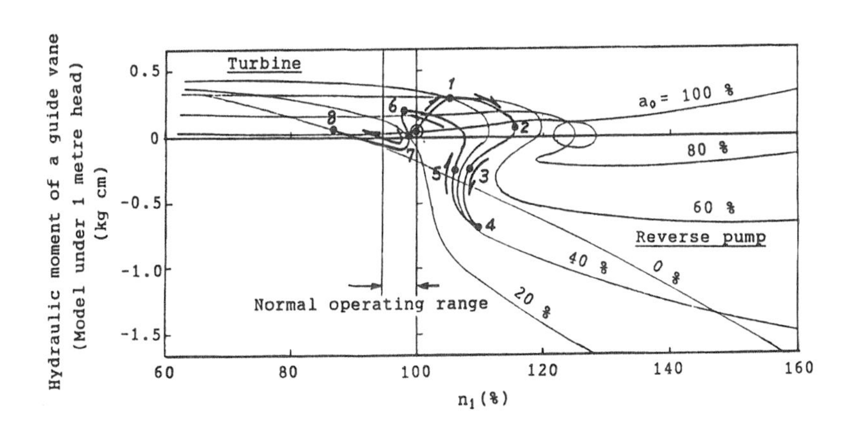

Figure 6.12 Hydraulic moment acting on guide vanes
in turbine and reverse pump ranges

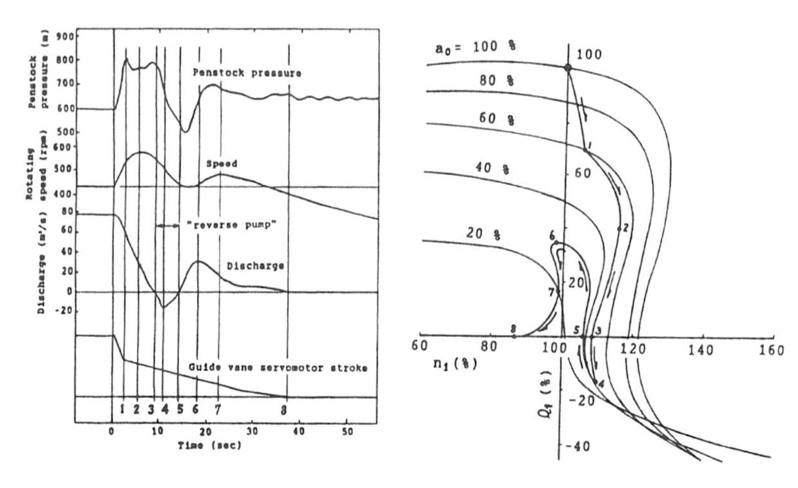

Figure 6.13 Transient locus of turbine load rejection on $n_{11} - Q_{11}$ diagram

The guide vanes are faced with very large hydraulic moment in the open direction under this condition as indicated by the transient loci *1* through *8* in Figure 6.12. In an extreme case, the closing motion of the guide vanes may stall or move in the reverse to open. Consequently, the capacity of the servomotors to operate the guide vanes must be determined with consideration of this hydraulic moment. The stall in guide vane motion can be alleviated to some degree by slowing down the gate closing speed in order to lessen the reverse pump flow in the transient condition.

Figure 6.14 shows the variation of hydraulic moment acting on guide vanes both in normal opening and closing, and in transient conditions of turbine load rejection and pump input power failure. In this diagram, the required servomotor force is that corresponding to the sum of both hydraulic and frictional moments. The capacity of the guide vane servomotor is designed to overcome these moments even when the operating oil pressure is at its allowable minimum.

When a pump–turbine is started as a pump and its guide vanes are about to open after priming (formation of pump shut–off head), self–excited vibration of the guide vanes may sometimes be induced, associated with heavily oscillating water hammer in the penstock. This phenomenon is caused by the following sequence of events:

1. When the guide vanes are slightly opened, positive waterhammer pressure is induced in the penstock.

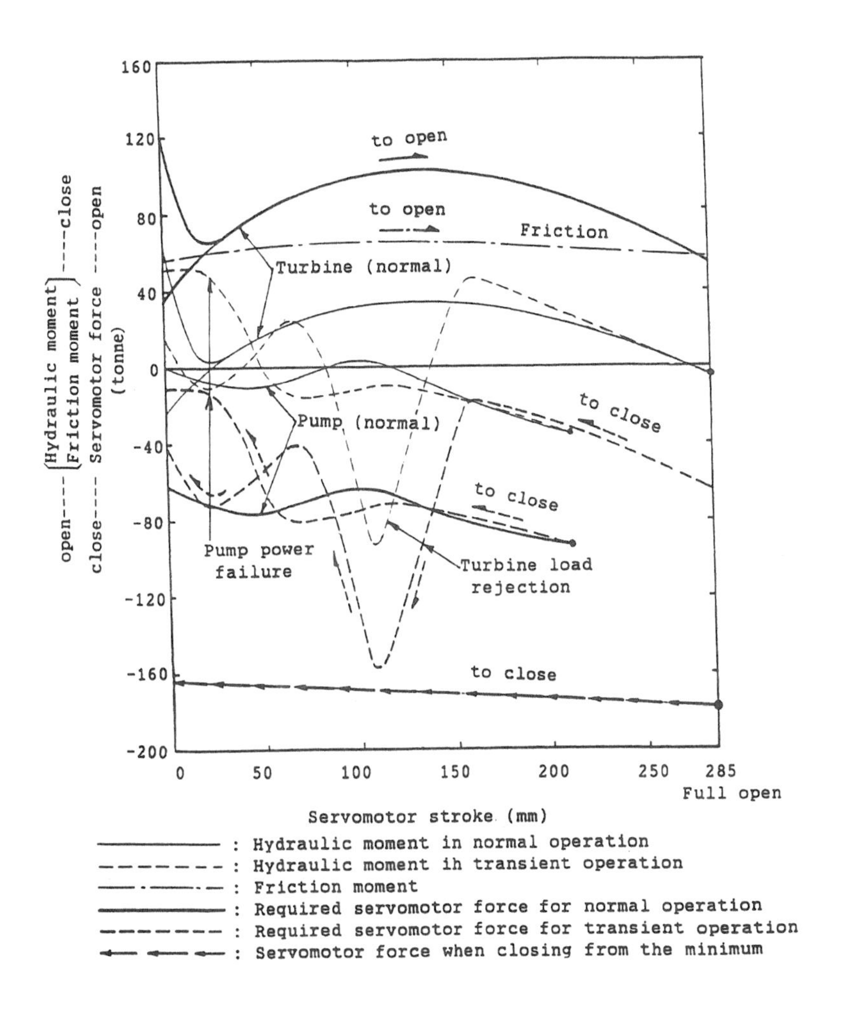

Figure 6.14 Hydraulic moment and friction moment in terms of servomotor force

2. This positive waterhammer wave is reflected at the upstream end of the penstock as a negative wave.

3. If the hydraulic moment of the guide vanes have a closing trend when the pressure in the spiral case decreases, the guide vanes will move slightly in the closing direction when the negative waterhammer wave reaches the pump–turbine.

4. The closing motion of the guide vanes reduces the pump discharge and hence amplifies the negative waterhammer.

5. Such amplification of the waterhammer wave is repeated at every reflection at the pump–turbine and the vibration of the guide vane is consequently amplified.

6. When positive waterhammer waves reach the pump–turbine, the guide vanes will move in the opening direction and amplify the positive water hammer pressure.

If such self–excited vibration takes place, it may grow to dangerous intensities within several reciprocations of the waterhammer wave.

This phenomenon is due to the characteristics of the hydraulic moment of the guide vanes as well as insufficient stiffness of the guide vane stems. Consequently, in order to prevent such vibration, it is required to alter the characteristics of the hydraulic moment of the guide vanes in pumping to have an opening (closing) trend when the pressure in the spiral case becomes lower (higher). Such characteristics can be achieved if the guide vanes are designed to have $L_1/L_2 > 1$ as shown in Figure 6.15

Pump–turbines are usually started in the pumping mode by depressing the water around the runner by compressed air to avoid excessive churning loss of the runner rotating in water. Guide vanes must be leak–proof at closed position to prevent leakage into the runner chamber from the spiral case. For this purpose, guide vane seals are usually provided on the facing plates on both the head cover and bottom ring where the top and bottom ends of guide vanes will rest at closed position. Two types of guide vane seals are commonly used as shown in Figure 6.16. One type is the simple rubber seal and the other is the metal seal backed by elastic material such as rubber for better contact. Metal seals are usually made of bronze to avoid seizure in sliding contact with the moving guide vanes, but are inferior to rubber seals in sealing results. Rubber seals, however, are often damaged by debris in the water and may cause serious leakage after years of operation.

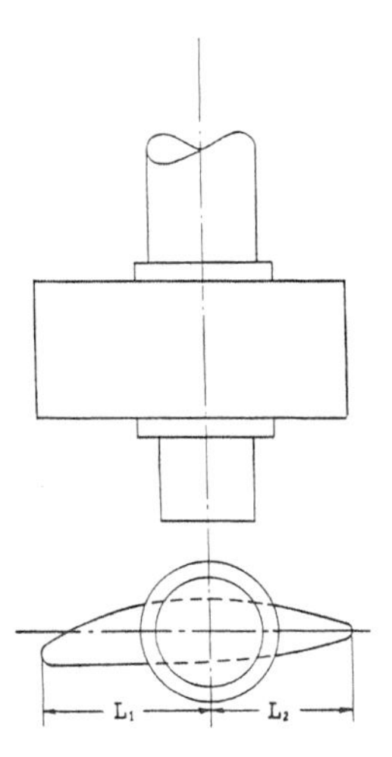

Figure 6.15 Position of the stem of a guide vane

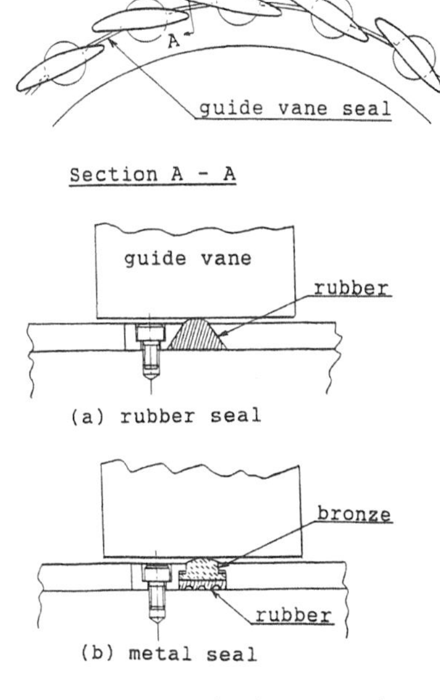

Figure 6.16 Guide vane seals

6.2.3 Draft tube

Draft tubes to be used for pump–turbines with less submergence are usually designed similar to those for conventional turbines. In case of high head pump–turbines, submergence often reaches 50 to 100m and the draft tubes are subject to high pressures. Such draft tubes are normally designed to have a circular cross section at the downstream end (turbine operation) to withstand the high internal pressure and for easy connection to the underground tailrace tunnel. An example of this draft tube design is compared with the conventional design in Figure 6.17.

As reversible pump–turbines are usually started as pump with the runner de–watered by compressed air, an air admission system with air compressors and air receivers, air admission valves, and water level sensors for detecting the draft tube water level are provided as shown in Figure 6.18.

In pump starting, the runner begins rotating after de–watering is completed. When the runner is accelerated to near the rated speed, a strong whirl is developed below the runner due to its rotating motion [6.3]. This whirl induces a revolving sloshing motion of the water surface in the draft tube as depicted in Figure 6.19. Since the circumferential velocity of high head runners is very large, the sloshing motion of the water surface is rather violent with the top of the sloshing wave blown off by the whirl. The water splash is brought into contact with the runner by an upward secondary flow in the central core of the whirl. If the water level in the draft tube is high, the amount of water splashing will become excessive and the input power required to drive the runner in air will increase greatly. On the other hand, if the water level in the draft tube is low, the bottom of the sloshing wave will reach the elbow section of the draft tube and some of the de–watering air will escape to the downstream, then a much larger air supply will be needed to make up the loss of air.

In order to maintain satisfactory operation, the vertical distance from the bottom of the runner to the depressed water level should be set at not less than the values indicated in Figure 6.20. The distance from the water level to the elbow section must be greater than Z.

The internal pressure of the draft tubes of high head pump–turbines may be quite high due to the very deep submergence, but it is not a critical problem in structural design of the draft tube liner as this pressure will be borne also by the surrounding concrete. Instead, the important question is the strength against external pressure. For machines with deep submergence, the external pressure acting on the draft tube liner by water seepage through the surrounding rock may reach nearly the value of the total submergence. When the draft tube is de–watered, it may cause separation of the liner from the surrounding concrete and in extreme cases, may result in buckling

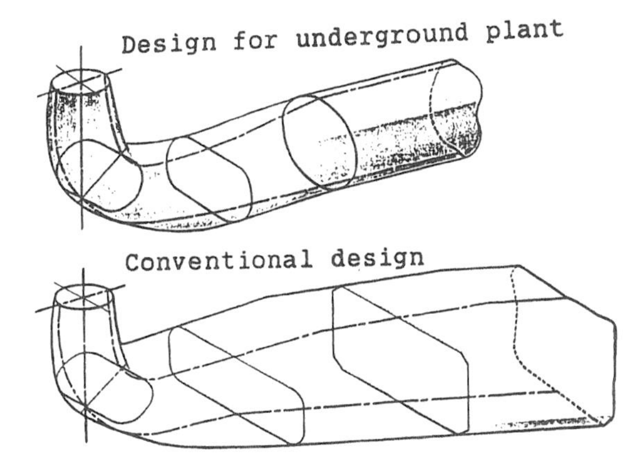

Figure 6.17 Draft tube design for underground power plant

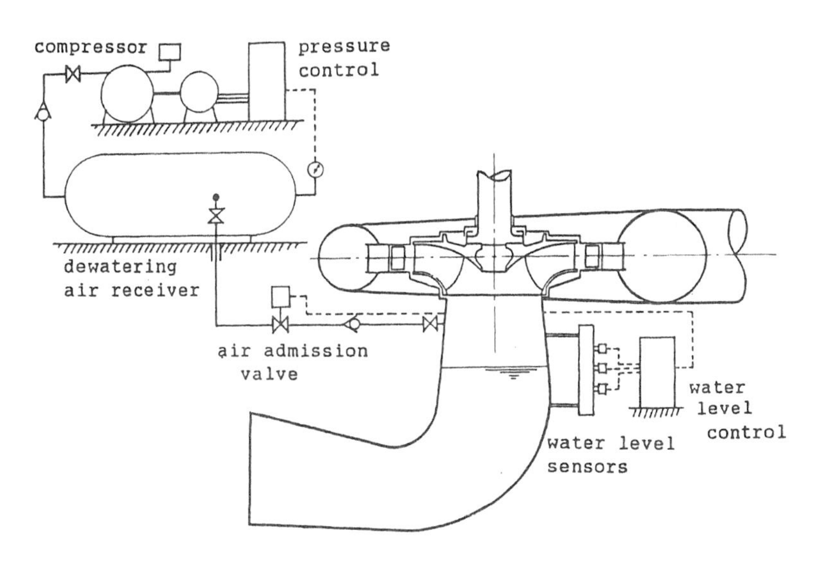

Figure 6.18 De–watering air admission system

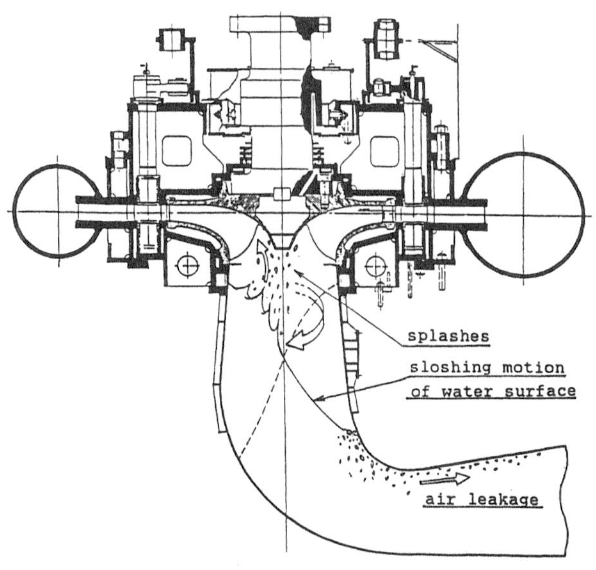

Figure 6.19 Behaviour of depressed water surface

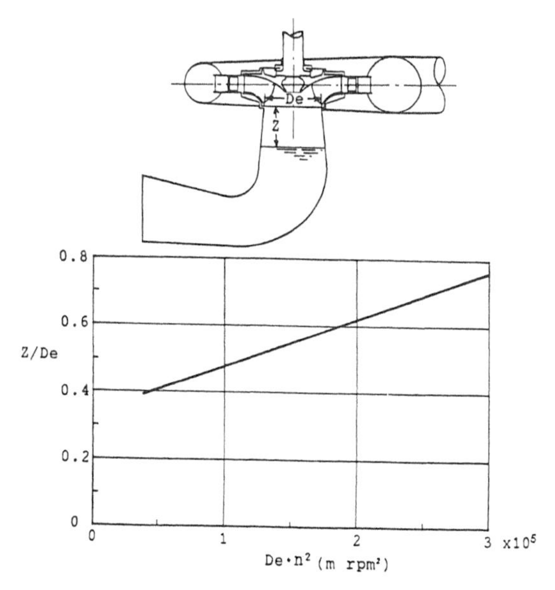

Figure 6.20 Depression of water level required in draft tube for de–watering operation

failure. Strength of the draft tube liner as well as the number and spacing of anchors attached to its outside must be determined with consideration of the external pressure if seepage water is expected.

6.2.4 Operation and control

(i) Normal starting and stopping

Starting and stopping of reversible pump–turbines in the turbine mode can be executed by the same procedures for conventional turbines, as shown in Figure 6.21.

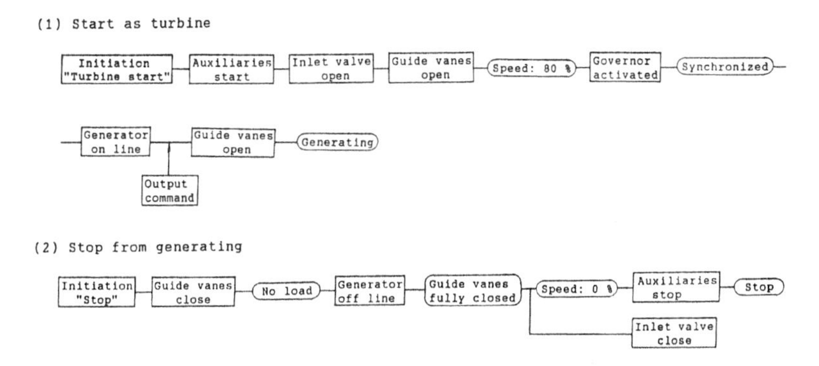

Figure 6.21 Start and stop procedures in turbine mode

In some cases, the turbine output may become negative, that is requiring input power to rotate the runner, immediately after the unit is connected to the line. This occurs when the guide vane opening is slightly smaller than the opening at speed–no–load or when the net head fluctuates downwards by water pressure surge after the unit is brought on line. As shown in Figure 6.22, the $'S'$ characteristics of high head pump–turbines appears in the vicinity of speed–no–load condition as indicated by the dotted line. Fluctuation or deviation of, guide vane opening, or the net head will cause shifting of the operating point from A (speed–no–load) to B (reverse pump) and the output torque of the unit is reversed. This negative input power may reach 40% of the maximum turbine output in an extreme case when the operating head is very low relative to the normal head. In order to avoid this reversal of output, the guide vanes must be further opened slightly immediately after the unit is connected to the line to ensure generation of positive power.

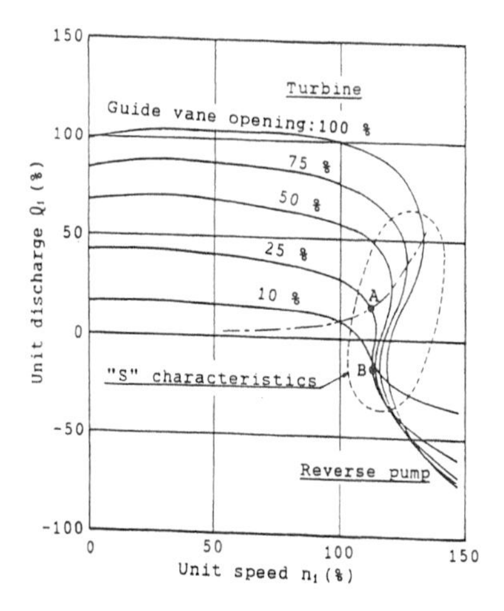

Figure 6.22 *'S'* characteristics of a high head reversible pump–turbine

To minimise the required power capacity for starting in the pump mode, the machine is usually started by prior de–watering of the runner, with exception of small capacity machines.

In general, the following starting procedures for the generator–motor are used to start the reversible unit in the pump mode

- Direct starting of the generator–motor as an induction motor
full voltage start,
reduced voltage start;
- Back–to–back synchronous starting with the aid of another generating unit;
- Pony motor starting with a small induction motor (pony motor) coupled to the main generator–motor;
- Synchronous starting by using a static converter consisting of thyristors to provide variable frequency power to the main generator–motor.

Whichever method is used, the torque capacity of the starting system must be larger than the sum of the accelerating torque and the power lost in de–watered operation as shown in Figure 6.23. Most of the power loss is produced by the rotating runner churning the water that comes from leakage through clearances around the guide vanes and from cooling water supplied to the runner seal clearances.

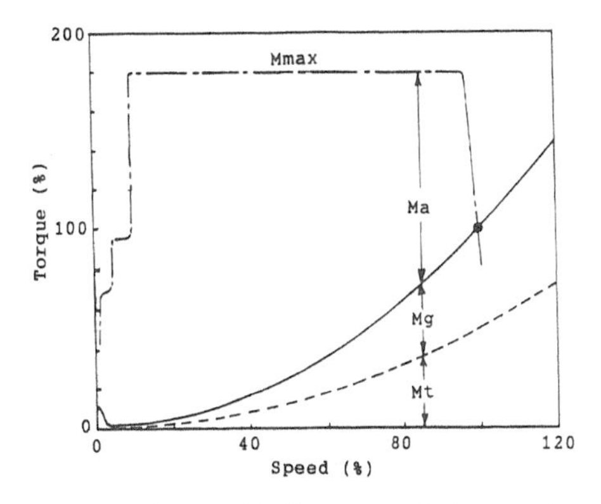

Mmax: Maximum available torque of the starting system
Ma : Torque available for acceleration of the rotating part
Mg : Torque corresponding to mechanical and windage loss of the generator-motor
Mt : Torque corresponding to mechanical and windage loss of the pump-turbine

Figure 6.23 Characteristics of starting torque (static converter) against lossed torque of generator–motor and pump–turbine

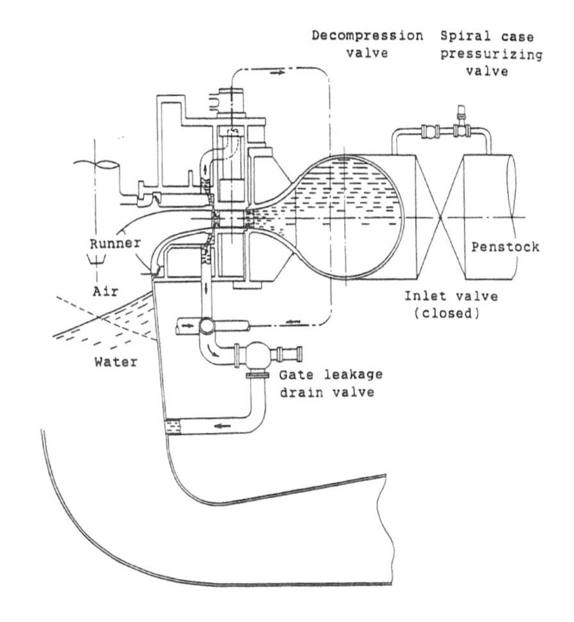

Figure 6.24 Guide vane leakage drain system

To reduce this lossed power, guide vane seals as shown in Figure 6.16 are generally used to reduce leakage and a leakage drain system as shown in Figure 6.24 is provided. To achieve efficient draining, the drain pipes are connected to the bottom ring at a location as close to the runner periphery as possible.

In the de–watering operation, usually the inlet valve upstream of the spiral case is closed and the spiral case is decompressed to reduce leakage through the guide vanes. However, if the pressure in the spiral case is below or equal to the pressure inside the guide vanes, air may leak outward into the spiral case and form an accumulation of air in the spiral case which will be discharged into the penstock when the inlet valve is opened. In practice, to prevent air accumulation during de–watering, the spiral case is pressurised slightly by penstock water through a decompression valve as shown in Figure 6.24 or by a small auxiliary pump.

An example of starting and stopping procedures of a reversible unit in the pump mode is shown in Figure 6.25. After the unit is connected to the line, the pump is primed by releasing the air through a pipe provided at the inner section of the head cover, until the runner is filled with water to develop the pump head. In this process, if the location of the air release pipe on the head cover, or that of the air vent through the runner crown is inadequate, stagnation of air release will take place resulting in failure of smooth filling. Behaviour of the air and water around the runner in the course of filling is illustrated in Figure 6.26.

When the outside section of the runner is mostly filled, more than half of the pump shut–off head is developed at the runner periphery and the water above the runner crown is compelled to flow inward by this pressure. If the runner seal clearance is too large and the drain pipe capacity is insufficient, this high pressure water will back up to the entrance of the air release pipes or the air vent on the runner crown and block them. Such conditions must be prevented by proper arrangement of air release pipes, and equalizer pipes, in order to obtain smooth pressure build–up in the filling process.

(ii) Emergency shut down

Figure 6.27 shows the complete characteristic curve of a reversible pump–turbine comprising all operating conditions from normal pump, brake, normal turbine to reverse pump, which is also known as the four–quadrant characteristics. In these diagrams, speed n_1, discharge Q_1 and torque M_1 are unit values under $1m$ head and are given a positive sign in the direction of turbine operation.

Both the discharge and torque characteristics of reversible pump–turbines in the normal turbine, turbine brake (the range between runaway and reverse pump) and reverse pump regions, are very much different from those

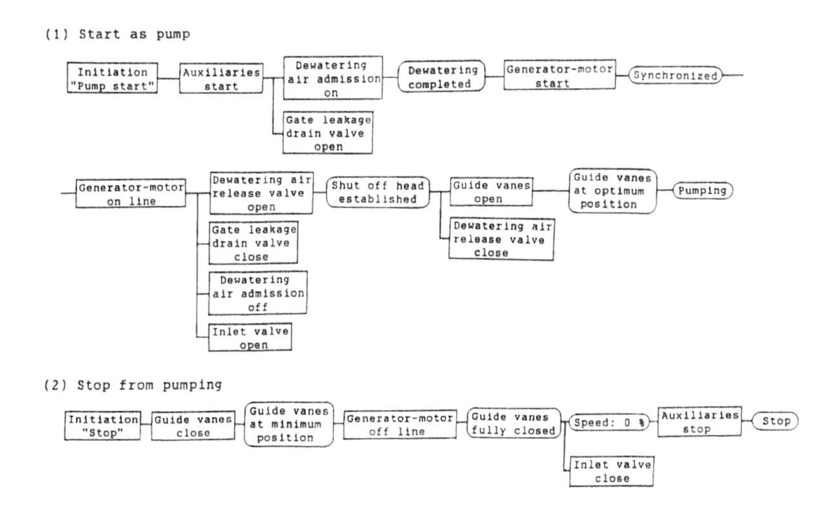

Figure 6.25 Starting and stopping procedures in pump mode

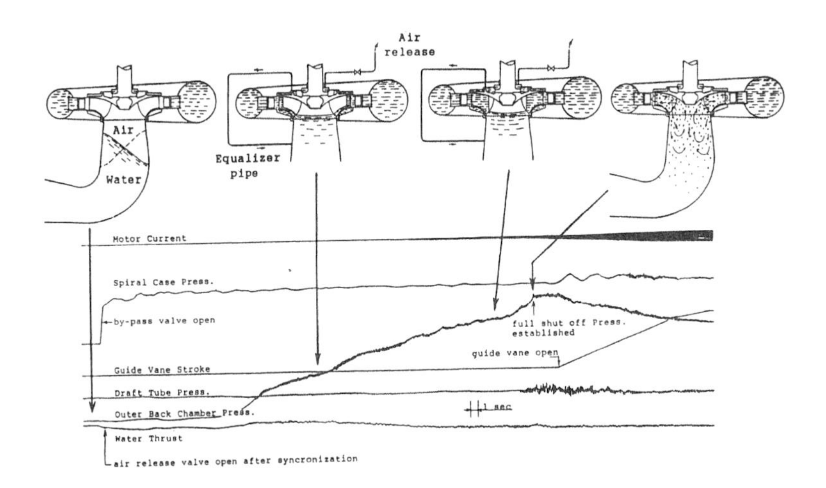

Figure 6.26 Filling process of a rotating runner

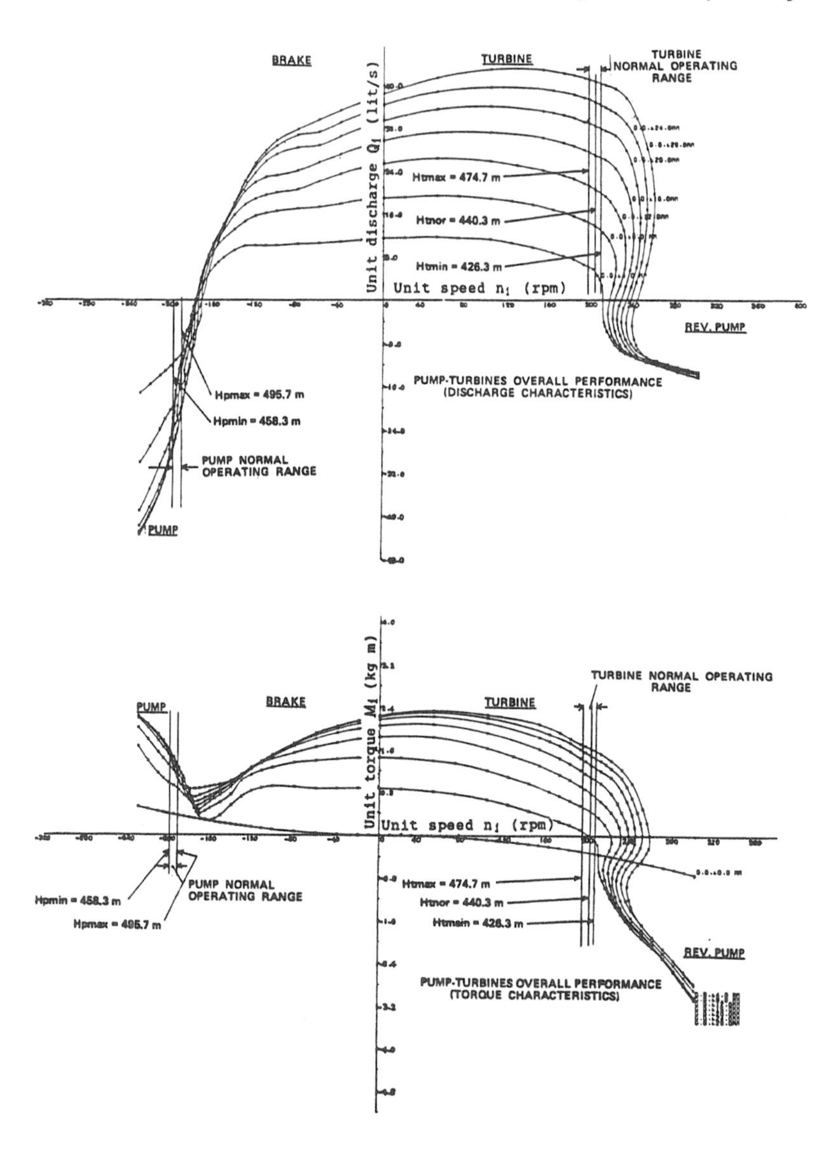

Figure 6.27 Complete characteristics (four–quadrant characteristics of a Francis reversible pump–turbine)

of conventional turbines. Simple formulae based on linear approximation of the turbine characteristics for the calculation of pressure or speed rise in turbine load rejection, as used for conventional turbines, can no longer be used. Therefore, computer simulation is necessary for study of the transient behaviour of reversible units, including pump power failure. Figure 6.28 shows results of such computations.

In turbine load rejection, a two–step closure of guide vanes as shown in Figure 6.28 (a) is often used both to reduce penstock pressure rise and to avoid excessive hydraulic moment on guide vanes in the reverse pump range. However, as there are cases when linear closure gives better transient response than two–step closure, selection must be made after careful comparative studies by these transient simulations. Figure 6.29 shows a comparison of speed variation Δn and pressure variation Δp between a reversible pump–turbine and a conventional turbine of low specific speed. In these diagrams, T_r represents the time constant of the rotating part, T_p represents the time constant of the pipeline water column and T_c is the closing time of guide vanes. They are defined as follows:

$$T_r = I\frac{\omega_0}{M_0} = I\frac{\omega_0^2}{1000P_0} = 2.74 \cdot 10^6 \left(\frac{GD^2 n_0^2}{P_0}\right)$$

$$T_p = \frac{\rho L Q_0}{A p_0} = \frac{L Q_0}{A E_0} = \frac{L Q_0}{g A H_0}$$

where I – moment of inertia $= GD^2/4 = WR^2$ (kgm^2)
ω_0 – angular velocity of rotation $= \pi n_0/30$ (rad/s)
n_0 – rated speed of rotation (min^{-1})
M_0 – torque at rated output $= 1000P_0/\omega_0$ (Nm)
P_0 – rated output (kW)
ρ – mass density of water (kgm^{-3})
g – acceleration of gravity (ms^{-2})
A – average cross sectional area of pipe conduit (m^2)
L – length of pipe conduit (m)
Q_0 – rated discharge (m^3s^{-1})
p_0 – pressure corresponding to rated head $=\rho g H_0$ (Pa)
H_0 – rated net head $= E_0/g = p_0/(\rho g)$ (m)
E_0 – rated net specific energy $= gH_0 = p_0/\rho$ (Jkg_{-1})

In case of turbine load rejection of a low specific speed pump–turbine, the discharge is reduced considerably in the overspeed period after rejection of the unit due to its flow characteristics, where the discharge decreases sharply as speed increases even without any closing motion of the guide vanes. Such decrease of discharge caused by speed increase develops positive waterhammer in the penstock. Consequently, as shown in Figure 6.29, pressure rise

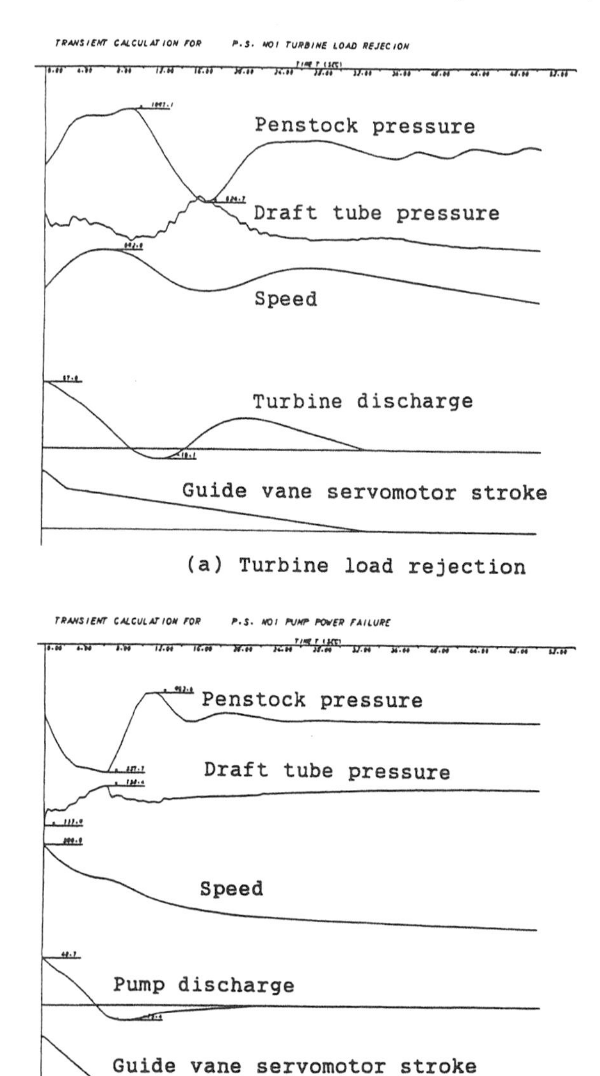

Figure 6.28 Examples of computer simulation of hydraulic transients of turbine load rejection and pump power failure

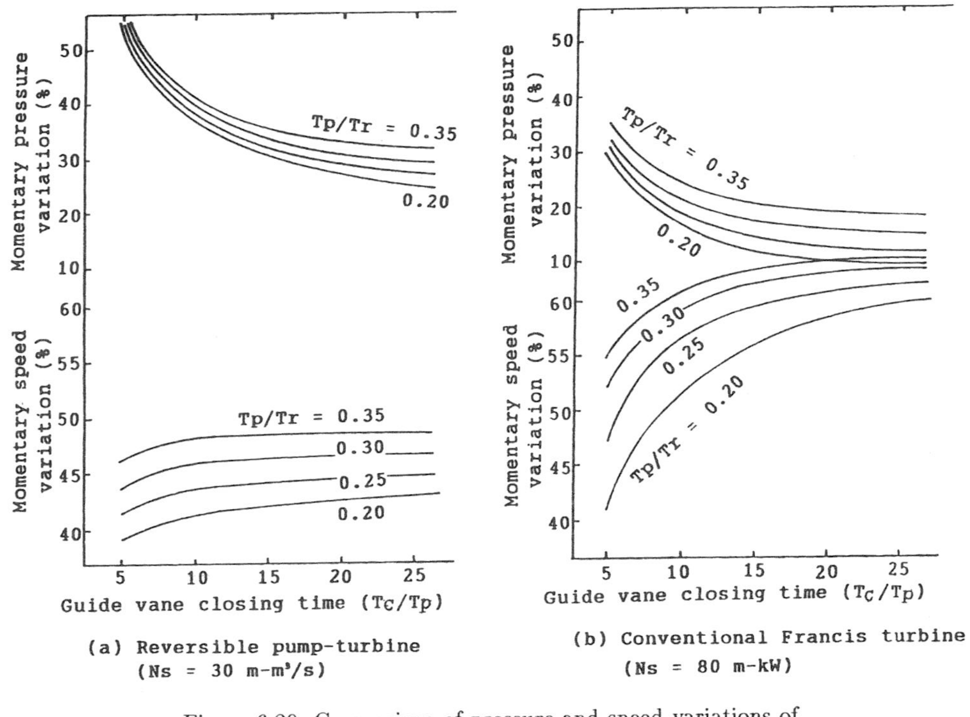

Figure 6.29 Comparison of pressure and speed variations of a reversible pump–turbine and a conventional turbine

in the penstock is higher than in conventional turbines and can hardly be reduced by extending the guide vane closing time. In such cases, not only the speed rise but also the pressure rise is governed by the flywheel effect of the rotating part.

When two or more reversible pump–turbines are connected to the common penstock and are shut down simultaneously in load rejection, they interfere with each other in going through the 'S' characteristics region and cause interactive waterhammer in the pipeline. Such interactive waterhammer results from the exchange of discharge by both units as shown in Figure 6.30, where one unit is going through the turbine brake range with discharge of $(Q_0 + \Delta Q)$ while the other is in the reverse pump range with $(Q_0 - \Delta Q)$. The transient process of this phenomenon is shown in Figure 6.31.

Abnormal waterhammer developed by this phenomenon will appear in both upstream and downstream sections of the pipeline. If the downstream manifold section has a large pipe factor $F = L/A$, this waterhammer may cause excessive vacuum in the pipeline and may result in water column separation which will lead to a subsequent pressure rise in the rejoining of the separated water column. Computer investigation for simulating such interactive phenomenon should be conducted in the design studies.

The magnitude of the pressure drop in the penstock following pump power failure depends mostly on the ratio T_p/T_r and very little on either the closing time or closing mode of the guide vanes. It depends also on the specific speed of the unit to a lesser degree. This is explained by the fact the pressure drop is primarily determined by the inertia of the water column and its deceleration. The inertia of the water column is represented by T_p and the deceleration, that is the decreasing rate of discharge due to the speed deterioration after tripping, is represented by $1/T_r$. Therefore, the larger the value of T_p/T_r, the larger will be both the deceleration of speed and the pressure drop in the penstock, after tripping.

The minimum penstock pressure takes place around the time when the pump discharge becomes zero. Following this, the pressure in the penstock recovers while the flow in the penstock reverses to the turbine direction (turbine brake region). So the value of pressure rise in this period is governed by the closing speed of guide vanes and a rapid closure will cause a high pressure rise. In general, the closing speed of guide vanes can be set faster for pump power failure than for turbine load rejection. Since strong turbulence is present in pump–turbines in the brake range, pressure fluctuation in the penstock increases correspondingly under these transient conditions.

(iii) Synchronous condenser operation

The pump–turbine unit is sometimes operated as a spinning reserve or synchronous condenser while connected to the line and with the runner spinning

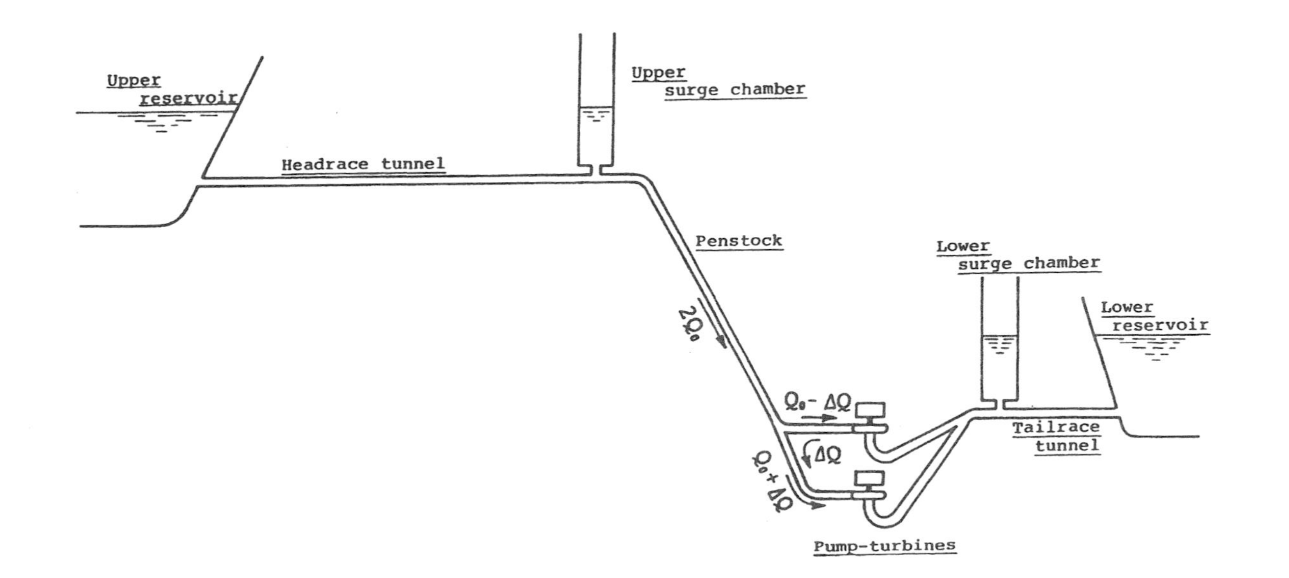

Figure 6.30 Reciprocating exchange of turbine discharge in case of interactive waterhammer

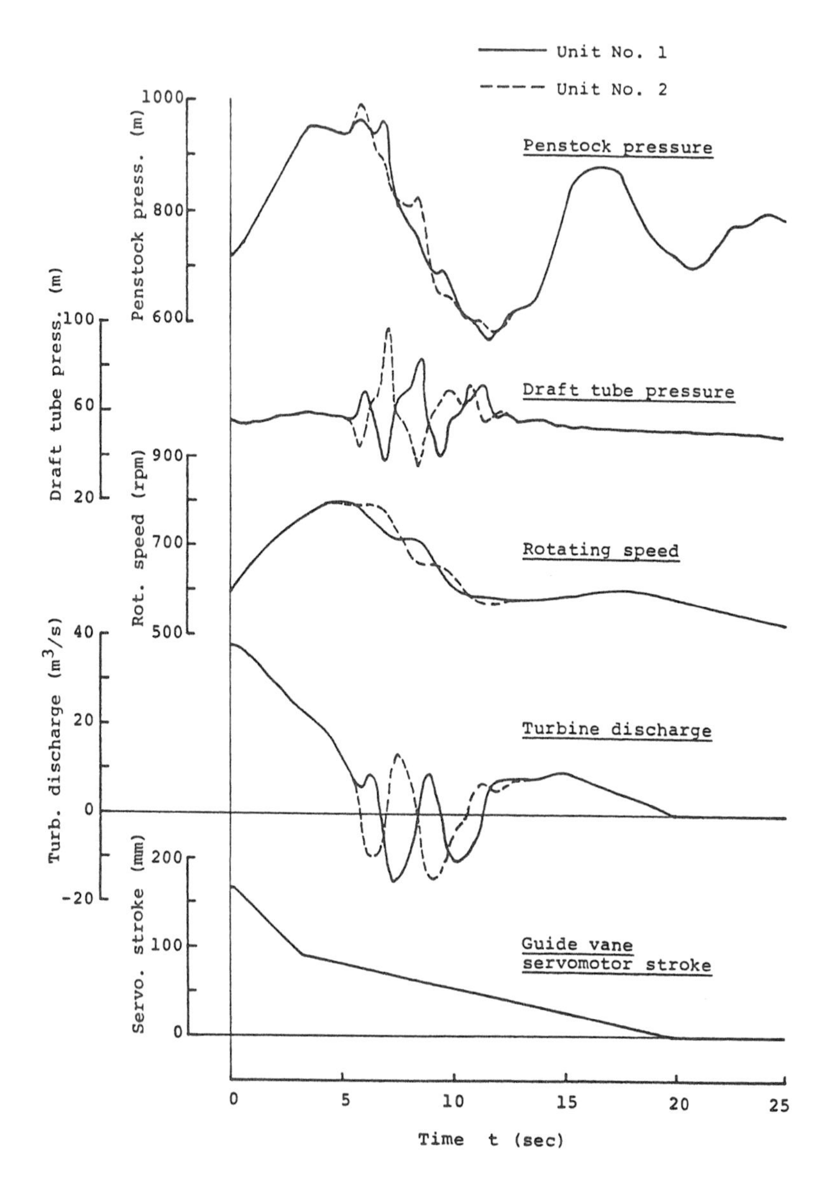

Figure 6.31 Interactive waterhammer in parallel turbine load rejection of two units

in air. Such an operation is done in either the turbine or the pump direction of rotation.

To engage the unit as a synchronous condenser in the turbine direction, the unit is started up as turbine, and following synchronisation the guide vanes are closed and the runner is de–watered. In this operation the following events must be taken into account:

1. If the guide vanes are brought to a smaller opening than the speed–no–load setting, the operation may go into reverse pump when the operating head is low. Should this happen, the unit will draw excessively large input power and may reach 30 or 40% of maximum pump input power in extreme cases.

2. If the de–watering air injection is insufficient, the injected air is carried downstream by the turbulent secondary flow below the runner and de–watering may fail altogether. Usually, for such dynamic de–watering with the runner in rotation, the air injection rate must be larger than a certain threshold value, which is approximately equal to the flow rate for de–watering the runner at standstill within 20 seconds or less.

3. The loss power in de–watered condition is due to the interference of water such as guide vane leakage with the rotating runner. When the unit is running in the pump direction, water is pushed out continually by the rotating runner blades, but when it is running in the turbine direction the runner blades scoop the water inward and then discharge it by centrifugal force. The turbulence of the water will thus become much greater than in the pump direction (Figure 6.32) and will cause much larger loss power with greater fluctuation, higher temperature rise and stronger vibration of the unit.

4. If the amount of guide vane leakage and runner seal cooling water are insufficient, the temperature of the water around the runner may rise due to heat from the loss power. Excessive temperature rise may cause thermal expansion of the runner and the reduction of runner seal clearances to a dangerous degree. With more guide vane leakage, the loss power will increase somewhat both the temperature of the water may however lower as shown in Figure 6.33. Therefore, the amount of guide vane leakage water and runner seal cooling water should be optimized to lower the temperature without unreasonably increasing the loss power. Installation of guide vane leakage drain pipe with sufficient capacity is effective in reducing the interference of water and to prevent both excessive temperature rise and loss power as shown in Figure 6.33.

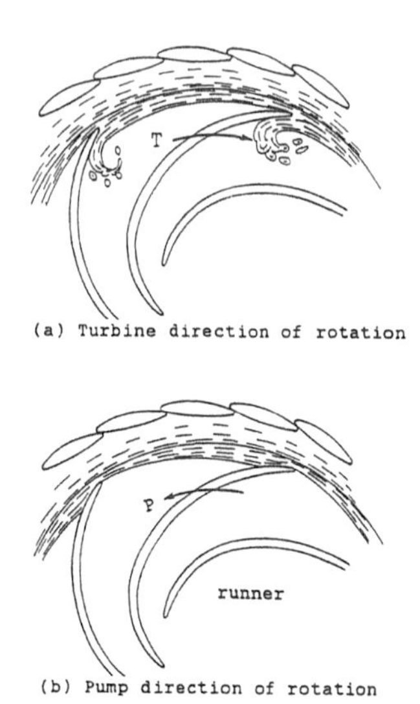

Figure 6.32 Interference of water at periphery of a rotating runner in de–watering operation

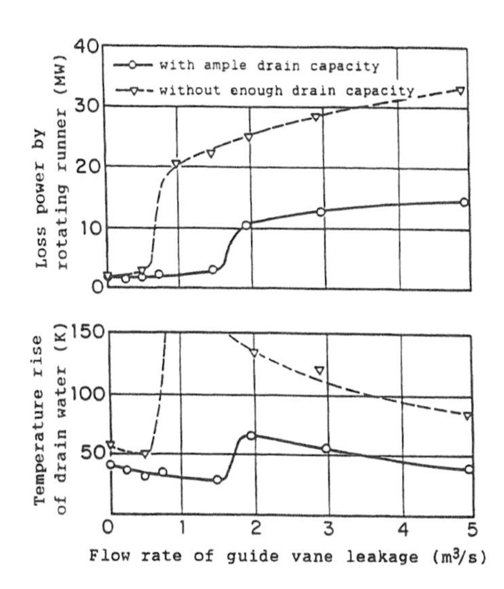

Figure 6.33 Power loss and temperature rise caused by interference of water at periphery in de–watering operation

It is imperative that when high head pump–turbines are operated as a synchronous condenser in the turbine direction of rotation, precautions must be taken for quick dewatering and sufficient drain capacity for removal of the water at runner periphery.

6.3 Multistage Reversible Pump–Turbines

6.3.1 Two stage pump–turbines

It is commonly accepted that the limit of the operating head of single stage pump–turbines would be somewhere around $900m$ considering the strength of the runner. In the case when higher operating head or less submergence is required, multistage pump–turbines may be used. When the unit capacity is small, multistage units may be utilized, for even lower head range, for the benefit of reduction in operating speed.

The number of stages currently adopted for large capacity multi–stage pump–turbines ranges from two to six. Among these, two stage machines fall in three groups, namely, the ones with movable guide vanes for both stages, the ones with movable guide vanes for the upper stage only and the ones with no movable guide vanes. Multistage machines with three stages or more are usually not provided with movable guide vanes. Difference in hydraulic performance among machines with and without movable guide vanes is shown in Figure 6.34. For better efficiency in part load operation in the turbine mode, machines with movable guide vanes for both stages are recommended. An example of such a machine is shown in Figure 6.35.

When the guide vanes of such machines are being closed, if a shear pin or breaking link of an upper guide vane is broken due to blockage by foreign substance and the lower guide vanes are in full close position, full penstock pressure will be imposed on the water passage between the upper and lower stages. For this reason, the guide vanes for the lower stage are sometimes designed not to close fully. In any case, the upper guide vanes must withstand full penstock pressure and are designed with very thick profile.

In general, a lower extension shaft through the draft tube and a bearing at the bottom end is needed to prevent excessive runout of the lower runner, as in Figure 6.35, although in other cases the lower runner has no support from below.

Two stage pump–turbines of small capacity can be started as pump without de–watering the runner, however, those of large capacity must have prior de–watering as with large capacity single stage machines. The process of air release and filling of two stage pump–turbines is much more complicated compared with single stage machines. Moreover, the method of air release and filling is different for those with movable guide vanes for both stages and

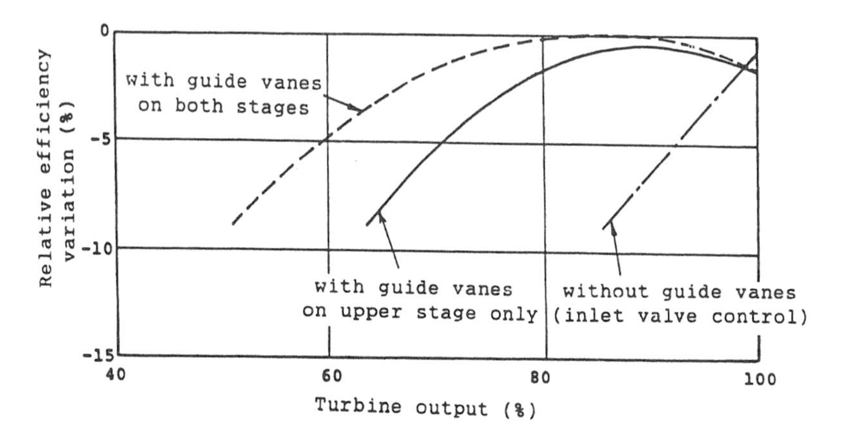

Figure 6.34 Comparision of turbine performance of two stage pump–turbine

for those for upper stage only. For the former, both stages can be filled in parallel in the same manner as single stage machines. For those with guide vanes on the upper stage only, a method of back filling from the penstock is used. In this process, the outer section of the upper runner is filled first by admitting water from the penstock, filling then proceeds to the inner section of the upper runner, return passage, outer section of the lower runner and finally to the upper portion of the draft tube and the inner section of the lower runner. Whatever the arrangement of guide vanes, the air vent for release of air in the filling process must be located as close to the runner centre as possible for effective air release without blockage by water. An example of the air vent arrangement is shown in Figure 6.36 for a two stage pump–turbine.

6.3.2 Multistage pump–turbines of three or more stages

Multistage pump–turbines with operating head exceeding $1000m$ may be designed to have three stages or more as shown in Figure 6.37. For these machines, movable guide vanes are normally not provided due to the difficulties in designing guide vanes able to withstand high penstock pressure without producing high hydraulic losses from their strong downstream wakes. Two main inlet valves are sometimes installed in series for multistage machines without guide vanes. The upstream valve serves as a normal inlet valve and the downstream one functions as a control valve to regulate the turbine flow.

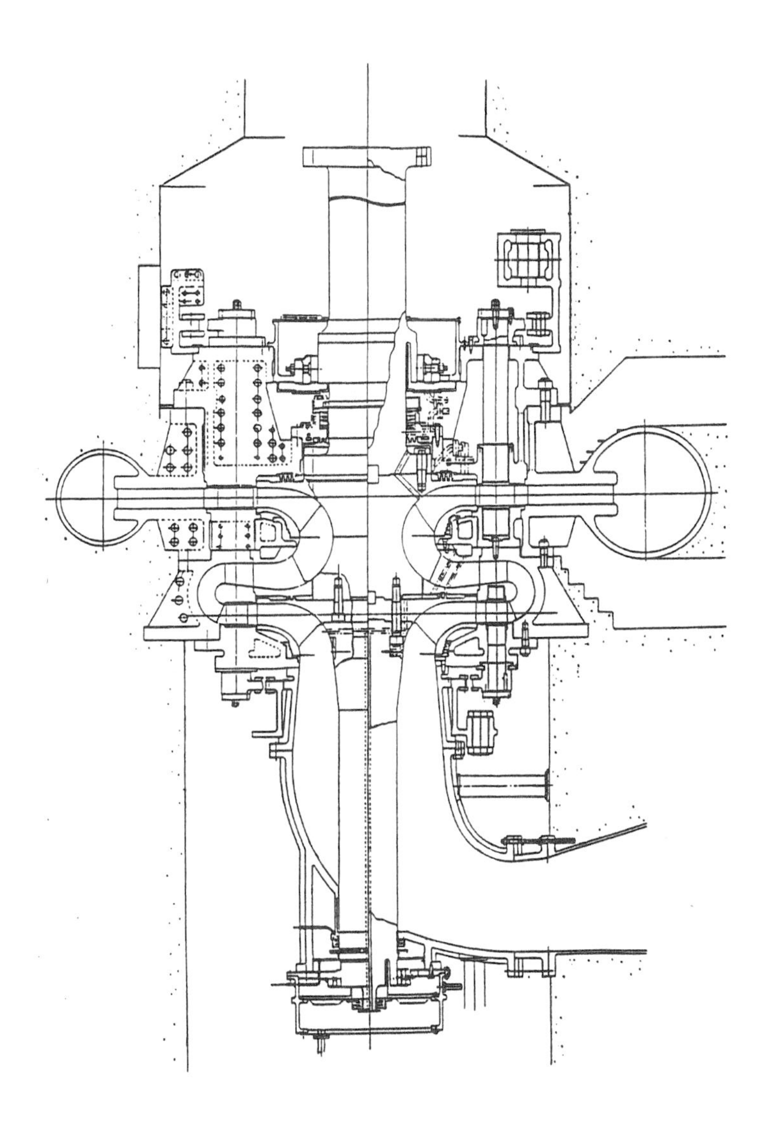

Figure 6.35 Two–stage reversible pump–turbine

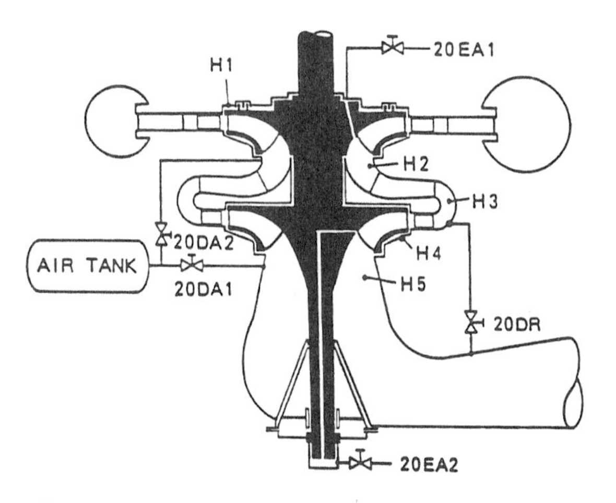

20 DA1: Dewatering air admission valve for lower stage
20 DA2: " " " for upper stage
20 EA1: Air release valve for upper stage
20 EA2: " " for lower stage
20 DR : Drain valve for upper stage

Figure 6.36 Schematic of de-watering air admission and release system

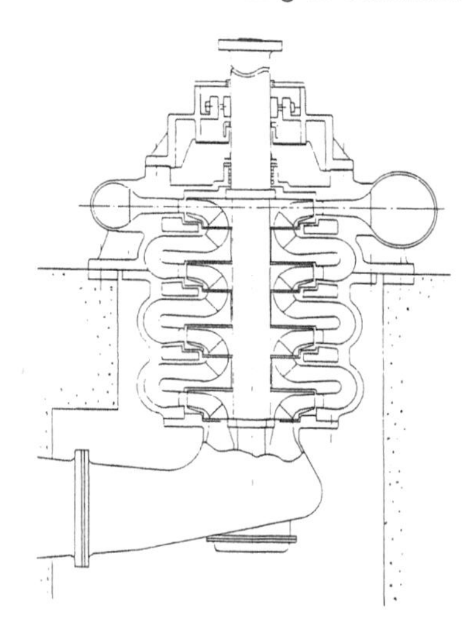

Figure 6.37 Example of multi-stage reversible pump-turbine

In turbine start, the downstream valve is used to regulate the speed of the unit and in emergency it closes rapidly to prevent overspeed of the unit. A bypass valve attached to the main valve or the seal ring of the main rotary valve may be used instead for speed regulation in place of the main valve.

Such methods of control, however, are used for speed regulation only and not for load regulation in generating operation. In generating, the inlet valve is normally kept full open throughout.

6.4 Tandem Pump–Turbines

6.4.1 Types of tandem pump–turbines

The term tandem type pump–turbine refers to a pump–turbine consisting of a conventional turbine and a storage pump connected to one common generator–motor. The cost of such a machine group is much higher than reversible pump–turbines but it has excellent operational advantages over reversible machines such as higher efficiencies and quicker and smoother changeover from one operating mode to another. Usually the following combinations of turbine and pump are used:

Head	Turbine	Pump
High head	Pelton turbine	Multistage single suction pump
		Multistage double suction pump
		Single stage single suction pump
Low head	Francis turbine	Single stage double suction pump

Tandem pump–turbines can be built with either vertical or horizontal shaft. Typical examples are shown in Figures 5.31 and 6.38.

Certain construction details are decided by the starting method of the pump–turbine including the following variants:

- Both turbine and pump are connected to the generator–motor by solid couplings. A de–watering system is provided to vacate the pump in generating operation, or the turbine in pumping operation, to reduce windage loss. This system is also used for de–watering the pump in pump starting.
- Clutches are provided to disconnect the pump in generating operation or the turbine in pumping operation. Two different kinds of clutches may be used: one makes connection or disconnection only with the unit at standstill; the other is able to perform these functions in rotation.

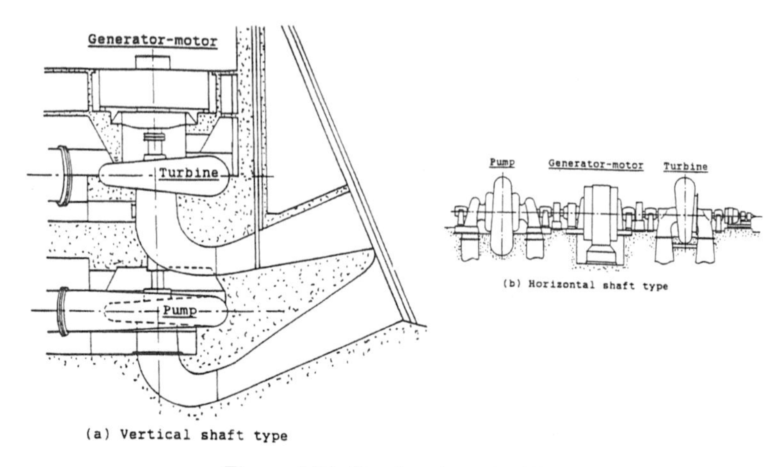

Figure 6.38 Tandem type turbines

- A small pony turbine is provided on the pump shaft for starting the pump.
- A torque converter with a clutch that is able to make connection while rotating is placed between the pump and the generator–motor.

The control technique for de–watering and filling while running, which has been well–developed in recent years, allows the use of a simple solid coupling connection that helps to save initial manufacturing costs. Large capacity tandem pump–turbines constructed in later years are often designed with this feature.

6.4.2 Features of tandem pump–turbines

In comparison with reversible pump–turbines, tandem units have the following advantages and disadvantages:

Advantages:

- Better efficiencies in both turbine and pump operations by 0.5 to 1.0%. Better efficiency in part load operation as turbine by several per cent may be expected.
- Reversal of direction of rotation in mode change from generating to pumping or vice versa is not necessary. Consequently, changeover times from one mode to another can be significantly reduced.

- By controlling the turbine output in parallel operation of both turbine and pump, it is possible to regulate the input power on pumping operation.

- Utilizing the same technique as above, smooth and fully controlled transition of power (input/output) in mode change from generating to pumping or vice versa can be accomplished.

- In most cases, such as with the solid coupling arrangement, pump starting may be accomplished by using the turbine without any other starting equipment.

Disadvantages:

- Large axial length of the pump–turbine unit, consequently greater dimension requirement for power house.

- More complicated control system for turbine and pump.

- Additional auxiliary equipment such as clutches, torque converters and a large de–watering system are required.

Tandem pump–turbines have great advantages in quick response and smooth transition of power. So, in spite of their higher construction costs, they are used where stability and reliability of the power network are of primary importance.

6.4.3 Operation and control of tandem pump–turbines

As explained before, there are many variants of tandem pump–turbines each with different methods of operation and control. Examples are given here for a vertical and a horizontal tandem unit with relatively simple arrangements as shown in Figure 6.38.

(i) Operating modes of the pump–turbine (Figure 6.39)

S – Unit at standstill. For quick start in generation, the pump is usually kept de–watered and the turbine filled.
G – Generation. The pump remains de–watered.
P – Pumping. The turbine is de–watered.
C – Synchronous condenser operation with both turbine and pump de–watered.

(ii) Changeover with turbine active

In advance of changeover to and from pumping, the turbine is filled with water and operates in parallel with the pump. The total input/output power to and from the unit is controlled by regulating the turbine output.

(iii) Quick changeover

The changeover operation is executed as quickly as possible without considering power fluctuation or influence to the network.

The above operating modes of the pump–turbine can be carried out by automatic sequence control according to the diagram in Figure 6.39. In these changeover operations, it is required to fill or de–water the machines while running.

(i) Turbine

(a) De–watering

- close guide vanes;
- air is injected preferably into the upper portion of the draft tube;
- open the drain valve for guide vane leakage water when most part of the runner is dewatered.

(b) Filling

- open the air release valve and discharge air through air vent near the centre of the runner;
- fill the runner from the draft tube upward;
- open the guide vanes upon completion of air release. (This last step may be done before the completion of air release if it is permissible to discharge large amounts of air to the downstream).

(ii) Pump

(a) De–watering

- close discharge valve;
- air is injected into the suction conduit;
- open the drain valve at the outer periphery of the impeller to drain the water accumulated at the outer section of the impeller and in the spiral case.

(b) Filling

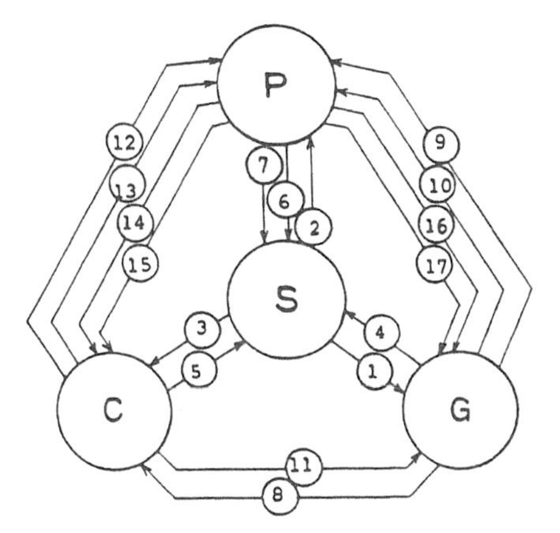


```
6, 9,12,14,16: Changeover with
               turbine active
7,10,13,15,17: Quick changeover
```

Figure 6.39 Mode changes of tandem pump–turbine

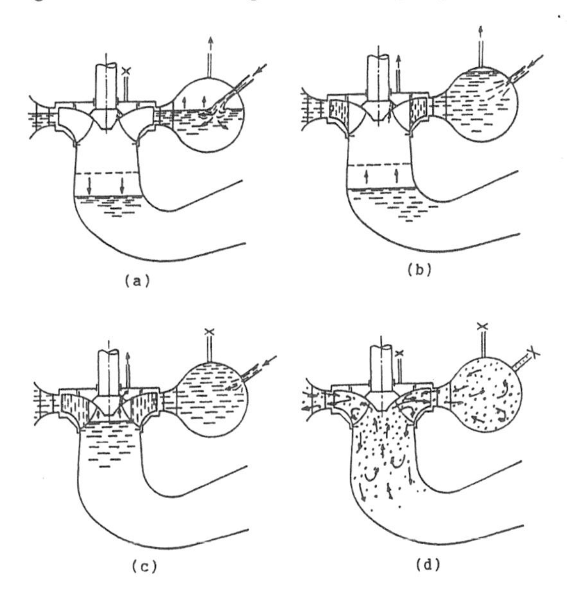

Figure 6.40 Filling of a vertical storage pump without movable guide vanes

- open bypass valve to fill the spiral case by admitting water from the penstock while the air vent on the top of spiral case is open;
- close bypass valve and air vent valve on spiral case when the outer section of the impeller is filled;
- open air release valve on the head cover and fill the impeller from draft tube upward;
- open discharge valve upon completion of air release from head cover (the above filling sequences are shown in Figure 6.40).

At the final stage of the pump filling, if the filling of the impeller from the bottom is started before the spiral case is completely filled, heavy shock may take place at the end of the filling of the impeller.

Among various mode changes, two examples of the changeover with turbine active are shown in Figure 6.41 for both pumping to generating and vice versa. In these examples, the total power of the unit including turbine output and pump input is controlled to change at a designated rate. When quick changeover is required in an emergency case, especially from pumping to generating, the changeover is performed without turbine output control.

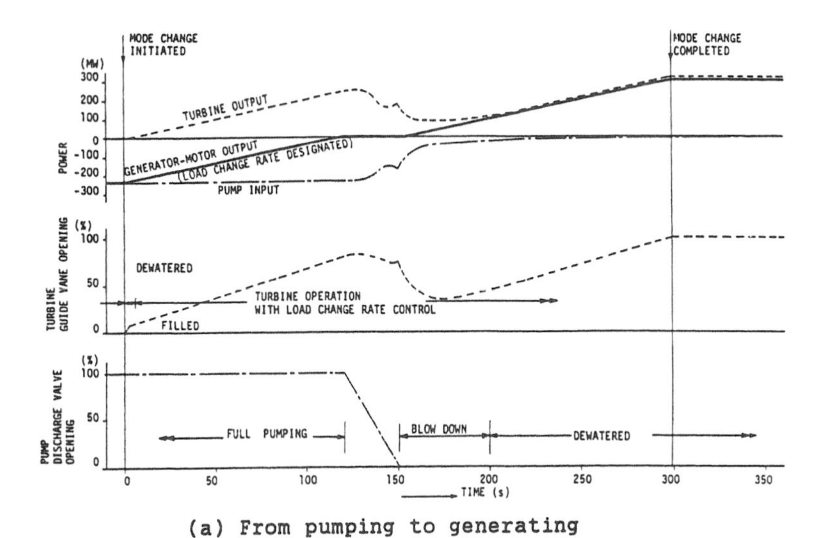

(a) From pumping to generating

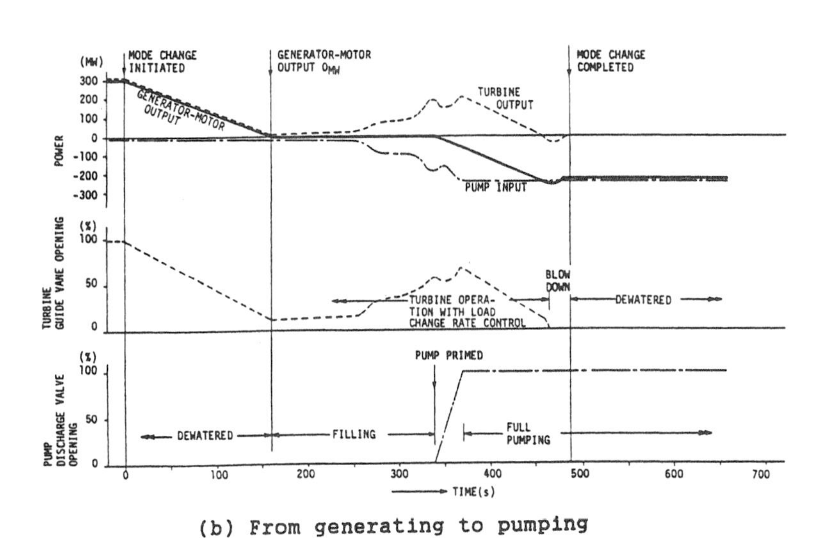

(b) From generating to pumping

Figure 6.41 Changeover of shaft operation between generating and pumping (with turbine active)

References

6.1 Kubota, Y. et al 'Vibration of a Rotating Disc with Blades Induced by Multiple Excitation Sources Distributed on Surrounding Stationary Part', Proc. JSME Symposium No. 824–9, 1982 (in Japanese)

6.2 Tanaka, H. 'Vibration Behaviour and Dynamic Stress of Runners of Very High Head Reversible Pump–Turbines', Special Book, IAHR Symposium, Belgrade, Sept, 1990.

6.3 Tanaka, H. 'Some Technical Review and Advanced Technologies Based on Operating Experiences of High Head Pump–Turbines', Int. Symp. on Large Hydraulic Machinery, Beijing, May 1989.

6.4 ASME Boiler and Pressure Vessel Code, Sect. XI, Rules for Inservice Inspection of Nuclear Power Plant, Appendix A, 1983 ed.

Chapter 7

Inlet Valves and Discharge Valves

Hiroshi Tanaka

7.1 General

For protection of a hydraulic turbine, an inlet valve is usually provided upstream of or at the inlet of the turbine. Rotary valves are normally used for medium and high head plants, and butterfly valves or through–flow valves are widely used for medium and low head plants.

Through–flow valves are an evolution from butterfly valves that have less hydraulic loss than butterfly valves. For cases of low head turbines with semi–spiral case or bulb turbines, gates of different constructions are generally installed in place of inlet valves. Figure 7.1 shows the approximate range of application of each type of valve.

In general, an inlet valve is provided with a bypass valve of smaller bore and is opened first to equalize the upstream and downstream pressure before the main valve is opened. During shutdown the main valve is closed first followed by the bypass valve. The size of the bypass valve must be large enough to compensate the leakage through the closed guide vanes of the turbine.

Inlet valves are strong and rigidly constructed so that they may cut off the flow in an emergency independent of the operation of the turbine. To prevent excessive waterhammer in this event, closing time of the valve is set to be longer than that of the turbine guide vanes. Since most valves are subject to large closing hydraulic moments on their rotors during shutdown (see Sec. 7.7), the closing motion during shutdown may become faster than in normal operation with pressure balanced on both sides. The normal closing time of the valve must be determined with consideration of the above effect to prevent excessive waterhammer during emergency operation.

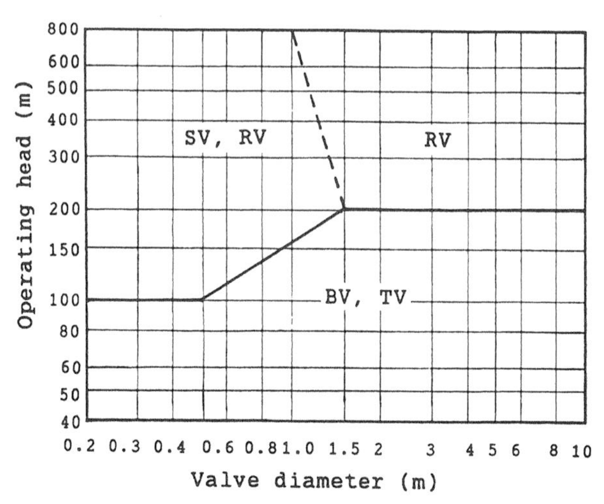

Figure 7.1 Range of application for each type of valve

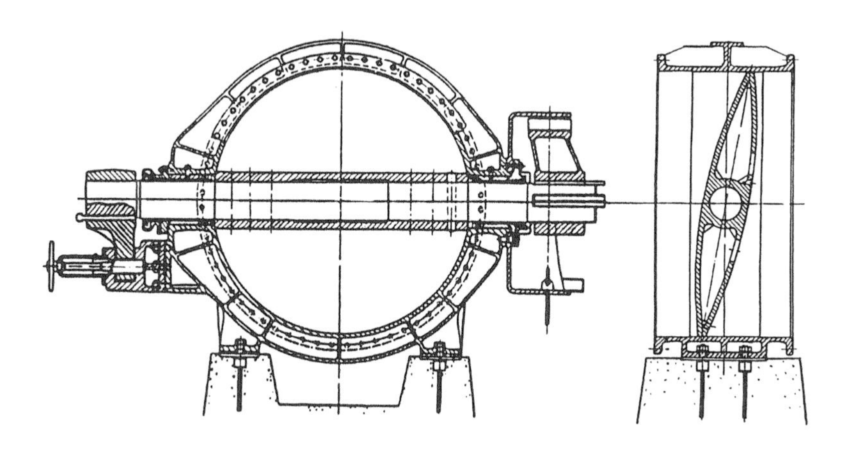

Figure 7.2 Typical construction of a butterfly valve

7.2 Butterfly Valves

Butterfly valves are the most commonly used of all valves. They are usually installed for operating heads up to $200m$. Figure 7.2 shows a typical construction of horizontal–shaft butterfly valves.

The hydraulic loss across a butterfly valve when fully opened is given approximately by the following formula

$$E_1 = \frac{t}{D} \cdot \frac{V^2}{2} \tag{7.1}$$

where E_1 – specific energy loss of fully opened valve (Jkg^{-1})
t – thickness of valve disc (m)
D – diameter of valve (m)
V – nominal flow velocity $= Q/(\pi D^2/4)\,(ms^{-1})$

The valve disc placed in the flow is normally subject to a hydraulic moment in the closing direction (self–closing) but the moment becomes zero at both the fully open or fully closed positions. If a self–closing moment is required at the fully closed position for security, the valve stem is made offset to the valve disc by the amount e_1 as shown in Figure 7.3 (a). Usually a rubber seal is provided on the periphery of the valve disc and is pressed against the valve seat to maintain watertightness at fully closed position. The compression of the seal at the closed position is the same on both sides of the valve disc when the valve stem is concentric to the valve disc. However, when an offset e_1 is intentionally incorporated in, the compression of the seal at the lower end of the disc becomes stronger than at the upper end. To make the compression at both ends equal, another offset e_2 must be provided as in Figure 7.3 (b).

Leakage through a closed valve is usually less than $2 \cdot pD$ $(lit \cdot min^{-1})$ where p is the differential pressure across the valve disc (MPa) and D is the diameter of the valve (m).

7.3 Through–flow Valves

When the thickness of the valve disc of a butterfly valve for high head application becomes too great and its hydraulic loss too large to afford efficient operation, a through–flow valve may be used. The valve disc of a through flow valve is of parallel thin–plate structure and is hollow in the centre so that water flows through it with fairly little loss at fully open position. It has however sufficient rigidity against bending due to upstream pressure at fully closed position. A typical construction of a through–flow valve is shown in Figure 7.4. The rubber seal mounted at the periphery of the valve disc,

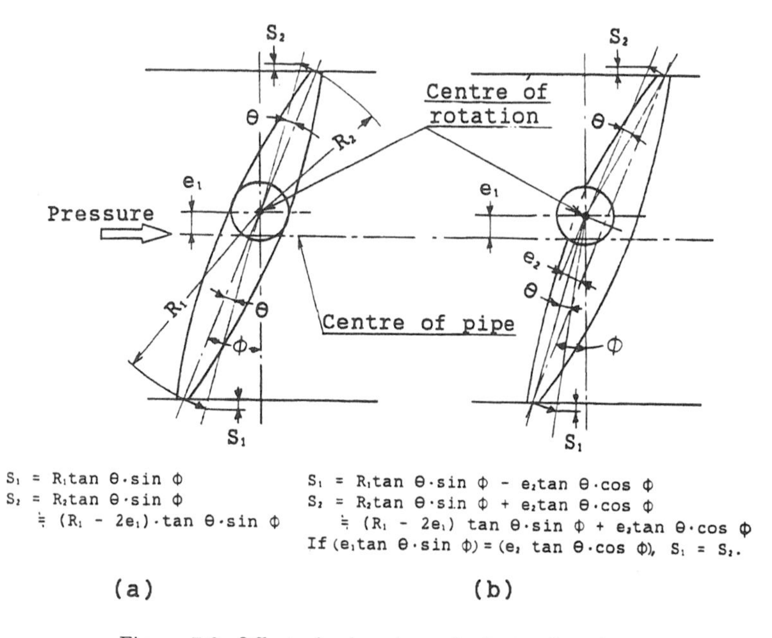

Figure 7.3 Offset of valve stem of a butterfly valve

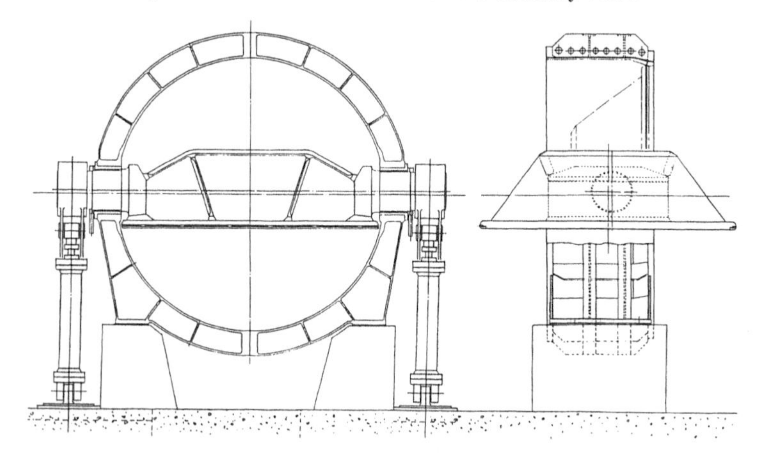

Figure 7.4 Typical construction of a through–flow valve

which is offset to the stem, may be made as one integral ring to give better water–tightness than those of butterfly valves which must have jointed ends at the roots of the valve stems.

Approximate values of the hydraulic loss coefficient ζ of through–flow valves at full open, are compared with those of butterfly valves in Figure 7.5. The loss of through–flow valves can be calculated by the following formula

$$E_1 = 0.8 \left[\frac{A_d}{\pi D^2/4} \right] \frac{V^2}{2} \tag{7.2}$$

where A_d is the projected area of the valve disc at opened position against a plane perpendicular to the axis of the pipeline.

A construction which has less frontal area of the disc is desirable in reducing losses. Some examples of construction of valve discs are shown in Figure 7.6. All the designs shown have the common feature of a main disc and a reinforcement structure made of thin plates. The construction shown in Figure 7.6 (a) has larger deformation than the other two types.

Leakage through a closed through–flow valve is usually less than $0.5 \cdot pD$. This is about one quarter of that for butterfly valves.

7.4 Rotary Valves

The valve body (shell) of a rotary valve is made spherical in shape as against other types of valves which are cylindrical. Consequently, the rotary valve is able to withstand higher internal pressure than other types and can be used for higher operating heads. A typical construction of a rotary valve is shown in Figure 7.7. Rotary valves for high heads are usually provided with movable metal seals which can withstand higher pressure than rubber seals. A movable metal seal operated by water pressure from the penstock is usually installed at each side of the valve rotor. The one at the downstream side is used for normal operation while the upstream one is used as an emergency seal and also for carrying out maintenance work on the downstream seal.

For long de–watered outage of the valve, a manual locking device is available for securing the seal. Rubber seals may be used for small size valves working under relatively lower heads. In such cases, the valve stems are fixed on the rotor with some offset similar to the amount e_1 in Figure 7.3 for butterfly valves. This offset is needed in order to make the angle of seating of the rubber seal larger when it approaches the fully closed position.

There are different constructions for the splitting of valve bodies as shown in Figure 7.8. These are determined by considering the allowable size and weight for transportation. If it is split as in (a), (b) or (d), the rotor can be made integral with the valve stems. When the valve body is split as (c), the valve stems must be made separate from the rotor. They are fitted on the

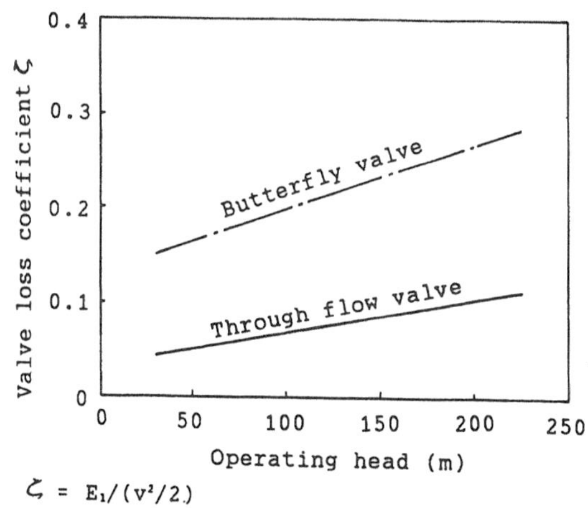

$\zeta = E_1/(v^2/2)$

E_1 : Loss specific energy of valve (J/kg)

v : Velocity (m/s) $= Q/(\pi d^2/4)$

Figure 7.5 Hydraulic loss coefficient of valves

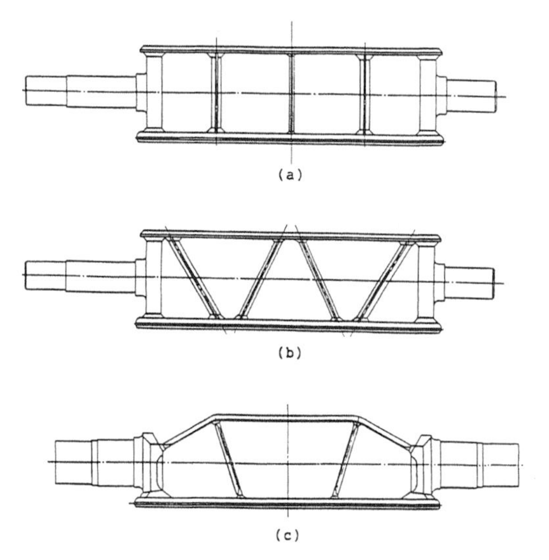

Figure 7.6 Various types of through–flow valve discs

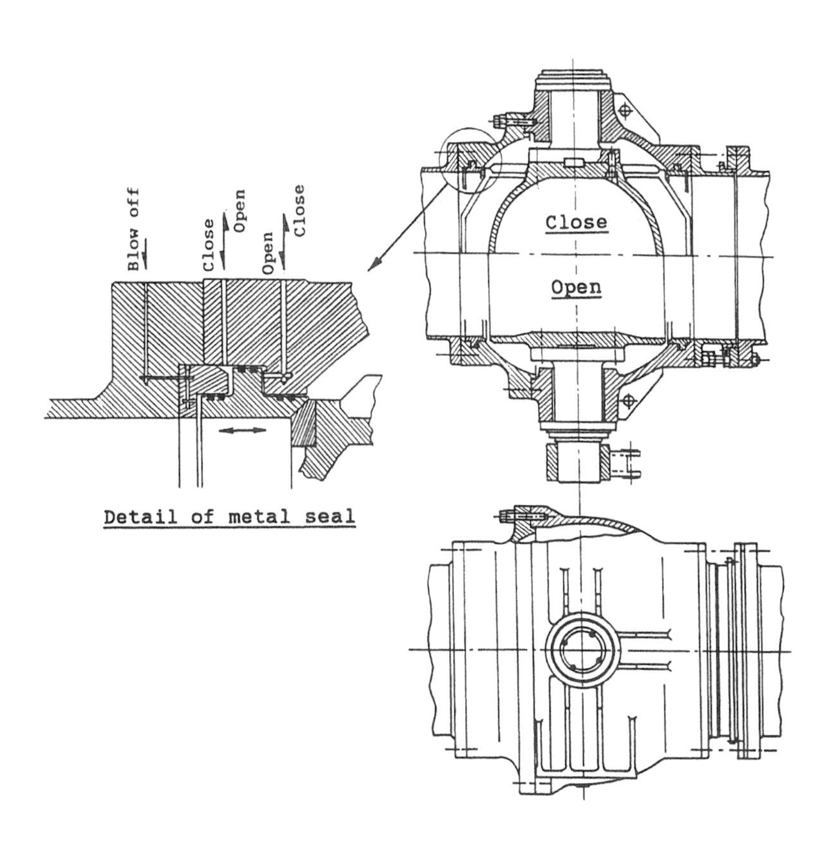

Figure 7.7 Construction of a rotary valve

rotor by bolts to be fastened from the inside of the rotor after being placed in the valve body.

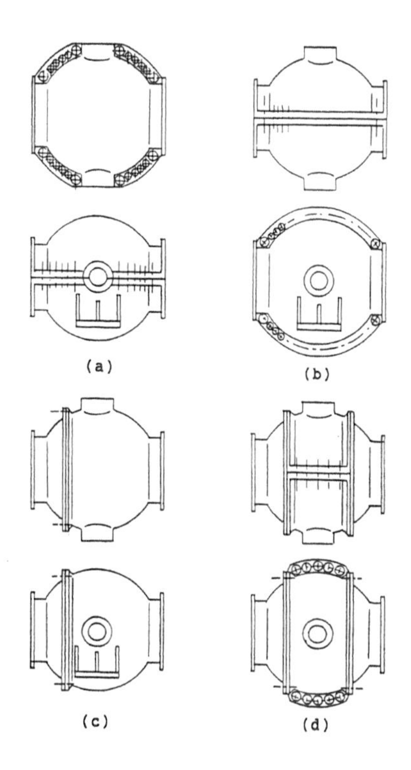

Figure 7.8 Ways of splitting the valve body

Rotors of rotary valves are made cylindrical in shape and supported by stems on both sides and are not as stiff as the valve bodies. When a lateral pressure force is applied on one side of the cylindrical rotor in the fully closed position, deflection of the rotor is considerable. To reduce this deflection to a tolerable degree, many ribs have to be provided around the rotor for reinforcement. An example of a high head valve rotor is shown in Figure 7.9. Metal seals must be flexible enough to easily bend with the rotor in case of deformation.

Examples of discharge coefficient C, hydraulic moment coefficient m and hydraulic force coefficient f are shown in Figure 7.10 (see Section 7.7 for the definition of these coefficients). The hydraulic moment acting on rotary

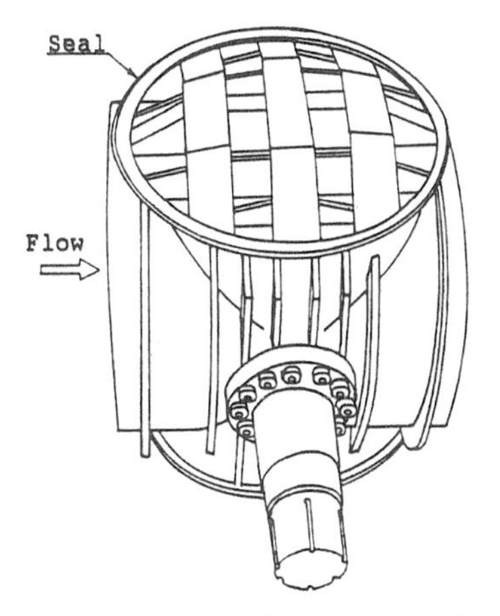

Figure 7.9 Rotor for a rotary valve

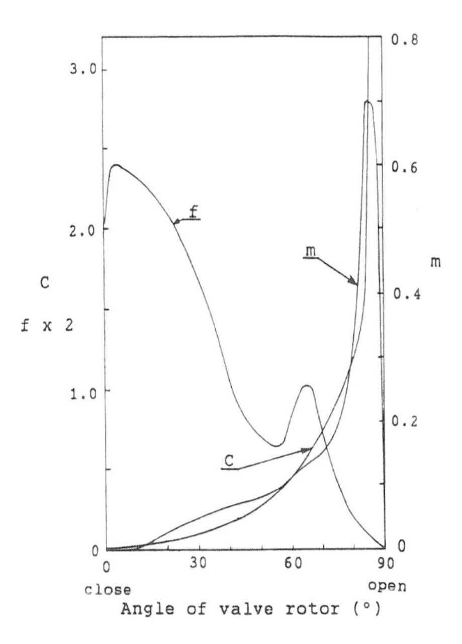

Figure 7.10 Discharge coefficient C, hydraulic lateral force coefficient f and hydraulic moment coefficient m of a rotary valve

valve rotors is also in the self–closing direction as other types of valves. The hydraulic loss at full open position is very small and is close to that of a cylindrical pipe of the same diameter.

7.5 Turbine Cylinder Gate

Instead of installing an inlet valve upstream of the turbine, a cylindrical shape gate may be built inside the turbine between the stay vanes and the guide vanes. This cylinder gate is lifted or lowered by several hydraulic or electrically driven screw jacks arranged around it.

Cylinder gates have an advantage over the conventional inlet valves in that they bring about a reduction of the space otherwise required by the inlet valve and hence is able to decrease the dimension of the power house. However, one drawback of it is that the inevitable leakage will spray around the guide vanes and the runner, and may hinder maintenance work in this area.

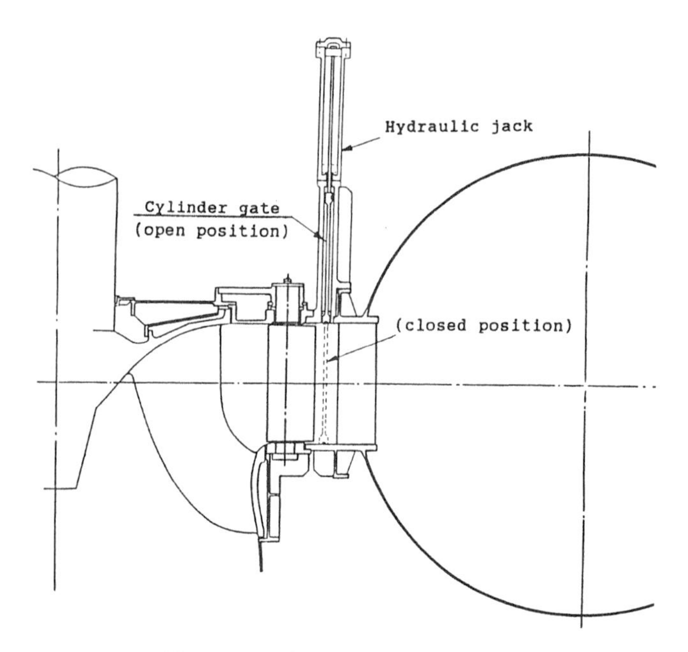

Figure 7.11 Turbine cylinder gate

7.6 Discharge Valves and Pressure Regulating Valves

Some power stations are required to discharge water to the downstream for irrigation and other purposes when turbines are not operated. For such cases, a discharge valve connected directly to the penstock or the turbine spiral case may be used.

For power stations having long penstocks, a long closing time for the guide vanes is necessary in order to avoid excessively high waterhammer pressure but the momentary speed rise will thus be quite high. In this instance, a pressure regulating valve may be attached to the penstock, and is designed to open in parallel with the closing motion of the guide vanes when turbine shutdown occurs. Thus the guide vane closure may be restored to the normal rate while both the waterhammer and momentary speed rise values are kept within allowable limits. An example of the installation of a turbine fitted with a pressure regulating valve is shown in Figure 7.12 and its function is illustrated in Figure 7.13. Pressure regulating valves which operate in coordination with turbine guide vane motion are sometimes called synchronous bypass valves.

Various types of valves are used as discharge valves or pressure regulating valves. Some of them are shown in Figure 7.14. These valves are required to meet the following

(1) Do not develop any serious vibration or cavitation when discharging high pressure water into the atmosphere.
(2) Energy of the water discharged is dissipated as fully as possible.
(3) Will cause neither detrimental wear by high velocity flow nor erosion by heavy cavitation in the conduit downstream the valve.
(4) Operating force of the valve is sufficiently small.

When a discharge valve is used as a synchronous bypass it must open without failure when the turbine guide vanes close. To accomplish this, an operating mechanism as shown in Figure 7.15 is used which is connected to the guide vane operating ring through a dash pot.

7.7 Operating Servomotors for Inlet Valves

Operation of inlet valves is achieved by one of the following methods

(1) Open and close by hydraulic servomotor using high pressure oil;
(2) Open by oil pressure servomotor and close by gravity through a counterweight;
(3) Open and close by hydraulic servomotor using high pressure water from

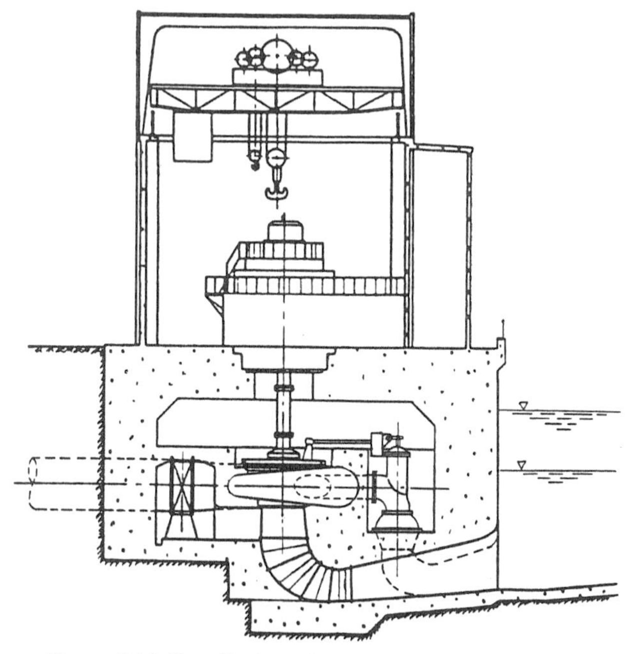

Figure 7.12 Installation of a pressure regulating valve

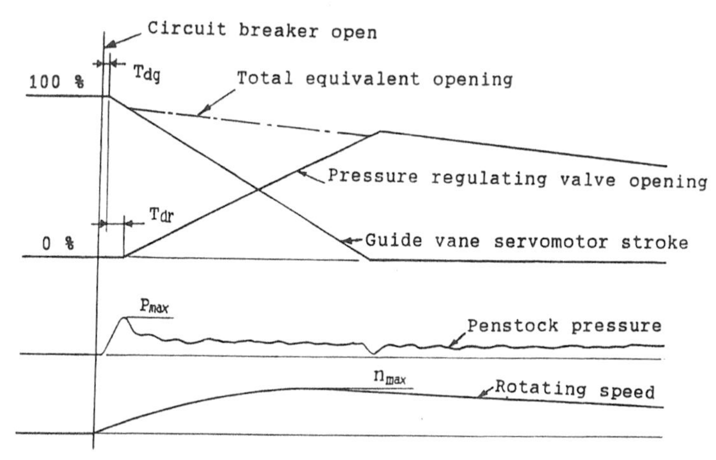

Figure 7.13 Function of pressure regulating valve

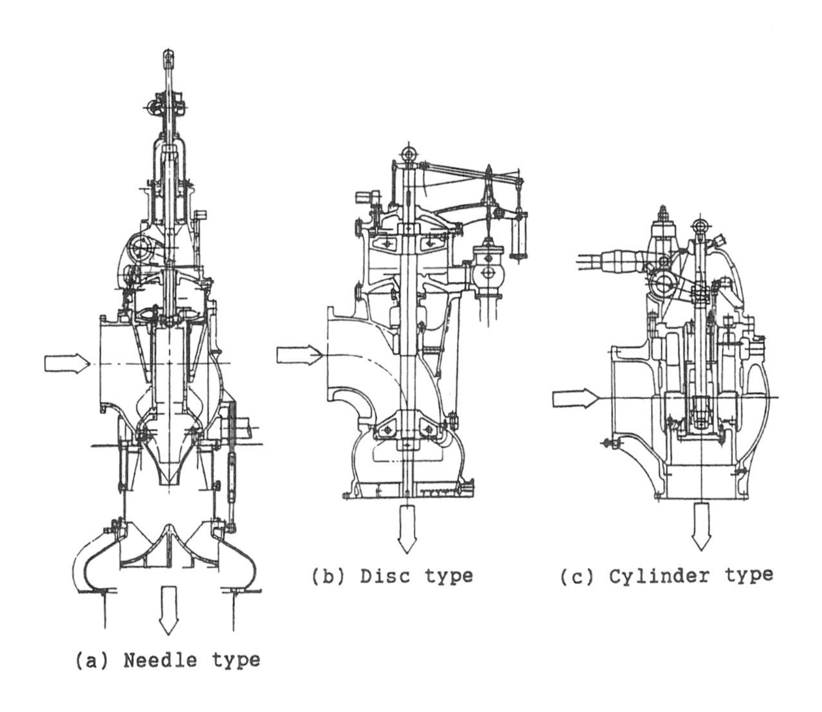

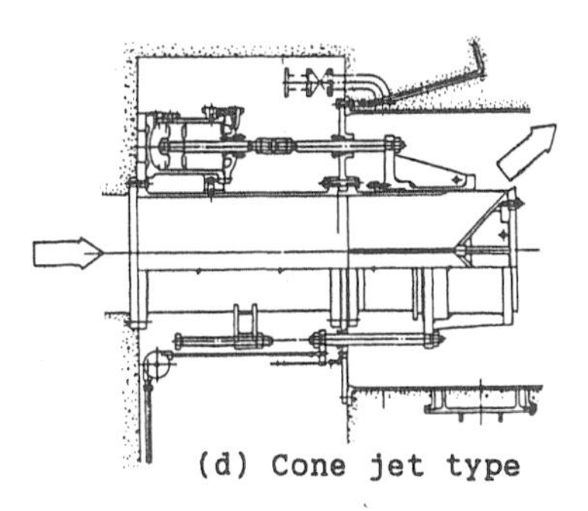

Figure 7.14 Various types of pressure regulating valves

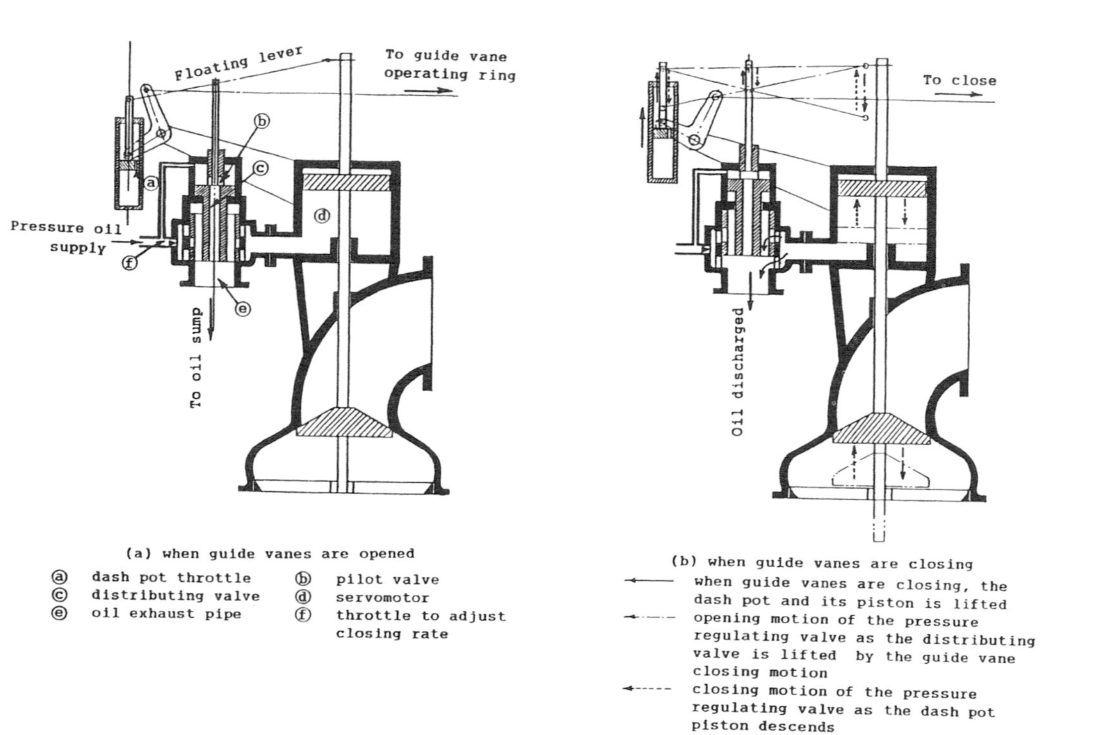

Figure 7.15 Operating mechanism of a pressure regulating valve

the penstock;
(4) Open by penstock pressure servomotor and close by gravity through a counterweight;
(5) Others, such as electric motor drive, etc.

When the gravity force of a counterweight is used to close the valve, the hydraulic servomotor, used to open the valve functions as a viscous damper for controlling the closing speed. Such servomotor(s) or counterweight(s) may be installed at one or both ends of the stems of a valve rotor.

In an emergency when the turbine guide vanes fail to close, the inlet valve must be capable of shutting off the flow regardless of the operating condition of the turbine.

Operating capacity of the servomotors to shut off the flow is determined under the condition that it closes the valve by overcoming the friction torque M_f acting on the valve rotor. The diameter D_s of the servomotor(s) is calculated from

$$np_s\frac{\pi}{4}D_s^2R = kM_f = k\mu p\frac{\pi}{4}D^2\frac{d}{2}$$

$$\text{or } D_s = D\left(\frac{k}{n}\,\frac{\mu}{2}\,\frac{p}{p_s}\,\frac{d}{R}\right)^{1/2} \tag{7.3}$$

where n – number of servomotor(s)
p_s – minimum operating pressure of servomotor (MPa)
D_s – diameter of servomotor (m)
R – radius of the operating arm of the valve stem (m)
k – factor of design margin = 1.2
μ – friction coefficient = 0.15–0.25
p – maximum static pressure difference across the valve (MPa)
D – diameter of the valve (m)
d – diameter of the valve stem (m)

When a valve is closed to shut off the flow, considerable hydraulic moment in the closing direction is developed. This closing hydraulic moment will cause the back pressure of the servomotor p_0 to reach the following value

$$p_0 = p_c + \frac{M_h - M_f}{n(\pi/4)D_s^2R} \tag{7.4}$$

where M_h is the hydraulic closing moment acting on the valve rotor, ($N.m$), and M_f is the friction torque due to lateral hydraulic force F_h working on the valve rotor, ($N.m$).

These two moments are calculated as follows

$$M_h = mp_vD^3\cdot 10^6 \qquad \text{and} \qquad M_f = \mu F_h\frac{d}{2} \tag{7.5}$$

where

$$F_h = f p_v D^2 \cdot 10^6, \quad p_v = \frac{p_c}{1-(CA/A_t)^2},$$

$$CA = \frac{Q}{(2p_v/\rho)^{1/2}}, \qquad A_t = \frac{Q_{max}}{(2p_t/\rho)^{1/2}}$$

and p_v – differential pressure across the valve during shut down operation (MPa)

m – coefficient of hydraulic moment

f – coefficient of hydraulic lateral force

Q – momentary flow rate through the valve during shut down $= CA(2p_v/\rho)^{1/2}$, (m^3s^{-1})

ρ – density of water, (kgm^{-3})

C – coefficient of discharge

A – cross sectional area (nominal) of the valve $= (\pi D^2/4)$ (m^2)

Q_{max} – maximum turbine discharge, $(m^3 \cdot s^{-1})$

p_c – total pressure difference across both turbine and valve (MPa)

p_t – total pressure difference across turbine (MPa) (corresponding to turbine net specific energy)

A_t – equivalent discharge area of turbine, (m^2)

The values of m, f and C are functions of the valve opening which depend on type and design of the valve, and must be determined by model tests. An example of these values for a rotary valve is shown in Figure 7.10.

An example of the hydraulic moment calculated for a rotary valve with a diameter of $3400mm$, which is to shut off the maximum turbine discharge of $124m^3s^{-1}$ under a head of $227m$, is shown in Figure 7.16. The calculations are based on the coefficients shown in Figure 7.10.

Due to the large hydraulic closing moment in this case, the differential pressure across the servomotor piston to withstand the closing moment reaches about 1.4 times the nominal operating pressure of the servomotor. When the valve is being closed, full oil pressure is applied to the closing side of the piston. Therefore, the pressure on the opening side of the piston (back pressure) reaches 2.4 times the nominal oil pressure. The cylinder and the piston of the servomotor as well as the throttle valve for controlling the closing speed must be designed to withstand this pressure.

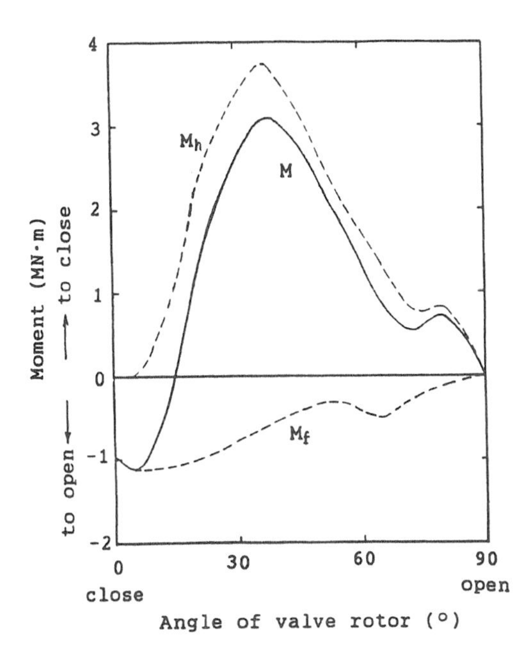

M_h : Hydraulic moment
M_f : Friction moment
M : Total moment to work on servomotor
$M = M_h + M_f$

Figure 7.16 An example of hydraulic moment of valve rotor when a rotary valve shuts off full turbine discharge

Chapter 8

Structural Design of Hydraulic Turbines

Hiroshi Tanaka

8.1 Materials for Hydraulic Turbines

8.1.1 The use of various types of materials

For many years the basic materials for construction of hydraulic turbines have been various grades of carbon steels for castings, and structural steels for welded parts. With the increasing growth of machine size and operating head, materials of greater strength have been developed for the fabricated parts of turbines, such as certain grades of low–alloy structural steels that have high strength and good weldability. With the increase of flow velocity through the machines, cavitation erosion becomes a dominating source of damage to turbines. It has become the major concern for many turbine designers and power plant operators. The common metallic material to resist cavitation erosion is stainless steel of various grades which are either used as protective overlays of the carbon steel parts, or to make castings for runner blades, guide vanes or complete runners, and in certain instances, the stay ring and spiral case as well.

Sand and/or silt erosion is a serious matter that limits the life and scope of application of hydraulic turbines. In recent years, erosion–resistant steels that have higher strength, greater hardness and higher resilience have been developed for turbine parts to work in silt–laden waters.

A large variety of materials are now used in the construction of hydraulic turbines. Some of the more commonly used materials are listed in Table 8.1.

Table 8.1 Materials used for manufacture of hydraulic turbines

Standards:	ASTM	DIN	BS	JIS
Rolled steel				
general use:	A283 Gr C/D	17100 St44-2	4360 Gr40A	G3101 SS41
for welded				
structure:	A284 Gr C/D	17100 St44-2	4360 Gr40A	G3106 SM41A
	A516 Gr 55	17100 St44-3	4360 Gr40C	G3106 SM41B
	A633 Gr D/E	17100 St52-2	4360 Gr50B	G3106 SM50A
	A516 Gr 70	17100 St52-3	4360 Gr50C	G3106 SM50B
	A678 Gr B/C			G3106 SM58
	A537 Gr B	17102 StE500		G3115 SPV50
	A517 Gr M	17100 St70-2		G3128 SHY70NS
for pressure				
vessel:	A612-796-55	17102 StE315	1501 Pt.1 223 Gr490	G3115 SPV32
Carbon steel				
forgings:	A668 Class E			G3201 SF50A
	A668 Class D			G3201 SF55A
Carbon steel				
pipes:	A53 Type E Gr A	2440 St33-2	1387 Bw22	G3452 SGP
	A53 Type S Gr A	1629 St37	3601 S 360	G3454 STPG38
Carbon steel				
for machine				
structure:	A576 Gr 1035	17200 CK35	970 Pt.1 080A35	G4051 S35C-N
	A576 Gr 1045	17200 CK45	970 Pt.1 080A45	G4051 S45C-N
	A576 Gr 1055	17200 CK55	970 Pt.1 070M55	G4051 S55C
Stainless steel				
plates, bars:	A167 Type 304	17440 X5CrNi189	970 Pt.4 304S31	G4303 SUS304
	A176 Type 403	17440 X10Cr13	970 Pt.4 403S17	G4304 SUS403
	A176 Type 410	17440 X10Cr13	970 Pt.4 410S21	G4304 SUS410
Cr-Mo steel				
bars:	A322 Gr 4137	17200 34CrMo4	970 Pt.1 708A37	G4105 SCM435
Carbon steel				
castings:	A27 Gr 65-35	1681 GS-45,2	1504-161 Gr480	G5101 SC46
	A27 Gr 70-36	1681 GS-52	1504-161 Gr540	G5101 SC49
	A216 Gr WCC	1681 GS-45,3	ZG20SiMn	G5102 SCW49
Stainless steel				
castings:	A743 CA-15	17445 G-X20Cr14	3100 St.410C21	G5121 SCS1
	A743 CA6NM	17445 G-X5CrNi134		G5121 SCS5/SCS6
Gray iron				
castings:	A48 Class 35	1691 GG-25	1452 Gr260	G5501 FC25
	A48 Class 40	1691 GG-30	1452 Gr300	G5501 FC30
Cupper alloy				
tubes:	B75 C12200	1787 SF-Cu	1527-79, T2	H3300 C1220T
Bronze				
castings:	B584 C90500	1705 CuSn10Zn	1176-74, ZQSn10-5	H5111 BC3
	B584 C83600	1705 CuSn5ZnPb	1176-74, ZQSn5-5-5	H5111 BC6
White metals :	B23 Gr 3	1703 LGSn80F	1174-74, ZChSnSb11-6	H5401 WJ2

8.1.2 Stainless steel

Of the common grades of stainless steel, both martensitic and austenitic are used in the manufacture of turbines. The martensitic stainless steel is mainly used for strength and the austenitic stainless steel essentially for corrosion resistance. Sometimes stainless steel may develop pitting corrosion in still water or in very low velocity flow. This is caused by insufficient growth of the passive layer on its surface due to deficiency of oxygen. Several kinds of passivation treatment may be taken for prevention of this pitting corrosion. One typical method is to have the steel surface cleaned, degreased and then soaked in 20–40% nitric acid for about 30 minutes to produce an oxidized layer on the surface (see Chapter 14).

Stainless steel, especially austenitic stainless steel, has higher electrolytic potential than carbon steel. When these two metals are placed in contact or very close to each other in water, the area of the carbon steel along the boundary with the stainless steel is likely to be corroded. If bolts made of chrome molybdenum steel are used in the vicinity of austenitic stainless steel, their screw threads may decay after several years of operation and cause structural trouble.

Erosion of material may come from two sources, one is by cavitation erosion and the other is by sand erosion. The two kinds of erosion are caused by different mechanisms but sometimes occur at the same time and may intensify each other. Cavitation erosion is mainly due to compression fatigue failure of the metal from high pressure shock wave or impact generated by collapse of cavitation bubbles, while sand or silt erosion is mainly caused by mechanical cutting or grinding action of solid particles.[1]

Stainless steel is much more resistant to cavitation than carbon steel. Therefore, areas of the turbine which are likely to be attacked by cavitation such as the runner, guide vanes and runner seal (wearing or labyrinth ring) are either made of stainless steel or overlaid by stainless steel weldment. In recent years, new protective overlay materials with good endurance against cavitation erosion have been developed. Some cobalt base alloys or those mixed with ceramic powder show better resistance to cavitation attack than ordinary stainless steel by several folds.

Endurance against cavitation erosion of various materials are compared on a relative basis by their amount of weight loss in cavitation erosion tests. Results obtained by ultrasonic vibratory cavitation test rig are shown in Fig. 8.1. Materials with high cavitation resistance are only practical in actual use only if they possess good weldability and resilience against deformation.

[1]The reader is referred to Volume 'Cavitation of Hydraulic Machinery' and Volume 'Erosion and Corrosion of Hydraulic Machinery' of this Book Series for more information on cavitation and sand erosion.

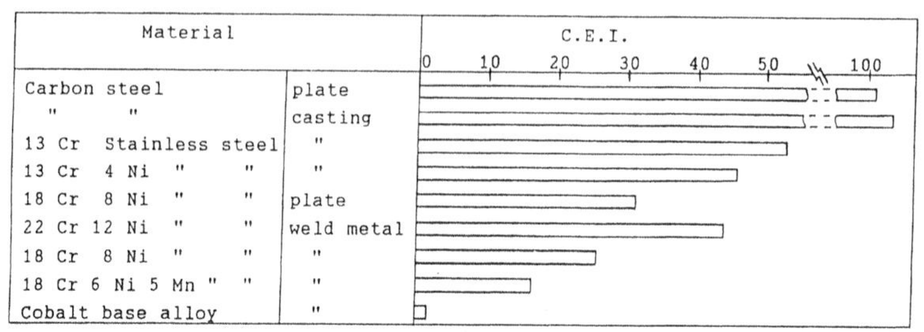

Note: C.E.I.: Cavitation erosion index
= [(Volume loss of metal, cm^3)/(Test period, min)]x10^4

Figure 8.1 Comparative test results of ultrasonic vibratory cavitation erosion test

8.1.3 Sand erosion resistant materials

Sand erosion is caused by a different mechanism from cavitation erosion. The resistance against sand erosion of a material is not necessarily related to its resistance against cavitation erosion. The degree of sand erosion of a material is affected by the physical properties of the sand, its grain size and shape, turbidity, velocity of the flow and the degree of flow turbulence as well as the angle of impingement of sand particles.

Consequently the abrasion resistance of a material cannot be defined by simple relations. For instance, a strong material against sand erosion in low velocity flow is not necessarily good in high velocity flow. A rubber lining on the surface of the water passages of spiral cases or inlet valves is effective only when the operating head and flow velocity are low.

For sand erosion in high velocity flow, no metallic material with satisfactory resistance has yet been found. At present, plasma spray coatings of some ceramics and welded overlays using special electrodes containing ceramic powder are being used on a trial basis. Solid ceramics are only practical for the parts not subjected to deformation because of their brittleness. For ceramic coatings, silicon carbide (SiC), silicon nitride (Si3N4), chrome oxide (Cr2O3), aluminum oxide (Al2O3) or tungsten carbide, (WC) may be used.

Fig. 8.2 shows an example of test results obtained by a comparative silt erosion test on a prototype turbine in operation [8.1].

8.1.4 Protective paints

Carbon steel is usually protected by painting to prevent corrosion. However, carbon steel parts embedded in concrete are not necessarily painted as the alkaline environment in concrete helps to prevents rusting of the embedded steel structure. On the other hand, for those structures embedded in concrete that are required to be painted, a paint that endures strong alkalinity must be used.

Some examples of paint systems are shown in Table 8.2.

8.2 Criteria of Structural Design

The structural design of hydraulic turbines can be determined by the criteria of either stress design or strain design depending on the requirement of each component. In stress design, the relation between the strength of material and the working stress is an essential criterion while in strain design it is essential to keep the deformation of a structure within an allowable value or to build up the stiffness of the structure above a given value.

Stress design is mostly related to fracture or failure of the structure and strain design is related to the functioning of the component such as the allowable runner deformation due to centrifugal force for maintaining runner seal clearances, deformation of the head cover under pressure to maintain top and bottom clearances of the guide vanes, deformation of the spherical valve rotor to ensure fitting and water–tightness of valve seals.

Structural design may be classified as follows:

Stress design:

Static strength – bolts, servomotor cylinder, pressure vessels, etc.
Low cycle fatigue – spiral case, stay vanes, guide vanes, etc.
High cycle fatigue – runner, horizontal shaft, etc.

Strain design:

Deformation – head cover, valve disc, runner (against centrifugal force)
Stiffness – shaft and bearing support (against vibration)

Since it is difficult to accurately calculate the deformation of a complicated structure by an analytical formula, numerical calculation such as the finite element method (FEM) as explained in Section 8.4 is used for strain design.

8.3 Stress Design

8.3.1 Allowable stress for static strength design

When stress design, according to static strength, is performed the maximum working stress of the structure, calculated for the most severe operating condition, must be kept under the allowable stresses $[\sigma]$ as given below. The

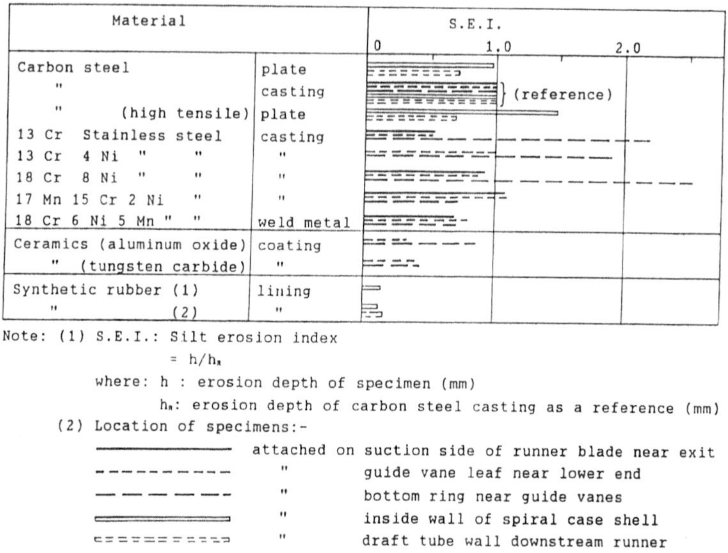

Material		S.E.I. (0 — 1.0 — 2.0)
Carbon steel	plate	
"	casting	(reference)
" (high tensile)	plate	
13 Cr Stainless steel	casting	
13 Cr 4 Ni " "	"	
18 Cr 8 Ni " "	"	
17 Mn 15 Cr 2 Ni "	"	
18 Cr 6 Ni 5 Mn " "	weld metal	
Ceramics (aluminum oxide)	coating	
" (tungsten carbide)	"	
Synthetic rubber (1)	lining	
" (2)	"	

Note: (1) S.E.I.: Silt erosion index

$= h/h_s$

where: h : erosion depth of specimen (mm)

h_s: erosion depth of carbon steel casting as a reference (mm)

(2) Location of specimens:-

———————— attached on suction side of runner blade near exit

- - - - - - - - - - " guide vane leaf near lower end

— — — — — " bottom ring near guide vanes

════════ " inside wall of spiral case shell

⊏=========⊐ " draft tube wall downstream runner

Figure 8.2 Example of comparative test results of silt erosion performed on a turbine in normal operation

Table 8.2 Examples of paint system

Area	Primer	Top coat
Embedded part in contact with concrete	(1) None	None
	(2) None	2 coats of tar paint
Water passages made of carbon steel plates	(1) None	2 coats of tar epoxy paint
	(2) 1 coat of zinc-rich primer	2 coats of epoxy paint
Non-machined surface exposed to atmosphere	(1) None	2 coats of vinyl paint or epoxy paint
	(2) Epoxy primer	1 coat of epoxy paint
Inside of the oil reservoir	None	3 - 4 coats of oil-resistant vinyl paint

normal operating condition refers to the most severe operating condition expected when all controls and protections function as designed, and the *extreme emergency condition* means the most severe operating condition anticipated when one or more of the control equipment fail to work as designed. In normal practice, runaway of the turbine with guide vanes fully open, or the unit shutting down by the inlet valve with guide vanes fully open are regarded as *extreme emergency conditions.*

Normal operating condition:

for runner, spiral case, stay vanes, pressure vessels, and so on.

$$[\sigma] \leq \frac{1}{4} \cdot \sigma_B \quad \text{or} \quad \frac{1}{(1.6 \sim 2)} \sigma_Y$$

for shafts, and so on.

$$[\tau] \leq \frac{1}{2} \cdot \tau_{max}$$

Extreme emergency condition:

for runner, and so on.

$$[\sigma] \leq \frac{2}{3} \cdot \sigma_Y$$

where σ_B is the ultimate strength, σ_Y the yield strength, and τ_{max} the maximum shear strength.

When the inlet valve is opened and closed at every starting and stopping of the turbine, the stresses in the spiral case and stay vanes will vary widely. If the unit is operated for peak load with frequent starts and stops it must be designed with consideration of low cycle fatigue.

If the stiffness of a structure fixed by bolts is small, the stress in the bolts will vary appreciably by the force applied on the structure. Line A in Figure 8.3 shows the relationship between the stress and strain of a bolt, where σ_i is the initial tightening stress and ξ_i is the initial strain of the bolt. Line B represents the relationship between the stress and strain of the structure held together by the bolt. If an external load corresponding to σ_e in terms of bolt stress is applied, this stress is shared by both increment of the tensile stress of the bolt σ_b and reduction of the compression stress in the structure σ_s. If the stiffness of the structure is low, line B will shift to the position of B', and result in a larger increment of bolt stress for the given load. The same situation is true for cases when tightening of bolts is insufficient and a small clearance is left between the fastened structures.

In cases like the aforementioned, if the load working on the structure is alternating, the bolt should be designed for additional consideration of fatigue.

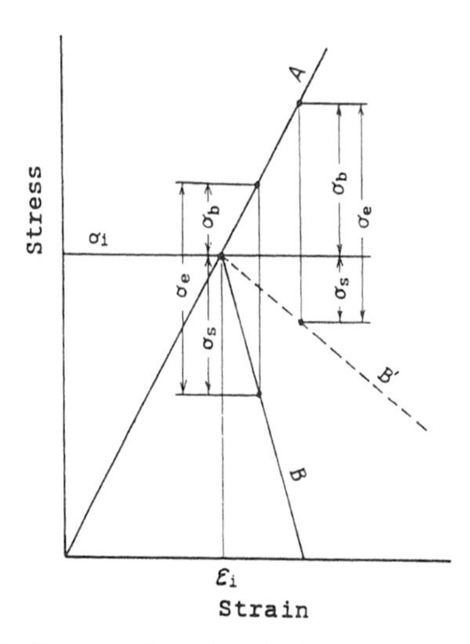

Figure 8.3 Stress and strain of a bolt and bolted structure

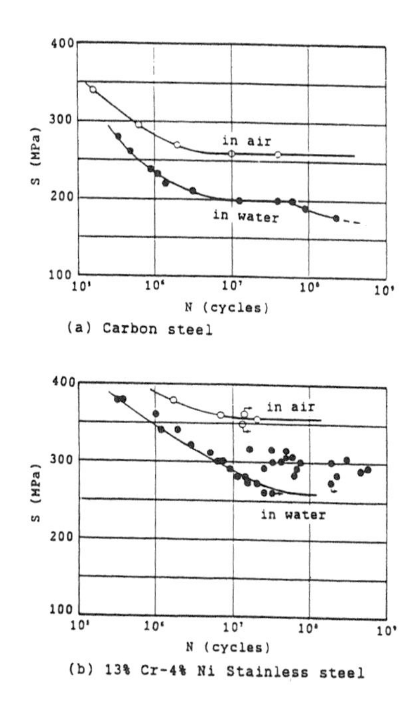

Figure 8.4 $S - N$ diagrams of carbon steel and 13Cr–4Ni stainless steel

8.3.2 Design with consideration of fatigue strength

Figure 8.4 shows the $S-N$ diagrams of some typical materials, where S is the amplitude of alternating stress (half amplitude, 0 to peak) and N is the number of cycles of the alternating stress up to fatigue failure. The fatigue strength of a material submerged in water will become lower than in air, so the fatigue strength tested in water must be used for designing structures operating in water.

It is worthy of note that the fatigue strength of mild steel declines appreciably in the high cycle range beyond 10^7-10^8 cycles. This is considered to be caused by heavy corrosion which proceeds nearly proportionately to the elapse of time. Fatigue strength declines further in acidic water or in sea water.

In general, cast steel contains many flaws such as microshrinkage and small blow holes. Welded joints also may have minor defects as insufficient fusion, blow holes and undercuts, which may affect the high cycle fatigue strength of the material considerably.

For assessment of the fatigue strength of the materials having these flaws, the values obtained from notched specimens with stress concentration factors $k_t = 2.5$ to 5 shall be used. An example of the test results with such notched specimens is shown in Figure 8.5.

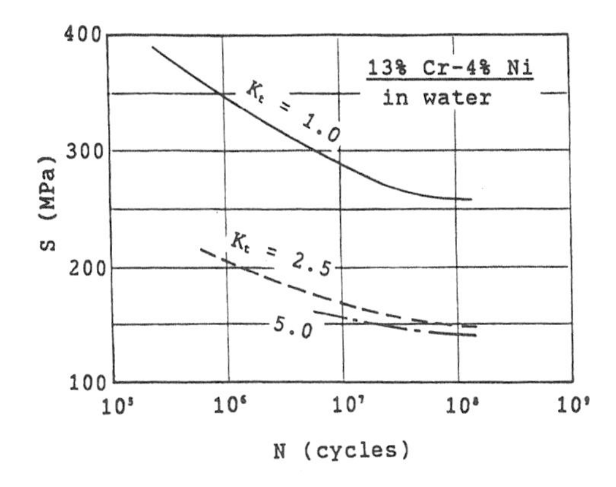

Figure 8.5 Fatigue strength tested with notched specimens

It is well known that the fatigue strength of steel in air usually becomes nearly constant beyond approximately 10^7 cycles. However, in corrosive environments such as water, it continually declines in the high cycle range beyond 10^7 cycles. Runner blades excited by Karman vortices, buckets of high head Pelton turbines, runners of high–head reversible pump–turbines and so on, are subject to alternating stresses with frequencies from tens to $200Hz$ and the cumulative total number of cycles of alternating stress in their service life may reach more than 10^{10}. For assessment of the fatigue strength of these components, sufficient margin of safety must be retained considering the decrease of fatigue strength in the high cycle ranges.

The decrease of fatigue strength in water in the high cycle range may be approximately estimated by the following relations

$$\log S = k \cdot \log N + \log C \tag{8.1}$$

where $k = -0.07$ to -0.10 for stainless steel
$k = -0.12$ to -0.14 for carbon steel

Spiral case, stay vanes, valve rotor, and so on are subject to large variation of stress at every starting and stopping, the anticipated number of cycles of alternating stress is somewhere around 10^4 to 10^5 cycles. In the range of these cycles, the fatigue strength varies significantly according to the value of N.

In design studies with respect to such low cycle fatigue strength, the number of cycles of alternating stress N during the whole life of the turbine is first estimated, and the fatigue strength S_N corresponding to the value of N is read from the $S - N$ curve for the material. If the value of alternating working stress S is less than S_N, it is concluded the component will be safe within the life of the turbine.

Generally, a structure is subject to both the stationary mean stress σ_m and alternating stress σ_a. For example, in the case of a runner in steady operation, the alternating stress due to pressure fluctuation is superimposed on the mean stresses from centrifugal force and hydraulic pressure. For design assessment of such combined stress conditions, the modified Goodman's diagram as shown in Figure 6.9 of Section 6.2 is used.

Also, under actual working conditions, the alternating stress applied to the structure is not uniform. For instance, when a runner is subject to regular alternating stresses due to hydraulic interference between runner and guide vanes in normal operation, it has to withstand also abrupt and intensified alternating stresses in the transient process following load rejection. For assessment of fatigue failure due to the cumulative effect of these different stress conditions, Miner's rule (cumulative damage rule) should be applied.

According to Miner's rule, the cumulative effect on a machine part will not cause fatigue failure if the value of D given by the following equation is less than 1.0

$$D = \sum_{i=1}^{m} D_i = \sum_{i=1}^{m} (n_i/N_i) \tag{8.2}$$

where

D – degree of fatigue damage
S_i – alternating stress at each operating condition
D_i – degree of fatigue damage due to n_i cycles of alternating stress S_i
n_i – expected number of cycles of repetition of S_i during the service life of the structure
N_i – number of cycles up to fatigue failure when S_i is applied alone
m – number of operating conditions in which different magnitude of alternating stress is expected

8.3.3 Growth rate of fatigue crack and allowable size of defects

After the initiation of a fatigue crack described in the above study, investigation should also be made to the length of the crack which will grow with every cycle of stress fluctuation. The rate of growth of a crack under alternating stress is estimated by the theory of linear elastic fracture mechanics [8.2]. According to this theory, the growth rate of the crack at each cycle of stress pulsation $\mathrm{d}a/\mathrm{d}N$ is given as a function of the variation range of stress intensity factor ΔK, which is defined by the following formula (Figure 8.6)

$$\Delta K = \Delta\sigma_m \cdot M_m\sqrt{\frac{\pi a}{Q}} + \Delta\sigma_b \cdot M_b\sqrt{\frac{\pi a}{Q}} \tag{8.3}$$

where

ΔK – variation range of stress intensity factor, $MPamm^{1/2}$
$\Delta\sigma_m$ – variation range of membrane stress (uniform stress across the section), MPa,
$\Delta\sigma_b$ – variation range of bending stress (stress distributed across the neutral line of section with linear gradient), MPa
M_m – correction factor for membrane stress for finite thickness of plate given as a function of a, ℓ and t,
M_b – correction factor for bending stress
Q – flaw shape parameter determined by aspect ratio ℓ/a of the flaw (crack)
a – depth of flaw, mm
ℓ – length of flaw, mm
t – thickness of plate, mm

For the calculation of M_m, M_b and Q, references may be made to [8.2] and [8.3].

The relationship of $\mathrm{d}a/\mathrm{d}N$ against ΔK is obtained experimentally for each material. Examples for carbon steel plate and 13Cr–4Ni stainless steel are shown in Figure 8.7.

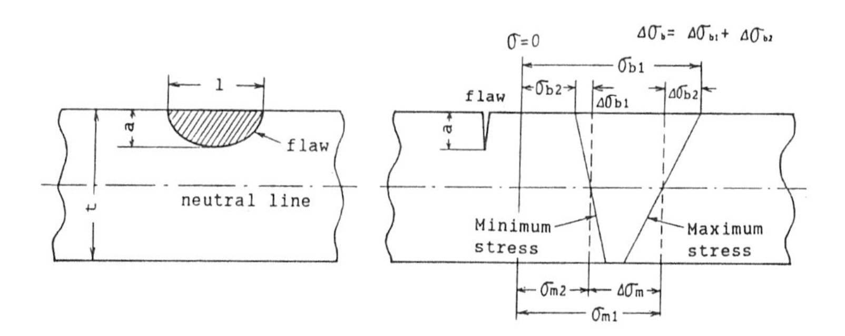

Figure 8.6 Stress condition and a typical flaw

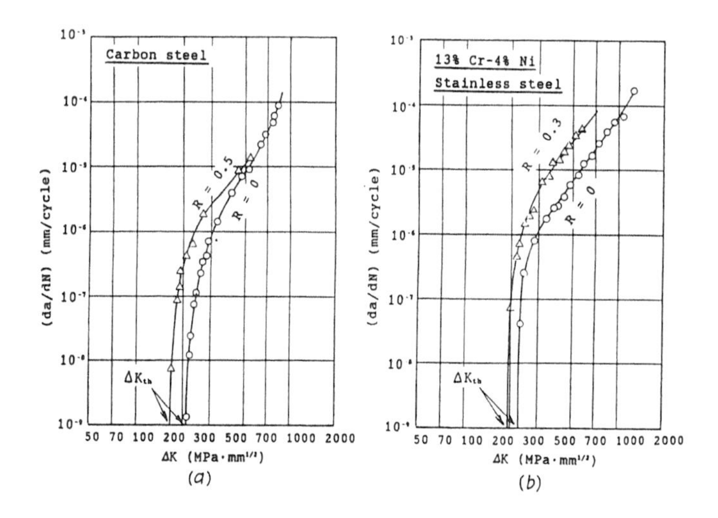

Figure 8.7 Crack growth rate of carbon steel and 13Cr–4Ni stainless steel

In these diagrams, curves for different values of parameter R are shown, where R is defined as

$$R = \frac{K_{min}}{K_{max}}$$

where

K_{max} – stress intensity factor for σ_{max}
K_{min} – stress intensity factor for σ_{min}.

For completely reversed alternating stress, R is -1, and for alternating stress pulsating between zero and peak, R is 0. When the mean stress is high the value of R approaches 1. From these diagrams it is seen that the growth rate increases with ΔK. If the ΔK value calculated from the expected working stress falls short of a threshold value ΔK_{th}, the flaw does not grow and will remain stable.

Inversely, from the threshold value ΔK_{th} of a material, it is possible to calculate the allowable size of defect from a given stress condition, $\Delta\sigma_m$, $\Delta\sigma_b$ and R. Examples of the allowable size of defect thus obtained, are shown in Figure 8.8, in which curves for surface defect and sub–surface defect are shown. If a linear defect exceeding the value shown in Figure 8.8 exists in castings or in weld joints, it may grow during the operation and cause serious fatigue failure eventually.

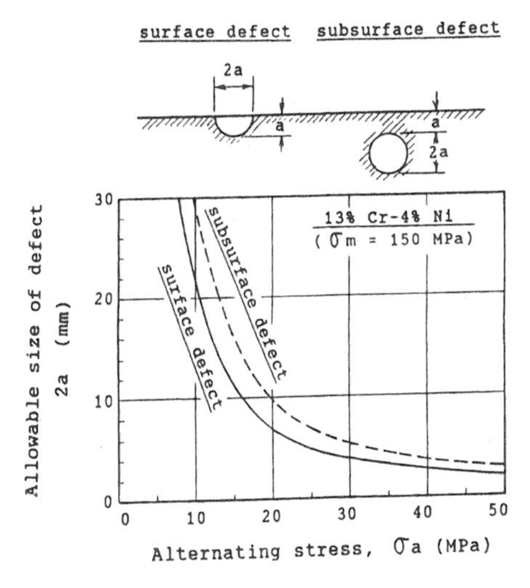

Figure 8.8 Allowable size of defect for 13Cr–4Ni stainless steel

As can be seen from these diagrams, the allowable size of defect is fairly large for low alternating stresses, however, it must be kept very small at high stress amplitudes. For alternating stress exceeding $\pm 40MPa$ the allowable size becomes smaller than the crack detection capability of commonly used non–destructive examination methods. In such cases, the design must be revised to decrease the stress amplitude in order to maintain the reliability of the structure.

8.4 Stress and Strain Analysis

8.4.1 Finite element method

The boundary problems of continuum are very important subjects of studies in the field of fluid dynamics, structural analysis, electromagnetism, and so forth. The finite element method (FEM) is a very powerful tool to solve these problems. At present it is widely used for analysis or design of hydraulic turbines, particularly in the flow analysis of flow components, stress and strain analysis of structural parts, eigenvalue analysis and dynamic response studies of structures. In this section, the application of the finite element method stress and strain analysis of structures will be described.

In the finite element method, a structure is divided into many small discrete elements as shown in Figure 8.9. The variables of the governing differential equations and their derivatives are specified at the nodes on the boundaries of these finite elements. In these elements, they are written as linear combination of appropriate interpolation functions of these variables. In stress and strain analysis, these variables represent the displacements of the nodes, which are shown as $\{U, V\}$ in Figure 8.9.

The governing differential equations are transformed by using the variational principles into finite element equations which represent dynamic equilibrium for individual elements. These finite element equations are collected together to form a global system of algebraic equations representing the whole structure in which boundary conditions are properly imposed as required. The displacements of the nodes are determined by solving this system of equations using matrix algebra. For details of the finite element method, references may be made to [8.4, 8.5]. The following should be noted when using the finite element method:

(i) Appropriate grid generation

Though a general rule for appropriate division of elements or grid generation has not been established, the decision as to the size and arrangement of grids is important in practice. Fine grids are desirable in areas of stress concentration, and elements of very slender proportions should always be

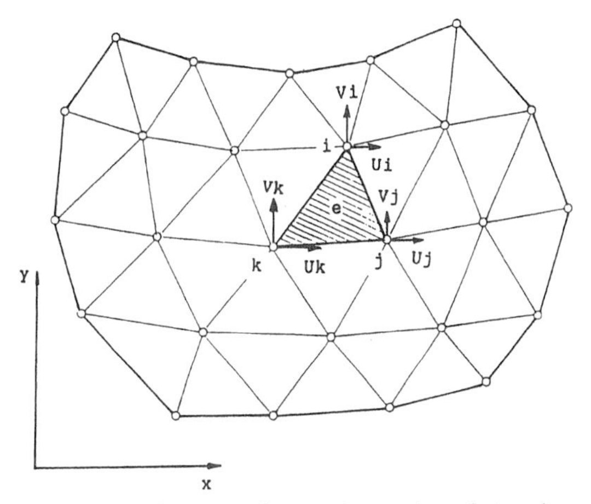

Figure 8.9 Division of a continuum into finite elements

avoided.

(ii) Correctness of boundary conditions.

When an engineering problem is to be solved by using the finite element method, the solution may involve serious error if the boundary conditions are given incorrectly. For example, when calculating the deformation of a head cover under internal hydraulic pressure, if the deformation of the stay ring is neglected and the periphery of the head cover is assumed rigidly fixed, the vertical displacement at the central part of the head cover calculated will be about one half of the value when taking the deformation of the stay ring into account.

Generally, when the finite element method is applied to three-dimensional problems, the number of elements will become very large and the required computer time will be considerable. But attempts to decrease the number of elements for the purpose of reducing computer time may affect the accuracy of the solution.

The following procedures will help to reduce computation time without significantly impairing the accuracy of the solution.

(i) Two-dimensional treatment of axisymmetric ring structure.

A radial section of an axisymmetric ring element is divided into two-dimensional elements. Each element represents an elementary ring around the axis of symmetry as shown in Figure 8.10. Finite element equation for dynamic equilibrium is established, considering the element as a ring though the variables are two-dimensional. The whole system of these equations for all finite elements is solved as a two-dimensional problem.

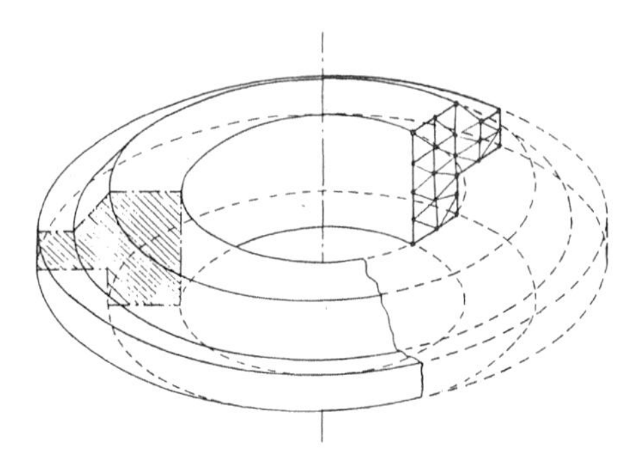

Figure 8.10 Axisymmetric ring structure

To properly use this method, it is essential that all the boundary conditions such as supports and external forces must be axisymmetric or uniform on the particular circle represented by a node. Structures such as pressure vessels of circular cross section and servomotor cylinder are constantly analyzed by this method.

(ii) Two-dimensional treatment of axisymmetric ring structure with evenly distributed radial members

A structure formed by axisymmetric ring elements with radial members can be treated also as a two-dimensional problem if the radial members are equally spaced around the periphery, as in the case of a typical head cover structure consisting of circular discs, rings, and cylinders with radial reinforcement plates.

Parts like radial membranes, which are not continual around the axis of symmetry but extend in equally spaced radial planes, are divided into elements in the radial section so that they may be distinguished from other parts. Young's modulus and Poisson's ratio in the directions of R and Z of these parts are given as (t/p) times the ordinary value and those in the θ direction are considered zero (here t is the thickness of plates and p is the circular pitch of membranes). Thus the radial member is regarded as a ring having lower stiffness in the R and Z directions.

By this approximation the deformation of the structure is first calculated, then the stress distribution of the radial members in the $R - Z$ plane is obtained by using common values of Young's modulus and Poisson's ratio in the $R - Z$ direction as of other parts. Deformations of head covers and

Francis turbine runners can be calculated fairly accurately by this method. Figure 8.11 shows an example of the computed stress distribution in the main parts of a pump-turbine.

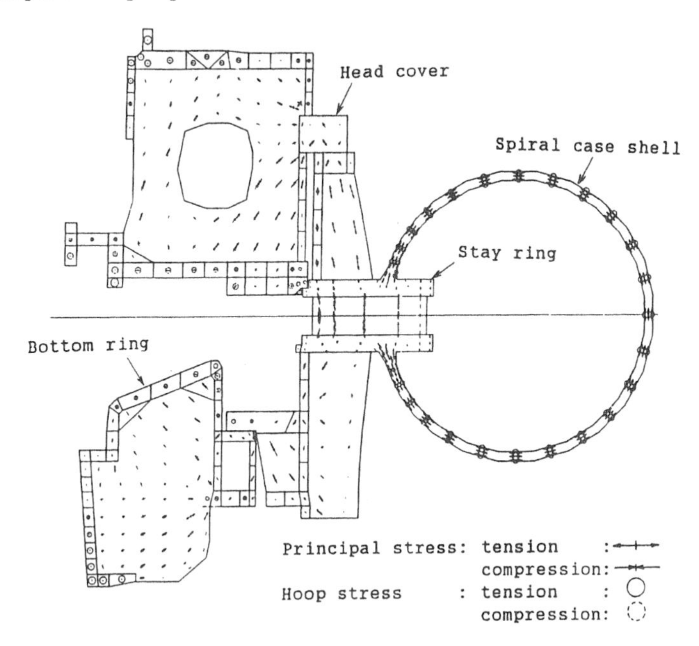

Figure 8.11 Stress distribution of main parts of a pump-turbine calculated by the finite element method as ring structures

(iii) Cyclic symmetry method

When the same shape is repeated around the axis of symmetry at a regular circumferential pitch, the structure for one pitch (substructure) is computed instead of the whole structure.

In this method, boundary conditions assigned to the nodes on the boundary facing the adjoining substructure are always kept consistent with those of the corresponding nodes of the adjoining substructure, namely, same displacements and inverse signs for values of force and moment. If the external boundary conditions such as external forces and displacements are of cyclic symmetry, boundary conditions between the adjoining substructures may be simplified. In this case, it is sufficient to keep consistency of boundary conditions on both sides of a substructure and to perform calculation on one substructure only. Figure 8.12 Shows an example of stress distribution of a pump–turbine runner computed by this method.

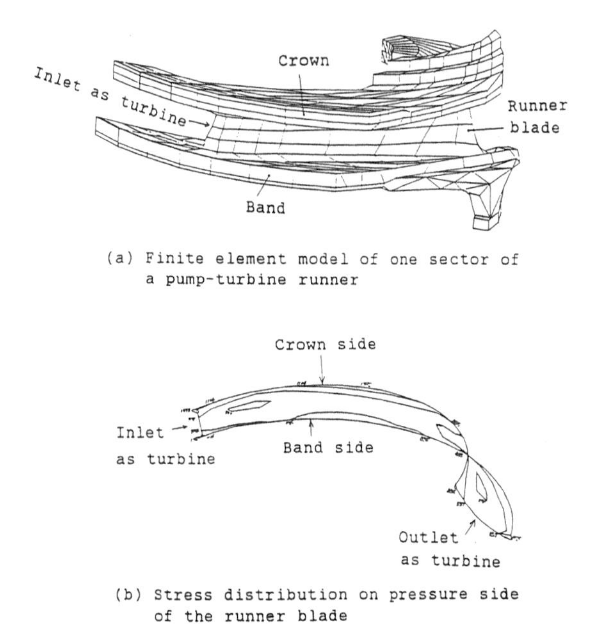

(a) Finite element model of one sector of a pump-turbine runner

(b) Stress distribution on pressure side of the runner blade

Figure 8.12 Stress distribution of a pump–turbine runner calculated by cyclic symmetry method

(iv) Substructure method

The cyclic symmetry method is a special case of the substructure method. While the application of the cyclic symmetry method is limited to a regularly integrated structure around an axis, the substructure method is applied to a structure of any shape. In this method, a structure is divided into several substructures and the boundary conditions between adjacent substructures are made consistent with each other as explained previously.

In computation by the finite element method, reduction of the size of the matrix contributes to the reduction of computation time considerably. Thus the division of a structure into substructures will reduce the total computation time for calculating the whole structure by a great amount, even though the computations to maintain the consistency of boundary conditions between adjacent substructures are necessarily increased.

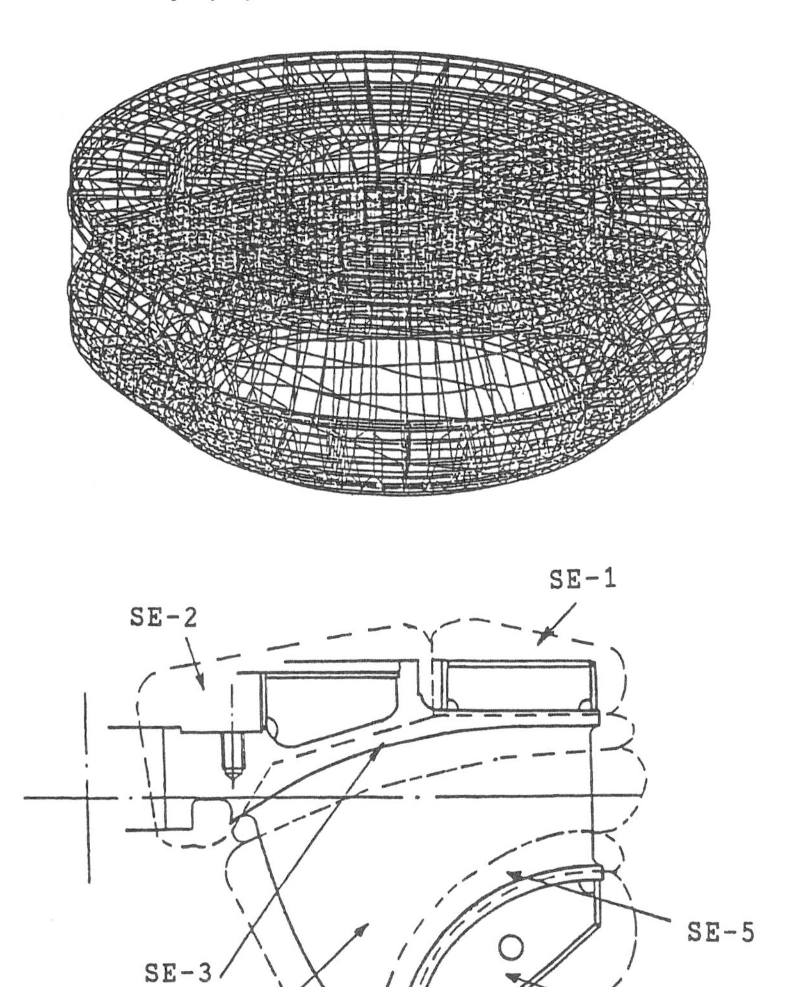

SE-1 to 6: substructures

Figure 8.13 Substructures of a split runner

Since the matrix for a slender structure can be transformed into a band matrix with narrow bandwidth, which may be calculated with less computation time, it is advisable that substructures be laid out slender when they are divided. Also the number of nodes on the boundaries between them should be as few as possible.

Figure 8.13 shows an example of the division of substructures for a split runner with cover plates. By this division, the total computation time can be reduced to less than one fifth of that for an integral structure.

8.4.2 Boundary element method

With a structure that is to be calculated by the finite element method the whole domain of the continuum is discretized into finite elements, but in the boundary element method (BEM) only the boundary of the domain is discretized into finite elements. In a three–dimensional problem only the surface, which is the boundary of the three–dimensional continuum, is divided into two–dimensional surface elements, and in a two–dimensional problem only the boundary curve of the two–dimensional area is divided into linear elements.

Both the finite element method and the boundary element method have similar approaches in numerical methods, in the use of discrete finite elements, but the principles of the calculations are quite different from each other. In the boundary element method, the governing differential equation defined in the domain is substituted by equivalent integral equations, defined at the boundary elements based on the Green theorem. For details of the basic theory of the boundary element method the reader is referred to [8.6, 8.7].

The integral equations defined on the boundary, which contains displacement u and surface force f as variables, are discretized by the boundary elements. Thus a boundary element equation is obtained for each element. A total system of algebraic equations is then established by collecting all these boundary element equations. If the number of boundary elements is m, this system will consist of m equations containing $2m$ variables that is one displacement and one surface force variable for each element. Therefore, by giving the values of m variables as boundary conditions, this system can be solved and the values of displacement and surface force of all boundary elements are determined.

As the boundary element method needs only the input data regarding the boundary elements, it is more advantageous over the finite element method in at least two respects, namely, the required amount of input data is less and it is more convenient to treat problems that involve infinite space such as the flow around a single aerofoil in air.

However, while the finite element method is based on simple dynamic equilibrium equations and uses rather common methods of calculation, the boundary element method deals with complicated algebraic equations and needs sophisticated numerical calculation techniques for solution. In addition, the size of the matrix can be reduced by utilizing the band matrix method in the finite element method but the full–size matrix must be computed in the boundary element method. Figure 8.14 shows an example calculated by the boundary element method for stress distribution along the surface of a T–shape welded joint.

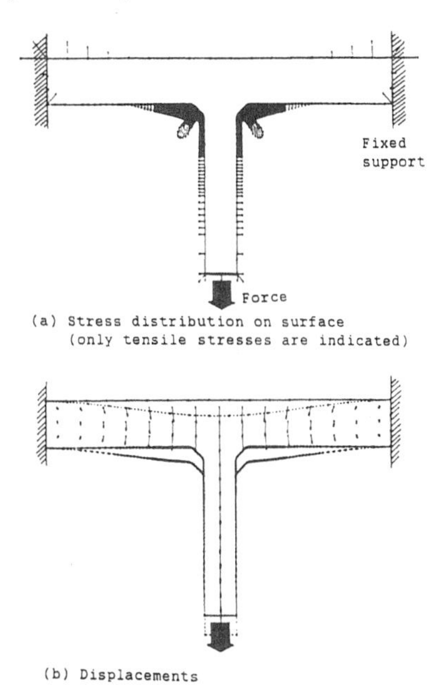

Figure 8.14 Stress distribution of a T–joint calculated by boundary element method

8.5 Computer–aided Engineering

8.5.1 General

Various techniques for design and engineering analysis which make full use of the computer have been developed and put into use in recent years. Such techniques involving repeated simulation studies on computers are aimed at reducing the cost and time required for the development of a new produ saving the time for testing on models or prototypes. They are compre sively known as computer–aided engineering or CAE. In general, the contains the following software tools:

1. Solid modeling to give visual image of the product.
2. Performance simulations such as flow analysis, stress and strain ana thermal analysis, dynamic simulation of motion including vibration, et
3. Computer–aided design or CAD to perform detail design calculations to prepare design drawings.
4. Computer–aided manufacturing or CAM which includes numerically trolled machining (NC) and robotics, etc.
5. Computer–aided testing or CAT for automated testing or detailed as ment of test results.
6. Other applications.

In the procedures above, items 1 and 2 are sometimes called comp aided engineering CAE in a narrow sense to separate from CAD, CAM CAT. In the process of design and manufacturing of hydraulic turbines, puter aid is used in the following aspects:

(i) CAE

- hydraulic design including performance prediction and loss analysis.
- numerical flow analysis.
- numerical stress and strain analysis.
- simulation of the dynamic response of shaft system against hydrauli bulence on turbines.
- simulation of hydraulic transients in pipelines.
- simulation of transients of power network systems including behavi governors and other control equipment.

(ii) CAD

- various design calculations.
- automatic drafting systems.
- interactive drafting systems.

(iii) CAM

- NC plate cutting.
- automatic welding.
- NC machining.
- robot applications.

(iv) CAT

- laboratory automation for model testing.
- computer–aided dimension checking system.
- computer–aided balance test of runners.
- computer–aided field performance tests.
- vibration analysis by FFT analyser.

(v) Others

- operation monitoring.
- computer–aided instructions (CAI) for operators and maintenance staff.
- training simulator for operators.

Some major items listed above are described in the following passages.

8.5.2 Flow analysis

For numerical flow analysis with the aim of obtaining velocity distribution in flow passages and pressure distribution on blades, various numerical methods are currently in use. The finite element method and boundary element method which are initially developed for use in structural analysis, are now extensively used in the field of flow analysis. The finite difference method which is also frequently used for flow analysis, solves the flow problem by replacing the partial derivatives in the governing differential equations by finite difference quotients [8.8, 8.9]. It can be applied to various problems including compressible flow and flow in boundary layer but it requires a regular grid formation of the calculation domain. For flow fields having complicated boundary shape, various techniques are developed for generation of regular grids [8.10, 8.11].

Other computation methods such as the finite volume method (FVM), discrete vortex method (for two–dimensional problems only) and the streamline curvature method are also used for flow analysis. Flow analyses conducted for ordinary hydraulic machines in most cases are based on the assumption of non–viscous flow, except for cases in boundary layer investigation. Flow analyses through runner/impellers using turbulence models have been performed by a number of researchers but no application in engineering design has been reported.

The flow around guide vanes of low specific speed machines can be treated as two–dimensional flow as shown by the example in Figure 8.15. Flows through turbine runners and pump impellers should correctly be treated as three–dimensional flow problems. In these cases, the quasi–three–dimensional analysis and the three–dimensional analysis are currently in use.

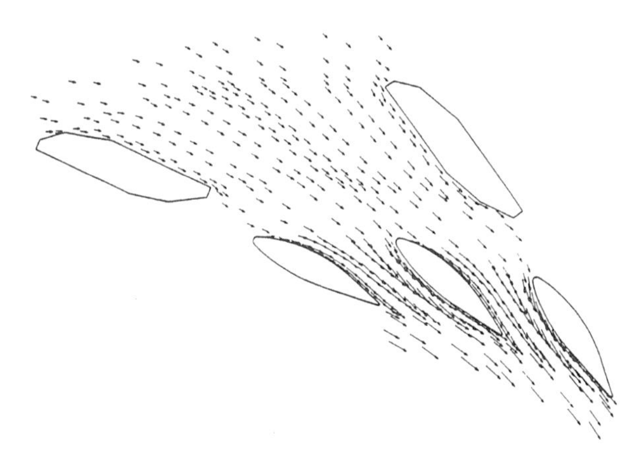

Figure 8.15 Two–dimensional flow field around guide vanes of Francis runner

(i) Quasi–three–dimensional analysis

The flow between the crown (hub) and band (shroud) in the meridional plane is analysed by using either the finite element method or streamline curvature method. The flow between the blades is then computed on a curved surface which is generated by one of the streamlines from the previous analysis around the axis of rotation as shown in Figure 8.16. By repeating the blade–to–blade analysis for all streamlines, the complete three–dimensional flow field is obtained. However, after the blade–to–blade calculation the stream surfaces have to be readjusted somewhat from their original positions, so iteration with the meridional plane calculation is necessary to obtain a more accurate picture of the flow.

(ii) Three–dimensional flow analysis

Some methods are designed to solve the three–dimensional flow problems directly, such as the finite element method and finite difference methods. The three–dimensional analysis itself utilizing the finite element method or finite difference method has no limitation in principle, but in practice, much difficulty is encountered in grid generation and in determination of boundary conditions.

Figure 8.17 shows an example of velocity distribution on the blade surface of a Francis turbine runner calculated by three–dimensional finite element method. Downward flow is identified in the vicinity of the blade leading edge on the pressure side while no such flow appears on the suction side.

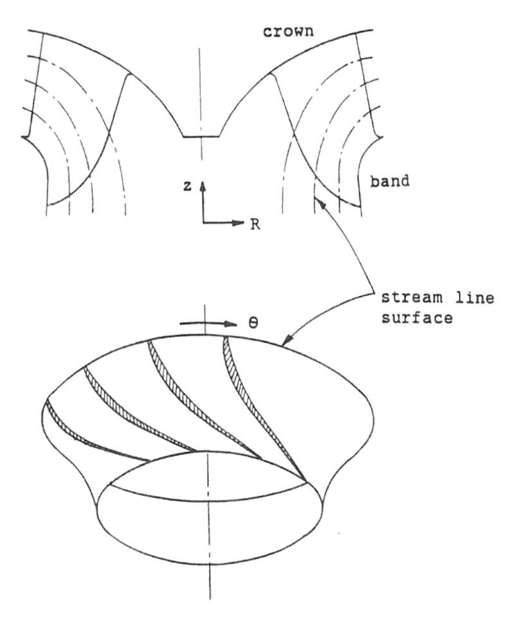

Figure 8.16 Quasi–three–dimensional flow analysis

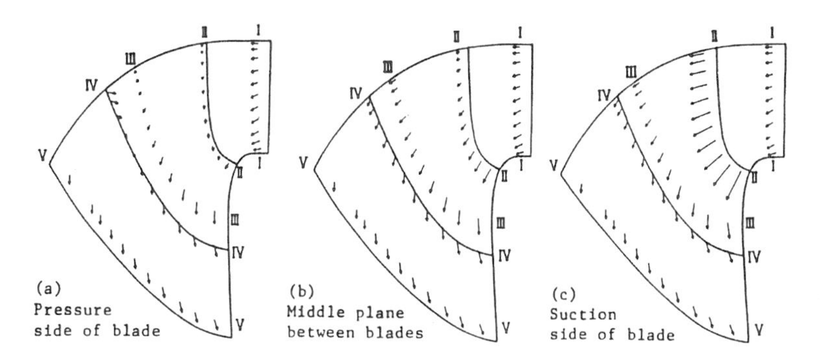

Figure 8.17 Velocity distribution of a Francis turbine runner calculated by three–dimensional finite element method

This flow pattern cannot be produced by quasi–three–dimensional analysis procedures.

8.5.3 Loss analysis of hydraulic turbines

A technique to evaluate the losses by utilizing known formulae has been developed. In this method, a numerical model is established in the computer consisting of various known theoretical and experimental formulae for assessing individual loss components under various operating conditions. By summing up these losses, it is possible to predict the performance of the turbine based on detailed design dimensions. The following loss assessment approaches , shown in Table 8.3, are in general used for evaluation of losses.

Table 8.3 Evaluation of losses in various components of a turbine

Component	Friction loss	Other loss
Spiral Case	Friction loss in a circular pipe or in a bend	
Stay vanes Guide vanes	Friction loss as plane plate or a pipe of rectangular section	Impingement loss at leading edges Wake losses downstream
Runner	Same as above, considering flow analysis results Disc friction on outside surface	
Draft tube	Friction loss as circular or rectangular pipe	Loss from runner discharge whirl Loss of diffusing flow
Runner seal clearances	Friction loss (tangential) in rotating concentric gap	Friction loss (axial) in rotating concentric gap

When a numerical model is composed from the above formulae, various corrections are necessary from experimental studies. Results of flow analyses are used in the determination of impingement loss (flow angle) and friction loss (flow velocity).

Figure 8.18 shows an example of component losses of a Francis turbine calculated by a computer model. Figure 8.19 indicates the variations of these component losses at the optimum operating condition for different machine specific speeds. It is seen for low specific speed turbines, significant

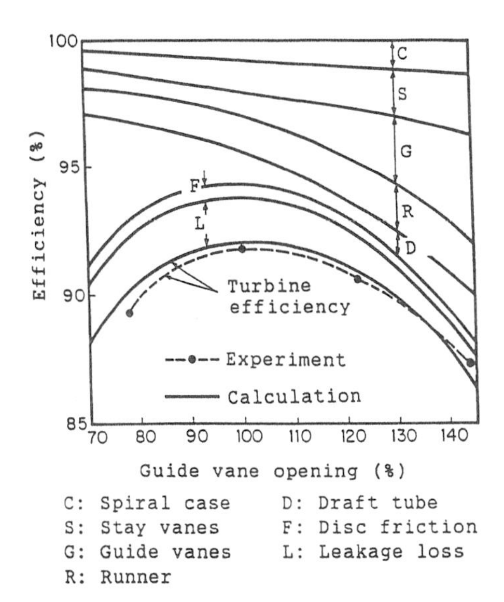

Figure 8.18 Loss distribution of a Francis turbine

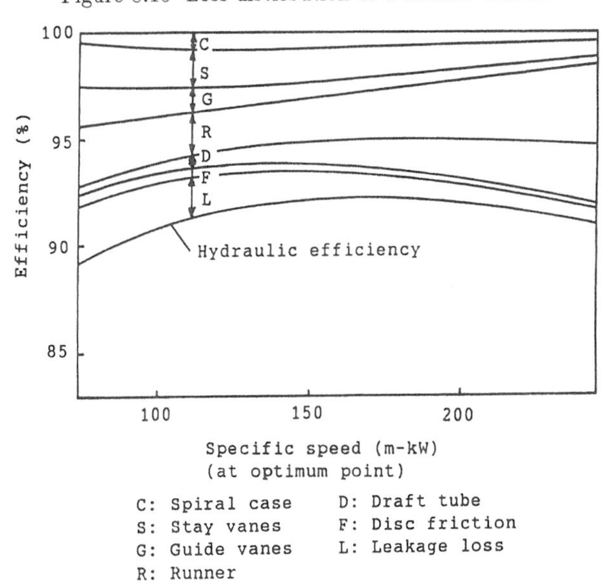

Figure 8.19 Variation of loss distribution against specific speed of Francis turbines

losses are developed in the stay vanes and guide vanes, while in high specific speed turbines losses occur mostly in the runner and draft tube. From these analyses, the extent of improvement of hydraulic performance expected by a certain design modification may be assessed in the design stage.

The conversion of hydraulic performance from model test results to prototype performance is conducted on the theoretical basis that friction loss in the prototype is reduced as Reynold's number increases. This is known as the scale effect for hydraulic machinery. For accurate assessment of this effect, the prerequisite is that the ratio of friction loss over total loss of turbines, that is the loss distribution coefficient V defined in IEC code, must be correctly known. By this loss analysis, the value of V can be ascertained and it becomes possible to estimate the amount of step–up of efficiency more correctly. The scale effects on other performance parameters such as head, discharge and power can likewise be assessed.

8.5.4 Analysis of dynamic characteristics of shaft system

The following numerical analysis for determination of critical speed of shaft systems is performed using a computer instead of using classical analytical methods

- Eigenvalue analysis calculation of critical speeds and their modes for shaft systems;
- Dynamic response of shaft system – dynamic response against hydraulic turbulence acting on the turbine runner.

For the latter, the magnitude and frequency characteristics of the hydraulic turbulence should be ascertained beforehand by measurement on prototype turbines or on models.

Approaches such as the transfer matrix method, direct integration method and finite element method are used for numerical analysis of vibration problems [8.12, 8.13]. In these analyses, the shaft system is divided into one–dimensional finite elements composed of mass, bending stiffness and damping effect. Masses which are not related to stiffness such as the generator rotor and turbine runner are treated as *added mass* to the appropriate element.

The transfer matrix method obtains solution by solving the matrix formed by transfer functions defining dynamic interaction between two elements in the shaft system. Less input data is generally required but sometimes in complicated systems difficulties may be encountered in determination of the transfer functions.

The direct numerical integration method solves the differential equations defined at the elements by step–by–step numerical integration. There are some variants of this method among which the well–known Runge–Kutta method is one. Generally, direct integration methods require longer computer time and may occasionally run into unstable solutions.

The finite element method is performed more easily as compared with the other methods. By assembling the dynamic equilibrium equations for all the finite elements of the shaft system, a global matrix system of equations is obtained. Eigenvalue analysis conducted on this matrix system gives critical speeds and their modes.

The modal analysis method [8.12, 8.13] can be performed with less computer time for obtaining response against external force like hydraulic turbulence in the turbine. Problems may be solved by modal analysis with the principle that any vibration mode of a structure can be represented by a linear combination of its eigenvalue modes. The response $\{X\}$ against steady harmonic excitation is represented by eigenvalue modes $\{\phi_r\}$ and coefficients β_r as

$$\{X\} = \sum_{r=1}^{N} \beta_r \{\phi_r\} \tag{8.4}$$

The values of β_r are coefficients called participation function which are obtained by modal analysis.

Calculation of the time history response against given time history excitation force is usually conducted in the following sequence. The response time at time (t) against unit impulse excitation at time (τ) is first calculated as $h(t-\tau)$. The unit impulse excitation is represented by a Dirac delta function which has infinite magnitude of excitation with zero bandwidth of time and its intensity as denoted by the product of magnitude and time bandwidth is unity.

Arbitrary excitation $x(t)$ may be regarded as a sum of impulses having an intensity of $x(\tau)\mathrm{d}\tau$ at time τ as shown in Figure 8.20. Therefore, the response of the shaft $y(t)$ against such arbitrary excitation is given by the following superposition or convolution integral

$$y(t) = \int_{-\infty}^{t} x(\tau) \cdot h(t-\tau) \cdot \mathrm{d}\tau \tag{8.5}$$

Figure 8.21 shows an example of modal analysis to simulate the shaft response against random hydraulic excitation which works on a turbine runner in speed–no–load operation. In the modal analysis for engineering purposes, higher eigenvalues can be practically neglected. In this example, eigenvalues up to sixth order are retained.

The magnitude of the response depends significantly on the magnitude of the damping effect in the system as well as on that of excitation force. Accordingly, the estimation of damping effect is of primary importance in the

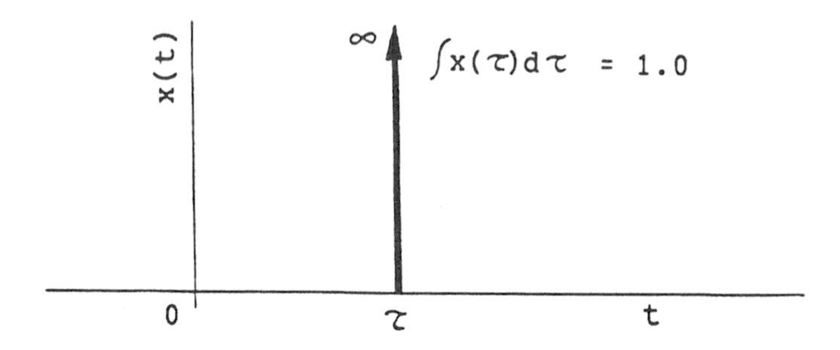

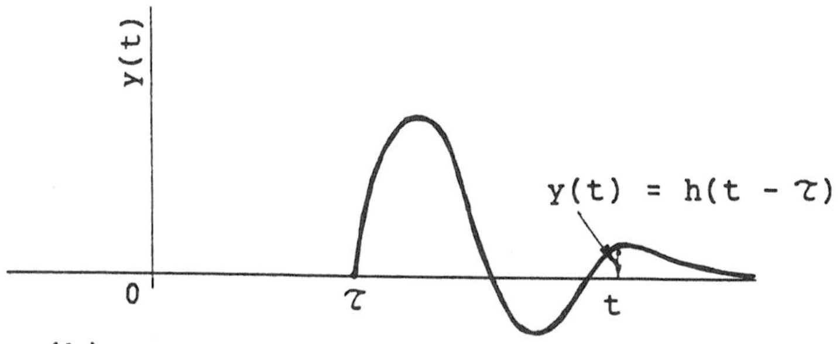

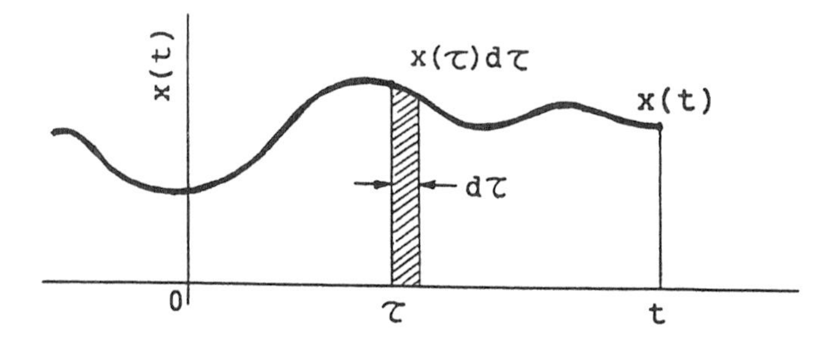

Figure 8.20 Arbitrary excitation $x(t)$ consisting of impulses

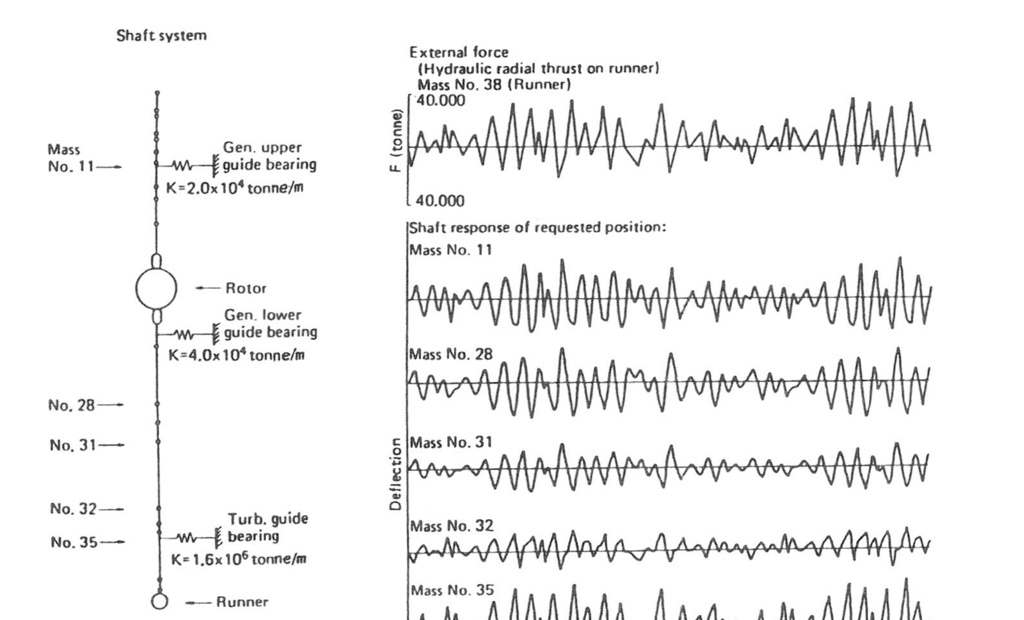

Figure 8.21 Dynamic response of shaft system against random hydraulic excitation

calculation of dynamic response although its accurate estimation is actually very difficult.

Assumptions of damping effect equivalent to $\zeta = 0.03 - 0.06$ will give reasonable response with ζ as the damping ratio. By this assumption, the damping effect for each bearing C_i can be determined by the following formula

$$C_i = 2\zeta(k_i/\omega) \tag{8.6}$$

where C_i – damping effect of $i-th$ bearing, Nsm^{-1},
k_i – stiffness of $i-th$ bearing, Nm^{-1},
ω – dominant natural frequency, $rads^{-1}$,
ζ – damping ratio $= C/2\sqrt{mk}$

8.5.5 Hydraulic transients in pipelines

Calculation of pressure variation in the pipeline and speed variation of the rotating parts in the transient process of turbine load rejection or pump input power failure are important items in the design stage of a hydroelectric power plant. For a conventional turbine connected to a relatively simple pipeline this calculation may be done by using approximation formulae which are based on linear relations of the turbine discharge and torque characteristics. For instance, pressure variation due to linear decrease of $q = Q/\sqrt{H}$ can be estimated by the well–known Allievi formula.

In case of reversible pump–turbines, discharge and torque characteristics in the operating ranges from normal turbine to reverse pump show much dependence on speed variation as shown in Figure 6.27. Also the characteristics in the ranges of normal pump and pump brake can hardly be approximated by simple curves. In these cases, pressure variation and speed variation are mutually influenced so no simple formula can deal with such complicated relations. Hence for the calculation of these transient processes, numerical simulation must be resorted to procedures such as the one in Figure 8.22. If more than one generating unit are connected to the common hydraulic conduit, the parts representing turbine and generator in Figure 8.22 will be duplicated as necessary and added to the hydraulic conduit in parallel. Water level variation in surge chambers, waterhammer in penstocks and speed variation of rotating parts are all represented in differential equation form.

The calculations to simulate time history transients are usually made by direct numerical integration methods (step–by–step integration). Since the variables are inter–related in a complicated manner in these calculations, iteration is necessary for each time step.

The values of n (speed) and Q (discharge) are initially assumed by extrapolation of past–time history. After a round of calculations for that time

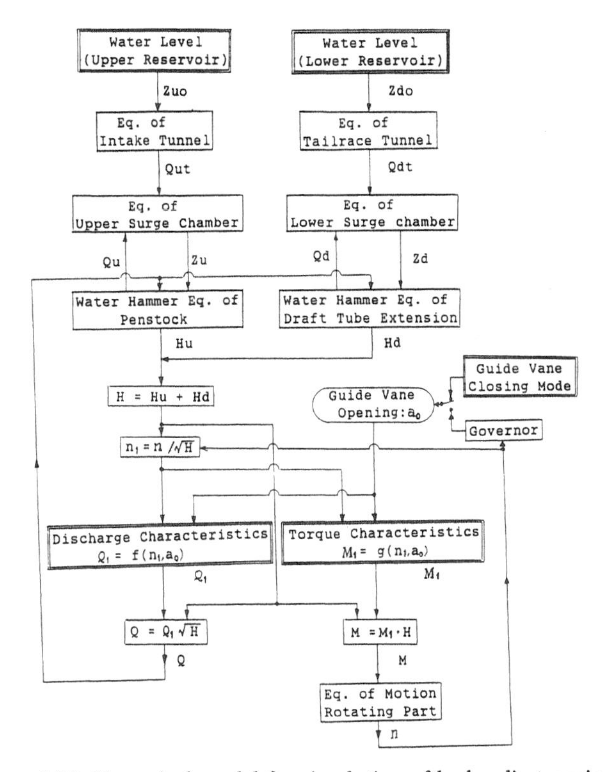

Figure 8.22 Numerical model for simulation of hydraulic transients

step, based on the assumed values, the results of n and Q are compared to the initially assumed values. The calculation is iterated to achieve satisfactory convergence. Details of calculation are referred to [8.14].

Diagrams in Figure 6.28 are examples of results of numerical simulation on turbine load rejection and pump input power failure.

8.6 CAD/CAM/CAT

8.6.1 Computer–aided design

Software packages which deal with detailed design calculations based on fundamental design parameters and produce drawings are known as computer–aided design (CAD). Many CAD systems applicable to various products have now been developed.

In the field of hydraulic turbines, the design work must cover varieties of types of turbines. Not only their ratings are different but also various special requirements and optional designs as required by purchasers have to be met. For this type of product it is difficult to make a CAD system completely automated, rather they are composed mainly of procedures involving interactive man–machine processes.

The design work which are done mostly by manufacturer's design criteria and are rarely affected by purchaser's optional requirements may be made up as fully automated systems, such as:

- Principal dimensions of flow passages.
- Flow passage configuration and blade profile of runner.
- Guide vane operating mechanism including simulation of guide vane motion.
- Servomotor design.
- Others.

Though most of the design work listed below can be done by automated process, the items indicated in brackets must be determined interactively:

- Spiral case layout (position and size of mandoor, supports, etc.).
- Draft tube liner (same as for spiral case).
- Head cover (access to bearing and shaft seal, internal piping).
- Others.

The prerequisite for making up a CAD system in any case is to standardize the design not only in design criteria but also in detail design of individual components. Figure 8.23 shows an example of the design of a Francis turbine runner. The drawing is done by semi–automated process. Figure 8.24 shows a cutting layout drawing or nesting plan drawing for a draft tube liner. This latter system first expands the three–dimensional curved surfaces to flat outlines automatically and then makes up an optimum cutting plan to minimize material waste by interactive man–machine process.

8.6.2 Computer–aided machining (CAM)

A CAM system includes the following techniques:

- Numerically controlled machining.
- Robot application.
- Factory automation.

In the field of hydraulic turbine manufacturing, the first two procedures are now widely used while factory automation is still too difficult to be put

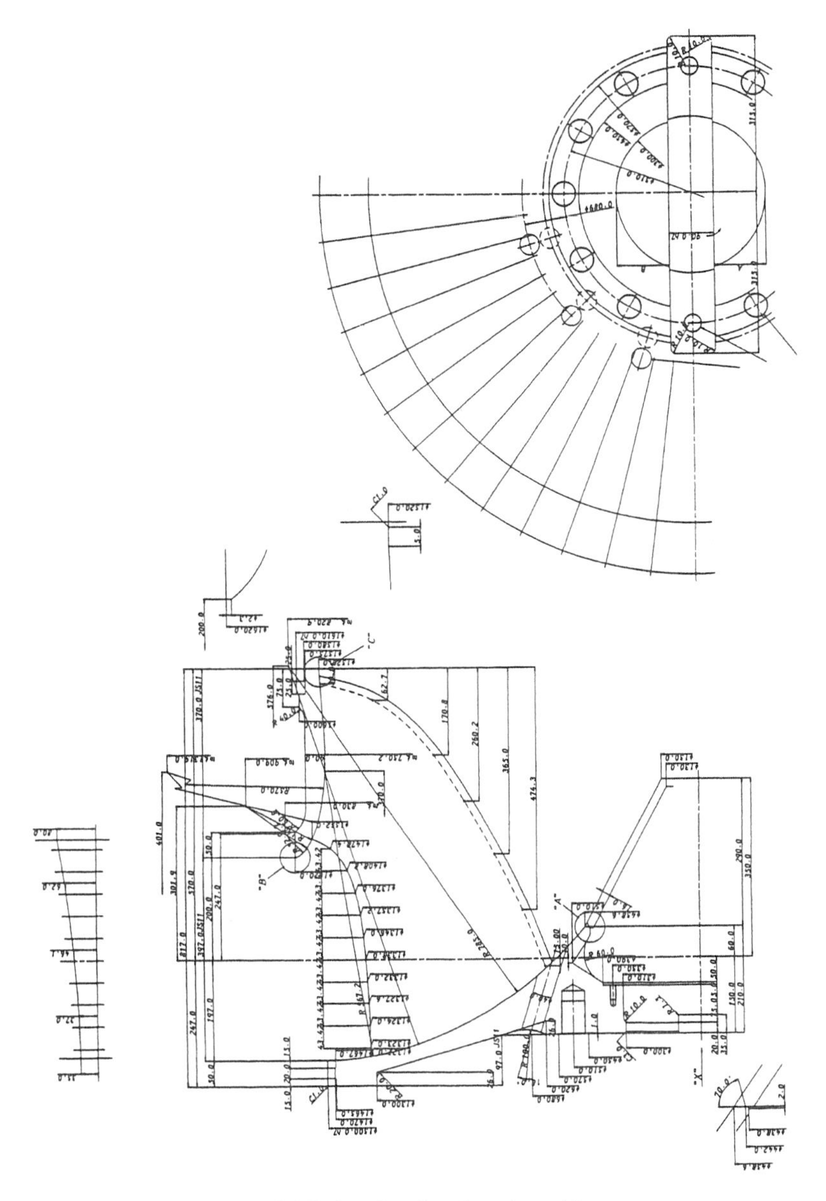

Figure 8.23 CAD drawing for Francis turbine runner

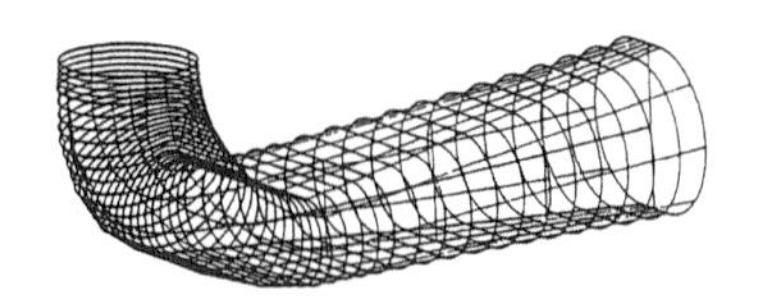

(a) Modelling of a draft tube liner

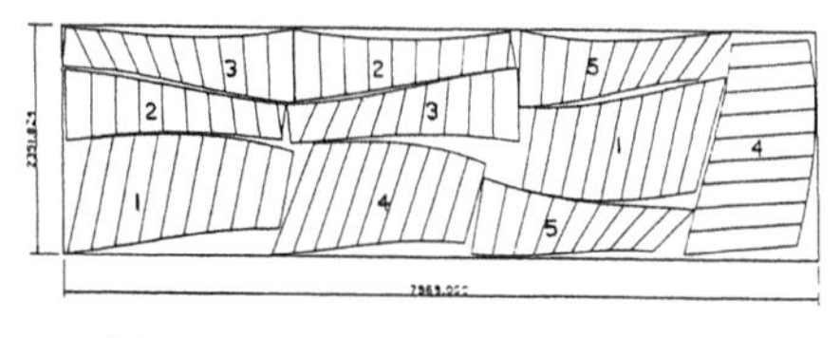

(b) Plate cutting plan (nesting)

Figure 8.24 CAD layout of draft tube liner

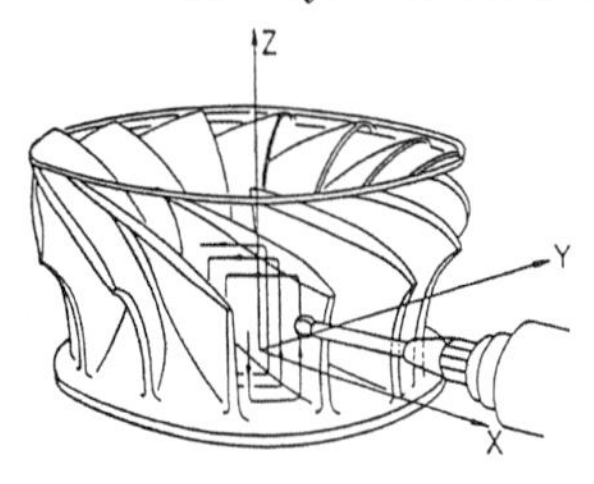

(a) Francis turbine runner

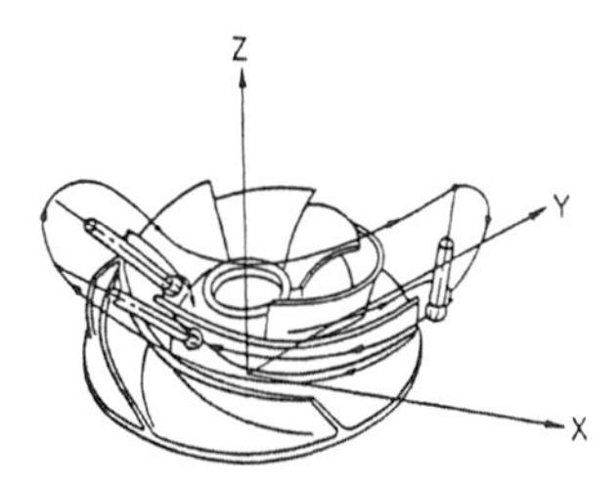

(b) Pump-turbine runner

Figure 8.25 Numercially controlled machining of model turbine runner

into practice.

There are a number of components of hydraulic turbines that have two– or three–dimensional curved surfaces, the manufacturing of which are handled by the very useful and powerful numerically controlled technology. The following are in use at the present:

- Cutting of steel plates for spiral case, draft tube and others.
- Machining of curved surfaces such as Francis runners, Pelton buckets, Kaplan runner blades and hubs, guide vanes, etc.
- Ordinary machining such as drilling, boring, cutting, milling as well as machining by so–called machining centre equipped with automatic tool changer.
- Others.

Figure 8.25 shows an application of numerically controlled machining to model turbine runners. The five–axis milling cutter is constantly approaching the curved surfaces of the model runner from the optimum angle for cutting while the tool holder is controlled from interferring with the adjacent runner blades.
Robots are also in use in various parts of the manufacturing. Examples are set out as follows:

- Cutting and bevelling of plate (S).
- Welding of plate structures (G), (S).
- Grinding of runners (S).
- Grinding of guide vanes (G).

where (G) indicates robots for general use and (S) means special robots designed for the specific work.

8.6.3 Computer–aided testing (CAT)

CAT systems used in the field of hydraulic turbines may comprise of:

- *Laboratory automation for model testing of hydraulic turbines*
  Most of present–day laboratories are equipped with automated instrumentation and data acquisition systems, some of which are furnished with automatic operation control.

- *Balancing test of runner*
  Most of the balancing test rigs are equipped with a computer for data processing which is able to output the amount and location of residual unbalance instantly.

- *Dimension check*
  Three–dimensional checking of the runner blade surface may be done

by three–dimensional measuring machines using either needle pointer or laser beam that are equipped with data processing equipment.

- *Non–destructive examination*
  Some of the test facilities are equipped with a computer for data and image processing that gives indication of the shape and location of the defects on graphic display and storing in memory for record.

- *Vibration analysis*
  Real–time analysis by Fast Fourier Transform (FFT) analyser is often used in field measurements of vibration which gives immediate indication of frequency spectrum or coherence and phase difference between two events.

- *Field efficiency tests*
  Field efficiency tests, with discharge measurement by, the acoustic method, the pressure time method, the current meter method, the thermodynamic method, etc. are all partly or fully automated with the aid of portable computers. Tests results are immediately displayed at site.

References

8.1 Duan, C. and Tanaka, H. 'Comparative Silt Erosion Test of Various Materials Using a Prototype Turbine', Proc. 2nd China–Japan Joint Conf. on Fluid Mach., Oct., 1987.

8.2 ASME Boiler and Pressure Vessel Code, Section XI, Rules for Inservice Inspection of Nuclear Power Plant, Appendix A, 1938 ed.

8.3 Shah, R. C. and Kobayashi, A. S. 'Physical Problems and Computational Solution of the Surface Cracks', ASME, 79, 1972.

8.4 Zienkiewicz, O. C. 'The Finite Element Method in Engineering Science', 2nd edn., McGraw–Hill, London, 1971.

8.5 Argyris, J. H. 'Recent Advances in Matrix Methods of Structural Analysis', Pergamon Press, Elmsford, New York, 1963.

8.6 Banerjee, P. K. et al. 'Developments in Boundary Element Methods,, Vol. 1, 1979, Vol. 2, 1982, Vol. 3, 1984, Applied Science Pub.

8.7 Hartmann, F. ed. by Brebbia C. A. 'New Developments in Boundary Element Methods', CML Publications, 1980.

8.8 Richtmyer, R. D. and Molton, K. W. 'Difference Methods of Initial–Value Problems', 2nd edn., Interscience Publishers, New York, 1967.

8.9 Roache, P. J. 'Computational Fluid Dynamics', Hermosa Publishers, Albuquerque, N.M., 1972.

8.10 'Numerical Grid Generation Techniques', NASA CP 2166, 1980.

8.11 Thompson, J. F. 'Numerical Grid Generation', North Holland, 1982.

8.12 Pestel, E. C. and Leckie, F. A. 'Matrix Methods in Elastomechanics', McGraw–Hill, New York, 1963.

8.13 Hurty, W. C. and Rubinstein, M. F. 'Dynamics of Structures', Prentice Hall, Englewood Cliffs, N.J., 1967.

8.14 Wylie, E. B. and Streeter, V. L. 'Fluid Transients', McGraw–Hill, New York, 1978.

Chapter 9

Manufacture of Hydraulic Turbines

Zheng Ben–Ying

9.1 Special Features of Hydraulic Turbine Manufacture

Hydraulic turbines are the kind of machinery that are manufactured singly or in small batches. They are characterised by the following features:

1. The parameters of hydraulic turbines are different in each hydroelectric power application, the structural design and production procedures including the making of castings and machining of individual parts may vary in each case.

2. For large pressure sustaining parts, high quality structural steels must be used; for parts that are to resist cavitation and erosion various grades of special steels are necessary.

3. The castings for certain component parts of large hydraulic turbines are quite large and heavy, their design is often limited by the capacity of the foundry, machine tools and transportation facilities. Therefore, coordination between firms is essential. Also due to their large sizes, some turbine parts have to be assembled and welded at the station site.

4. Conventional machine tools become inefficient when used to machine large component parts of turbines, or are not capable of producing the required degree of accuracy. Special purpose machine tools must then be produced, or special machining processes devised.

5. Certain component parts have intricate shapes that require complicated manufacturing procedures, some even involve large amounts of

inefficient manual labour, consequently the production cycle is usually fairly long.

6. Because of the dimension of the turbine parts, special large size measuring tools of high accuracy, including digital instruments, must be provided to assure accurate machining.

To modernise the manufacturing of hydraulic turbines, it is essential to improve the production techniques and quality of workmanship, closely following the development trends of the general machinery trade. Important measures may be: (a) the use of new steel grades with superior overall properties and durability; (b) the elimination or reduction of machining after fabrication and factory assembly; (c) the use of high precision working methods such as numerically controlled cutting and machining; (d) the improvement of installation, welding and machining methods at power station sites.

9.1.1 Production features of Francis turbines

For turbines of higher than $100m$ head, runners are usually made of martensitic stainless steel because of its strength and cavitation resistance. Runners for high head Francis turbines traditionally made from steel castings are now produced more often in fabricated (cast–weld) constructions. Runner blades for these turbines generally have limited curvature and less variation in thickness which make them adaptable to forming by hot pressing. Bands of such runners may be made by rolled steel plates or welding from cast parts.

Runners for turbines of lower head are usually made of carbon steel or low–alloy steel castings, some have stainless steel overlays applied at positions where cavitation is likely to occur. Many runners are welded from dissimilar materials such as blades of martensitic stainless steel with crown and band of carbon steel.

Runners larger than five metres in diameter are sometimes made in segments in order to meet transportation restrictions and are then welded together at the station site. For very large size runners, there have been successful cases where the runner blades, the crown and band are all welded at site.

For Francis turbines operating at around $40m$ head, welded spiral cases are generally used. With medium and small size turbines, some spiral segments are welded to the stay ring, up to transportation limitation, while other segments are later welded at site. With large turbines, all segments are transported separately and welded to the stay ring at site. With the increase of machine size and capacity, high strength steels are used with the purpose of reducing the consumption of material and welding volume.

Stay rings are often made with parallel upper and lower rings instead of the conventional flared rims mainly for ease of fabrication. For split stay rings, the massive flanges for bolts are now less common, instead, direct welding is applied to joint the segments. Special provisions are made in the design to help eliminate the necessity of machining after welding.

Main shafts for large turbines are now being made from rolled plates of considerable thickness. This fabrication technique is quite representative of the present trend of development in the production of large turbine parts.

Many other new production methods have been introduced to enhance the technique in manufacture of turbine parts, such as the elimination of simultaneous boring of head cover and bottom ring, the use of high precision rotary cutters to machine guide vanes (wicket gates), the use of pressings to form runner blades and the use of electroslag welding of runners and other heavy parts.

9.1.2 Production features of Axial–flow turbines

There are two main types of axial–flow turbines, the ones with adjustable blades (Kaplan) and the ones with fixed blades (propeller). The latter kind of construction is mostly used in small size turbines and are comparatively less complicated in manufacture.

Runner blades of Kaplan turbines are traditionally made of castings, but such techniques cannot always assure uniform material properties and strength throughout the blade, due to the variation of blade thickness. Casting alone is not able to produce the blade profiles up to the specified accuracy, so generous allowances must be made on the blade surfaces for subsequent milling or grinding. In recent years, cast blade blanks are being finished by numerically controlled machine tools. A number of attempts have been made to form runner blades completely from welded steel plates.

As the size of turbines grow larger, the runner hubs are becoming more difficult to cast and create problems in subsequent machining. Endeavours to replace casting with welding or cast–weld structures are being made in several countries. One successful example is the welding with electroslag and submerged–arc automatic welding of a large runner hub of over $3m$ in diameter (9.2.2).

Welded construction for runner chambers is generally accepted as common practice. In order to resist cavitation attack, a stainless steel middle section is sometimes put in between the top and bottom carbon steel sections. A new technique is the use of composite steel plates initiated in the USSR that allows no machining after welding and maintains very good clearance with the runner (9.5.3).

9.1.3 Production features of tubular turbines

Large size horizontal shaft tubular turbines, though highly efficient in performance, bring problems to manufacture because of their large size and complicated structure.

The stay ring of a tubular turbine with runner diameter of $7.5m$ is so large it has to be split into eight segments that makes machining and heat treatment difficult. One solution to this is to divide the stay ring into an inner ring and an outer ring. The inner ring is made as an ordinary machined part while the outer ring is checked only for dimensions during factory assembly. The guide vanes are then machined to heights according to the space provided by the outer ring.

Conical guide apparatus present challenges to machining and fitting, such as the finishing and inspection of the guide vanes with their spherical surfaces, the sealing between guide vanes, machining of the spherical inner and outer rings and the simultaneous reaming of guide vanes bores.

9.1.4 Production features of impulse turbines

For an impulse (Pelton) turbine, the major problem is in the manufacture of the runner and the distributor piping.

The Pelton runner can be either integrally cast or cast–welded. The shape of the buckets, their surface finish and positioning are all subject to very stringent tolerances. For integrally cast runners, great reliance is placed on manual grinding and fitting to obtain the accuracy in dimension and finish. In case of cast–weld runners, the buckets are individually cast and may be processed by various means to greater dimensional accuracy and surface finish. Also cracks in the roots of buckets are avoided. However, this will entail much larger amounts of welding work and will require complicated fixtures to guarantee a constant value for the blade pitches.

Distributor pipes must assure that the discharge from the several nozzles falls into one plane with their centre lines tangential to the pitch circle of the buckets. Some pipes are cast in segments and joined by flanges, others are welded at site from pipe segments.

9.2 Manufacturing Technique of Turbine Runners

Structural design and manufacturing techniques vary greatly from different types of machines and their runners. Only general descriptions can be given here on the representative types, the Francis runner, the Kaplan runner and the Pelton runner.

9.2.1 Runners for Francis turbines

(i) Fabricated runners

Runners welded from cast or formed parts are called welded runners or cast–weld runners, and sometimes fabricated runners. Runners made this way have the advantage of higher precision in the profiles and dimension control, possibility of machining of flow surfaces as well as the use of dissimilar materials for runner blades and crown/band. For runners that are too large for production in one piece in the factory this is the only way of fabrication.

Some large turbine runners are completely welded at the station site. An outstanding example of this is the large Francis runners (diameter nearly $10m$) for Grand Coulee III in the USA [9.3]. There are two types of turbines at Grand Coulee, the $600MW$ and $700MW$ units, which have runners fabricated in different ways. The former has stubs cast on the crown and band to enable butt welding with the blades, while in the latter case the blades butt directly against the crown and band to form T–joints, as shown respectively in Figures 9.1 and 9.2.

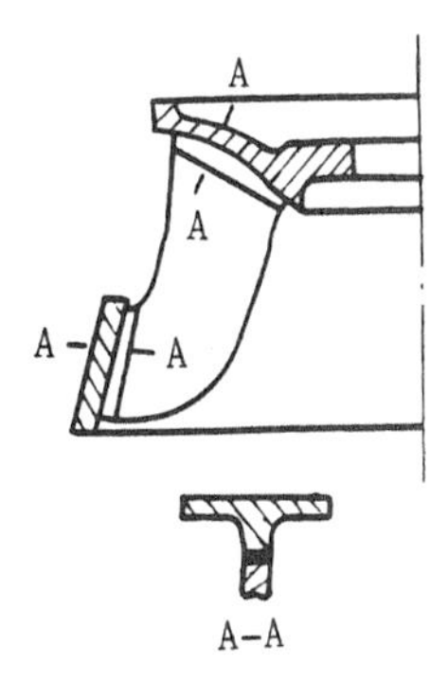

Figure 9.1 Butt welding of blades to crown and band [9.4]

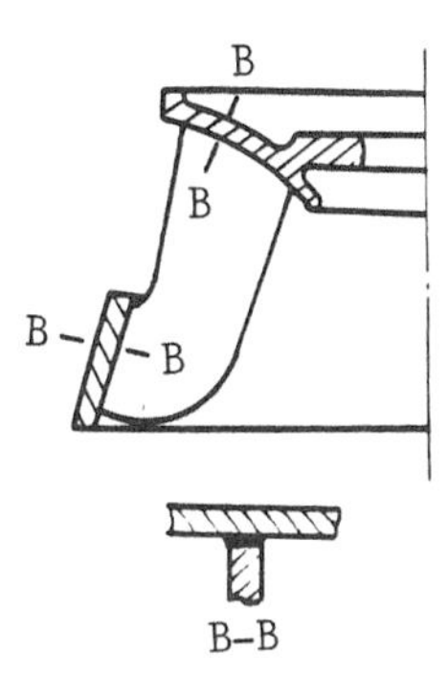

Figure 9.2 T–welding of blades to crown and band [9.4]

To meet the requirements of transportation, large runners often have to be split into parts. There are many ways of splitting a Francis runner. With some manufacturers, the runner is welded complete in the shop and then cut into three large pieces for transportation and finally welded together at the station site as is the case shown in Figure 9.3 (a). There are other instances where the blades are welded to each of the halved runner bands and finally welded together with the crown at site, such as Figure 9.3 (b).

For high head runners with long blades, the runner may be cut into

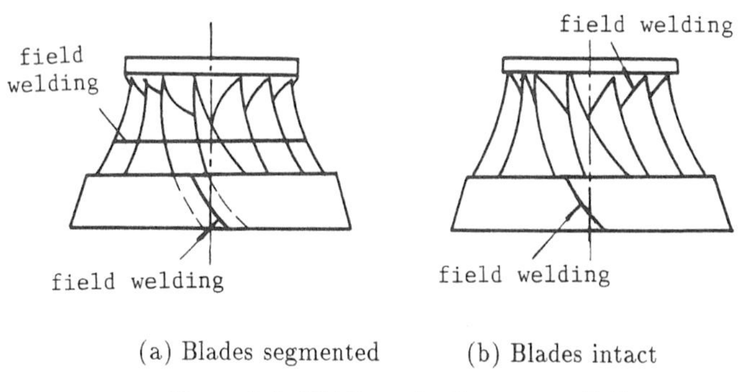

(a) Blades segmented (b) Blades intact

Figure 9.3 Welding of split runners [9.5]

two concentric pieces so as to facilitate subsequent welding and grinding, as shown in Figure 9.4. For low head runners where there are more blades with smaller wrap angles, the splitting is often made in the radial direction. The segmented crown is normally joined by bolting and the band either by welding (Figure 9.5) or by sunken–head bolts inserted in the tangential direction (Figure 9.6). The machining of runners with joints is comparatively difficult so the current practice is to use welding as much as possible.

For radially split runners, when the number of blades is even, the parting lines through the crown are invariably diametrically opposed. The parting lines through the band are so aligned as to keep all blades intact so they must be in a skewed plane, as shown in Figure 9.3. For runners with odd numbers of blades, two or more blades have to be cut. The Longyang turbines in China ($320MW$) which have a runner diameter of $6m$ and 17 blades are split $250mm$ off centre into two unequal parts in order to limit the number of blades that have to be cut into two [9.7].

The crowns of welded runners are usually made of steel castings while the runner blades and the bands are either cast or formed from steel plates. When the batch of production is large and the variation of blade thickness is slight, it pays to use pressed blades from steel plates or cast blanks. For blades with greater variation in thickness, a two–stage pressing process may be used where the steel blank is first pressed into an intermediate curved shape and then machined by planing or milling the excess metal on the concave side. A second hot pressing will shape the blank into the desired profile (Figure 9.7) which requires very little further finishing. When using hot pressing to form runner blades, large hydraulic pressure (water or oil actuated) are necessary. Presses of $4{,}500tonne$ capacity are sufficient to

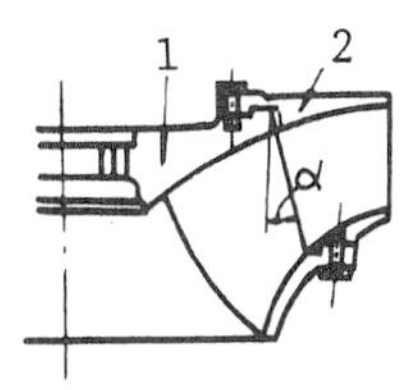

Figure 9.4 Concentrically split runner

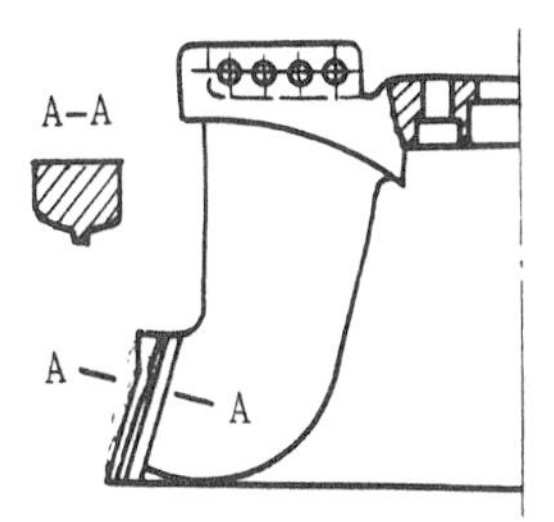

Figure 9.5 Radially split runner–joining of band by welding

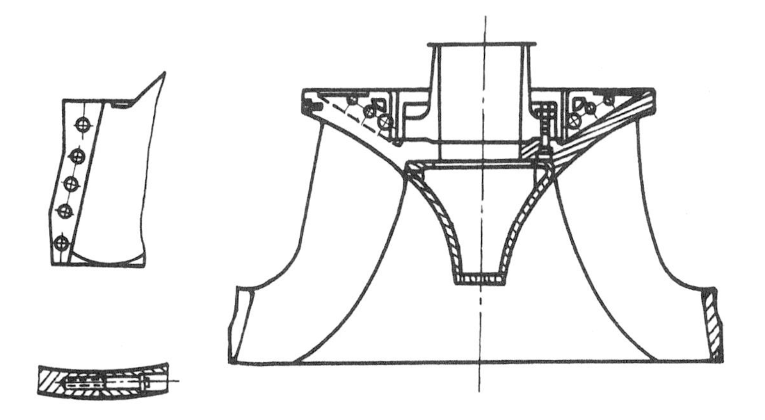

Figure 9.6 Radially split runner – joining of band by bolts

form blades for most Francis turbines [9.1, 9.8].

Numerically controlled milling machines are becoming more widely used to machine pressed or cast blade blanks such as the example shown in Figure 9.8.

(ii) Integrally cast runners

For many years, integrally cast runners are mainly used in medium/small turbines, but the size of such runners is increasing steadily with the advancement of casting techniques, up to now stainless steel runners of over $7m$ have been successfully cast. An outstanding example is the runners for Guri II in Venezuela cast by the Kobe Steel Works in Japan. Typical specifications for integrally cast runners are:

(a) Dimension tolerances:

- Spacing of blades: 3% of blade pitch.
- Thickness of blade: 10–15% of design value.
- Angular variation of blade: 1° - 1.5^o.
- Unevenness of surface: 3–$5mm$ in length of $200mm$ [9.1].

(b) Defects:

- No hot cracks, shrinkage cavity and inclusions allowed in joints between blades and crown/band.
- No cold shut allowed over the whole flow surface, especially in the blades and the transition sections from blade to crown/band.

When castings for runners meet the above requirements, they have to go through the very laborious process of cleaning and grinding in addition to machining. These are some of the reasons why the use of integrally cast runners has become somewhat limited.

(iii) Welding of runners

The assembly of runner parts and the welding technique determine, to a great extent, the quality of the machine and the required fabrication time cycle. The alignment of blades to crown and band is done in various ways with different manufacturers, one of the two common techniques is the upright method where the blades are first fixed to the band and then have the crown mounted after the upper edges of the blades are trimmed. The other is called the inverted method where the blades are first aligned with the crown (placed upside down) and then have the band mounted. In cases where it is necessary, the half assembled runner is stress relieved before final welding, hence there are the terms one–step or two–step assembly.

There are two essential ways of welding the runner, by electric–arc and

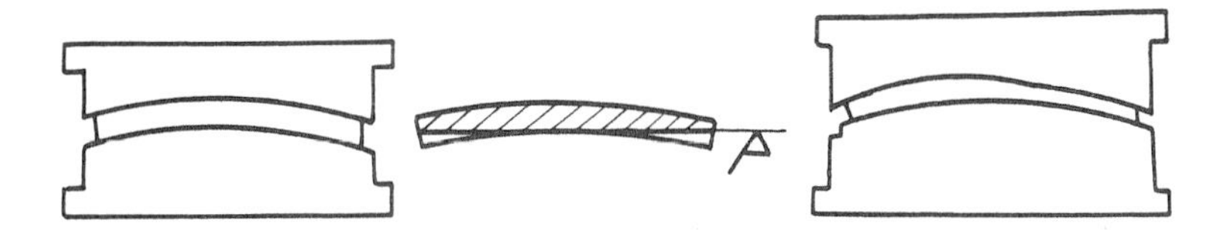

Figure 9.7 Pressing of runner blades from steel plates

Figure 9.8 Francis runner blade machined by computer numerically controlled milling cutter (Photograph courtesy Sulzer Escher Wyss)

by electroslag welding (consumable–guide ESW). The gas–shielded welding has largely replaced hand welding for both higher efficiency and quality. Semi–automatic carbon dioxide shielded welding is said to have three times the efficiency of hand welding and reduce the cost of welding by over 60%. It requires lower preheat temperature and reduces welding stresses. Argon (or argon + carbon dioxide) shielded welding is used in joining stainless steel parts or parts made of dissimilar materials.

Electroslag welding has been developed for thirty years in hydraulic turbine production and is now the most advanced technique in making large runners. It is the only means to bring about automatic fabrication since it is able to weld in one pass, joints of any shape and depths up to several hundred millimetres.

When welding together turbine runner parts of carbon steel and martensitic stainless steel, aside from the common rules in welding practice, attention should be paid to the properties of materials such as the use of high nickel stainless steel to prevent cracks after welding. In some cases, high nickel material 8–10*mm* thick is first built up on carbon steel parts before welding to stainless steel. This method of using a transition layer, though able to produce good welds, makes firstly the alignment difficult because the overlay thickness cannot be controlled very accurately, then it may bring about additional deformation of the parts, and lastly the overlay requires additional grinding and dressing that would prolong the production cycle. The Dongfang Electrical Machinery Works (DFEM) in China has produced welded runners for nearly twenty years without this transition layer.

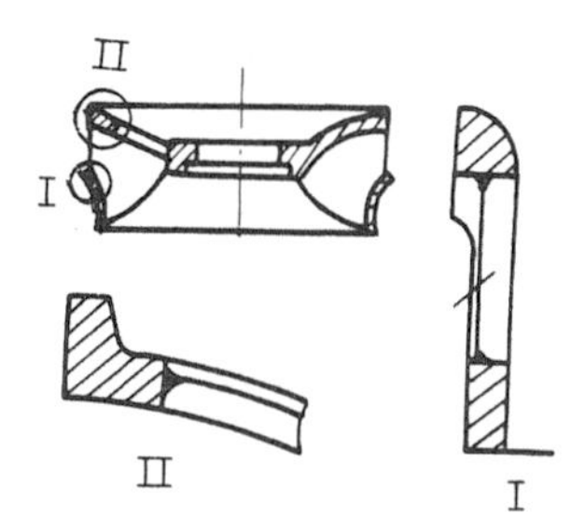

Figure 9.9 Welding scheme for blades with tenons [9.12]

For high head Francis runners, due to the long blade lengths and small passage heights, the accessibility for welding is very poor, especially along the joints between the blades and the band. Special measures such as that shown in Figure 9.9 are taken to overcome this handicap. Slots accommodating the tenons built up on the blades are milled out in the crown and band. After insertion of the tenons which have bevels for welding into the slots, joints inaccessible on the inside can be reached from the outside [9.10].

For very high head runners, the band may be made from two concentric rings as shown in Figure 9.10. The outer (upper) band is first welded to the blades and the crown under more favourable conditions. After machining of the lower ends of the blades the inner (lower) band is joined to the assembly. The weld length along the joints between the blades and each band is thus reduced to about one–half of the total length.

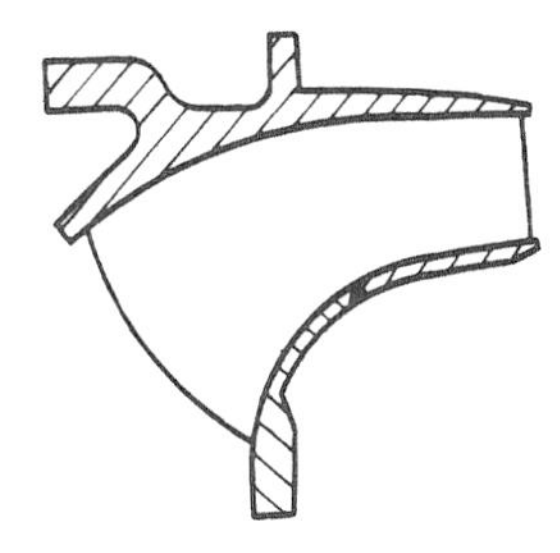

Figure 9.10 Welding scheme for runner with two–piece band

In the assembly of split runners, the welding of the band must be strictly controlled in order not to cause any deformation of the main shaft flange on the crown since it cannot be re–machined. The two halves of the flange must have no upward bulging, but a depression of less than $0.05mm$ is allowed.

Runner bands to be welded in the field may be preheated by portable electric heaters to a temperature of $100 - 150°C$. The heater can be used again for local stress relieving after welding.[1]

Then a specially designed rotating grinding device is used to true the labyrinth ring. As the deformation here is usually not large, less than $0.8mm$, the trueing operation is relatively light [9.1, 9.6].

Application of the electroslag welding requires a higher level of technique and longer preparation time. As it needs greater space to operate, only runners with diameters larger than $5m$ are justified to be welded this way. The larger the runner, the easier is the operation, so is the effectiveness of welding. This is because the welding is done in one run regardless of the cross

[1] According to reports from the USSR, if the band joint is welded by austenitic electrodes in narrow passes (each pass not wider than 2.5 times electrode diameter and thickness less than $5mm$) and the temperature kept to $100 - 120°C$, then local stress relieving will not be required [9.12].

section of the weld. But electroslag welding will cause greater deformation to the runner due to the larger amount of heat generated.

One thing peculiar to the electroslag welding is that the joints must be in a near vertical position and welding is done from the lower end up. Welding positioners or specially designed rotating fixtures that are required to continually adjust the position of the parts to obtain equal fusion depth on both sides of the weld [9.11, 13, 14, 16].

Well known examples of large runners welded by electroslag are the ones at Grand Coulee III in the USA and at Itaipu in Brazil–Paraguay. Both types of runners are made of carbon steel and welded by electroslag, the former were assembled, welded, heat treated and finished all at the station site while the latter were made complete in the factory and transported to the site by highway [9.11]. Typical quality control for welded runners are as follows [9.1]:

1. For castings to be used on runners, no cracks or chain–like defects are allowed. Other defects smaller than $5mm$ and spaced greater than $50mm$ apart are permitted, so are accumulated minor inclusions with a total area less than $20cm^2$ and spacing between groups greater than $100m$.

2. For steel blanks, scales smaller than 50 x $50mm$ and spaced more than $500mm$ apart may be allowed.

3. No cracks or chain–like defects are allowed in runner welds as inspected by ultrasonic and magnetic particle methods; only defects smaller than $3mm$ spaced greater than $10mm$ apart or of sizes smaller than $5mm$ spaced greater than $50mm$ apart are permitted with the condition that there are less than five such defects in each metre length of weld.

(iv) Machining of runners and static balancing

The machining of assembled runners is comparatively simple because of its regular shape. Vertical lathes and boring machines are the basic tools used. The flange surface on the crown and bolt holes require higher degrees of accuracy than other parts. The flange surface must be ground after finish turning to a run–out of less than $0.02mm$. The bolt holes when bored singly must have an allowance 3–$5mm$ for reaming unless they are centred and worked by numerically controlled machines. When the runner is being bored simultaneously with the main shaft, the flange surface of the crown must be adjusted to within $0.05mm$ from the vertical. Boring jigs of $150m$ depth may be used in place of numerically controlled machining which can have an accuracy of 0.01–$0.02mm$ and would save the simultaneous reaming.

Holes to fit dowel pins must be reamed at the same time with the main shaft holes.

For very large and heavy runners, that are greater than 200t in weight, it is difficult to align it with the shaft in a horizontal position. So the bolt holes are often reamed initially by a vertical boring mill, then a special-purpose tool is employed to bore the bolt holes together with the main shaft in the vertical position.

Only the blade inlet edges of integrally cast runners require trimming and milling. Blades for welded runners are dressed to the desired size and shape before welding and generally required no further work on the vertical lathe.

Francis runners whether cast or welded in the factory or fabricated at site inevitably require static balancing to assure stability in operation. Runners lighter than 15t are usually balanced horizontally on parallel rails, others are balanced on vertical rigs utilizing a hardened sphere for support, as shown in Figure 9.11. For high head medium/small turbines, the runners are sometimes checked for dynamic balancing with the complete rotating system.

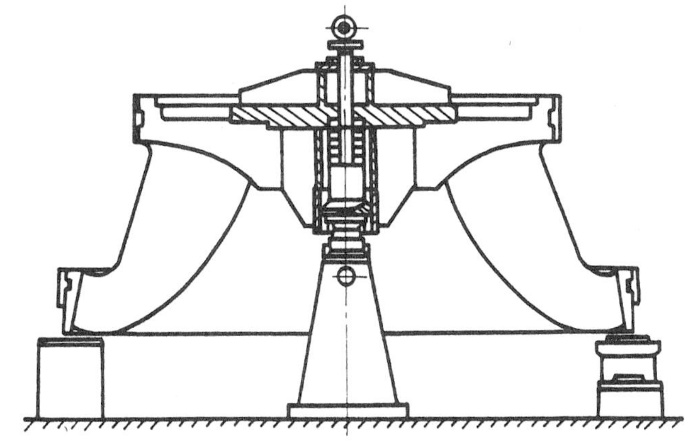

Figure 9.11 Static balancing of runner in vertical position [9.1]

For the very large runners at Grand Coulee III (weight 450t), where support by a sphere was impossible, the runner was placed on four specially made hydraulic jacks located on two centrelines perpendicular to each other.

The runner was raised alternatingly by two opposite jacks and the required weight for balancing recorded. The amount and location of unbalance was determined after repeated trials [9.4].

9.2.2 Runners for axial–flow turbines

(i) Manufacturing of runner blades

Fixed blade propeller turbines usually have the runner blades and the hub cast separately and assembled by direct welding. A novel way of manufacture is to cast a radial segment of the hub together with one complete blade; after the contact surfaces are machined, all segments are assembled into one complete runner and secured by a cold shrunk collar. Some propeller runners have blades adjustable when the machine is at standstill, the blades are then made similar to Kaplan blades but are secured to the hub by a number of bolts.

Kaplan blades are traditionally cast because of their complicated shape, stainless steel is often used for resistance to cavitation. As these blades are separate parts from the hub, they may be cast from martensitic stainless steel of low nickel content both for economy and ease of casting.

The stem of a blade may be cast integral with the blade or as a separate piece and fastened to the blade by bolts. The composite construction has the advantage of allowing the use of different materials for the blade and the stem, for instance, stainless steel for the former and carbon steel for the latter (see 3.2.2).

The runner blades are generally cast in a sand mould in the upright position with the flange facing up. The cast has to be of such quality to guarantee high accuracy of blade profile and sound internal structure and mechanical properties. Samples are taken from the blade flanges for checking the mechanical properties of large sections. Ultrasonic inspection is generally specified along with magnetic particle detection.

Various types of copying type tools have been used to machine the curved surfaces of runner blades. Now numerically controlled milling machines are able to shape the blade profiles to a high degree of accuracy and good surface finish (Figure 9.12).

Because of the difficulty of casting very large and heavy turbine blades, attempts have been made to fabricate blades from steel plates. In the USSR, the six runner blades of a $9500mm$ diameter $178MW$ Kaplan turbine, each $3m$ by $5m$, were welded from cast steel pieces (leading and trailing edges of blade and blade flange) and plate steel (pressure and suction faces of blade). A lattice frame is formed inside the blade to strengthen the face plates, as shown in Figure 9.13. To prevent the blade from distorting in welding, the blade frame is first mounted to a fixture for the welding of one side, and is tempered together with the fixture. The frame is removed after cooling and fixed again for welding the other side. Blades fabricated this way is lighter by $15t$ than cast blades and is produced in half the time. Most parts of the blade surface are within $10mm$ of the design profile as checked by templates

[9.1].

Runner blades for medium/small turbines are normally checked by composite templates (cage templates). But for large blades this will become inconvenient or impossible because of the deformation of the templates themselves, so checking by coordinate methods are mostly used.

(ii) Manufacturing of runner hub

Runner hubs are the most massive and complicated of all parts of a Kaplan turbine. They are traditionally cast in carbon steel and machined on vertical lathes. Kaplan runner hubs have a spherical section at the centreline of the blade bores. This spherical contour may be machined by electric copying machines from templates or by numerically controlled vertical lathes. The blade bores are machined on horizontal boring machines or by milling cutters of vertical lathes. As soon as the bore is finished, a chilled bronze bushing for the blade stem is inserted into the hole and then machined at the same setting to finish dimensions. This operation will guarantee the maximum degree of concentricity and avoid the tideous work of fitting by hand.

The size of hubs increase with the growing capacity of Kaplan turbines. It comes to a point where the casting and handling of large hubs (over $100t$ in weight) have become a problem in manufacture. One solution of this is to use a welded construction as in the case of the $140MW$ Ice Harbour turbines [9.17]. The all-welded hubs have diameters of $3m$ and heights of $2.5m$ and accommodate six blades each. The upper circular plates (each made in two halves) and the hexagonal body are all made of high quality remelt steel plates $330mm$ in thickness and electroslag welded (Figure 9.14). A cost saving of 15% was attained by the change to welded construction.

(iii) Preassembly and balancing

Runners of Kaplan turbines must be preassembled in the shop to check the motion of blades, the relationship between blade angle and opening, and also leakage from blade seals with the blades rotated to all angles. Whenever possible, the runner assembly should be transported to the station site intact.

The main steps of runner assembly is as follows:

1. Hydrostatic testing of runner hub with all openings sealed by temporary bulkheads and plugs. Cast steel hubs are tested to 1.5 times working pressure and welded hubs to 1.2 times for 15–30 minutes.

2. Fitting of bronze bushings chilled in liquid nitrogen into blade stem bores. The amount of nitrogen used must be controlled to 4–5 times the volume of the bushing to prevent cracking and deformation.

3. The runner blades are fitted to their respective bores considering the

Figure 9.12 Kaplan runner blade machined by gantry type five–axis numerically controlled miller (Photography courtesy J. M. Voith)

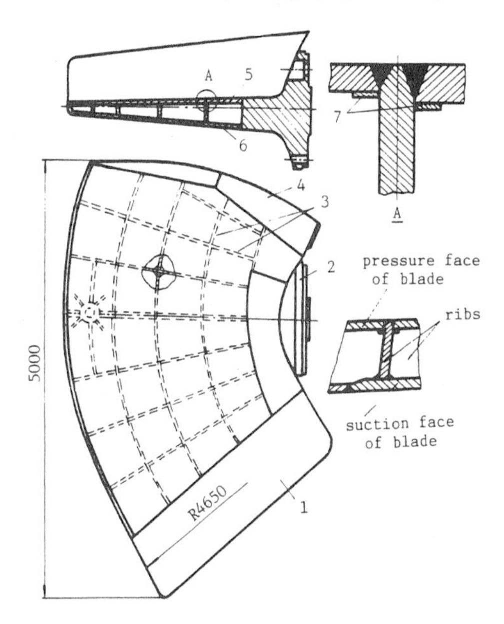

Figure 9.13 Welded runner blade made up of cast parts and steel plates [9.1]

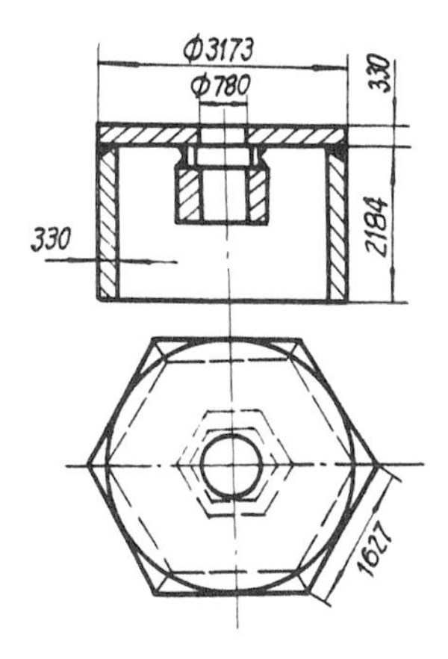

Figure 9.14 Schematic layout of all-welded runner hub [9.2]

static moment they create and the unbalance of the blades themselves. Each blade stem is fitted with a counter-weight when being inserted into the hub.

4. In order to reduce the amount of balance weight for the runner assembly, the blades, hub, bottom cover, draft cone, servomotor piston and cover should all be balanced individually first and corrected as much as possible.

9.2.3 Runners for Pelton turbines

Pelton turbine runners are subjected to very large pulsating loads during operation, and the runner buckets have not only to be precise in shape but also sturdy in construction. Pelton runners may be integrally cast or with buckets individually cast and then welded to the runner discs. Previously individually cast buckets were secured to the runner by bolting but this construction is seldom used today. Better accuracy in casting may be obtained from individually cast buckets than from integrally cast wheels. High strength martensitic stainless steel, such as OCr13Ni6Mo, are often used to make integral or welded runners, while in some countries stainless steels of low Ni content, such as OCr12NiCu are used.

Accuracy required for integrally cast Pelton runners (pitch diameter 2–2.5m) may be in the order of:

1. Working surfaces of buckets to within 5mm of design profile when

checked with solid templates;

2. Displacement of buckets to be less than $\pm 2mm$;

3. Pitch between buckets to be within $\pm 3mm$.

The inside surfaces of buckets are the most difficult to machine, previously electrochemical processes have been used to finished the buckets. This technique is fairly efficient, being able to bring the work time per bucket to several hours and attain surface finish to 2.5 (ISO) [9.1].

For a fabricated Pelton runner, the buckets are fixed to the runner wheel in selected positions according to their weights, and are then welded to the runner each in the sequence $N_3 - N_4 - N_3 - N_4 - N_1 - N_3 - N_2 - N_5 - N_6$ as labelled in Figure 9.15.

With the development of numerically controlled machining techniques, complicated curved surfaces of Pelton runners can be finished to higher degrees of accuracy. This gives an impetus to the use of integrally cast runners which may have many advantages over the welded construction (see 5.2.1). Figure 9.16 shows such a runner being machined by a numerically controlled milling cutter.

9.3 Manufacturing of Main Shafts

Manufacturing of main shafts for hydraulic turbines fall mainly into two categories: forging and welding. Figure 9.17 shows a typical construction of a forged shaft. The shaft may have an attached sleeve (a) made of stainless steel for use with water–lubricated guide bearing, or it may have a skirt (b) for use with oil–lubricated guide bearing.

A welded shaft may be made up of a forged barrel welded to forged or cast flanges, or it may be made up of a rolled steel hollow shaft with welded flanges. For very large shafts, the inside of the barrel is quite spacious so the flanges, also made of plate steel, can be fitted inside the shaft.

A saving of up to 20% in material may be obtained from a welded shaft from forged barrel and cast flanges against an all–forged shaft. It may also save several hours of work time on the hydraulic press and many hours in machining, the production cycle may be reduced by 20%.

9.3.1 Forged shafts

Integrally forged shafts are mostly used for medium/small turbines and made from medium carbon steels, for example 0.35 or 0.45% carbon content, which are not as strong as low–alloy steels. Shafts of these turbines are normally designed according to critical speed, and not strength, so the actual stresses

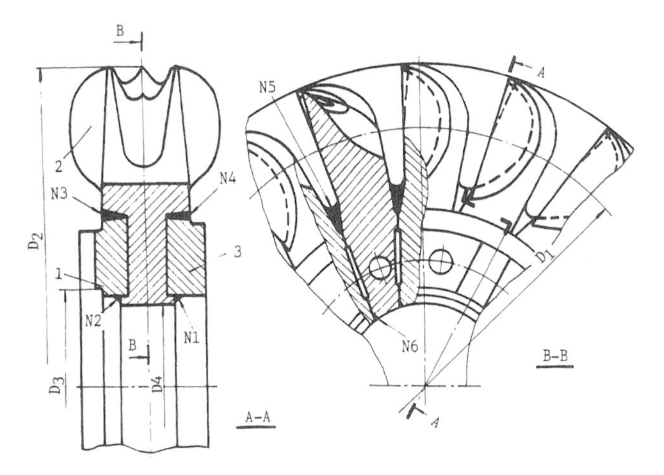

1 – front collar 2 – bucket 3 – rear collar

Figure 9.15 Construction of welded Pelton runner [9.1]

Figure 9.16 Pelton runner machined by numerically controlled milling cutter (Photograph courtesy Sulzer Escher Wyss)

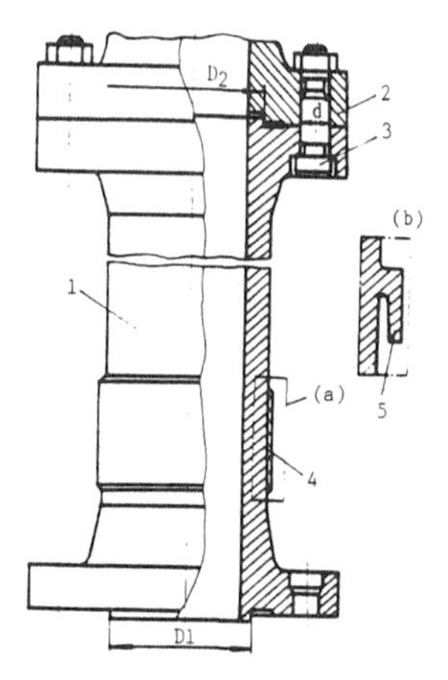

1 – shaft barrel 2 – mating flange 3 – tight–fit bolts
4 – shaft sleeve 5 – shaft skirt

Figure 9.17 Construction of forged main shaft [9.1]

in them are quite low. Low–alloy steels are mainly used in cast–weld or forge–weld shafts because of their good welding properties but they are much more expensive than medium carbon steels.

The steel ingot is heated to 1250^oC and forged to shape before it is cooled to 800^oC. Normally the blank is first forged into a near round stock and is reheated before extrusion of the centre hole and forming to the final dimension. After the forging process, two sections not less than $0.5m$ in length are cut off from each end of the shaft for checking of the mechanical properties and residual stresses, as in Figure 9.18.

In the case of large forged shafts, the finish–machined shaft may have a weight of only 25% of the original steel ingot, which indicates that this production method is rather wasteful.

9.3.2 Welded shafts

Welded shafts are mainly made from forged shaft barrels with forged or cast flanges. The joints between the barrel and the flanges are formed by either electroslag welding or submerged–arc welding. Forged shaft barrels are made in similar manner to integrally forged shafts and must have test sections each $150mm$ long provided at both ends.

Fabricated shaft barrels are either made from cold– or hot–rolled plate

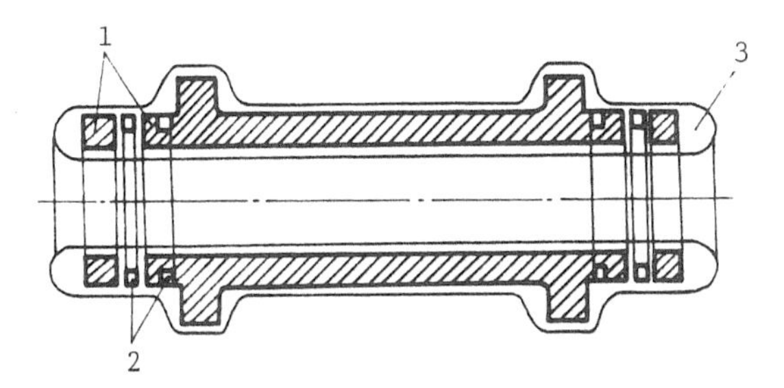

1– ring shape test pieces
2– rings for determinajtion of residual stress
3– left–over stock

Figure 9.18 Test pieces taken from main shaft forging [9.1]

cylinders or from pressed half cylinders. The longitudinal joints are mostly electroslag welded and tempered for stress relieving. Main shafts made of rolled steel plates will need no machining on the inside.

The largest welded shafts manufactured so far are for the $700MW$ turbines at Grand Coulee III made by the Allis–Chalmers Corporation. The shaft barrels are made of hot–rolled cylinders of thickness $178mm$ and from three longitudinal sections with outside diameter of $3200mm$. The sections are welded on the inside of the barrels. Shafts formed by $220mm$ thick hot–rolled plates have been made in France.

The skirt on the shaft is usually made of two halves and attached to the shaft after it is welded, heat–treated and rough machined. Alternatively the skirt may be made integral and slid on the main shaft before the end flanges are welded, such procedure was used to make the 170 MW Kaplan turbines at Gezhouba. The skirt, after welding, must be heat treated to ensure stability of dimensions.

Stainless steel sleeves for use with water lubricated bearings, usually made of Cr18Ni10 plates 10–$12mm$ thick, are welded onto the main shaft after it is finish machined and require no further heat treatment.

9.3.3 Machining of main shafts

Main shafts are mostly machined on horizontal lathes, with the exception of very large shafts which are machined on vertical lathes. Because of the rather large diameter of the flanges, the centre height of many lathes are insufficient so that special measures must be taken to raise the centre height. Static bearings have been introduced in the design of headstock of large lathes to reduce their runout to $0.01mm$. The location of the work pieces on the lathes can be determined to $0.01mm$ by use of optical–electronic measuring systems. Figure 9.19 shows a large Pelton turbine shaft being finished on a grinding machine.

Figure 9.19 Pelton turbine shaft on a large grinding machine (Photograph courtesy Sulzer Escher Wyss)

The tight–fit bolt holes connecting the main shaft with the runner and the generator shaft must be in perfect alignment which is done by reaming simultaneously the holes of the two flanges temporarily held together (pull–up pressure $200MPa$). Specially designed machines for reaming these bolt holes are available, but with the advancement in numerically controlled machine tooling, simultaneous reaming may become unnecessary.

9.4 Manufacture of Guide Apparatus

The guide apparatus constitutes the largest subassembly of the many parts that make up a hydraulic turbine. The production quality of the guide apparatus directly affects the performance of the turbine, so a high degree of manufacturing accuracy is required of these parts. The head cover and bottom cover, though rather large structural members, are traditionally regarded as part of the guide apparatus since among other functions they serve to support the guide vanes.

9.4.1 Head cover

The head cover is one of the largest components of a turbine. Head covers for large turbines may be over $10m$ diameter and over $200t$ in weight. Except for medium/small turbines, the head covers are commonly split into two, four or even eight parts and joined by bolting. Head covers are mainly made from welding steel plates. In order to ensure rigidity of the cover, very heavy sections are used in the structure, for large turbines the plate thickness may go up to $200mm$.

For protection against cavitation damage to the head cover lower plate and the guide vane end faces, type 18:12 stainless steel overlays are applied to these surfaces to a thickness of $3mm$ or more. Another way is to use replaceable stainless steel wear plates attached to the head cover by fasteners as is the common practice in China.

Machining of the head cover is mainly performed on vertical lathes for its inner and outer diameters. Holes for guide vane stems may be bored on horizontal or vertical boring mills, as shown in Figure 9.20.

Concentricity of the guide vane bores in the head cover and the bottom ring is very important. These bores may be machined by simultaneous reaming or machined successively with the finished one acting as guide for the next one. But when these parts become very large they are difficult to be worked together. One of the many alternatives is shown in Figure 9.21. The finished head cover is placed on top of the bottom cover and properly aligned.

Positioning rings are inserted into the bores on the head cover with clearance of less than 0.15–0.2mm and tag–welded. The bottom ring is then machined with the positioning rings as guides when the head cover is removed. In very well equipped factories, by use of large composite vertical boring mills fitted with precision indexing heads, the guide vane bores in these two parts may be machined separately to the same degree of accuracy.

9.4.2 Bottom ring

The bottom ring, like the head cover, is mostly made from welding of steel plates, a typical construction is shown in Figure 9.22. It is essentially made up of a top plate *1* a ring wall *2* plus guide vane receptacle *3* and an ample number of stiffening ribs. For axial–flow turbines the skirt *4* is quite large which has a double curvature that has to be pressed in sections and joined together. In other cases, the skirt may be made of straight cylindrical sections and joined into a polygon, only the transition portions of the skirt with the upstream and downstream structures are machined to a smooth contour.

To prevent deformation of the bottom ring during fabrication, as con-

Figure 9.20 Machining guide vane bores of a head cover on horizontal boring mill (Photography courtesy J. M. Voith)

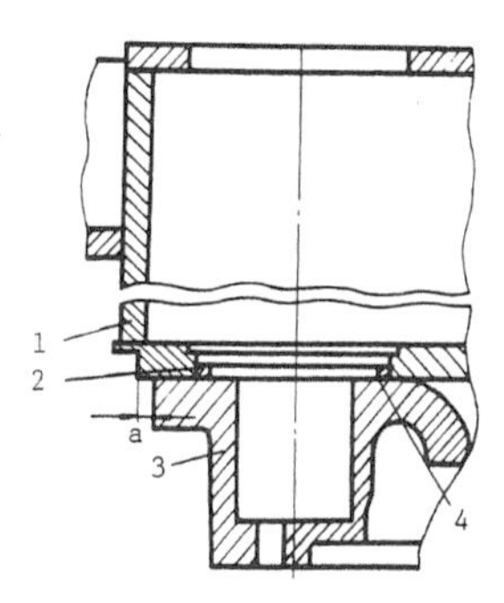

1 – head cover 2 – positioning ring
3 – bottom ring 4 – tag weld

Figure 9.21 Machining of guide vane bores on bottom ring positioned by the head cover

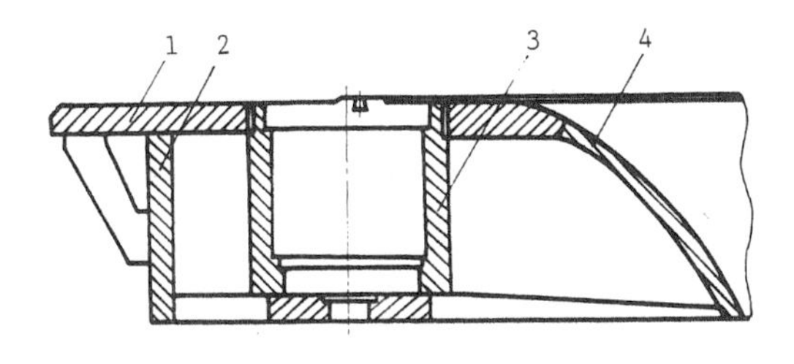

1 – top plate 2 – ring wall
3 – GV receptacle 4 – skirt

Figure 9.22 Section of typical bottom ring construction

siderable welding is done on the underside of the ring, two circumferential segments of the ring are joined temporarily *top–to–top* and stiffened by supports during welding (Figure 9.23). Machining procedures for the bottom ring is essentially the same as for the head cover.

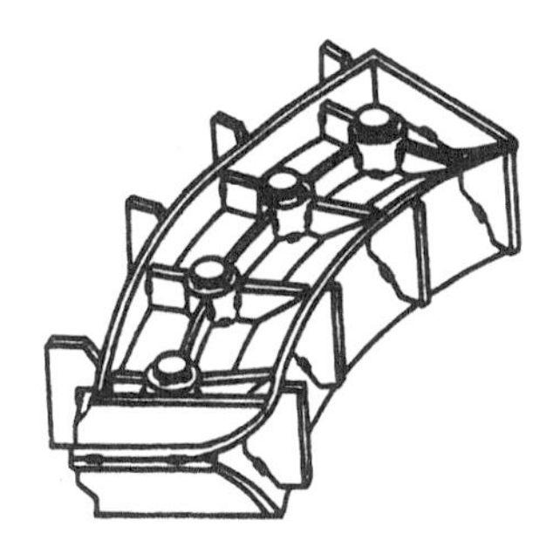

Figure 9.23 Welding of underside of bottom ring

9.4.3 Guide vanes

Guide vanes for high head turbines may be made either from integral castings or from the welding of cast vanes and forged stems (mainly the long stems). For medium/low head turbines where the vanes are much larger, it is common to fabricate the vanes from pressed–steel plates and forged stems, as shown in Figure 9.24.

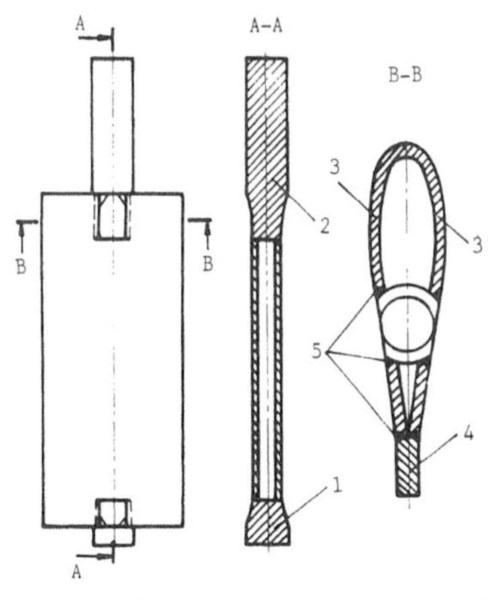

1 – short stem 2 – long stem 3 – profile plates
4 – tail piece 5 – vertical seams

Figure 9.24 Construction of welded guide vane

There have been certain difficulties in casting guide vanes because they are long and slender objects with considerable changes in section. Longitudinal deflections for guide vanes $2m$ long may reach $10–15mm$. These vanes must be corrected before machining.

Typical requirements for unmachined surfaces of cast guide vanes are

1. Deviation in vane profile less than $4mm$.
2. Longitudinal deflection, including torsional deflection, less than $3mm$ per metre length.
3. No non-metallic inclusions on the surfaces.
4. Tool marks from cleaning to be less than $1mm$ deep.

Guide vanes of welded construction can attain better precision in shape and surface finish. Typical requirements are

1. Deviation in vane profile less than $3mm$.
2. Longitudinal deflection less than $4mm$ in length of $3.5m$.
3. Selective inspection of 10–15% weld joints by non-destructive methods.

One shortcoming of welded guide vanes is the inherent low stiffness due to its hollow construction. This has been overcome by welding longitudinal ribs inside the vanes. A special technique known as *tunnel* welding is used to weld the ribs.

For protection against cavitation damage, the trailing edges of welded guide vanes (part 4 of Figure 9.24) may be made of stainless steel. For the same purpose, stainless steel cladding or overlay are applied to the end faces of the vanes. For protection against wear on the guide vane stems, stainless steel sleeves are fitted on the stems at the bearing positions, sometimes overlays are applied directly to the stems themselves [9.1,9.4].

For machining of the guide vane stems, special purpose lathes employing rotary cutters are the most effective tools as the vanes are held stationary and the tool(s) rotate, most of the out-of-roundness of the stems is thus eliminated. Figure 9.25 shows a guide vane being set up on a machining centre with rotating cutters.

Figure 9.25 Guide vane stem being machined with rotary cutter (Photograph courtesy J. M. Voith)

9.4.4 Preassembly of guide apparatus

Most manufacturers specify shop assembly of the guide apparatus for

1. Checking the dimension of the moving parts and the end clearance of the guide vanes.
2. Checking the movement of all parts of the control mechanism and openings of the guide vanes.
3. Checking the tightness between guide vanes when they are fully closed.

In case the stay rings are transported to the site ahead of the control mechanism, preassembly must be done on a mock-up tool, or the mechanism is preassembled at the site before erection. Only when very good manufacturing accuracy is assured can the shop assembly be waived or only the first machine in the batch shop assembled.

9.5 Manufacturing of Embedded Parts

By embedded parts it is generally understood to mean the spiral case, runner chamber (for axial-flow turbines), draft tube liner and the distribution piping (for Pelton turbines). Stay rings for reaction turbines have been classified as part of the guide apparatus from the hydraulic point of view, although they are also truly embedded parts. Descriptions are given here on the spiral case, stay ring, runner chamber and draft tube liner.

9.5.1 Spiral case

Welding is the predominant method used in fabrication of spiral cases. With the ever increasing machine size and/or operating head, the steel plates that make up the spiral case have reached such a thickness that they are difficult to roll, assemble, weld and handle. Consequently, high strength steel has been introduced to the manufacturing, in order to bring the plate thickness down to a practical level. Some better known grades of medium strength steels are ASTM A516 Gr70 (USA), CtB25A and BCt3Mn (USSR) and TTStE36 (FRG). In the USSR, the turbine industry is required to limit the plate thickness to $40mm$ [9.1], this is well below the greatest thickness now used ($73mm$). For very large turbines, high strength low-alloy steels with yield strength of $500\text{–}700MPa$ are employed.

A typical construction of welded spiral case is shown in Figure 9.26. The various sections of the spiral case are sometimes made of different grades of steel and in different thickness because of their varied dimensions and loading conditions. Spiral case sections are now commonly laid out by computers and cut by numerically controlled flame cutters (with edges bevelled for welding) to an accuracy of 1 mm. The ends for some spiral case sections are first pressed to the required curvature before rolling, while for others the blanks are directly rolled to shape.

Stay rings of the parallel-plate design are becoming more common. Welding of spiral case sections to this kind of stay ring is easier than for cases with flared rims. The purpose of shop assembly of the spiral case is to check the accuracy of formation of the sections, and the fitting between sections, and with the stay ring. The tail section of the spiral case is of such intricate shape that it has to be formed by gauging with templates.

Spiral cases must be welded in strict conformity to welding codes, like

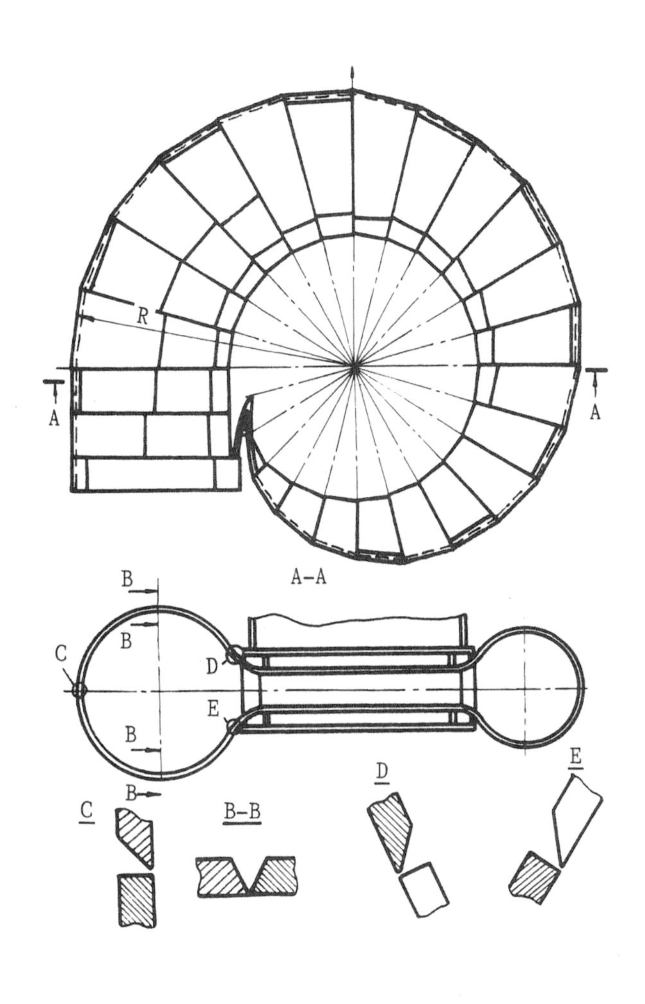

Figure 9.26 Typical construction of welded spiral case [9.1]

pressure vessels, by qualified welders. X–ray inspection must be applied to all welds of the spiral case. Spiral cases required to operate under high heads are required to be pressure tested after assembled at site, in which case the inside of the stay ring and the spiral case inlet are temporarily blocked off by bulkheads.

Although spiral cases of many large turbines are not shop assembled, installation time may be saved by welding two or more sections of the case to one segment of the split stay ring outside the power house and then lowered together into the turbine pit for final assembly welding. An example of spiral case made in two halves being welded to a stay ring of parallel–plate design is shown in Figure 9.27.

Figure 9.27 A split spiral case being welded in the factory
(Photograph courtesy J. M. Voith)

Turbines of high operating head or of small capacity have spiral cases of such size that are easily welded complete in the factory. They are then cut into two halves along with the stay ring for transportation to the site, and then assembled by bolting.

Some high head turbines of small size have cast spiral cases. They are made of split construction with two, three or four segments mainly from casting consideration.

9.5.2 Stay rings

Stay rings of hydraulic turbines used to be exclusively made from castings many years ago, but welded stay rings, all-welded or cast-welded, has increasingly replaced castings except for small size turbines.

Stay rings fall into two major types of construction: with flared rims and parallel plates. The flared rim structure may be all cast (in split segments) or a combination of welded plates and cast components. One of the latter is shown in Figure 9.28. The flared rim is actually formed by a large number of straight sections and pieced into a polygon. The division of sections usually corresponds with those of the spiral case sections.

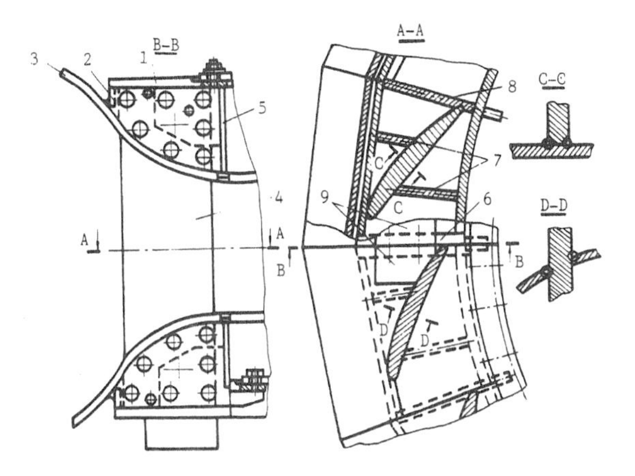

Figure 9.28 Construction of cast–weld stay ring with flared rim

Stay rings of the parallel–plate design are much simpler in structure, as the upper and lower rings are merely of box or triangular construction (Figure 9.29). The points of contact with the spiral case will vary from the large to the small sections by design consideration. The square corners are covered with segments of curved plates for better flow conditions.

Because the upper and lower rings and the stay vanes of the stay ring have very large welding sections, measures must be taken to prevent laminar tearing. In recent years, high quality steel plates with very low sulphur content ($<0.008\%$) and cross section contraction ratio (in direction of thickness) larger than 25% that have good anti–tearing properties have become available, for example TTStE36–Z3 steel from Germany [9.11] and WELTEN62CF steel from Japan [9.19].

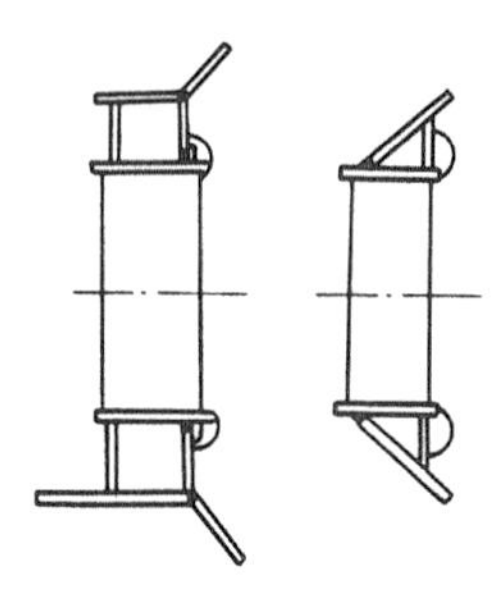

(a) Box construction (b) Triangular construction

Figure 9.29 Stay ring of parallel–plate design

When high strength steel plates are used to make the spiral case, there emerges the problem of unequal strength between the spiral case and the stay ring. They may be treated by:

1. Using also high–strength steel for the portion of stay ring that is joined to the spiral case.

2. Building up the cross section of the carbon steel member of the stay ring by welding in the contact region.

3. Employing a transition band also of high–strength steel between the stay ring and the spiral case.

The first measure is the most desirable but is more expensive; the second one is not highly recommended since it requires a larger welding volume and brings in additional deformation; the third measure has applications in stay rings with flared rims (Figure 9.30) as the transition band may be made with a taper that also offers a gradual transition in thickness between the two parts. All three measures have been used in large hydraulic turbines built in recent years.

The stay rings, like the runners, may be hand–welded, by carbon dioxide semi–automatic welding or by electroslag welding. Electroslag is most suitable for welding the butt or T–joints between the stay vanes and the upper and lower rings because of their large cross sections. Greater difficulties are

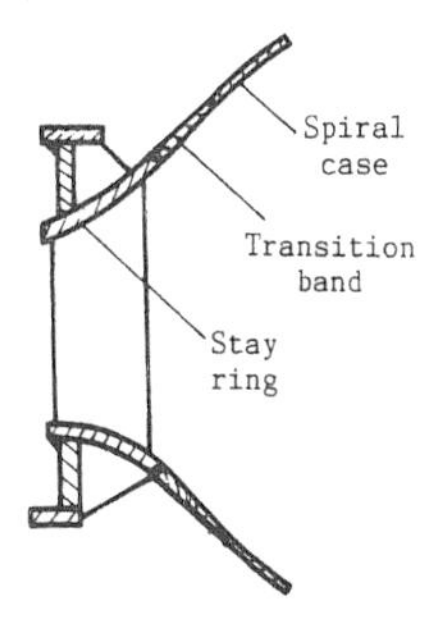

Figure 9.30 Transition band between stay ring and spiral case

encountered in welding stay rings with flared rims where the T-joints have three-dimensional curvatures.

Machining of the stay rings is similar to all ring-shaped parts of the turbine and is not complicated in itself. But the size and weight of present-day stay rings have increased so much that they far exceeded the capacity of ordinary machine tools. For example, the Grand Coulee III $600MW$ turbines have stay rings of $14m$ outside diameter, $5.9m$ height and $450t$ weight that would present difficulties to most conventional manufacturing methods. One solution to this problem is to devise new techniques utilizing existing machine tools. An example of this is the conversion of a vertical lathe by putting one post in the centre of the faceplate and the other outside the work which will enable the machining of parts up to $12m$ in diameter, as shown in Figure 9.31. Another example is to put the cutting tool of a vertical boring mill on the faceplate with the stay ring around it and held stationary (the inside bore to be not less than $8m$), the flanges are machined with the tool revolving. The limiting size of work for this is nearly $17m$ which is sufficient to handle the largest turbines envisaged today.

A different approach to this problem is to design the turbine that will require no finishing of the stay ring flanges. Figure 9.32 shows details of the joining of an unmachined stay ring with the head cover. The level of the head cover is adjusted by the mounting screws and its position fixed by final welding of the two retaining rings. Experience gained in the USSR with this design shows the guide vane end clearances can be adjusted to within $0.2mm$ [9.1].

The advantages of using unmachined stay rings for large machines are

1. Reduction of stay ring weight by eliminating the massive flanges.

Figure 9.31 Numerically controlled machining of large stay ring on vertical boring and turning mill (Photograph courtesy of Sulzer Escher Wyss)

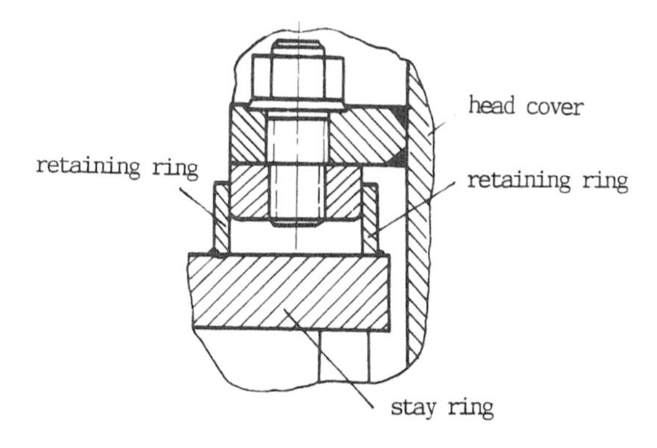

Figure 9.32 Joining of unmachined stay ring with head cover [9.1]

2. Reduction of factory welding and machining work load.

3. Removal of limitation on size of the stay ring.

4. Improved accuracy of installation due to the flexibility of site adjustment.

The obvious drawback of this design is the necessity of special provisions for removal of the runner and the guide vanes for repair. Another example of employing unmachined stay rings is the case of the 730MW Itaipu turbines made by J. M. Voith [9.11]. The clearances at A, B, C and D in Figure 9.33 are all adjusted by shims and fixed by screws and dowels.

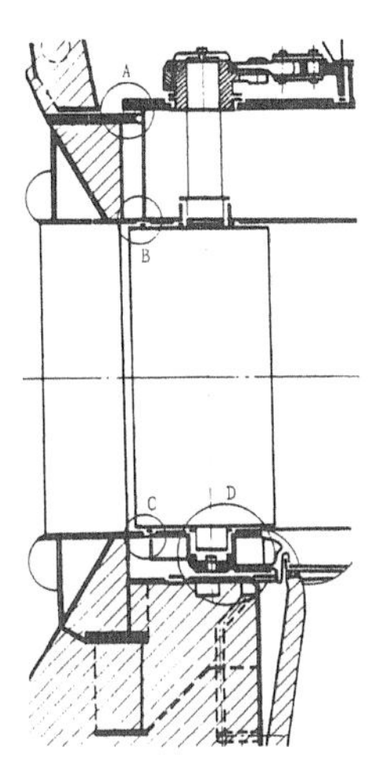

Figure 9.33 Joining of head cover and bottom ring with unmachined stay ring flanges (Drawing courtesy J. M. Voith)

9.5.3 Runner chamber

Runner chambers are required by axial–flow, diagonal–flow and tubular turbines but differ greatly in shape according to the type of machine. Runner chambers may be made of steel casting or welded from steel plates. The cast runner chamber is limited to use in small size turbines since it is a thin–shell structure that is difficult to cast in large sizes. The majority of designs today are of welded construction, or partly with cast pieces.

Depending on its size, the runner chamber may be made integral or split into several pieces, either in the radial direction or into upper and lower sections, or in both ways. Most runner chambers are machined on vertical lathes, the internal curvature shaped by numerically controlled machine tools or by template copying tools.

As the runner chamber is subjected to low pressure during operation of the turbine, cavitation is likely to occur in a band that corresponds to the position of runner blades. Stainless steel overlays are used to protect this specific region of the runner chamber. In large machines, the middle or lower sections of the runner chamber may be wholly made of stainless steel.

Band–electrode submerged–arc or multi–electrode automatic surfacing machines are used to assure even and smooth surfacing.

Different segments of the runner chamber used to be joined by flanges which became very massive in large turbines, therefore, all–welded constructions are now being used to save these flanges. But difficulties exist in the welding of dissimilar materials such as uneven thermal expansion and different heat treatment behaviours. The situation becomes worse when there are radially–split joints.

A new type of composite steel plate is now used to make the curved sections of the runner chamber without the need of subsequent over–laying. The composite plate is $30mm$ thick with a layer of Cr13 stainless steel 3–$4mm$ thick on one side. A typical example of using this material is shown in Figure 9.34.

According to experiences in the USSR, the runner chamber is fabricated in the machine shop by welding the composite steel plate and reinforcements in a predetermined order with measurement by solid templates at every stage of work. When transported to the site, the dimensions of the runner chamber is likewise carefully controlled when the components are assembled and welded. With this technique, it is able to control the clearance between the runner blades and the chamber to within $0.0005D_1$, this is better than that attainable by machining because deformation in transport is always unavoidable. For runner chambers larger than $7.5m$, the design is changed to a combination of a conical section and a cylindrical section in order to save the work load of pressing curved parts [9.9].[2]

[2]The latest technique in fabricating a runner chamber with composite steel plates is to first weld the external longitudinal and lateral reinforcements and machine after heat treating. The composite plates after pressing to shape are then welded on with no further machining. This technique has been successfully applied in producing runner chambers of 7.5m diameter in large batches [9.12].

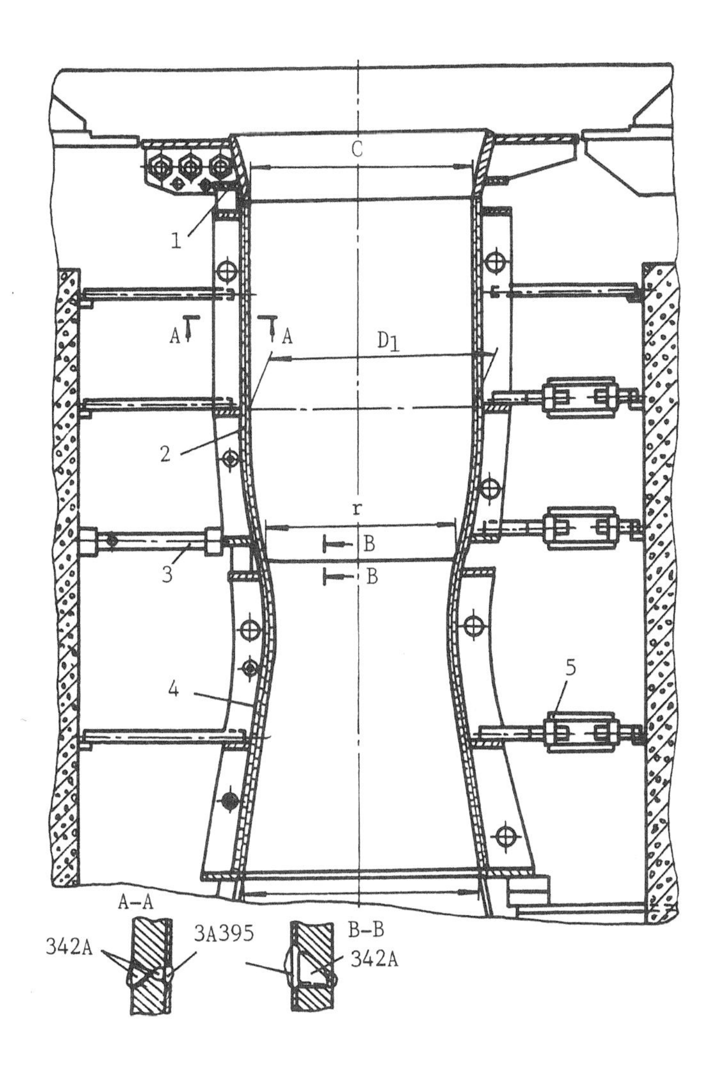

Figure 9.34 Runner chamber with no machining and welded at site [9.1]

9.5.4 Draft tube liner

The draft tube liner involves a great deal of work in manufacture because of its complicated shape. For medium/small machines, or those operating under high head, a full draft tube liner is normally supplied (see Figure 2.16), but for large turbines usually only the straight conical liner and occasionally also the elbow are supplied, leaving the rest of the draft tube passage to be formed by concrete.

Draft tube liners are always made of welding from steel plates. As they are large in size and low in stiffness, measures to prevent deformation in transport and erection are important. Except for medium/small turbines where the liners may be made integral in the shop, they are invariably assembled and welded at station site. The draft tube liners may be erected, whether in the shop or at site, either from the bottom up or from the top down. Temporary supports and braces are necessary to prevent deformation and serve as safety measures.

The draft tube liners are subject to quite a large pressure difference in operation, as the water pressure built up from seepage through the concrete outside of it may be several times the pressure inside of it, so the liners must be designed against collapsing by providing sufficient reinforcements and securely anchored to the foundation concrete.

The largest draft tube liners ever manufactured are the ones for Grand Coulee III ($16.9m$ in height and $167t$ in weight) and the ones for Sayan–Shushensk in the USSR ($16.9m$ in height and $27.3m$ in length) [9.18].

The draft tube liner is also exposed to cavitation at the upper part especially with Francis turbines. These vulnerable areas are sometimes also protected with stainless steel. Various piping and devices are fitted inside the draft tube for admission or injection of air during operation. Some are nozzles fitted on the wall of the liner, others may be three or foyr–legged crosses placed directly in the flow passage. The devices have proved to be effective in reducing noise and vibration at part load operation of the turbine, but the sturdiness of construction cannot be over–stressed because numerous cases of these devices being washed away have been reported.

If the turbine is to operate in silt–laden waters especially when containing hard particles, the draft tube liner may have to be made of erosion–resistant material such as 1Cr18Ni3Mn3Cu stainless steel.

References

9.1 Bronovski, G. A. et al 'Technology of Hydraulic Turbine Manufacture', Machinostroenia, Leningrad, 1978 (in Russian).

9.2 Rowlands, E. W. 'Welded Turbine Hub Improves Quality and Reduces Cost', *Welding Journal*, Vol. 53, (1974), No. 7.

9.3 Rowlands, E. W. 'World's Largest Turbine Runners', *Welding Design and Fabrication*, 1977, No. 10.

9.4 Sirman, W. R. 'Turbines for the Grand Coulee Hydro Powerplant', *Water Power and Dam Construction*, 1977, No. 4.

9.5 Casacci, S. et al. 'Modern Trends in Large Hydraulic Turbines', *Water Power and Dam Construction*, 1975, No. 10.

9.6 Bronovski, G. A. Article in 'Problems in Turbine Manufacture', edited by G. A. Drobielka, *LMZ Publication Vol.* 7, 68–77, Moscow, 1960 (in Russian).

9.7 Zheng, B.Y. "Manufacture of 6–metre Diameter Dissimilar Material Runner of Special Design', *Dongfang Electrical Machinery*, 1984, No. 3 (in Chinese).

9.8 Bronovski, G. A. et al. 'Special Features of Manufacture of Turbines for Krasnayarsk Hydro Power Station', *Energomachinostroenia*, 1965, No. 3 (in Russian).

9.9 Kovalev, N. N. (editor), 'Hydraulic Turbine Handbook', Machinostroenia, Leningrad, 1984 (in Russian).

9.10 Kirov, S. M. 'Cast–weld Runner for High Head Francis Turbine', *Energomashinostroenia*, 1973, No. 11 (in Russian).

9.11 Hardt, E. 'Manufacturing a Spiral Casing and a Turbine Runner of a Francis Turbine', Hydro Power Project: Three Gorges, Presentation, May 1985 by J. M. Voith, GmbH.

9.12 Babanov, O. S. 'Hydraulic Turbine Manufacturing at LMZ for 12th Fifth–year Plan', *Energomachinostroenia*, 1986, No. 1 (in Russian).

9.13 Voloshkevski, G. E. 'Manufacture of Cast–weld Runner of Hydraulic Turbine', *Automatic Welding*, 1968, No. 2 (in Russian).

9.14 Paton, B. E. et al. 'Electroslag Welding and Surfacing', MIR Publisher, Moscow, 1983.

9.15 Beltran, G. 'The 100 MW Hydraulic Turbines for the Guri Project', Toshiba Publication.

9.16 Zheng, B. Y. 'Construction and Manufacturing Techniques of Francis Turbine Runners Abroad', *Dongfang Electrical Machinery*, 1981, No. 3 (in Chinese).

9.17 Siebensohn, R. B. 'Trends in U.S. Hydro Equipment Design', Water Power, 1974, No. 3.

9.18 'Cooperation in Hydro Power: China and Allis–Chalmers', Allis–Chalmers Engineering, October, 1979.

9.19 Zheng, B. Y. 'Development of Use of Low–alloy High–strength Steel for Hydraulic Turbine Spiral Case', *Dongfang Electrical Machinery*, 1982, No. 1. (in Chinese).

Chapter 10

Construction of Small Turbines

Lu Chu–xun

10.1 General

The hydro power potentials for small scale developments are indeed very abundant and widely distributed all over the world. In China alone, the theoretical small hydro potential amounts to $160GW$ with the exploitable about $70GW$ and the developed so far about $10GW$. Small hydro power developments that are well adapted to local conditions are very effective in satisfying the ever–increasing demand for electric energy in rural areas. Especially for developing countries, harnessing of small hydro power has been one of the most challenging undertakings in their reconstruction programmes.

Small hydro power stations may vary over a very wide range of ratings such as from low to high head, and from small to very large discharge. Consequently, turbines designed to operate in these small hydro stations must cover a wide range of specifications although the quantity required in each category may not be very large. Therefore, such turbines belong to the kind of product that is varied in rating and small in production batches. For this reason, government agencies and manufacturers in many countries have standardized the construction of small turbines into a number of design series typified by simplification of construction, standardization and interchangeability of parts, with the aim of reducing production costs and facilitating the building of small hydro plants.

10.1.1 Range of application of small turbines

The range of application of small turbines is classified according to the power output and size. Different standards in designating the range of application are in use in different countries as the stage of technical and economical development are not the same.

Since the energy crisis in the late 1970s, greater attention was given to the development of small hydro power resulting in the convening of the First International Small Hydro Conference held in Katmandu in 1979 [10.1]. Different views on the definition of range of application has been raised in several international conventions since. The following suggestions were made by the Preparatory Committee for the United Nations Conference on New and Replenishable Energy Source in 1981:

Small hydro	(small installation, plant or unit)	$< 10MW$
Micro hydro	(micro installation, plant or unit)	$< 1\ MW$
Large hydro	(large installation, plant or unit)	$> 10MW$

It is also commonly accepted that installations with capacities less than $10MW$ are called small hydro, less than $1MW$ mini hydro and less than $100kW$ micro hydro. General head ranges applicable to the above categories are given in Table 10.1.

Descriptions on small turbines in this chapter will be confined to the above power and head ranges. Figure 10.1 shows the range of application for machines smaller than $10MW$ with working head from 1 to $400m$ and discharge from 0.05 to $100m^3s^{-1}$.

10.1.2 Types of turbines and their installations

In the general realm of small turbines, the following types are now in use:

Reaction turbine
- Axial–flow turbine
- Tubular turbine
- Mixed–flow (Francis) turbine

Impulse turbine
- Cross–flow (Banki) turbine
- Inclined–jet (Turgo) turbine
- Tangential–jet (Pelton) turbine

The recommended specific speeds and head ranges of the above types are given in Table 10.2. Installations of the above turbine types may be classified as:

1. Alignment of shaft: horizontal, vertical or inclined.
2. Type of inlet chamber: open–flume, spiral case (semi– and full–spiral case), can–type, shell–type.

Table 10.1 General head ranges applicable to small turbines [10.5]

	Head, m		
Unit output, kW	Low	Medium	High
5–50	1.5–15	15–50	50–150
50–500	2–20	20–100	100–250
500–5000	3–30	30–120	120–400

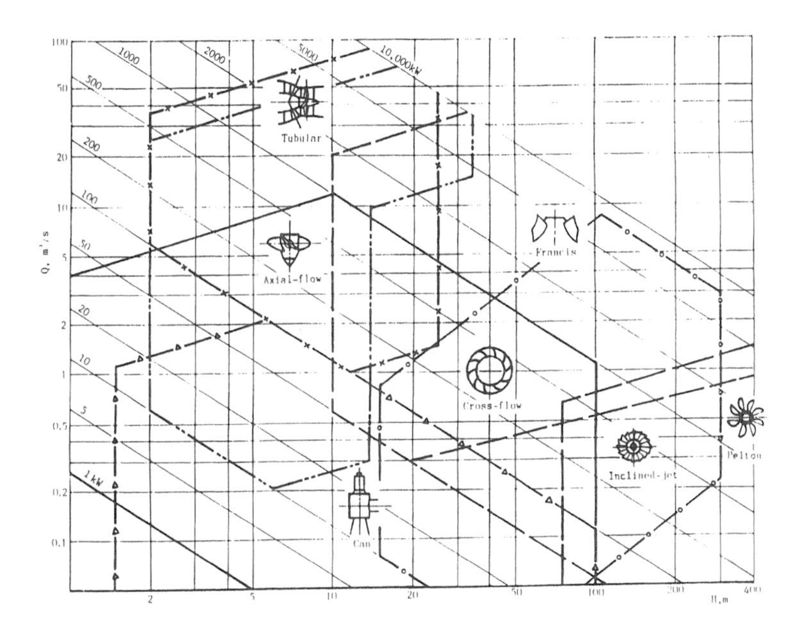

Figure 10.1 Range of application of small turbines

3. Regulation mechanism: fixed runner blades, adjustable runner blades, fixed guide vanes, adjustable guide vanes.

Schematic arrangement of these installations are shown in Figure 10.2.

Table 10.2 Specific speeds and head ranges for various types of small turbines [10.4]

		n_s $(m \cdot hp)$	Ω	H, m
Reaction turbines	Axial–flow	1100/350	4.89/1.55	2/25
	Francis	450/250	1.99/1.11	25/100
		250/150	1.11/0.67	100/250
		150/60	0.67/0.26	250/600
Impulse turbines	Cross–flow	300/30	1.32/0.14	20/200
	Pelton	70/30	0.31/0.14	100/400
		30/10	0.14/0.04	400/1800

10.2 Small Reaction Turbines

10.2.1 Tubular turbines

Tubular turbines of small sizes may be built in the open–flume, (a) of Figure 10.2, siphon (b), pit (c), bulb (d) or S–tubular (e) types. As with other types of small turbines, the open flume is used for low–head, small size turbines with fixed guide vanes and directly driven generators.

A siphon built in the inlet flow duct serves to control the functioning of the turbine: the turbine is started by drawing the air out of the top part of the siphon and is stopped by letting air into the duct to break the vacuum.

The bulb turbine is composed of stay ring, conical guide apparatus, runner and draft tube. The stay ring is a hydraulic component and also a structural member that supports the bearings and the rotating parts. Spatially curved vanes provide the required circulation for the runner and shuts tightly to cut off the water flow at machine standstill. The bulb turbine normally uses a straight conical draft tube (see 4.2.3).

S–type tubular turbines of small sizes may be of several types of construction. Standardized products are available for head range of 2–15m, power output up to 5,000kW and runner diameters up to 3.0m. They have proved to have better performance than vertical axial–flow units of comparable parameters and require appreciably less space and height in the power house.

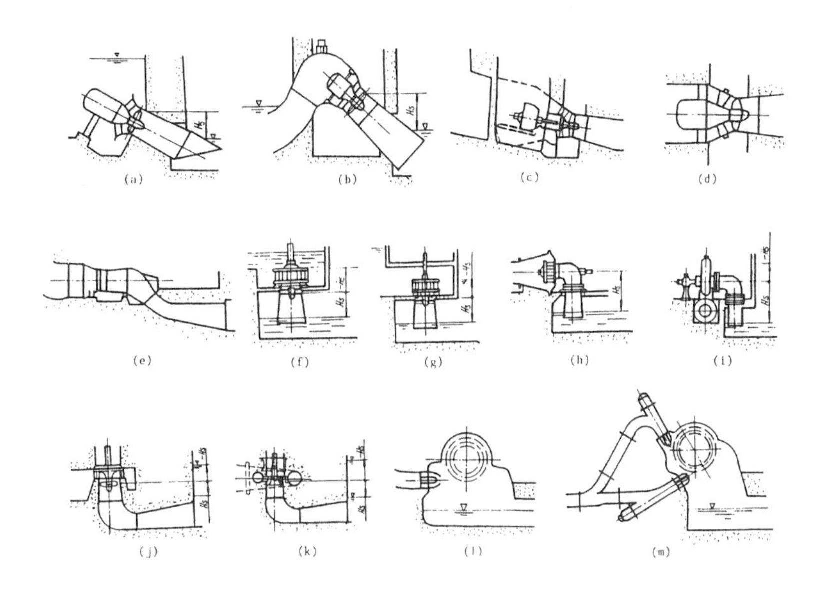

(a) open–flume tubular
(b) siphon tubular
(c) pit tubular
(d) bulb tubular
(e) S–type tubular
(f) open–flume vertical
(g) pressurized open–flume vertical
(h) can–inlet horizontal
(i) metallic spiral case horizontal
(j) concrete spiral case vertical
(k) metallic spiral case vertical
(l) one–jet impulse
(m) two–jet impulse

Figure 10.2 Installation arrangements for small turbines [10.7]

The S–type tubular turbine may have its shaft extend from the curved inlet passage and connect to a generator upstream of the turbine as shown in Figure 4.10 or through the curved draft tube and connect to a generator downstream of the turbine as shown in Figure 4.11.

The turbine may be regulated by various means, such as with adjustable runner blades and guide vanes, adjustable runner blades and fixed guide vanes, fixed runner blades and adjustable guide vanes or both blades and vanes fixed. When the guide vanes are fixed, an additional shutdown device is necessary.

Depending on the operating head, the runner may have four or five blades that are controlled by a servomotor working through control rods, levers, a crosshead and arms. The servomotor may receive signals from a speed governor, a load–control device or a water level transmitter. A butterfly valve is usually installed at the turbine inlet and is closed by oil pressure and/or a counter–weight.

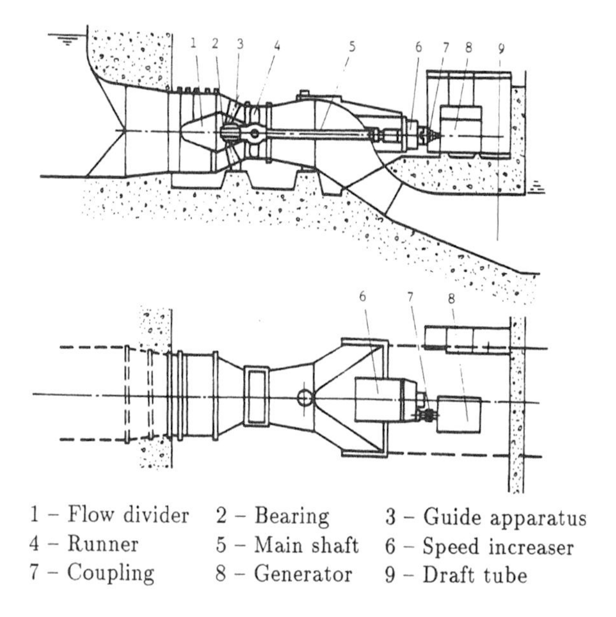

Figure 10.3 S–type tubular turbine [10.4]

For machines working in fairly low heads, the turbine speed will become quite low and it is difficult to fit a generator with matched characteristics. By using a gear speed increaser, a standard-line generator of much higher speed may be used that has better performance and of smaller size and lower cost. Figure 10.3 shows an installation of a downstream S-type tubular turbine equipped with a speed increaser.

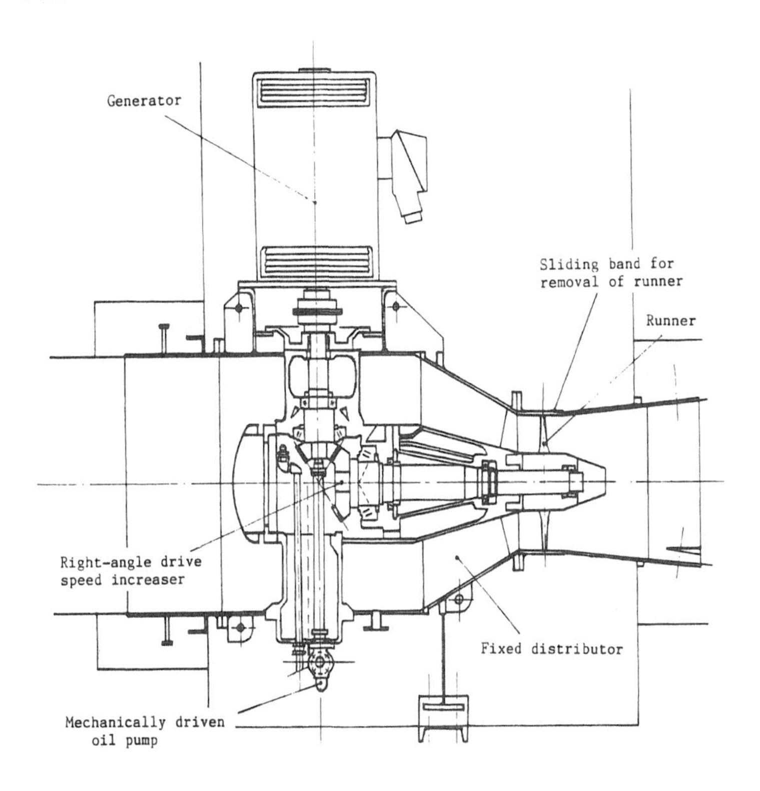

Figure 10.4 Tubular turbine with right-angle drive
(Drawing courtesy Neyrpic)

The arrangement shown in Figure 10.4 has the torque from the turbine transmitted upward through a pair of right-angle gears (equi-speed or increaser) to drive a vertical generator that sits on the turbine casing. The complete unit may be assembled in the factory and transported to the site intact. By placing the generator in a more accessible position and by eliminating the generator foundation, a further reduction in power house excavation is attained.

10.2.2 Axial–flow turbines

Vertical shaft small axial–flow turbines of either the propeller or Kaplan types may be used for the head range of 2 to $30m$, with propeller turbines exclusively in runner diameters less than $1.2m$. Open–flume inlet passages (Figure 10.2 (f)) are used for heads less than $6m$ and runner diameter smaller than $1.2m$. Pressurized open flume inlets (Figure 10.2 (g)) are sometimes used for heads over $6m$. Concrete and metallic spiral cases are used with axial–flow turbines of higher ratings (Figure 10.2 (j) and (k)).

The open–flume propeller turbine shown in Figure 10.5 is simple in construction in that it does away with the spiral case and the associated stay ring. The guide apparatus including guide vanes, linkages and the control ring are all immersed in water. The turbine shaft uses a guide bearing with pads made of rubber, plastic or hardwood submerged in the flume and running in water. To reduce wear of the bearing pads, a filter screen is sometimes fitted over the bearing. If the water contains much sediments, specially filtered water must be provided or an oil–lubricated bearing is used instead.

Low head axial–flow turbines of reasonable sizes are connected directly to the vertical shaft generator. Small output turbines are sometimes connected via belts or speed increaser gearing to generators of higher speed. Component parts for low head axial–flow turbines are mostly made of cast iron with runners either cast integral or welded from cast steel pieces. Alternatively the blades may be fastened to the runner hub by collars (see 9.2.2). The drive shaft is usually made from bar stock with flanges welded on both ends.

10.2.3 Francis turbines

Small Francis turbines are predominantly of horizontal type, the standard–line turbines are fitted with spiral cases (Figure 10.2 (i)) and multi–vane guide apparatus. But for small–size units the can–type (Figure 10.2 (h)) and single guide vane casings may be used.

Figure 10.6 shows a horizontal–shaft Francis turbine with cast spiral case which forms the basic structure housing the runner and the multi–vane guide apparatus. The main shaft is supported by two sliding bearings mounted on a base plate with the flywheel in between them. The spiral case unit which is separately mounted from the bearing unit must be assembled with great care in order to assure good alignment of the runner which is fastened to the shaft by taper fit. Since the guide vane height is relatively short, two–pivot overhung layouts are used for ease of maintenance.

For smaller turbines the spiral case and the bearings may be mounted together on one common base plate. To prevent the machine from running

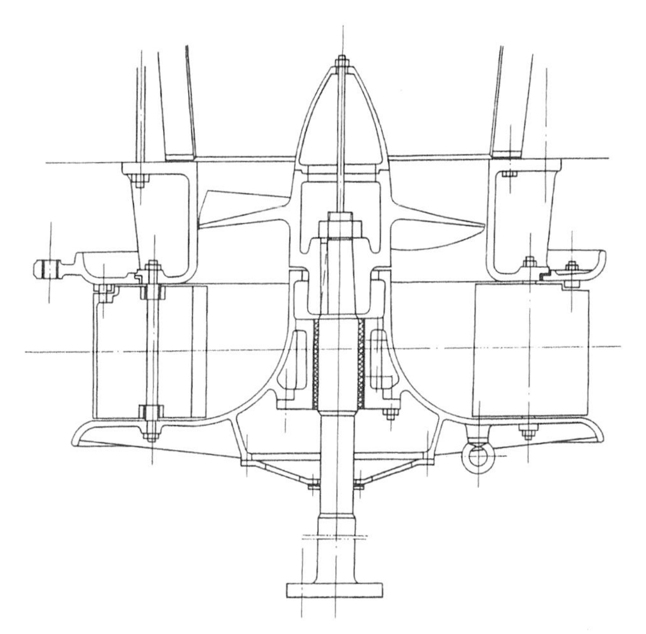

Figure 10.5 Section of open–flume propeller turbine
(Drawing courtesy TRIED)

at low speed for any length of time, brakes are provided that act on the sides of the flywheel. Spiral cases for small turbines may be integral castings from iron or steel or welded from steel plating. Cast spiral cases not only have greater strength but also help to absorb noise from water flow, specially for high head turbines.

For high head low specific speed turbines, the runner diameter is comparatively small so the guide apparatus may be mounted on the draft tube side to facilitate assembly and repair. The free hanging control ring which connects with the guide vanes is pulled or pushed sideways by a control rod to regulate the opening of the guide vanes. Governors for small turbines are generally built integral with the servomotor and oil reservoir in one unit.

Runners for small turbines are commonly made of cast–weld construction, that is, the blades, the runner crown and band are cast separately and then welded together. With good control in welding technique, runners thus made have more accurate profiles and better surface finish than cast runners. The blades may also be pressed from steel plates for even better shapes

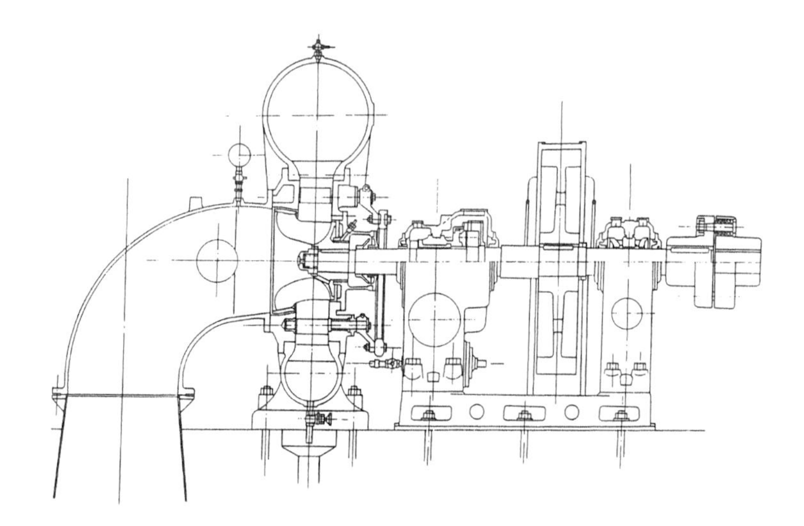

Figure 10.6 Section of horizontal shaft Francis turbine (Drawing courtesy TRIED)

and finish (see 9.2.1). Static balancing is required after final finishing of the runner.

The main shaft may be fitted with gland packing, mechanical seals or clearance seals. For units installed below the tail water level, an additional standstill seal is sometimes used.

Main shafts for small horizontal turbines are usually made from bar steel stock, with forgings used only for larger units. Small turbines usually run at higher speeds so they require fairly small shaft diameters, but stiffness calculations are necessary in the design to assure safe operation of the turbine, such as by the formula

$$d = K\sqrt{P/n} \tag{10.1}$$

where d – shaft diameter (cm)
P – power (kW)
n – speed (min^{-1})

Factor K may be taken as 11.3–14 corresponding to an allowable stress of 35–18MPa (K = 11.3–12.6 for shaft diameter 50–25cm, K = 12.6–14 for 25–10cm).

Flywheels are made of cast iron or steel, or fabricated from steel plates in occasional cases. The linear speeds for flywheels are:

Cast steel	$V < 100 - 110$	ms^{-1}
Cast iron	$V < 50 - 60$	ms^{-1}
Fabricated	$V < 120 - 150$	ms^{-1}

Flywheels are standardized in design and may be selected according to the required rotational inertia (GD^2).

Oil lubricated sleeve bearings are designed with a clearance of $\delta = d/1000$ as a rough rule, but for large shafts it must be kept much smaller such as according to

$$\delta = 0.15 + (0.2 - 0.3)\frac{d}{1000} \quad mm \tag{10.2}$$

To ensure sufficient load capacity of the bearing, the following factors are important for design and operation

1. Sufficient oil supply must be ensured by either thrower rings or pressure oil sources. The load factor expressed in specific load $P(MPa)$ and linear speed $V(ms^{-1})$ is used as a criterion

$\sqrt{PV^3} = 6$–50	- lubrication by thrower rings or thrust plate, no cooling required
$= 50$–1000	- same as above, with cooling
> 100	- lubrication by forced circulation

2. Well designed route for oil recirculation. The hot oil coming out of the bearings must be separated from the returning cooled oil.

3. The bearing assembly must be located in a position convenient for maintenance and repair, especially for readjustment of clearance.

10.2.4 Flow passage components

(i) Inlet passages and their application range

Aside from the various types of inlet passages used in medium/large turbines, the following types are peculiar to small turbines:

(a) Open–flume inlet

Open–flumes are easy to build and offer excellent inflow conditions when properly designed. They are used for the head range of 2–6m and for turbine diameters smaller than 1.5m. In order to reduce the hydraulic losses, the inlet velocity of open–flumes are limited to

$$V = (0.5 - 0.7)\sqrt{H} \approx 0.7 - 1.7 \ ms^{-1} \tag{10.3}$$

The following dimensions are recommended for open–flume inlet passages

Horizontal–shaft installations (Figure 10.7)

$$B = (3.5 - 4.0)D_1 \quad \text{and} \quad L = (3.5 - 5.0)D_1 \tag{10.4}$$

with higher values for front inlet and lower values for side inlet, also $h_1 = 2D_1$ and $h_2 = 1.5D_1$.

Vertical–shaft installations (Figure 10.8)

$$B = (3 - 4)D_1; \ \ h = (0.9 - 1.0)D_1; \ \ \text{and} \ \ L = (3 - 4)D_1 \tag{10.5}$$

Better performance may be obtained when the outer walls of an open–flume inlet is made in the shape of a spiral case.

(b) Can–type inlet casing

The can–type inlet casing is used for very small turbines for its simplicity and ease in manufacture. But its use is limited because of the poor flow conditions in the can (Figure 10.9). The inlet flow velocity is determined by $V = \alpha\sqrt{H}$, with flow coefficient $\alpha = 0.6 - 0.8$, the can–inlet diameter D_e is given by

$$D_e = \sqrt{\frac{4Q}{\pi V}} \tag{10.6}$$

Recommended values of major dimensions are

$$D_k = (2.6 - 3.5)D_1; \quad L_k = (2.0 - 2.5)D_1$$

The ratio of flow cross sections is

$$\frac{F_k}{F_e} = 1.3 - 1.5$$

where

$$F_k = \frac{\pi}{4}(D_k - d_k) \quad \text{and} \quad F_e = \frac{\pi D_e^2}{4}$$

(c) Full spiral case with circular cross sections

Spiral cases of small dimensions are usually made of cast iron and laid out with circular cross sections as shown in Figure 10.10. Calculation of the cross sectional areas follows essentially the same methods for large turbines, and can be found in various references, such as [10.3].

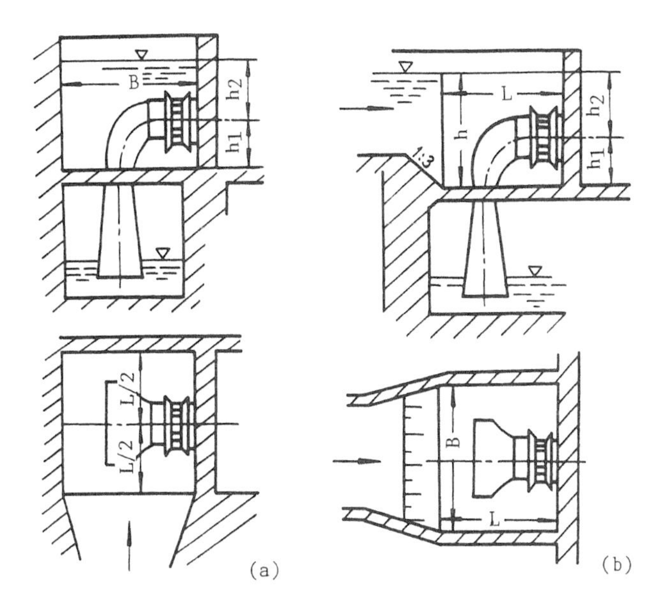

Figure 10.7 Open–flumes for horizontal–shaft turbines
(a)–side inlet; (b)-front inlet

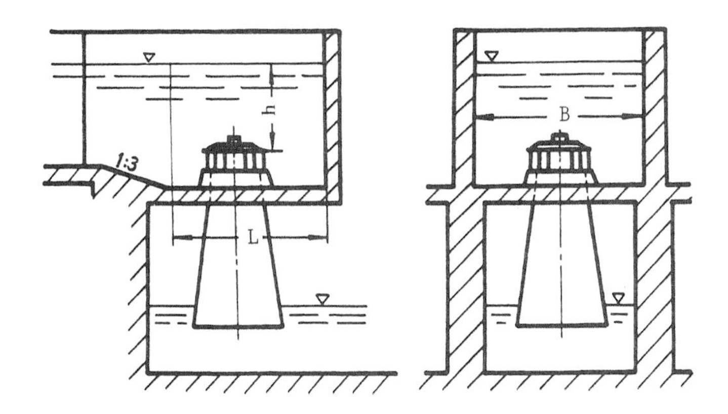

Figure 10.8 Open–flumes for vertical shaft turbines

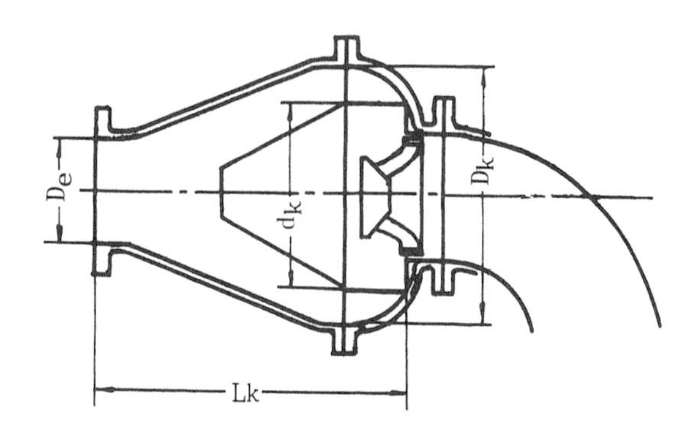

Figure 10.9 Can–type inlet casing

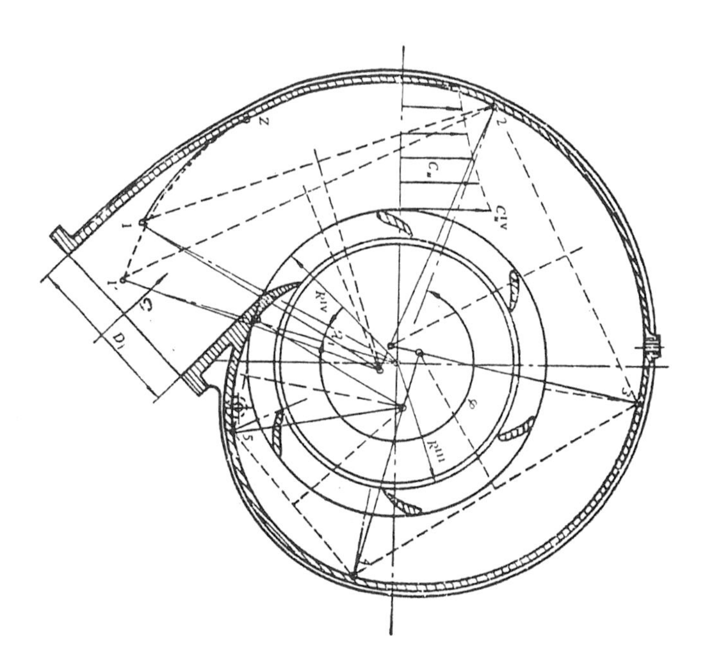

Figure 10.10 Metallic spiral case with circular cross sections [10.3]

(d) Single–guide vane spiral case

A simplified form of the spiral case has only one large guide vane at the inlet in place of the multi–bladed guide vanes, so the circulation required by the runner must be formed solely by the spiral case.

The spiral case may be laid out with circular cross sections, rectangular cross sections or a combination of a rectangular inlet and the other parts circular, as in Figure 10.11.

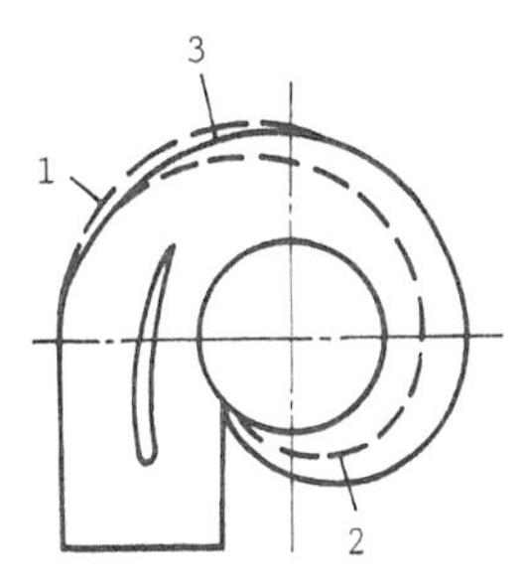

1 – circular cross section
2 – rectangular cross section
3 – transition from rectangular to circular cross section

Figure 10.11 Spiral case with single guide vane

(ii) Design of guide vanes

When the prototype turbine is analogous to the model turbine, the following relation exists for the guide vane openings

$$a_0 = a_{0m} \frac{D_0 \cdot z_{0m}}{D_{0m} \cdot z} \tag{10.7}$$

Small turbines, because of their particular requirements in manufacture, generally have fewer guide vanes, such as 12 to 16, and larger D_0/D_1 ratios than the models which are mostly intended for scaling up to larger machines. In this case, the above similarity relation no longer holds true and will result in calculated vane outlet angles different from the model. Hence for small turbines the guide vane outlet angles must by adjusted from model data by calculating

$$\tan \alpha = \frac{K n_{11} Q_{11}}{60 \alpha \eta g (B_0/D_1)} \tag{10.8}$$

where K – factor to compensate for displacement by GV, (=1.07)
α – factor to compensate for losses in circulation in spiral case and at runner exit, $\alpha = 1.15 - 1.2$.

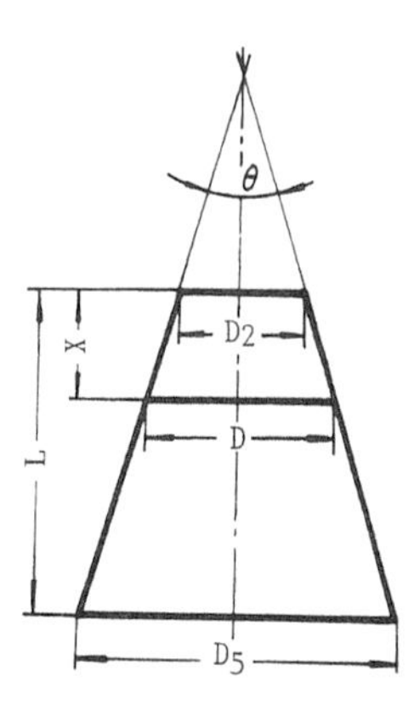

Figure 10.12 Straight conical diffuser

(iii) Design of draft tube

Straight conical diffusers are the simplest type of draft tubes and are widely used in small turbines. The straight conical diffuser has good energy recovery since it has even flow distribution and low hydraulic losses and may reach an efficiency of 83%. The outflow velocity is given by:

$$V_5 = 0.008H + 1.2 \ (ms^{-1}) \tag{10.9}$$

The hydraulic losses in a straight conical diffuser are shown in Figure 10.13 where the tolal loss is lowest at a diffusing angle of $12^o - 16^o$. The length of the diffuser may be calculated from the inlet area A_2, outlet area A_5 and cone angle θ. The relation between L/D_2 and the best cone angle with efficiency is shown in Figure 10.14.

Small turbines with horizontal shaft generally have an elbow of constant cross sectional area fitted before the vertical conical diffuser. Pertinent dimensions are given in Figure 10.15

$$L_0 = 0.4D', \qquad r = 0.6D'$$

and $\gamma \leq$ conical angle of runner band, therefore,

$$D''' = D' + 2L_0 tg\gamma, \quad D'' = D'''/\cos\gamma$$

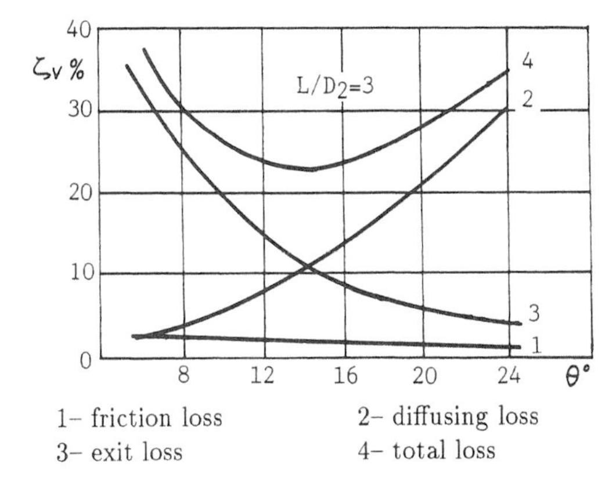

Figure 10.13 Relative losses in diffuser [10.6]

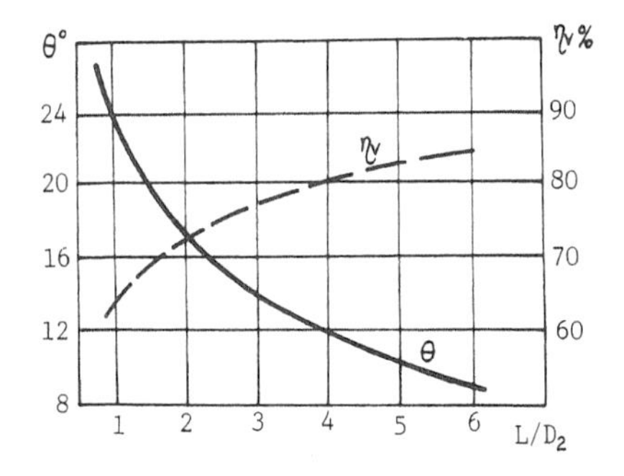

Figure 10.14 Relation of L/D_2 to cone angle

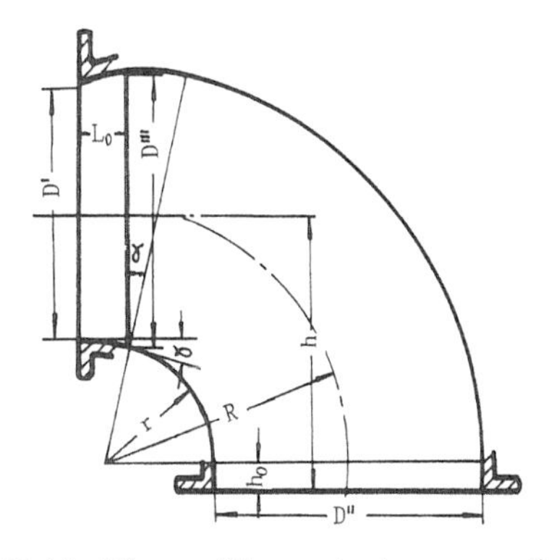

Figure 10.15 Elbow with constant cross sectional area

$$\text{and} \quad h = (r + D''/2)cos\gamma + h_0 \tag{10.10}$$

Elbows with constant cross sectional areas usually have lower efficiency than elbows specially designed for hydraulic turbines, so the latter type should be used for small turbines of higher specific speeds.

Some types of small turbines (can–type, S–type tubular) have the main shaft passing through the draft tube, a sleeve over the rotating shaft is sometimes added to reduce hydraulic interference.

10.3 Small Impulse Turbines

Small tangential–jet impulse (Pelton) turbines are an important category in small turbine application. Since their descriptions are well covered in Chapter 5 along with large Pelton turbines, space here will be given only to the other two types of impulse turbines, the cross–flow (Banki) and the inclined–jet (Turgo) turbines which are used exclusively in small hydro developments.

10.3.1 Cross–flow turbines

The cross–flow turbine gets its name from the fact that water flows through the blades of the cage–like runner in the upper part first and again in the lower part.

(i) General description

Figure 10.16 shows a section of a cross–flow turbine. The runner *3* is fitted inside the casing *1* and rides on ball bearings *4* on each end of the shaft. Water enters from the inlet, *8* at the top and flows around the guide vanes *2* twice through the runner, and leaves downward by way of the discharge duct *7*. The turbine may be connected to the generator directly or through increaser gears. For low head machines, a draft tube may be used to help recover part of the exit flow energy. In this case, an air valve is installed to control the pressure in the casing to keep the runner from being submerged.

(ii) Analysis of flow through the runner

In a cross–flow turbine, the water flows through the blade passages twice, that is from point *1* to *2*,and again from point *3* to *4*, as shown in Figure 10.17.

The basic equation of one–dimensional flow is

$$P = \frac{Q\gamma}{g}[(U_1C_1\cos\alpha_1 - U_2C_2\cos\alpha_2) + (U_3C_3\cos\alpha_3 - U_4C_4\cos\alpha_4)] \tag{10.11}$$

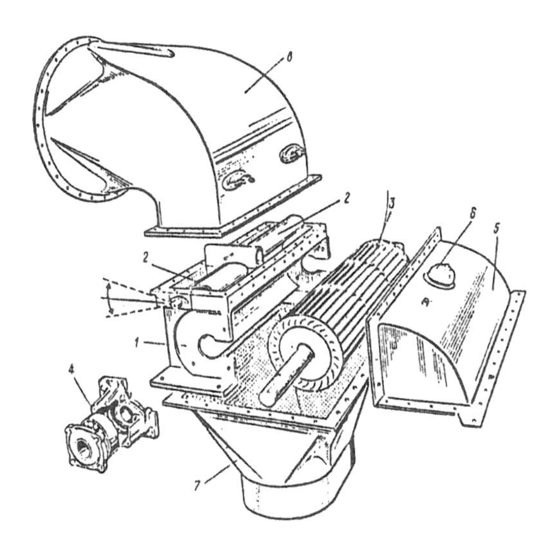

1- runner chamber 2- guide vanes 3- runner
4- bearing assembly 5- cover 6- air valve
7- discharge duct 8- inlet duct

Figure 10.16 Component parts of a cross–flow turbine
(Drawing courtesy Ossberger)

From the diagram, $U_2 = U_3$, $U_4 = U_1$ and $\alpha_3 = \alpha_2$, so the equation becomes

$$P = \frac{Q\gamma}{g} U_1 (C_1 \cos \alpha_1 - C_4 \cos \alpha_4) \qquad (10.12)$$

Expressing the absolute velocities C in terms of relative velocities W, and from the following relations

$$\beta_4 = 180^o - \beta_1, \quad \beta_3 = 180^o - \beta_2, \quad \beta_3 = \beta_2 = 90^o, \quad W_3 = W_2 \;\; W_1 = W_4$$

the equation becomes

$$P = \frac{\gamma Q}{g} 2 U_1 W_1 \cos \beta_1 \qquad (10.13)$$

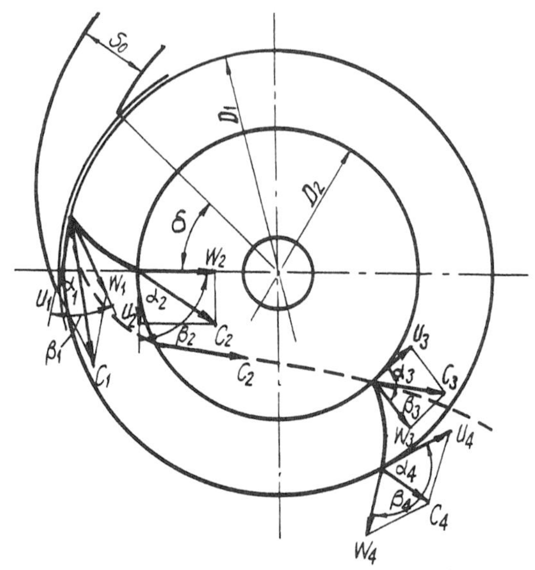

Figure 10.17 Schematic of flow through cross–flow turbine

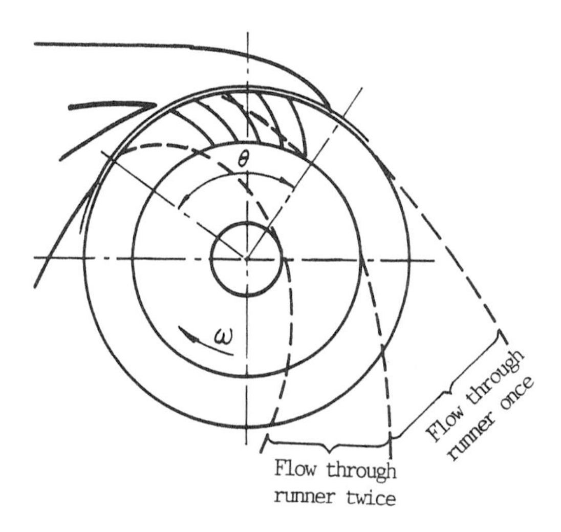

Figure 10.18 Actual flow through a cross–flow runner

The efficiency of the turbine is then

$$\eta_h = \frac{4U_1W_1\cos\beta_1}{2U_1W_1\cos\beta_1 + U_1^2 + W_1^2} \tag{10.14}$$

By putting $d\eta_h/dU = 0$ the following expression is obtained, after simplification

$$\eta_{h_{max}} = \frac{2\cos\beta_1}{1+\cos\beta_1} \quad \text{at} \quad U_1 = W_1 \tag{10.15}$$

It is found from the flow the triangles and further substitution, that

$$\alpha_4 = \beta_4/2 = \frac{1}{2}\,(180^o - \beta_1) \tag{10.16}$$

It is difficult to find the best value of β_1 by analysis since in theory the efficiency is 1 at $\beta_1 = 0$ which is obviously impossible as β_1 cannot be zero. From experience, β_1 is taken to be 30°, the maximum efficiency of the turbine is found to be 0.928 by neglecting the exit loss. With $\alpha_1 = \beta_1/2 = 15^o$, the optimum runner speed is found to be $U_1 = 0.518C_1$.

(iii) Design considerations

Research made on cross–flow turbines indicate that

- A certain amount of pressure must be maintained at the nozzle outlet so the runner inlet velocity C_1 will decrease somewhat with an increase in S_0/D_1 and create a higher pressure at the inlet, therefore, the cross–flow turbine is not truly an impulse turbine.

- As shown in Figure 10.18, the main body of the flow passes through the runner twice but a certain part passes through only once and is really *non–cross–flow.* This flow which is found to exist at the runner outlet rotates with the runner for about 90° before leaving the runner [10.9]. Other secondary flows are suspected to exist also in this region.

- Flow in the central core of the runner tends to converge so an acceleration exists in the space between the two stages.

- Maximum efficiency of the cross–flow turbine is obtained only when the inlet flow spans over about 90° at the inlet. This and the above phenomena cannot be explained by theory, so far, and is treated only from experience [10.8].

(iv) Application

Cross–flow turbines may be used under heads up to $100m$, their specific speed being in the range of $\Omega = 0.28 - 1.04$ ($n_s = 55 - 200$ $m.kW$). The largest cross–flow turbine in operation rates at $1000kW$ under $172m$ head.

The German firm Ossberger has many years of experience in manufacturing cross–flow turbines, a typical performance curve is given in Figure 10.19. The runner is partitioned into two parts, so 1/3, 2/3 or the full length may be used at a time depending on the load condition. The machine is able to attain efficiencies of 83–84% over the flow range of 25–100%.

10.3.2 Inclined–jet turbines

(i) General description

The inclined–jet turbine is an impulse turbine where the jet hits the runner at an angle ($22.5^o - 25^o$). It aims to avoid the losses occurring from the back side of the buckets being hit by the outflowing water. Also due to its geometry, the inclined–jet turbine can be applied to much higher specific speed (up to $\Omega = 0.31 - 0.36$, or $n_s = 60 - 70$) than the Pelton turbine (usually less than $\Omega = 0.16$ or $n_s = 30$), that is, the jet to runner diameter ratio may go as high as 1:3.57 (Figure 10.20) as against 1:(6–7) or much smaller for Pelton turbines (see 5.2.1).

(ii) Analysis of flow through runner

The flow through the runner is represented by the diagram shown in Figure 10.21. From the basic equation of flow through the turbine

$$\eta_h gH = U_1 C_1 \cos\alpha_1 - U_2 C_2 \cos\alpha_2 \tag{10.17}$$

it is seen that $U_1 = U_2 = U$ as points 1 and 2 are on the same radius. The terms $C_1 \cos\alpha_1$ and $C_2 \cos\alpha_2$ in the equation can be changed into functions of W_1, W_2 and β_1, γ_2 as

$$C_1 \cos\alpha_1 - C_2 \cos\alpha_2 = W_1 \cos\beta_1 + W_2 \cos\gamma_2 \tag{10.18}$$

From the Bernoulli equation for relative motion, it can be shown that:

$$W_1^2/2g = W_2^2/2g + h \tag{10.19}$$

where h is the head loss between the two points and may be expressed as $h = \zeta W_2^2/2g$, so that $W_2 = W_1/\sqrt{(1+\zeta)}$

If the velocity coefficient of the jet is φ, then

$$gH = C_1^2/2\varphi^2$$

and the basic equation may be further expressed as

$$\eta_h = \varphi^2 \frac{2\sin\alpha_1 \sin(\beta_1 - \alpha_1)}{\sin^2\beta_1}\left(\cos\beta_1 + \frac{\cos\gamma_2}{\sqrt{1+\zeta}}\right) \tag{10.20}$$

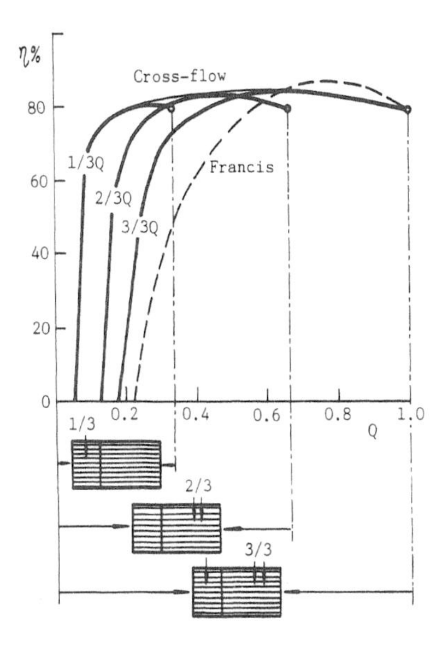

Figure 10.19 Performance curve of cross–flow turbine [10.5]

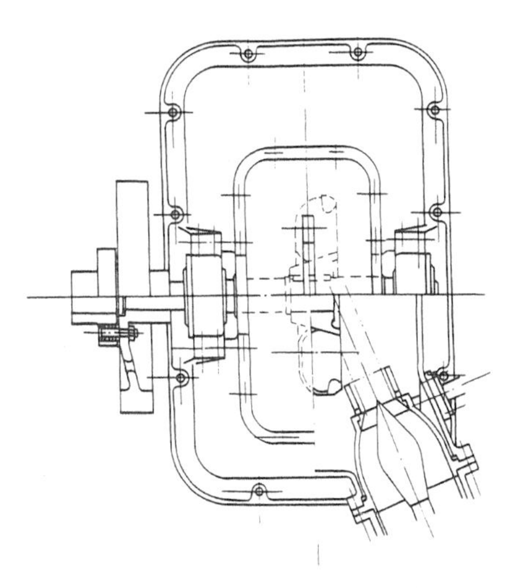

Figure 10.20 Construction of inclined–jet turbine
(Drawing courtesy of TRIED)

η_h is the hydraulic efficiency of the turbine, therefore it follows that the hydraulic efficiency of the runner is

$$\eta_{hr} = \frac{1}{\varphi^2 \cdot \eta_h} \tag{10.21}$$

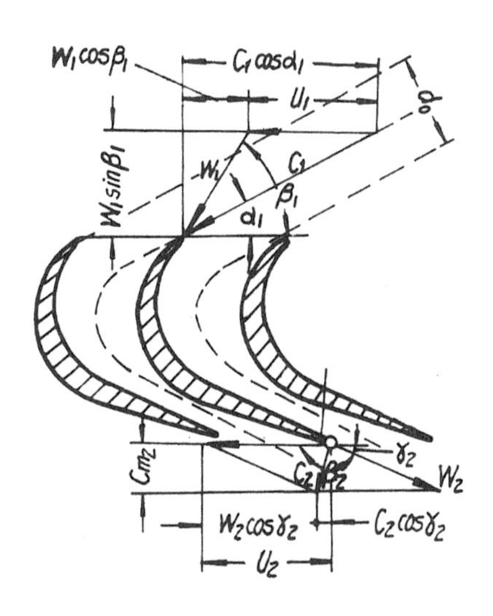

Figure 10.21 Flow through inclined-jet turbine [10.10]

It is known that the runner efficiency η_{hr} is related to the inlet angle α_1, inlet blade angle β_1 and outlet blade angle γ_2. Assuming no friction loss in the runner passage, then $h = 1$, or $U_1 = W_2$. The condition for η_{hr} to be a maximum is

$$\frac{\partial \eta_{hr}}{\partial \alpha_1} = 0 \quad \frac{\partial \eta_{hr}}{\partial \beta_1} = 0 \quad \frac{\partial \eta_{hr}}{\partial \gamma_2} = 0 \tag{10.22}$$

From these, the following solutions are obtained

$$\gamma_2 = 0^0, \quad \alpha_1 = \frac{1}{2}\beta_1$$

The inlet flow triangle must be an isosceles triangle, so the optimum peripheral velocity is

$$U_1 = \frac{C_1}{2\cos\alpha_1} \quad \text{or} \quad ku_1 = \frac{kc_1}{2\cos\alpha_1} \tag{10.23}$$

When $kc_1 = 1$ (that is $\varphi = 1$), $ku_1 = 0.542$ (for $\alpha_1 = 22.5^o$), it is seen the optimum linear velocity of an inclined–jet turbine is slightly higher than that for a Pelton turbine.

(iii) Design considerations

The hydraulic characteristics of an inclined–jet turbine depends on the position of the jet ellipse, the number of runner blades, the shape of inlet edges, incident angle of the jet, distance between the nozzle and runner, and the shape of the casing, and so on. [10.10].

(a) Influence of number of blades

Although the number of runner blades has little effect on the unit speed n_1, it does influence the unit flow Q_1 and the efficiency η greatly. From model tests, it is found with Z decreasing from 28 to 22, the value of η drops from 85.6 to 85% and Q_{11} increases from 70 to 85 ℓs^{-1}. When Z decreases further to 20, η drops to 84% and Q_{11} rises to $100 \ell s^{-1}$.

(b) Influence of incident angle of jet

Good characteristics is obtained when the peripheral velocity of runner equals to the relative inlet velocity flow, $U_1 = W_1$, that is the incident angle $\alpha_1 = \beta_1/2$. As shown in Figure 10.22, for a blade angle of $\beta_1 = 45^o$, the machine has the best efficiency with incident angle $\alpha_1 = 22.5^o$. At all other incident angles the performance is lower.

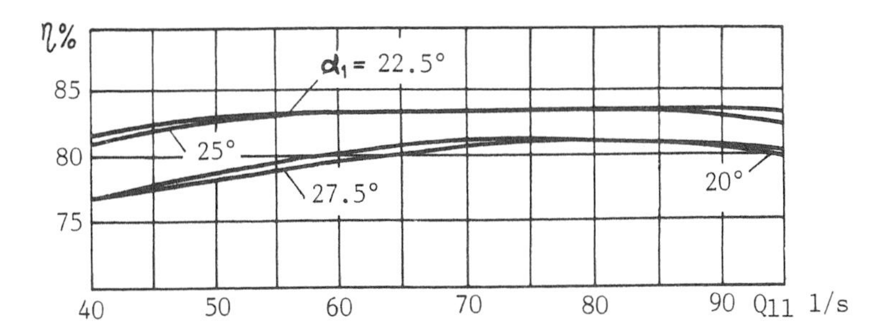

Figure 10.22 Relation of efficiency and unit flow of inclined–jet turbine

(c) Influence of blade inlet profile

With the same number of blades, $Z = 20$, and the same incident angle $\alpha_1 = 22.5^o$, runner 2 (Figure 10.23) with a more curved blade and changing blade angle along the radius has higher efficiency than runner 1 with constant blade angle.

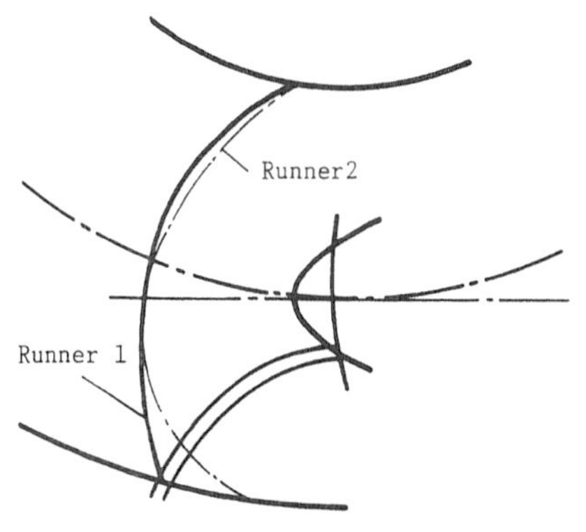

Figure 10.23 Comparison of inlet blade shape

(d) Influence of distance between nozzle and runner

Figure 10.24 shows with the increase of the distance between the nozzle and the runner, efficiency of the turbine decreases. Therefore, the shortest distance practical should be used in designing the turbine.

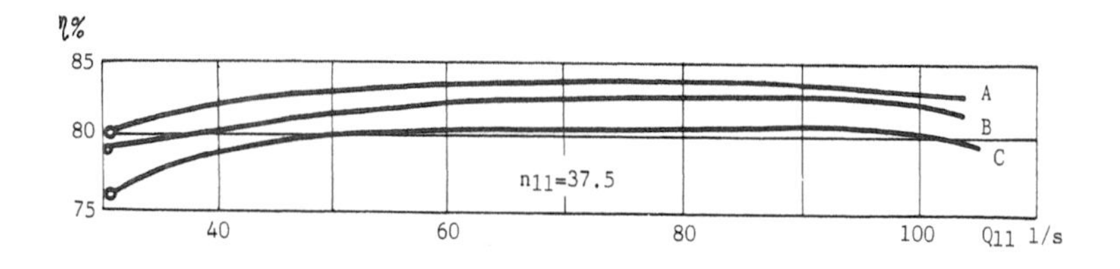

A – 100mm, B – 17mm, C– 258mm

Figure 10.24 Effect of distance between nozzle and runner

(e) Influence of casing shape

The turbine casing is installed off the runner central plane since the water discharges on one side only (see Figure 10.20). A deflector on the casing periphery must be provided to scrape off the splashing water from the outflow which would otherwise interfere with the runner. It is found that the water rotates about 120° with the runner from inflow to outflow. For horizontal-shaft units, the jet ellipse should be advanced 50° against the direction of rotation (from 0 to 0′ in Figure 10.25) to allow effluent discharge of the runner.

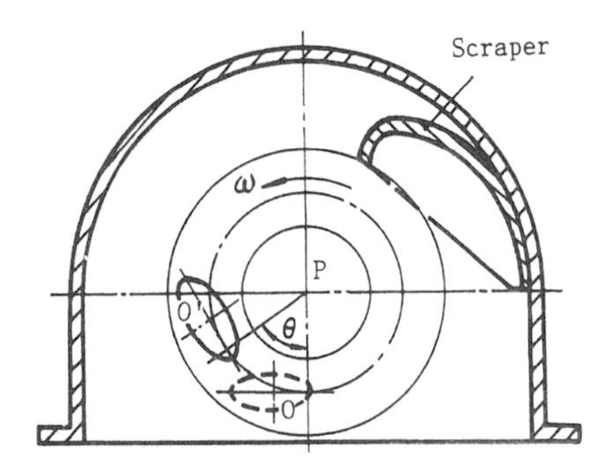

Figure 10.25 Runner casing of inclined-jet turbine

(iv) Specific speed of inclined-jet turbine

The specific speed is defined as:

$$n_s = \frac{n\sqrt{P}}{H^{5/4}} \quad \text{in engineering units}$$

Since

$$P = \frac{\eta\gamma HQ}{102} \quad \text{and} \quad Q = \frac{\pi}{4} d_0^2 \varphi \sqrt{2gH}$$

where d_0 is the diameter of the jet and φ the coefficient of flow, the specific speed is then

$$n_s = A\frac{n\sqrt{\eta} \cdot d_0}{\sqrt{H}} \quad \text{with} \quad A = \left(\frac{\pi\varphi\gamma\sqrt{2g}}{4 \cdot 102}\right)^{1/2} \tag{10.24}$$

With $m = d_0/D_1$, then theoretically the specific speed becomes

$$n_s = 490 \cdot ku_1 \cdot m\sqrt{\eta} \tag{10.25}$$

and unit flow

$$Q_1 = \frac{Q}{D_1^2\sqrt{H}} = \frac{\pi}{4}\varphi\sqrt{2g} \cdot m^2 = 3.48\varphi m^2 \tag{10.26}$$

(v) Application

The inclined–jet turbine is suitable for a specific speed range of $\Omega = 0.13 - 0.36 (n_s = 25 - 70mkW)$. The first such machine built by Gilbert Gilkes & Gordon in England in 1919 was rated at $40h.p.$ under $60m$ head and attained an efficiency of 81%. The earliest jet/runner diameter ratio was 1:5.25 and was later raised to 1:4.5 with increased discharge flow and efficiency. From 1939 to 1945, inclined–jet turbines reached ratings of $4300h.p.$ $(H = 275m, \quad n = 1000min^{-1})$ with the largest one of $6000h.p.$ $(H = 244m, \quad n = 600min^{-1})$. By 1960, the products known as *high capacity Turgo impulse turbines* have jet/runner diameter ratio of 1:3.75 and specific speed of $\Omega = 0.34$ $(n_s = 66)$. The advancement of efficiency in the course of development is shown in Figure 10.26. At present the maximum working head for inclined–jet turbines is $273m$, the maximum output being $10{,}090h.p.$

10.4 Serialisation and Selection of Small Turbines

10.4.1 Standardized turbine series

Serialisation of small turbines is to decide on a number of designs whose parameters can best cover the very wide range of head and flow (output) required by hydro power developments. The gist of the matter is to select the proper range of a limited number of machine specifications to satisfy the most applications.

(i) Items of Serialisation

- Type of machine and the runner series of each.
- Machine specifications of each runner series.
- Arrangement of installation (alignment of shaft. type of inlet).

(ii) Guide lines for establishing the runner series

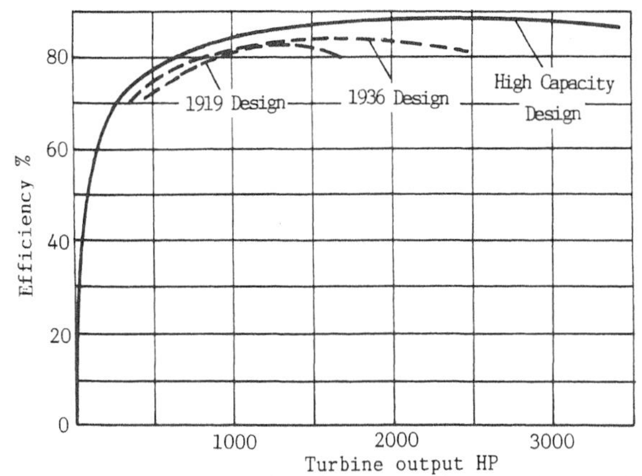

Figure 10.26 Evolution of performance of inclined–jet turbine (Drawing courtesy Gilbert Gilkes & Gordon)

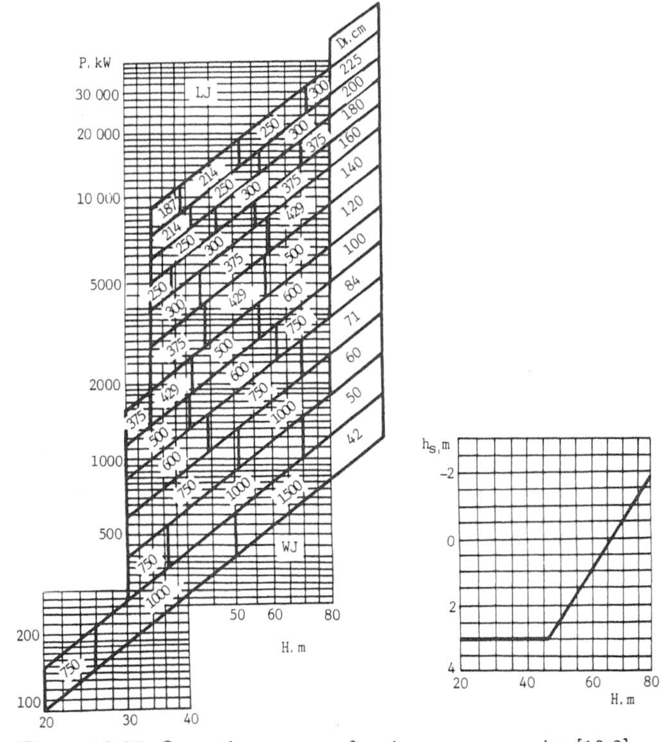

Figure 10.27 Operating ranges of a given runner series [10.2]

- Division of the total head range into suitable sectors by the principle that the increase in n_s derived from each sector corresponds to one increment in the runner diameter series. The runner selected for this series must keep the required suction height within a certain range, for example not less than -$2m$ for small turbines. For horizontal units with output of less than $1000kW$ the H_s must be positive.

- Runners are selected under the condition that the drop in efficiency within the operating range must not exceed an allowable amount. A preferred ratio of runner diameter between two adjacent grades is decided and applies to all increments.

(iii) Operating range of turbines

To facilitate selection of turbines from the serialized ratings, several kinds of performance charts are available, such as the ones showing the applicable operating range on logarithmic H and Q scales (Figure 10.1). A chart showing the applicable ranges of a given runner series is shown in Figure 10.27. The recommended speeds for each runner diameter are marked in the parallelogram bound by the upper and lower limits in output of adjacent products and on either side by the head range. The small diagram gives values of h_s for calculating the required suction height H_s:

For vertical units $H_s = h_s - \nabla/900$

For horizontal units $H_s = h_s - \nabla/900 - D_2/2$

where ∇ is the elevation of the hydroelectric station above sea level.

10.4.2 Selection of small turbines

(i) Original data

The following information must be supplied by the prospective user or the consulting agency:

- Maximum head, design head, minimum head and weighted average head, water levels of upper and lower pools, maximum flow, design flow.

- Total installed capacity, number of units.

- Hydrological data of the station, for example salinity of water, sediment contents (average and short–time peak values).

- Operation mode of the station, for example isolated or tie–line operation, base or peak load, synchronous compensating operation.

- Maximum allowable excavation depth of station.

(ii) Selection of machine parameters

The machine selected must have:

- Not only a high peak efficiency but also good average efficiency under variation of head and load, that is a turbine with the flattest possible performance curve.
- The smallest possible dimensions, i.e. of highest possible specific speed and operating speed.
- The best possible cavitation characteristics.

When more than one variant is found applicable to the project at hand, which may be from different series of the same type of machine or from different choices of the same series, or even from different types of machines, such as axial–flow versus Francis, studies of comprehensive nature must then be carried out to critically judge the merits of each.

Axial–flow turbines have good regulation characteristics and are superior to Francis turbines in adjusting to variation of head and flow. Francis turbines are applicable to a very wide range of heads and are noted for their rugged construction and trouble–free operation. Impulse turbines have the flattest performance curve and is best suited for stations with great variation of load. When used in silt–laden rivers, the Pelton turbine has the advantage of ease of replacement of worn parts.

(iii) Calculation procedure for selection

The reader is referred to Volume *Hydraulic Design of Hydraulic Machinery* of this Book Series for detailed procedures for selection of turbine parameters. Only outlines are given here that pertain to the selection of small turbines

- select the type of turbine and runner series;
- calculate the required diameter and adjust to the – next larger standardized size;
calculate the rotating speed from the recommended unit speed and design head, adjust to the nearest synchronous speed;
- convert model efficiency to prototype by recommended formula;
- correct unit flow and unit speed values when necessary;
- draw up the projected operational performance curve;
- calculate the required suction height;
- calculate the runaway speed;
- calculate the hydraulic axial thrust.

References

10.1 Ministry of Water Resources and Electric Power. *Proceedings of International Symposium on Small Hydro Power*, Water Conservancy Press, 1982 (in Chinese).

10.2 Tienjin Design and Research Institute of Electric Drive. *Design and calculation of Hydraulic Turbines*, Science Press, 1971 (in Chinese).

10.3 Nechleba, M. *Hydraulic Turbines, Their Design and Equipment*, Artia Press, 1957.

10.4 Karelin, V. Y., Volshnik, V. V. 'Construction and Operation of Small Hydropower Stations', *Energoatomizdat*, 1986 (in Russian).

10.5 Moniton, L., Lenir, M., Roux, J. 'Micro Hydroelectric Power Stations', John Wiley & Sons, Chichester, 1984.

10.6 Cheng Liang-jun. 'Hydraulic Turbines', *Machinery Industry Press*, 1981 (in Chinese).

10.7 Tienjin Design and Research Institute of Electric Drive, *Atlas of Hydraulic Turbine Construction*, Science Press 1978(in Chinese).

10.8 Arter, A. 'The Splitflow Turbine–Crossflow Development Results in a New Design', Proceedings of 3rd International Conference on Small hydro Power, Hangzhou Regional Centre for Small Hydropower, Hangzhou, China, 1986.

10.9 An Li-rang, Lu Li. 'Investigation of Hydraulic Characterisitics of Cross-flow Turbines', *Hydro Power Equipment*, 1986, No.1 (in Chinese).

10.10 Shupulin, E. F. 'Inclined-jet Turbines', Proceedings of VEGM, No.19, Mashgiz, 1956. (in Russian)

Chapter 11

General Concepts of Pump Construction

R. K. Turton

11.1 Fields of Application

Pumps are the workhorses of many industries. This volume is concerned with pump applications ranging from city water supply to hydroelectric storage systems.

The commonly used arrangements of machine flow paths are shown in Figure 11.1. Single stage axial–flow pumps are designed for low heads and high flows, a typical duty being $10m$ and $14m^3s^{-1}$. They are used for land drainage and irrigation, for dock de–watering and large flow transfer duties. Single stage centrifugal pumps, used in many water supply facilities, are used to produce heads in the range of 5 to $300m$ with corresponding flow rates of 0.05 to $0.5m^3s^{-1}$. Single stage mixed–flow pumps may have flow rates up to $1.7m^3s^{-1}$ with the same typical heads for pumping duties.

Typical impeller profiles, diameter ratios and pump characteristics are illustrated in Figure 11.2 which indicate that the flow path tends to go from radial to axial as the characteristic number Ω increases, where Ω is defined by

$$\Omega = \frac{\omega\sqrt{Q}}{(gH)^{3/4}}$$

If higher heads are required, multi–stage pumps are available, as shown in Figure 11.3. These machines may be built in either vertical or horizontal arrangements.

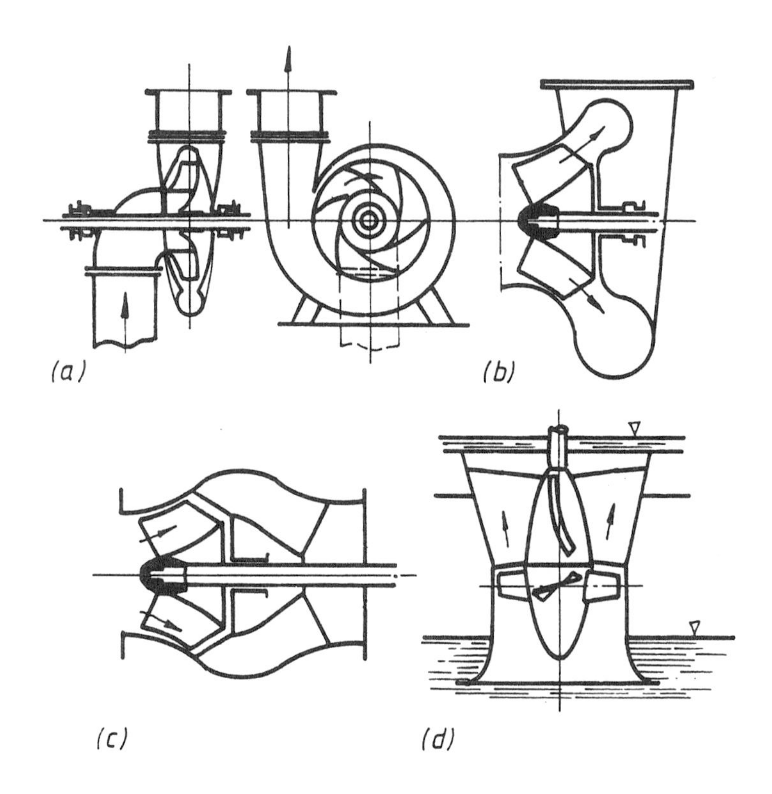

(a) Centrifugal pump (b) Mixed–flow volute pump
(c) Mixed–flow pump with return passage (d) Axial–flow pump

Figure 11.1 Arrangements of different types of single–stage pumps

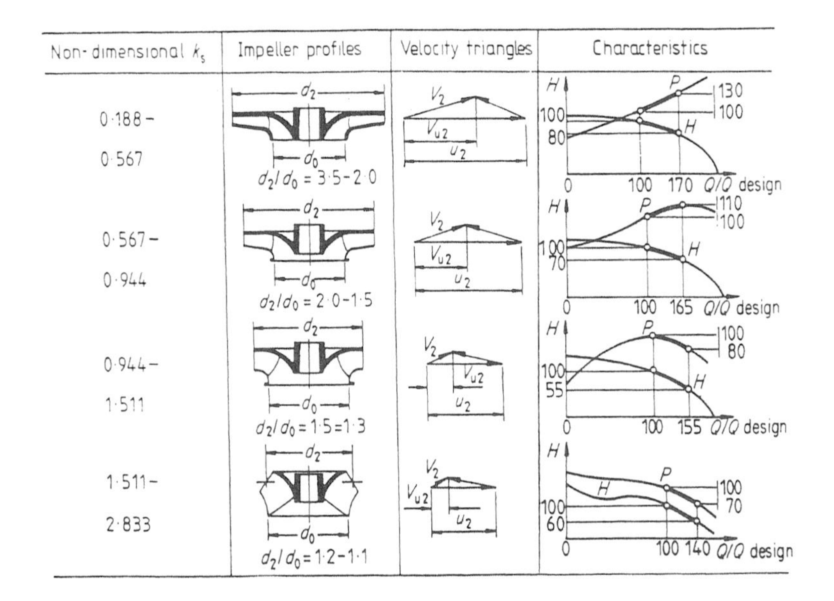

Figure 11.2 Typical impeller profiles and characteristics of single-stage pumps (Drawing courtesy Sulzer Escher Wyss)

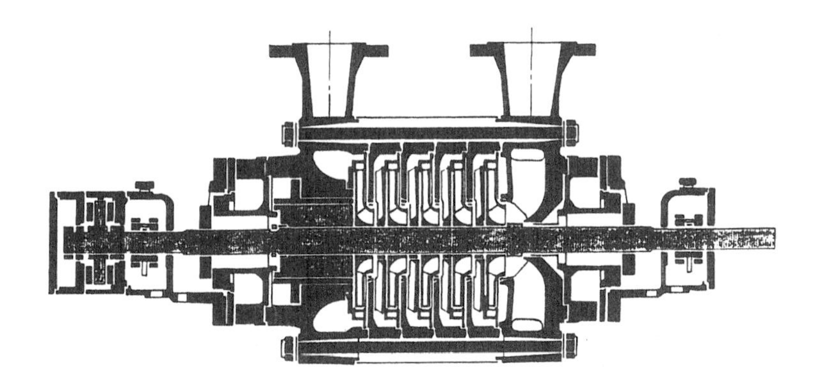

Figure 11.3 Section of a multi-stage centrifugal pump

11.2 Typical Pump Designs

11.2.1 Axial–flow pumps

Most axial–flow pumps are constant speed fixed–blade machines, and may have vertical or horizontal shafts. Figure 11.4(a) shows a vertical pump with metallic casing. The larger machines have concrete casings, as in Figure 11.4(b), and may be motor or engine driven through a gearbox.

The fixed blade axial–flow pump has a comparatively narrow range of application. If greater variations in duty curves are needed, the pump can have a variable speed drive, or be provided with variable–pitch impeller blades, as shown in Figure 11.5.

11.2.2 Single stage centrifugal pumps

Figures 11.6, 11.7 and 11.8 show typical single–stage pump designs. In the small pump in Figure 11.6, the impeller is fitted on the motor shaft or an extension to the shaft, so that the motor bearings are required to withstand the hydraulic loads imposed. These machines deliver heads up to $200m$ and flows up to $0.3m^3s^{-1}$. Figure 11.7 shows a typical modern single–stage pump of back pull–out design, arranged so that the bearing housing, seal, shaft and impeller assembly may be removed for maintenance without the need to break the casing to pipe joints. As can be seen a muff coupling is needed to provide the axial movement necessary for removing the impeller, bearing and seal assembly. Such machines give flows in the range of 0.005 to $0.1m^3s^{-1}$, with a head range of 10 to $240m$.

If twice the flow rate at the same head or better cavitation characteristic at the same flow rate is required, a double suction pump may be used, as Figure 11.8. In this design the pump shaft passes through the flow passages at both eye sections of the impeller. This type of pump is made with a two–part casing where the joint is usually horizontal through the centreline, so the pump may be opened for inspection and repair without disturbing the main pipe connections. These pumps are often used in town water supply systems and deliver flows up to $8.5m^3s^{-1}$.

11.2.3 Single stage mixed–flow pumps

Mixed–flow (Francis) and bulb pumps are normally used to deliver higher flows than by centrifugal pumps. The flow path in the impeller is three–dimensional. Figure 11.9 shows a volute type mixed–flow pump which has construction features similar to centrifugal pumps. Figure 11.10 shows a multi–stage design where the flow is directed back to the inlet of the next

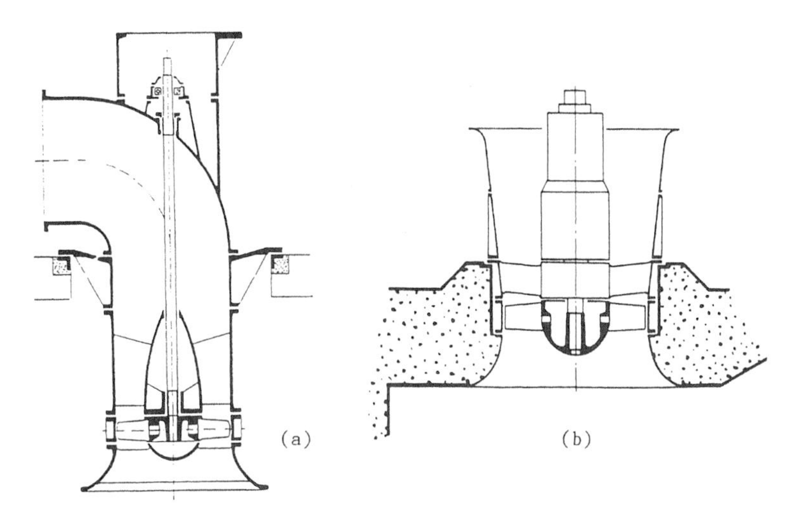

(a) Pump with metallic casing (b) Pump with concrete casing

Figure 11.4 Vertical axial–flow pumps

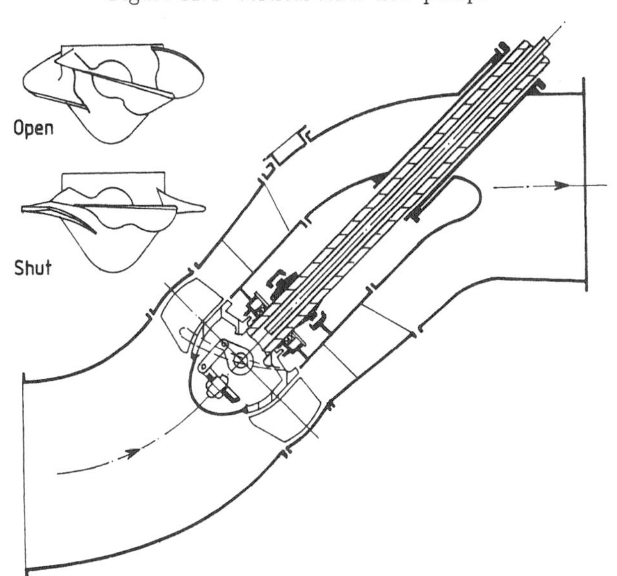

Figure 11.5 Schematic diagram of a variable–pitch axial–flow pump

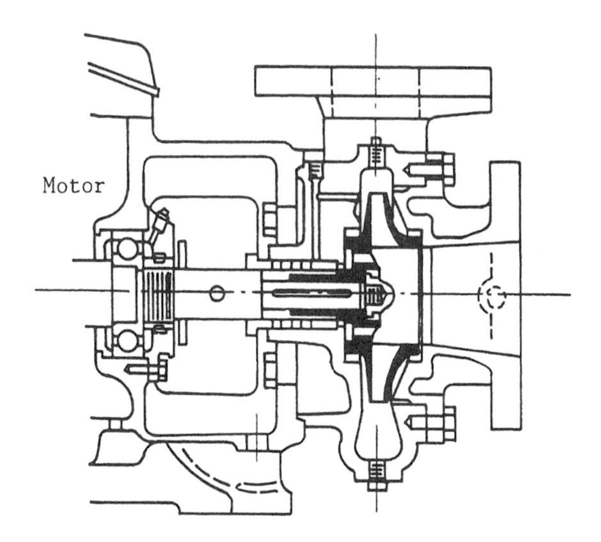

Figure 11.6 Single-stage centrifugal pump mounted integral with driving motor (monobloc)

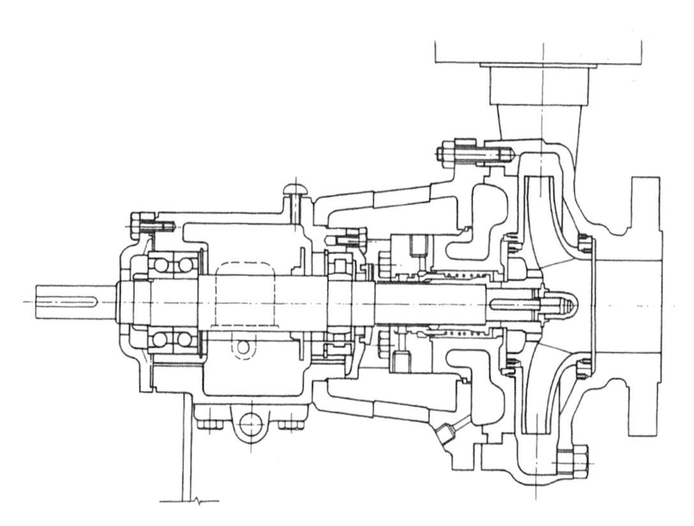

Figure 11.7 Centrifugal pump of back pull-out design

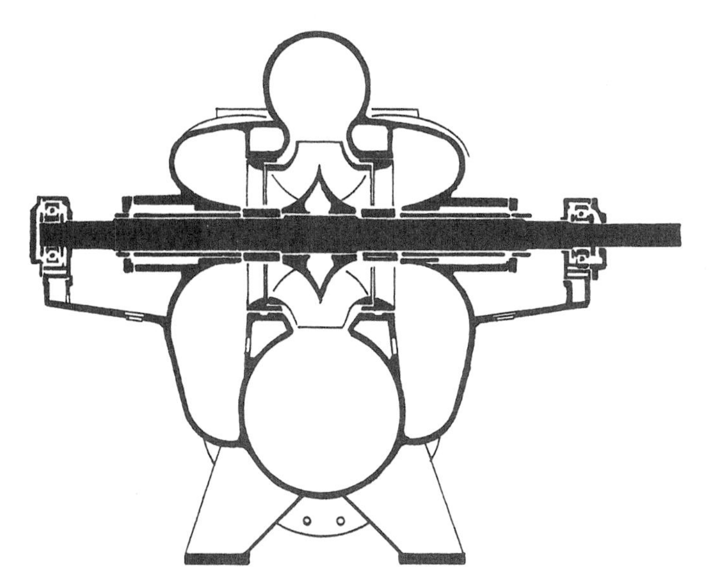

Figure 11.8 Double suction centrifugal pump

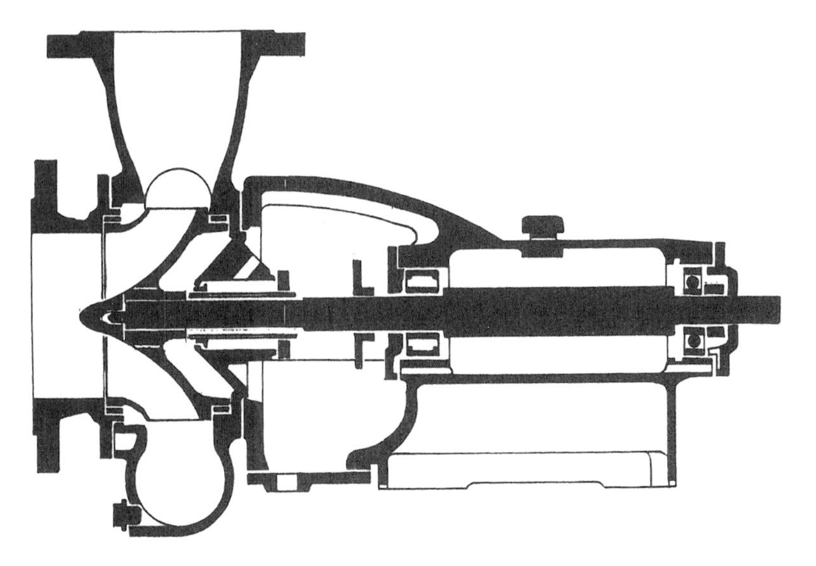

Figure 11.9 Volute type mixed–flow pump

stage via a return passage. This type of construction is most commonly used in bore hole pumps.

11.2.4 Multi–stage pumps

Where required delivery heads are greater than a single–stage pump can handle, several stages of impellers are arranged in series to form a multi–stage pump. The conventional horizontal shaft layout for boiler feed pumps and similar high pressure duties is shown in Figure 11.2. A typical bore hole (deep well) design is shown in Figure 11.10.

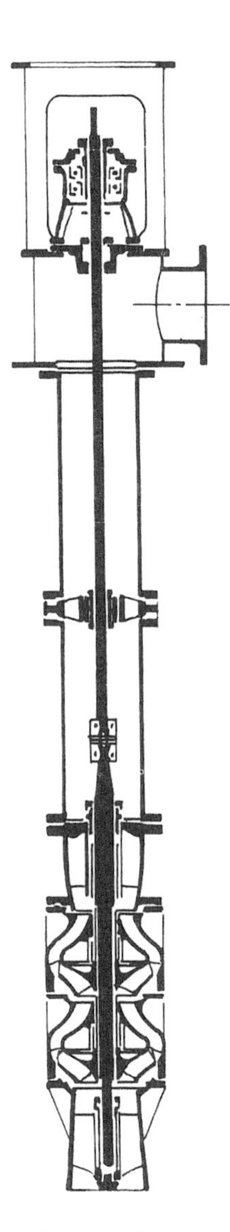

Figure 11.10 Vertical mixed–flow multi–stage pump

Chapter 12

Construction of Centrifugal Pumps

Huang Jing–guo

12.1 General Trends of Impeller Design

The centrifugal pump is the most widely used type of all rotodynamic pumps today. The name *centrifugal* originated from the very early days of pump history. Judging from the direction of flow through the pump, *radial–flow* would be more fitting for this type as compared with other types of pumps described in this volume.

When analysed from concepts of fluid motion, centrifugal force is not the only force exerting on the fluid in such a pump, nor are the other types of pumps free from centrifugal forces. However, as the name centrifugal pump is so predominantly used in the industry today, this more conventional name is retained in the text of this chapter.

12.1.1 Specific speed

The specific speed represents a similarity criterion for all rotodynamic pumps. When two impellers are geometrically and dynamically similar, they will have the same specific speed. The specific speed is the condition of similitude for enlarging or reducing a known impeller design. From dimensional analysis, the expression

$$n_s = 3.65\frac{n\sqrt{Q}}{H^{3/4}} \tag{12.1}$$

is designated specific speed, where speed n is in min^{-1}; flow rate Q in m^3s^{-1}; and head H in m. In most European countries, the specific speed is expressed as

$$n_q = \frac{n\sqrt{Q}}{H^{3/4}} \tag{12.2}$$

Obviously, there is the relation $n_s = 3.65 n_q$.

The numerical value of the specific speed will vary greatly with different definition formula and different units used. The following table shows the conversion constants between a number of commonly used expressions.

Table 12.1 Conversion constants for specific speed expressions

	Non-dimensional	European	China/USSR	Japan	U. K.	U. S. A.
Formulae	$\Omega = \frac{\text{rad/s} \cdot \sqrt{\text{m}^3/\text{s}}}{(981 \cdot \text{m})^{\frac{3}{4}}}$	$n_q = \frac{\text{min}^{-1}\sqrt{\text{m}^3/\text{s}}}{\text{m}^{\frac{3}{4}}}$	$N_s = \frac{365 \cdot \text{min}^{-1}\sqrt{\text{m}^3/\text{s}}}{\text{m}^{\frac{3}{4}}}$	$N_s = \frac{\text{min}^{-1} \cdot \sqrt{\text{m}^3/\text{min}}}{\text{m}^{\frac{3}{4}}}$	$N_s = \frac{\text{min}^{-1} \cdot \sqrt{\text{l/s}}}{\text{m}^{\frac{3}{4}}}$	$N_s = \frac{\text{min}^{-1} \cdot \sqrt{\text{gal/min}}}{\text{ft}^{\frac{3}{4}}}$
Conversion Constants	1	52.9	193	410	1671	2732
	0.0189	1	3.65	7.75	31.6	51.7
	0.0052	0.274	1	2.12	8.66	14.15
	0.0024	0.129	0.471	1	4.08	6.67
	0.0006	0.0316	0.116	0.245	1	1.63
	0.00037	0.0194	0.0707	0.150	0.612	1

It should be pointed out that impellers of the same specific speed may differ in geometric shape because of different hydraulic considerations assumed in the designs. The designer may select design elements such as the blade inlet angle β_1, discharge blade angle β_2, number of blades z and discharge width b_2, for designs with different emphasis in hydraulic performance and arrive at impellers with varied geometric dimensions, and consequently different head discharge characteristics, efficiency and suction performance.

Stepanoff proposed the impeller discharge specific speed as a criterion of pump performance, and is expressed as [12.1]

$$S_2 = \frac{n\sqrt{Q}}{h_{v2}^{3/4}} \tag{12.3}$$

where h_{v2} is defined by $C_{m2}^2/2g$, where C_{m2} is the meridional component of the discharge flow velocity. For impellers of different specific speed but with the same flow Q, speed n, and blade discharge angle β_2, the outlet velocity C_{m2} will be the same and is independent of head H. Therefore, the impeller–discharge specific speed S_2 is a constant and is not effected by n_s, H or D_2 for the same β_2. This is true for all impellers with β_2 smaller than 90° [12.1].

For the condition $S_2 =$ const, the following is obtained

$$\left[\frac{\phi_2}{\phi_1}\right]^2 = \left[\frac{n_{s2}}{n_{s1}}\right]^{4/3} \left[\frac{\tan\beta_2 - \phi_2}{\tan\beta_1 - \phi_1}\right] \tag{12.4}$$

where ϕ is the flow coefficient, and

$$\phi_1 = C_{m1}/U_1 \quad \text{and} \quad \phi_2 = C_{m2}/U_2 \tag{12.5}$$

For a given impeller, if the ϕ value is known for a certain specific speed, the values at other specific speeds may be determined from equation (12.4), which has been proved by practice.

Flow coefficient ϕ and head coefficient ψ values for standard designs at best efficiency are given in Figure 12.1. Every pair of points on the chart is derived from the optimal b_2/D_2 ratio for the given specific speed. By relating different β_2 values for the same specific speed, impellers of different dimensions, efficiency and different inclination of head–discharge performance curves may be obtained.

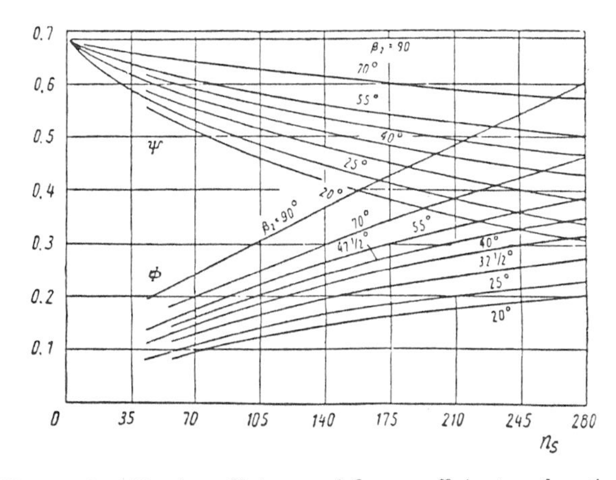

Figure 12.1 Head coefficient and flow coefficient as functions of specific speed [12.1]

Normally, centrifugal impellers are considered best for applications in the range of $n_s = 35 - 280$. But from the point of view of hydraulic designs, an impeller may be laid out as *centrifugal* for a specific speed as high as 450, that is with reference to the direction of flow being perpendicular to the rotating axis. Advantages in removing the *hump* in the head–discharge curve at low flows for high specific speed impellers may be obtained after *centrifugal design* treatment. The complete characteristics of such a design will be noticeably different from that of the standard mixed-flow impeller design, with the $H = 0$ line in the third quadrant of the complete characteristics moving to the fourth quadrant. Pumps designed with this special method have been successfully used in main coolant circulating pumps for pressurized water nuclear power stations.

12.1.2 Suction performance

The Thoma number that has been very widely used as a cavitation criterion for hydraulic turbines is also used to show the suction or cavitation characteristics of centrifugal pumps, expressed as the ratio of net positive suction head ($NPSH$) to total head H

$$\sigma = \frac{NPSH}{H} \tag{12.6}$$

In actual practice, designers may prefer to use another criterion, the suction specific speed, defined by

$$C = \frac{5.62n\sqrt{Q}}{NPSH^{3/4}} \tag{12.7}$$

where Q is in m^3s^{-1}, n in min^{-1} and $NPSH$ in m.

The $NPSH$ can be shown to be a function of absolute velocity C_1 and the relative velocity W_1 at the impeller inlet:

$$NPSH = \frac{\mu C_1^2}{2g} + \frac{\lambda W_1^2}{2g} \tag{12.8}$$

The suction specific speed C can better represent the cavitation characteristics of an impeller since it contains the terms C_1 and W_1 that are closely related to the flow Q and is commonly used to replace the Thoma number σ in the field of pump design.

The suction specific speed was first proposed as a general criterion for inlet similarity conditions, however, it was found this relation does not hold true under different inlet operating conditions as the suction specific speed will vary with the speed, specific speed and efficiency of the pump [12.2].

It is difficult to accurately predict the $NPSH$ value of a centrifugal pump at different operating speeds and is even more difficult when dealing with different designs of pumps. Yedidiah analysed the experimental data of some 600 pumps produced by ten manufacturers throughout the world and proposed an approximate statistical formula as

$$\frac{NPSH_p}{NPSH_m} = \left(\frac{n_p}{n_m}\right)^{1.424} \left(\frac{D_p}{D_m}\right)^{1.272} \tag{12.9}$$

which shows the statistical regularity for variation of $NPSH$ with linear dimension and speed [12.3]. Therefore, prediction of $NPSH$ under condition of increasing speed according to formula (12.7) is on the safe side as it is reduced to the form

$$\frac{NPSH_1}{NPSH_2} = \left(\frac{n_1}{n_2}\right)^2 \tag{12.10}$$

A formula is available to estimate the *NPSH* value at any operating point from the *NPSH* at the optimum point (best efficiency point), as from [12.4]

$$NPSH = \frac{1}{3}\,NPSH_{opt}\left(\frac{Q}{Q_{opt}}\right)^2 + \frac{2}{3}\,NPSH_{opt} \tag{12.11}$$

where the flow Q must be within the normal operating range.

The suction specific speed is also expressed in a form similar to the specific speed as

$$S = \frac{n\sqrt{Q}}{NPSH^{3/4}} \tag{12.12}$$

Naturally, there is the relation $C = 5.62 \cdot S$ as compared with equation (12.7). Comparison of values of suction specific speed when applied in different expressions and units similar to those for specific speed expressions is given in Table 12.2.
Referring to the μ and λ values in equation (12.8), Pfleiderer in early researches gave the values for μ as $1.1 \sim 1.3$ and for λ as $0.25 \sim 0.35$ [12.5]. More recent results obtained by Oshima [12.6] show the μ value represents the non–uniformity and loss factor of the inlet flow and is independent of the flow and may be taken as $1.1 \sim 1.2$, whereas the cavitation coefficient λ varies with the flow over a wide range and may be taken from the curve in Figure 12.2.

There are four factors, μ, λ, C_1 and W_1 that determine the magnitude of the *NPSH*, from equation 12.8. These values are in turn related to the following elements:

1. Shape of suction chamber – prerotation of inlet flow.
2. Design of impeller – curvature of blade passage, location of inlet edge, angle between inlet edge and axis of rotation, sharpening of inlet edges.
3. Angle of attack α of flow at blade inlet.
4. Area of suction eye and number of blades.

In the design process, the impeller eye area is first determined to ensure good cavitation characteristics. The coefficient K_o is convenient for this purpose, as given in:

$$K_o = \frac{\sqrt{D_o^2 - d_h^2}}{\sqrt[3]{Q/n}} \tag{12.13}$$

Table 12.2 Conversion constants for suction specific speed

	China/USSR	U. K.	Japan	U. S. A.
Formula	$C=\frac{5.62\,(r/min)\,(m^3/s)^{\frac{1}{2}}}{(m)^{\frac{3}{4}}}$	$N_{ss}=\frac{(r/min)\,(l/s)^{\frac{1}{2}}}{(m)^{\frac{3}{4}}}$	$S=\frac{(r/min)\,(m^3/min)^{\frac{1}{2}}}{(m)^{\frac{3}{4}}}$	$N_{ss}=\frac{(r/min)\,(USgal/min)^{\frac{1}{2}}}{(ft)^{\frac{3}{4}}}$
Conversion Constants	1	5.63	1.38	9.2
	0.178	1	0.245	1.63
	0.725	4.08	1	6.67
	0.109	0.612	0.15	1

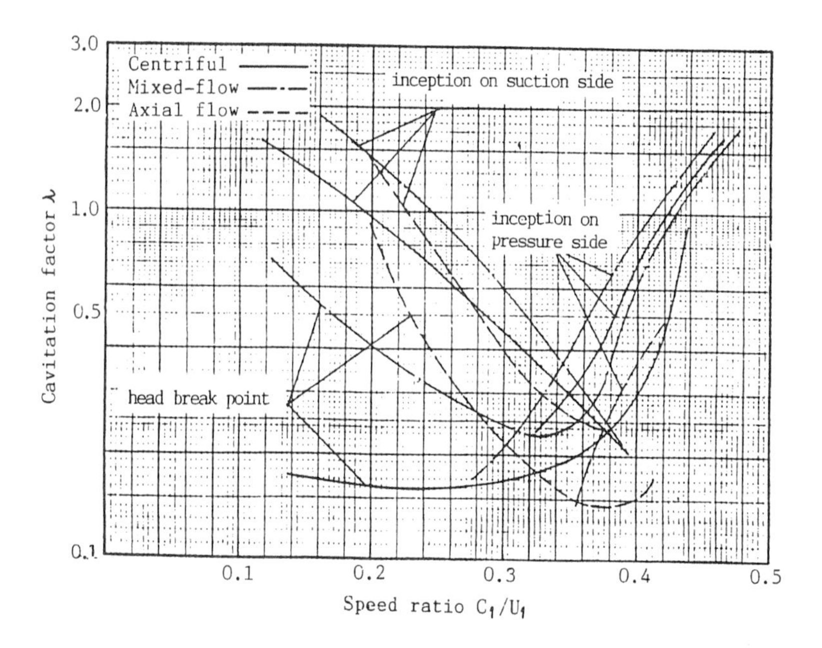

Figure 12.2 Cavitation coefficient for centrifugal mixed-flow and axial-flow pumps [12.6]

where D_o – diameter at suction eye (m)
d_h – diameter of hub (m); (for end suction pumps, $d_h = 0$)
Q – inlet flow accounting for leakage loss, $(m^3 s^{-1})$
n – speed of revolution, (min^{-1}).

Both the absolute value of Q and the velocity ratio C_1/U_1 (proportional to Q/n) are contained in the expression of K_o. Values of K_o may be selected for various applications according to Table 12.3.

Table 12.3 Guide for selection of K_o

K_o	Application of Design
3.1 ~ 4.0	Mainly in application for high efficiency and low cavitation requirement, like in second and higher stage impellers of multi – stage pumps
4.2 ~ 4.5	High efficiency and moderate cavitation requirements, as for single stage pumps of standardized design
5.0 ~ 5.5	For good cavitation characteristics, impellers used in combination with conical contour inducers
5.5 ~ 6.0	For combination with various types of inducers to obtain very good cavitation characteristics

The prerotation of the inlet flow is very important to the cavitation characteristics of a pump. From the inlet velocity diagram (Figure 12.3), where W_1' and β_1' represent the conditions at normal inlet and W_1 and β_1 represent those under prerotation, it is seen that the prerotation will reduce the relative inlet velocity W_1. Stepanoff proposed an empirical constant R_1 for use in deciding the inlet blade angle which may be decided by considering the C_{m1} in normal design and multiply by R_1. For the condition of no prerotation, R_1 is simply an expression of the inlet angle of attack. For a given design, the R_1 value may be easily converted to get the angle of attack.

$$R_1 = \tan\beta_1 \frac{U_1}{C_{m1}} = \frac{\tan\beta_1}{\tan\beta_{1'}} \tag{12.14}$$

It must be pointed out that prerotation of the inlet flow cannot be caused by the impeller itself. The amount of prerotation necessary to improve the cavitation characteristics is different from the prerotation at the impeller inlet as induced by the viscosity of the fluid from the rotation of the impeller, the required prerotation can only be supplied by properly designed inlet passage shapes.

The common procedure is to use a suction chamber of spiral or semispiral shape and for multi–stage pumps the outlet of the return passage can be designed to form the necessary prerotation for the next stage. In pumps for

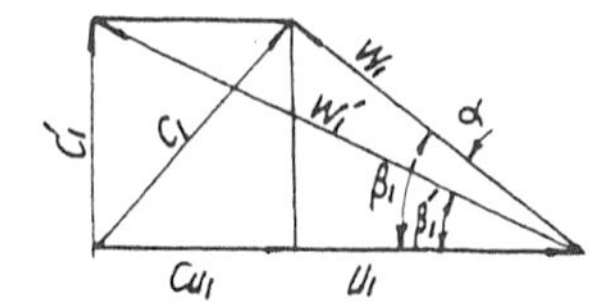

Figure 12.3 Velocity triangle at impeller inlet

rocket engines, a portion of the pump discharge flow is diverted through the throttle ring to inject circumferentially on the upstream flow of the inducer for formation of the prerotation. When the pump is operating at low flow, about 10% of the return flow is able to produce marked improvement of cavitation performance, the required *NPSH* may decrease by 50% [12.7]. This method of creating prerotation is equally applicable to ordinary centrifugal pumps, the simplest way is by directing the flow from the axial–thrust balancing pipes to the upstream flow of the pump inlet.

The angle of attack of the inlet flow α is of utmost importance in predicting the maximum operating capacity Q_{max} under given available *NPSH* and in changing the slope of the *NPSH* - Q curve.

The amount of prerotation that can be formed by the inlet passage should be taken into consideration in deciding the value of (α). Since distance between the suction chamber and the impeller inlet is quite small, it is proper to estimate the component C_u by assuming $C_u R = const.$ The applicable range for angles of attack is rather large, from 3° to 15°, therefore different values of α may be used for different streamlines along the blade inlet edge.

For first stage impellers of high–capacity high–speed pumps such as in boiler feed pumps, the angle of attack must be carefully selected so as to prevent damage to the impeller inlet edges due to too large an incidence. One recommended procedure is to observe the local cavitation conditions at the impeller inlet by visualization methods during model testing, the extent of cavitation bubbles is used as a measure to decide the blade angle. The best impeller design is the one with no local cavitation. The required *NPSH* for safe operation may also be determined by observation in such tests. Figure 12.4 shows the relation between the development of cavitation bubbles, the weight loss per unit time, and the cavitation coefficient at constant head.

Experimental investigations have proved the influence of impeller design on cavitation performance. Schöneberger gave the following expression for evaluation of cavitation performance [12.8]

$$K_A = \frac{NPSH}{U_1^2/2g} \tag{12.15}$$

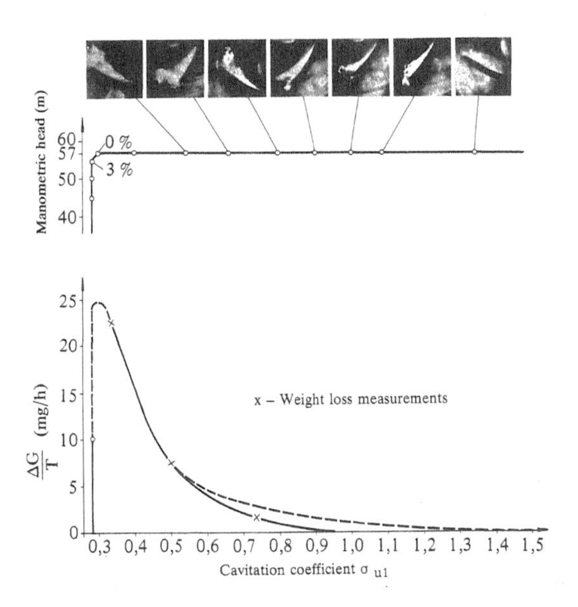

Figure 12.4 Development of cavitation bubbles and unit weight loss as function of cavitation coefficient at constant head [12.8]

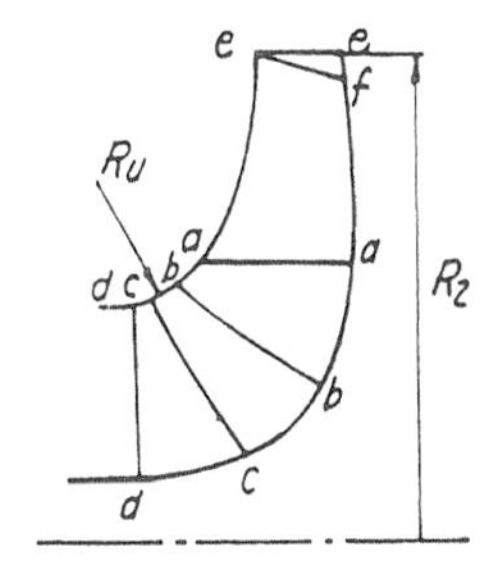

Figure 12.5 Impeller geometry

which may also be expressed as:

$$K_A = C\left[\frac{1}{1-\zeta}\left(\frac{C_{m1}}{U_1}\right)_{st}\right]^{3/2} \tag{12.16}$$

where

ζ – displacement coefficient at the impeller inlet
C – constant related to position of blade inlet edge, being 1.33–1.73.

Tests have shown that by shifting the blade inlet edge from a position parallel to the axit. (a–a in Figure 12.5) to one perpendicular to the axis (d–d), the *NPSH* value is reduced by 20–25%. Curvature of the flow passage may be represented by radius ratio R_u/R_2, where R_u is the radius of curvature of the front shroud and R_2 the outlet radius. The tested impellers have $R_u/R_2 = 0.08 - 0.15$. Actually, when the radius ratio is increased to 0.10–0.15, the cavitation coefficient K_A is reduced by 18%.

12.1.3 Stability considerations

Stability of a pump is normally considered from two aspects:

1. Hydraulic design of the impeller and stability of its characteristic curve.
2. Stability related to varying operating conditions.

The characteristic curve of a centrifugal pump is said to be stable when it has a constantly drooping tendency over the complete operating range starting from zero flow. For a given pump design, the slope of the characteristic curve may be expressed as

$$\frac{Q_N}{H_N}\cdot\left|\frac{\Delta H}{\Delta Q}\right|$$

where Q_N and H_N are values at best efficiency point; ΔH and ΔQ are variation in head and flow rate respectively within the operating range, the characteristic curve is said to be flat for

$$\frac{Q_N}{H_N}\cdot\left|\frac{\Delta H}{\Delta Q}\right| \leq 0.2$$

and is not flat for

$$\frac{Q_N}{H_N}\cdot\left|\frac{\Delta H}{\Delta Q}\right| \geq 0.2$$

Factors effecting the stability of the characteristic curve are inlet angle β_1, outlet angle β_2 and number of blades z, and so on. When the shutoff head is specified for the duty, the outlet diameter D_2 is also an important factor. Schröder proposed a chart to evaluate the stability regions of a pump on the basis of β_2 and z, as shown in Figure 12.6.

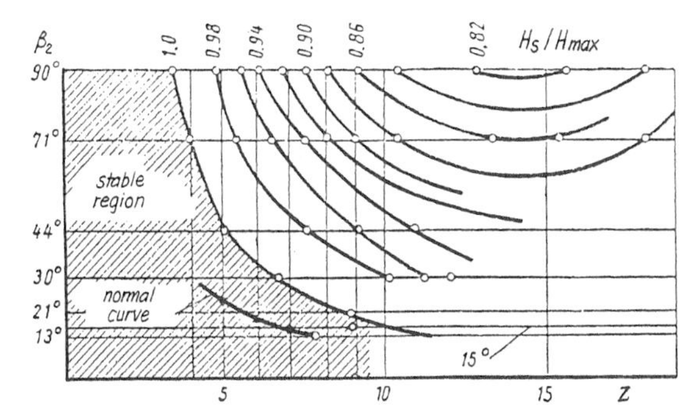

Figure 12.6 Stability condition for head–discharge curve [12.10]

It has been proved by practice that extension of the impeller inlet edge into the suction eye will increase the stability of the characteristic curve besides improving the suction performance, although the inlet blade angle will change somewhat with the shifting of the inlet edge. Inclination of the impeller outlet edge from the front shroud to the back shroud, as e–f in Figure 12.5, will also have the effect of improving stability of the characteristic. However, impellers are usually not designed this way, the outside diameter is sometimes machined to a slant only for the purpose of reducing the head or preventing instability.

When the application specifies a certain shutoff head of the pump, the following formula

$$H_0 = 0.585\frac{U_2^2}{g} \tag{12.17}$$

may be used to estimate the necessary outside diameter D_2 [12.1]. This dimension must be checked by accurate calculation after the outlet blade angle β_2 is fixed. Figure 12.1 may be used as a guide for selection of β_2. The D_2 satisfying the best efficiency point head H_N obtained from ψ value corresponding to the predetermined β_2 should be checked with the D_2 satisfying the shutoff head H_o.

By selecting volutes of different sizes to match a given impeller it is possible to change the slope of the head–discharge curve and position of the best efficiency point. This matching technique cannot remove the hump in the head–discharge curve but is able to change the slope of the curve. Figure 12.7 shows the performance of an impeller fitted with three different volutes of varying throat cross–sectional areas, the larger volute will give greater best efficiency point flow rate. The head at best efficiency point for all conditions will fall on the common tangent of the three head–discharge curves.

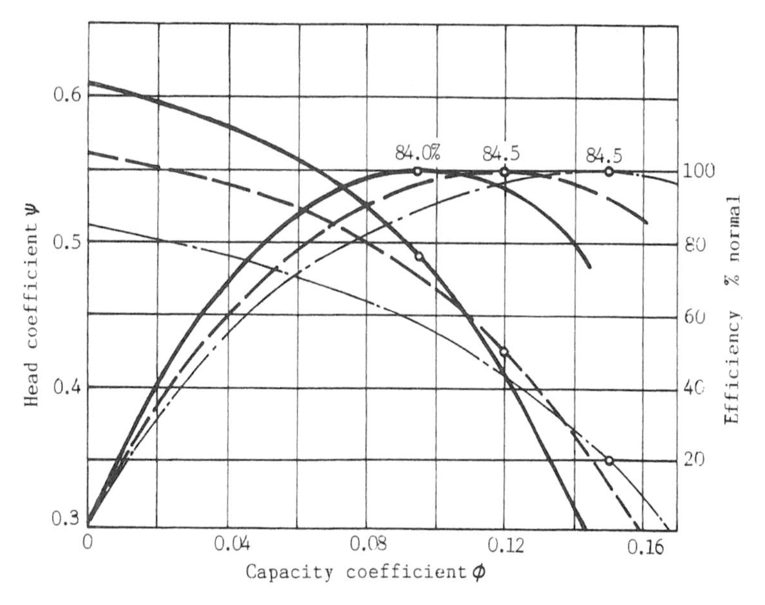

Figure 12.7 Performance curves of an impeller fitted in three different volutes [12.2 from Krisam]

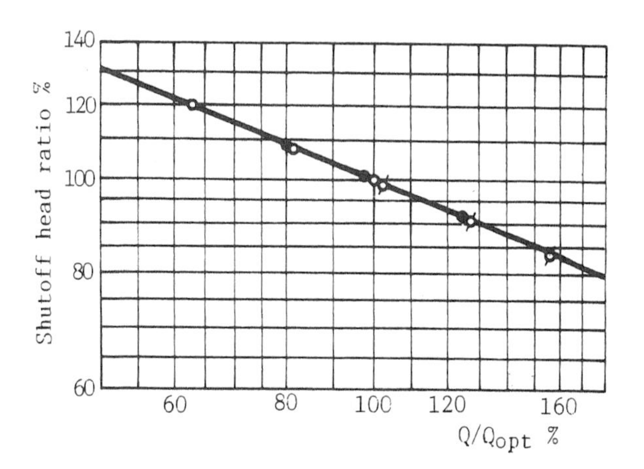

Figure 12.8 Ratio of shutoff heads of an impeller in three different volutes [12.2]

Figure 12.8 shows the ratio (percentage) of shutoff heads for an impeller fitted in different volutes to that fitted in the standard volute. All tested points fall along one straight line, which indicates it is immaterial which volute to pick as the reference. The shutoff head is closely related to the impeller design and is further shown in the example of Figure 12.9 where the shutoff head of an impeller of smaller flow is changed nearly linearly in ratio to the reduction of impeller outlet area from a standard impeller.

The pump operation in the high flow rate region, when driven by sufficient power, is usually limited by the available *NPSH* of the system. When the maximum discharge determined by the required *NPSH* is surpassed, the pump may run into unstable operation and vibration induced by cavitation.

In the low flow rate region, the operation may be limited by two factors: one is the temperature rise due to energy dissipation within the system; the other is the reverse flow at the impeller inlet (in some cases also at the impeller outlet).

In certain applications, the energy of internal losses will be converted into heat and cause the fluid temperature to rise. When the vapour pressure of the fluid under this temperature exceeds that of the pump pressure, the fluid may partially or totally vaporize. To prevent this undue vaporisation, the pump must discharge a certain amount of flow to carry away the heat, thus requiring a limit on the minimum operating flow Q_{min} to be imposed.

The temperature rise within the pump may be calculated by

$$\Delta t_R = \frac{0.00981}{C} \cdot \left(\frac{1}{\eta} - 1\right) \quad {}^oC \tag{12.18}$$

where C is the specific heat of the fluid, in $kJ \cdot kg^{-1} \cdot K^o$. The temperature rise in the clearance of axial-thrust balancing devices is

$$\Delta t_D = \frac{0.00981H}{C \cdot \eta} \tag{12.19}$$

The total temperature rise is

$$\Delta t_E = \Delta t_R + \Delta t_D$$

The minimum flow is then

$$Q_{min} = \frac{3600P_Q}{\rho C(t_E - t_s)} \quad m^3hr^{-1} \tag{12.20}$$

where

P_Q – liquid power kW

t_s – temperature of fluid in suction pipe

t_E – admissible temperature of fluid behind the balancing device – normally, (t_E - t_s) is $17^o - 20^oC$.

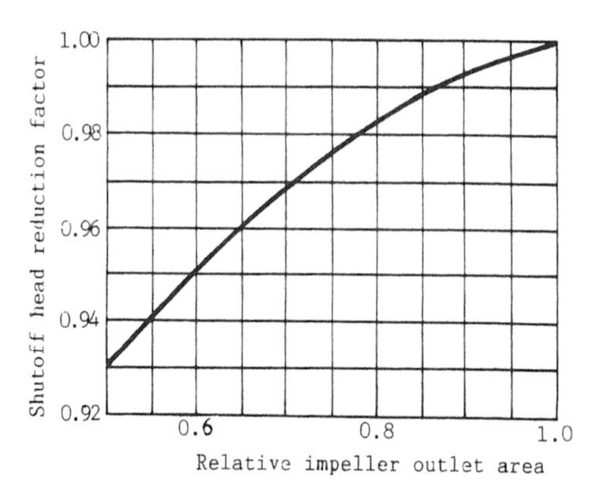

Figure 12.9 Ratio of drop in shutoff head as function of impeller outlet area [12.2 from Esman]

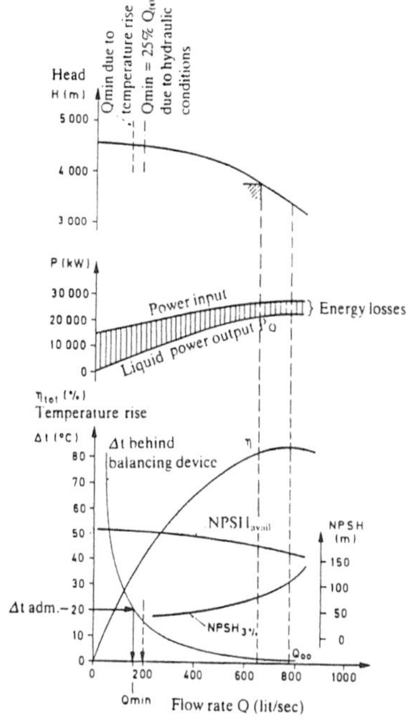

Figure 12.10 Allowable minimum flow as determined from temperature rise of fluid [12.11]

Figure 12.10 shows a calculated example, the allowable minimum flow rate is about 150 ℓs^{-1}, determined by an admissible temperature rise of 20^oC. It should be pointed out that although formula (12–20) is independent of the driving power, Q_{min} values thus calculated hold true only for pumps of smaller ratings, say up to $100kW$. For large power inputs (over $1000kW$) and high specific speed pumps, the minimum flow must be increased by 25% to 35% of rated flow rate or even more, otherwise, the pump and the piping system may be subjected to severe vibration. Makay found that for pumps of large power inputs and high specific speed the determination of Q_{min} is related to the specific speed and suction specific speed. As can be seen from Figure 12.11, the Q_{min} value increases with suction specific speed, and for impellers of the same specific speed, those with higher C values have the higher allowable Q_{min}. The highest value of Q_{min} may reach 40–60% of Q_N.

The other limitation to safe operation of pumps is due to hydraulic conditions, that is, reverse flow generated at the impeller inlet below a certain flow rate which may become the cause of unstable operation of the pump and vibration of the system. When reverse flow is developed in the suction cone and at the impeller inlet, the inlet flow passage is divided into two concentric areas: the through flow of the pump passing in the centre space while the reverse flow returning in the outer space. The amount of back flow Q_r measured for a pump of $n_s = 190$ is as follows [12.13]:

Flow rate Q, % of Q_N	78	60	53	45	31	0
Back flow intensity Q_r/Q_N	0	0.19	0.44	0.56	1.14	∞

When sufficient reverse flow is developed, it consumes so much input power that a drop in discharge head occurs which may fall continuously to shutoff or form a depression in the head–discharge curve (the cause of the well–known *saddle* or *hump*). Critical point for the appearance of reverse flow is 60–80% of rated flow depending on the specific speed and design of the suction passage of the pump, but detrimental effects from reverse flow will come at much smaller flow rate.

By observation of the flow phenomenon at the pump inlet in model testing, the best values for the pump suction eye and inlet blade angle may be found from conditions of minimal local cavitation and the least reverse flow.

The optimal combination of width and clearance between the impeller outlet and discharge chamber (volute or diffuser vanes) are also factors governing the stability of pumps. Florjancic proposed criteria for hydraulic stability for different combinations of impellers and diffusers and also for diffuser vane stress, as shown in Figure 12.12.

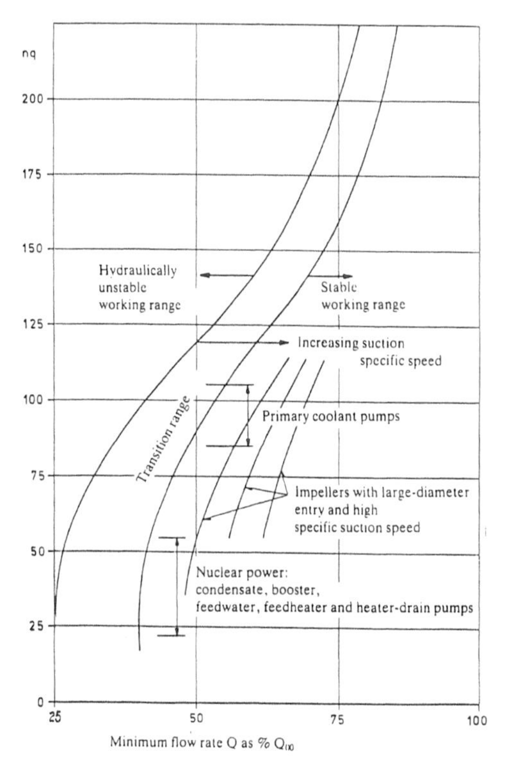

Figure 12.11 Minimum flow rate as function of n_s and pump type for power inputs exceeding 1,000 kW [12.12]

12.2 Construction of Suction and Discharge Chamber

The suction and discharge chambers are designed to conduct the fluid flow with the least losses while satisfying the overall layout requirements of the pump. In centrifugal pump design procedures, the fluid entering and leaving the impeller are assumed to be axisymmetric steady flow. The following assumptions are used under various conditions:

$$C_m = const, \quad C_u = const, \quad C_u R = const \tag{12.21}$$

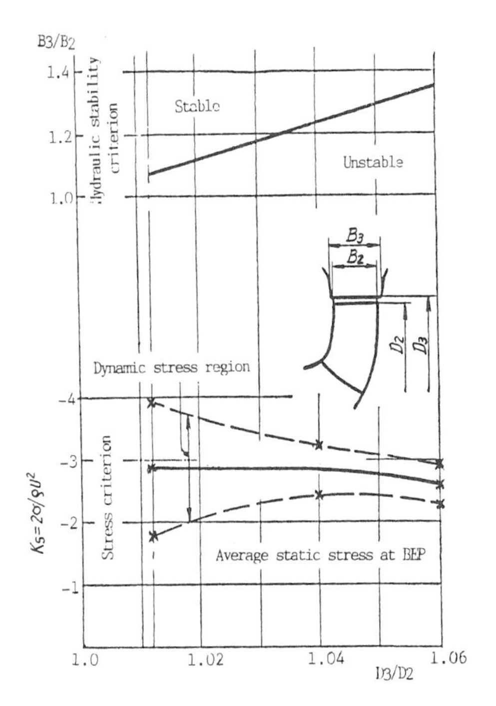

Figure 12.12 Criteria for hydraulic stability and diffuser vane stress as function of dimension ratio [12.14]

12.2.1 Suction chamber

For end–suction pumps, good suction flow conditions can be readily provided by a reducing conical pipe section before the impeller inlet. Swirl flow caused by successive bends in the inlet piping may be straightened out by short axial guide vanes in the inlet pipe.

For volutes, semi–volutes and circular suction chambers, it is important to keep the losses low where the flow changes from the radial to the axial direction. This portion of the passage, such as the collector pipe, can best be laid out as a lemniscate

$$r^2 = a_\Lambda^2 \cos 2\alpha \tag{12.22}$$

where, $a_\Lambda = (0.6 - 0.8)D_1$, and D_1 is the inside diameter of pipe.

For simplicity, the lemniscate curve may be replaced by two connecting arcs, such as with $r_1 = 0.5D_1$ and $r_2 = 1.4D_1$. By use of a portion of

$r_{min} = (0.15 - 0.2)D_1$ it is possible to fit well with the conical reducer and obtain uniform flow and a high flow coefficient when the space is limited by construction considerations.

It has been proved by practice in multistage pumps that the geometry at the inlet turn and the length of the inlet inducer has a decided effect on the efficiency and cavitation characteristics of the pump, this is more apparent in pumps with a double–suction impeller as the first stage. Noticeable improvements have been made in recent years in the design of inlet chamber of this type of pump. Figure 12.13(b) shows new types of suction chamber design which reduce the required *NPSH* and lower the cost of pumps.

12.2.2 Discharge chamber

Centrifugal pumps generally have discharge chambers of the following types with the exception of certain special purpose pumps such as ash pumps and slurry pumps:

- volute
- diffuser vanes plus volute
- diffuser vanes plus circular discharge chamber.

The function of the discharge chamber is to convert the kinetic energy of the fluid into pressure, the matching of its geometry with the impeller determines the position of the best efficiency point on the head–discharge curve. The area ratio rule proposed by Anderson [12.15/16] is well proven in practice for designing the discharge chamber. Figure 12.14 shows the correlation between the area ratio and specific speed while Figure 12.15 illustrates the prediction of pump performance from the throat velocity determined by the area ratio rule.

From the stand–point of energy balance, a centrifugal pump with an open impeller chamber is able to recover part of the disc friction loss from the front and rear shrouds and help to improve the efficiency. But with high speed and high head volute pumps, in small discharge rates far removed from the best efficiency point, pressure fluctuation induced by exit reverse flow may cause severe vibration of the pump and even the discharge piping system. By reverting to a closed impeller chamber with clearances between impeller outlet diameter and volute wall controlled to 1.3–1.9mm and by providing ample overlap of impeller shrouds with side walls, vibration and noise from the pump and piping system can be subdued [12.12]. Low discharge rate operation may then be extended from $50\%Q_N$ down to $18\%Q_N$.

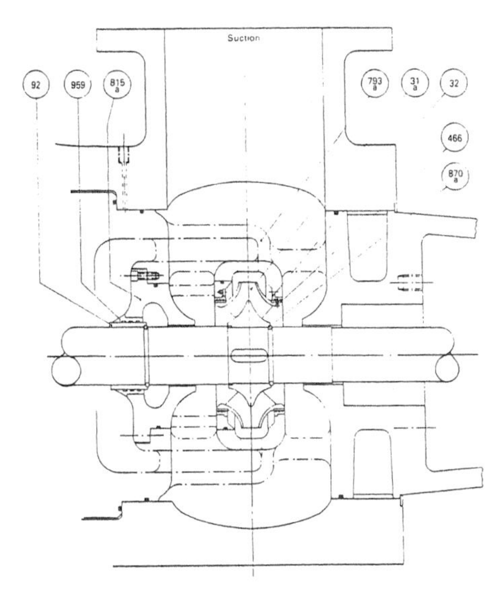

(a) Double–suction impeller (Drawing courtesy Mather & Platt Ltd.)

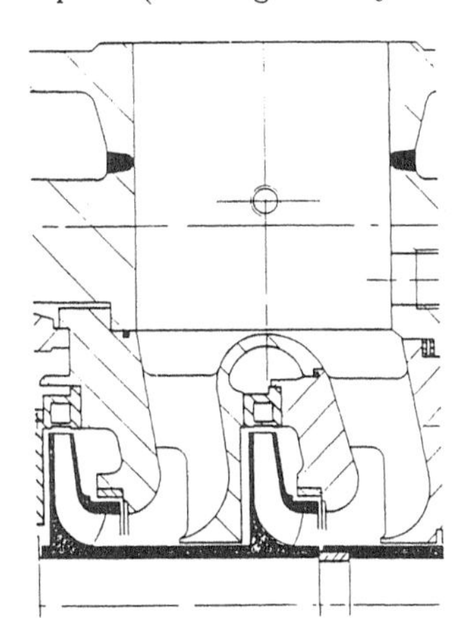

(b) Tandem impellers (Drawing courtesy KSB Ltd.)

Figure 12.13 Suction chamber of multistage pump

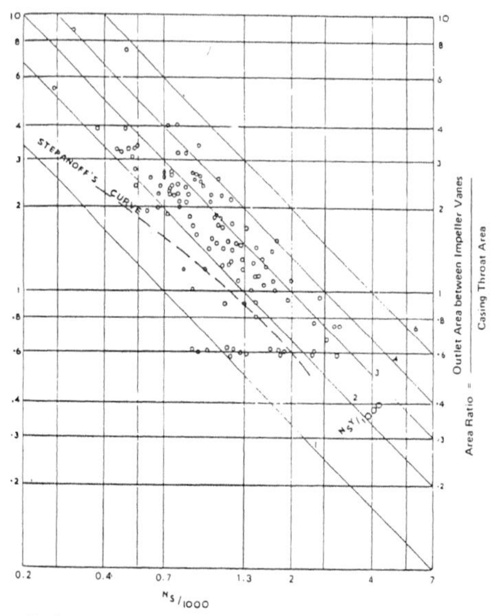

Figure 12.14 Relation between area ratio and specific speed [12.15]

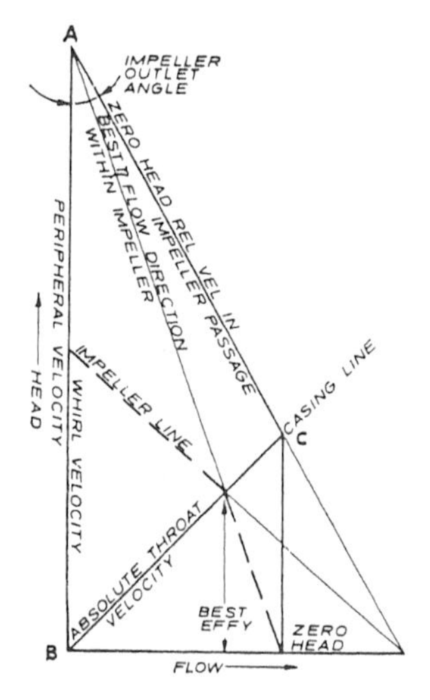

Figure 12.15 Determination pump performance from discharge velocity triangle [12.16]

12.3 Diffuser and Return Passage

For a centrifugal pump, the volute and diffuser vanes serve essentially the same function of diffusing the discharge flow except that the best efficiency may be 1–2% higher with the use of diffuser vanes and a wider high efficiency range may be obtained with the use of a volute. Selection is made according to specific requirement of the design.

As in the design of the volute, the area ratio Y is also a very important factor in the design of diffusers. The sum of all the throat areas of vanes is used in place of the throat area of the volute in the calculations. Normally, to prevent pressure fluctuation, the number of guide vanes is taken to be the number of impeller vanes plus one. Radial dimensions of guide vanes may be D_4/D_3 = 1.35-1.5 (D_3 being the diameter of the guide vane inlet circle and D_4 the outlet circle). Shorter vanes are sometimes used to reduce the radial dimension of the pump by using 11–13 guide vanes and D_4/D_3 = 1.1–1.15.

In multi–stage pumps, cross–over vanes are used to direct the flow back to the inlet of the next stage impeller. The angle of the cross–over vanes may be set to give either axial entrance or with some prerotation into the next stage. The guide vanes and the cross–over vanes together make up the interstage flow passage. When the diffuser guide vanes and cross–over vanes are connected into one spatially curved surface, it is known as a continuous passage type diffuser (Figure 12.16); when there is a vaneless space between the sets of vanes it is known as a sectionalized diffuser (Figure 12.17). The two types of construction have both been widely used in present day pump designs. Because the continuous passage diffuser is costly both in the process of casting and machining, an intermediate design between the two types has been in use in large pumps and is called the semi–continuous passage diffuser. Outer boundary of the diffuser passage is formed by the cylindrical casing of the pump (Figure 12.18), *windows* are opened in the walls of the guide vanes to facilitate cleaning and grinding of the passages.

For high speed (greater than $6,000 min^{-1}$) and high power input boiler feed pumps, diffuser guide vanes starting from an axial direction are used to prevent pressure fluctuation and erosion by the high velocity outflow (Figure 12.19). The guide vanes are impact–forged and attached to the diffuser frame by electron–beam welding to give high degrees of dimensional accuracy and surface finish.

The inclination of the head–discharge curve may be altered by changing the distance between the outlet of the cross–over vanes and the next stage impeller inlet, typical spacing is shown in Figure 12.17. When the cross–over vanes extend into the eye of the next stage impeller, as in Figure 12.19, it will contribute to steepen the head–discharge curve and improve stability.

For the purpose of balancing the axial thrust, multi–stage pumps of sym-

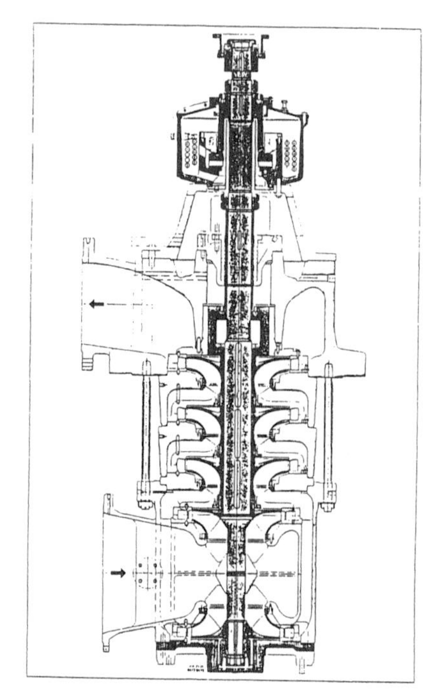

Figure 12.16 Continuous passage type diffuser for vertical pump (Drawing courtesy Sulzer)

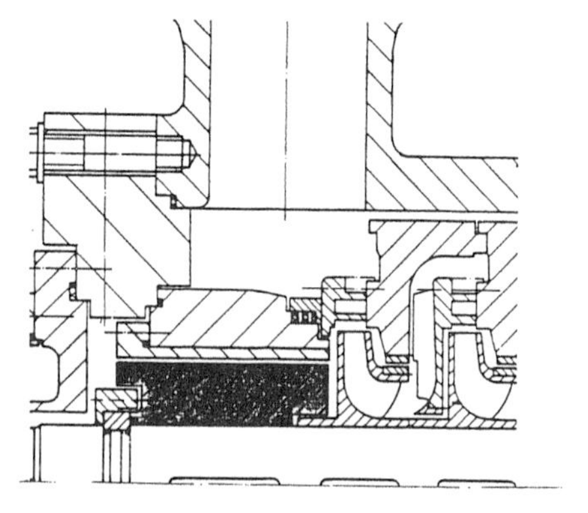

Figure 12.17 Sectionalized diffuser for horizontal shaft pump (Drawing courtesy KSB)

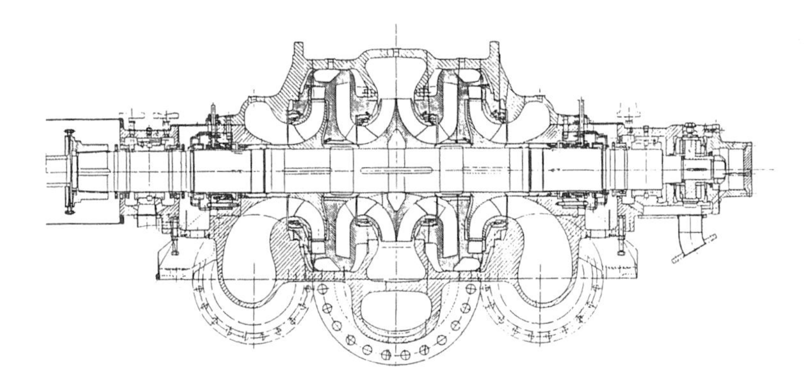

Figure 12.18 Large transfer pump with semi–continuous passage type diffuser (Drawing courtesy Sulzer)

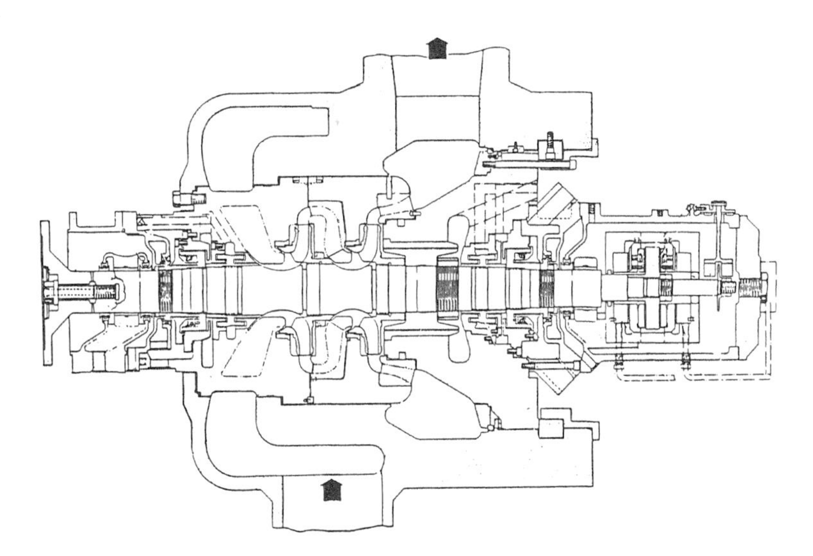

Figure 12.19 Boiler feed pump with axial diffuser vanes (for $660MW$ generating unit) (Drawing courtesy Weir Pumps)

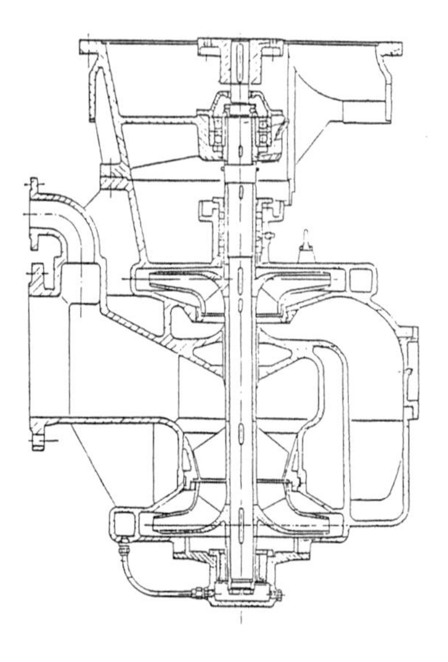

Figure 12.20 Vertical condensate pump with partly external return flow passage (Drawing courtesy Shanghai Pump Works)

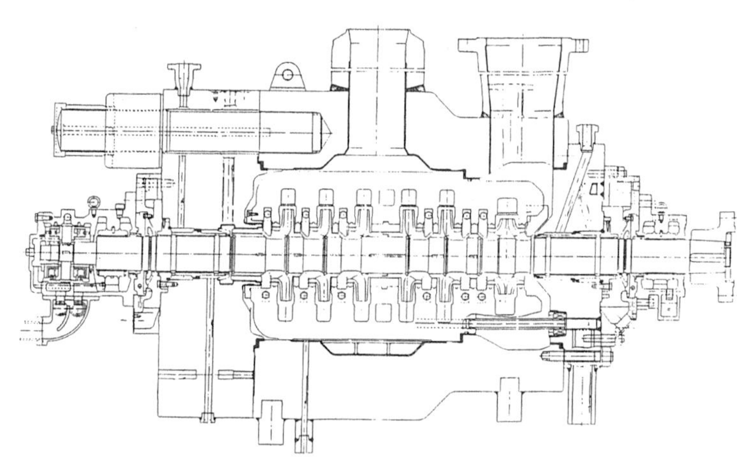

Figure 12.21 Boiler feed pump with twin–volute and internal return flow passage (Drawing courtesy EBARA)

metrical arrangement are used which utilize external pipe passages to divert the flow between stages. Designs with round external pipes are becoming infrequent nowadays and have given way to a special kind of interstage flow passage arrangement. When enough radial space is available, the construction shown in Figure 12.20 is often used both in vertical and horizontal layouts. In cases where the radial space is limited, the discharge chamber may be designed as a twin–volute with the cross–over passage located in the space near the small end of the volute, as shown in Figure 12.21. Requirements on castings are very stringent in this design since the flow passages are very intricate and are very difficult to clean when cast integrally.

12.4 Construction of Pump Casing

12.4.1 Horizontally split casings

The horizontally split casing is the earliest design used in centrifugal pumps. Advantages in installation and maintenance of this type of construction have led to the continual use in large–size, high–performance pumps. The example shown in Figure 12.18 shows application of a large pump with $Q = 10,800 m^3h^{-1}$, $H = 420m$, $n = 1,790 min^{-1}$ and $P = 13,990 kW$. The pump shown in Figure 12.21 is a high power input boiler feed pump also with horizontally split inner casing, the discharge pressure is $38 MPa$ [12.17].

The high pressure safety injection pump (nuclear class II) with horizontally split casing as shown in Figure 12.22 is a design with distinctive features in nuclear application. The upper and lower pump casings are made of stainless steel forgings of rectangular shape which facilitates working in the forging process and non–destructive inspection as well as in cost reduction.

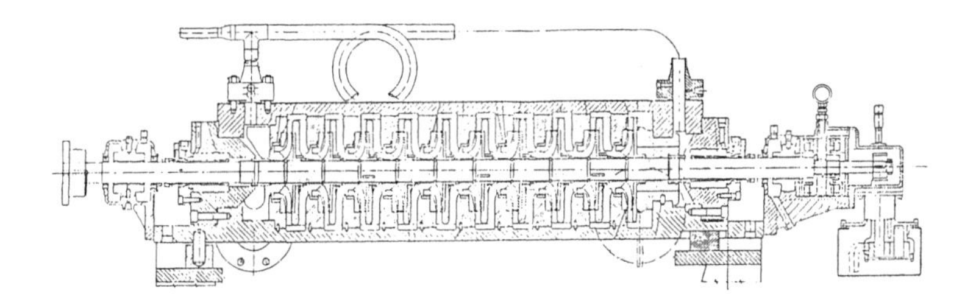

Figure 12.22 Nuclear class II pump utilizing horizontally split casing (Drawing courtesy Pacific Pumps)

The sealing of flat surfaces between the split casings and flanges for locating hold–down bolts are of utmost importance for high duty pumps. For inner casings, hard seals with metal–to–metal contact that require high degrees of surface accuracy are often used. For outer casings, dechlorinated asbestos gaskets reinforced with fine metallic mesh have proved to be the best choice.

12.4.2 Ring section casings

The ring section type of casing is used very early for multi–stage centrifugal pumps, a typical construction is shown in Figure 12.23. The ring section design is now used in equally wide applications as the horizontally split casing, each for its irreplaceable advantages. The ring section design is also used in vertical multi–stage pumps, an example of a vertical condensate pump is shown in Figure 12.24.

In the design of multi–stage pumps with ring section casing, the sealing between the casings of stages and the design of tie bolts are of great importance. For sealing between stages, metal–to–metal seals have now replaced gaskets. Rubber 0–rings are used as auxiliary seals in applications of high safety requirements.

For multi–stage pumps operating at temperatures above $120°C$, the tie bolts must be designed to ensure the interstage sealing when exposed to heat transient conditions.

12.4.3 Barrel type casing

For multi–stage pumps with discharge pressure higher than $15MPa$, the barrel type double casing design is often used to provide added safety, typical constructions are shown in Figure 12.21 (inner casing horizontally split) and Figure 12.25 (inner casing ring section type).

The axisymmetric geometry of the barrel casing makes it easier to perform finite element analysis of the stress–strain field and temperature field of the pump. For high speed pumps the barrel casing must have sufficient rigidity as well as satisfactory heat transient behaviour. The end covers for the barrel casing and holding down bolts must also be checked in the same analysis as the barrel casing. A spiral wound gasket with sufficient back–elasticity is to be used at the sealing between the barrel casing and the cover to ensure effective sealing under heat transient conditions.

Tightening of the holding down bolts is currently done with a hydraulic tensioner instead of expansion by electric heating in order to shorten the time required for replacement of the pump cartridge. This practice is especially important for high–pressure water injection pumps used in off–shore oil fields.

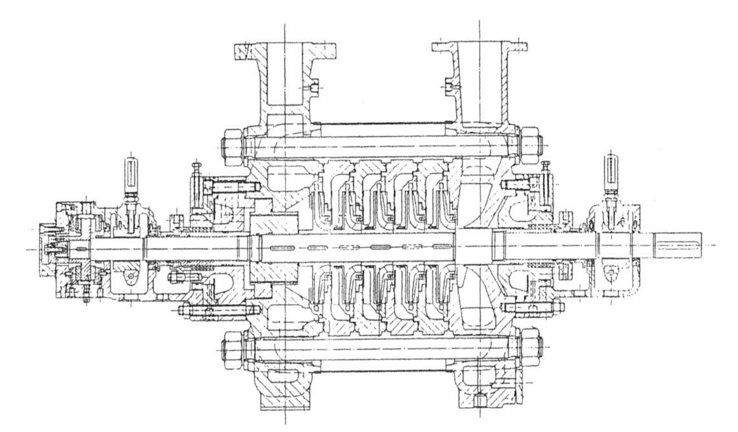

Figure 12.23 Typical multi-stage pump with ring section casing (Drawing courtesy Sulzer)

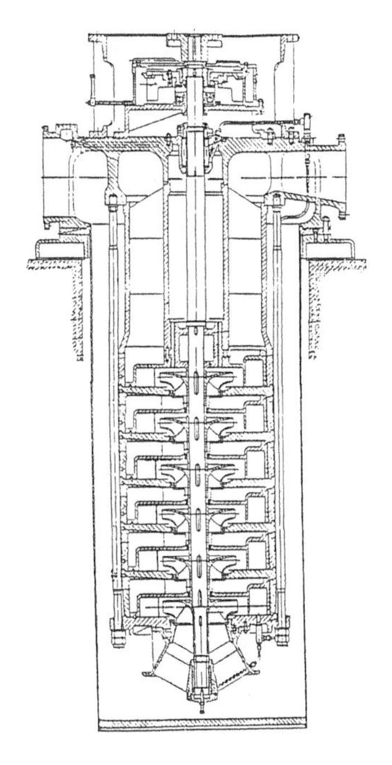

Figure 12.24 Vertical condensate pump with ring section casing (Drawing courtesy Shanghai Pump Works)

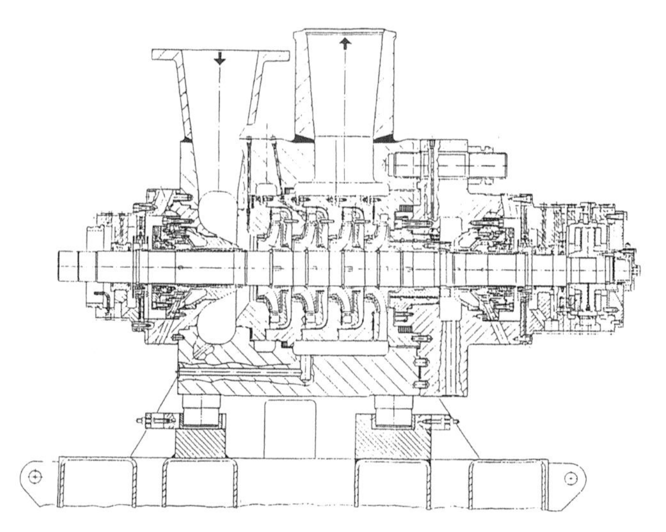

Figure 12.25 Typical double–barrel casing boiler feed pump (Drawing courtesy Sulzer)

The construction of the pump cartridge has improved to the extent that the bearing housing at the driving end need not be disassembled, the complete pump cartridge can be pulled out after loosening the nuts of the end–cover holding down bolts. The procedure enables the replacement of the pump cartridge within eight hours.

12.5 Rotating Components

12.5.1 Shaft

The pump shaft is the most important element among the rotating parts. The stiffness of the shaft of a multi–stage pump has the greatest effect on reliability of the pump. A factor K is used to describe the shaft stiffness

$$K = \frac{\sqrt{W} \cdot L^{3/2}}{D^2} \tag{12.23}$$

where L is the distance between centre lines of bearings; W the weight of rotor between bearings and D the average diameter of rotor shaft between bearings. Recommended values of K for application under different operating conditions are given in the chart in Figure 12.26, which is derived from many years of operating experience [12.18].

In order to improve the reliability and availability of the pump, the following points in shaft design should be observed:

- Avoid details that may induce stress concentration, such as by omitting screw threads on shafts, replacing impeller snap rings of square section with ones of round section.
- Avoid mounting of impellers on shaft by shrunk-fit whenever possible, sliding fit of impeller on smooth shaft should also be used on high-performance pumps.
- For high speed pumps, use round-rooted spline shaft for fitting impellers instead of the standard two-key design for torque transmission.

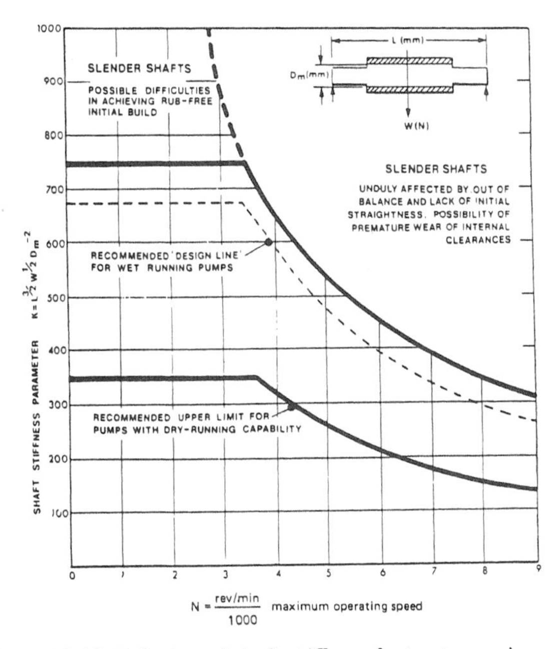

Figure 12.26 Relation of shaft stiffness factor to maximum operating speed [12.18]

12.5.2 Bearings

Anti-friction roller bearings are extensively used in standard series centrifugal pumps, the problem facing the designer is the selection of bearings with sufficient load capacity and the way of lubrication. Sliding bearings are more often used in high performance pumps because of the high shaft speed and load rating. Figure 12.27 is useful in selecting and designing of sliding bearings.

Bearings for centrifugal pumps belong to the high speed and light load category. From the point of view of rotordynamics of the pump, the important question is to maintain the stability of the bearings and prevention of whipping of the oil film. Multi-lobe bearings are commonly used in boiler feed pumps. In cylindrical journal bearings, the pressure dam type is always preferred. The tilting-pad type bearing which has the highest stability is seldom used in centrifugal pumps except for some special cases such as in main coolant circulating pumps in nuclear power stations and in super high-speed (over $10,000 min^{-1}$) multi–stage pumps.

In large vertical centrifugal pumps and high performance multi–stage pumps, double-action thrust bearings are often used. In the former case the auxiliary bearing is of smaller size and is used to prevent upward thrust at pump start, in the latter case the bearings for both directions are of the same size. When double-action bearings and thrust balancing devices are working in combination, load transducers are mounted between the main thrust bearing pedestal and the bearing housing to monitor the effectiveness and reliability of the balancing device under all operating conditions.

The effectiveness of lubrication has a great deal to do with the load capacity and service life of a bearing. For journal bearings, the manner of lubrication is decided by the value of $\sqrt{pV^3}$

$\sqrt{pV^3}$ < 15 lubrication with oil ring but no oil cooler
$\sqrt{pV^3}$ $=$ 15-30 lubrication with oil ring and oil cooler
$\sqrt{pV^3}$ > 30 lubrication with forced oil circulation

where p is the specific pressure of bearing (MPa) and V is peripheral speed of shaft (ms^{-1}).

A lubrication oil supply system consisting of oil pump, oil cooler, strainer, electric heater, regulating valve and control relay is required for the case of forced oil circulation. As greater degrees of reliability are required of oil pumps, shaft driven pumps are being replaced by motor driven pumps with standby units.

The oil-bath type lubrication is often used in vertical pumps, the oil is circulated either by the centrifugal action of the thrust collar or by specially provided screw type oil pumps. The oil cooler is directly submerged in the reservoir. Double-action thrust bearings generally have external forced oil supply systems except the ones with low parameters where it is sufficient to have a viscosity pumping ring to supplement the thrust collar.

The general principle for a thrust bearing lubrication is to supply sufficient quantities of fresh oil into each bearing pad in all operating conditions

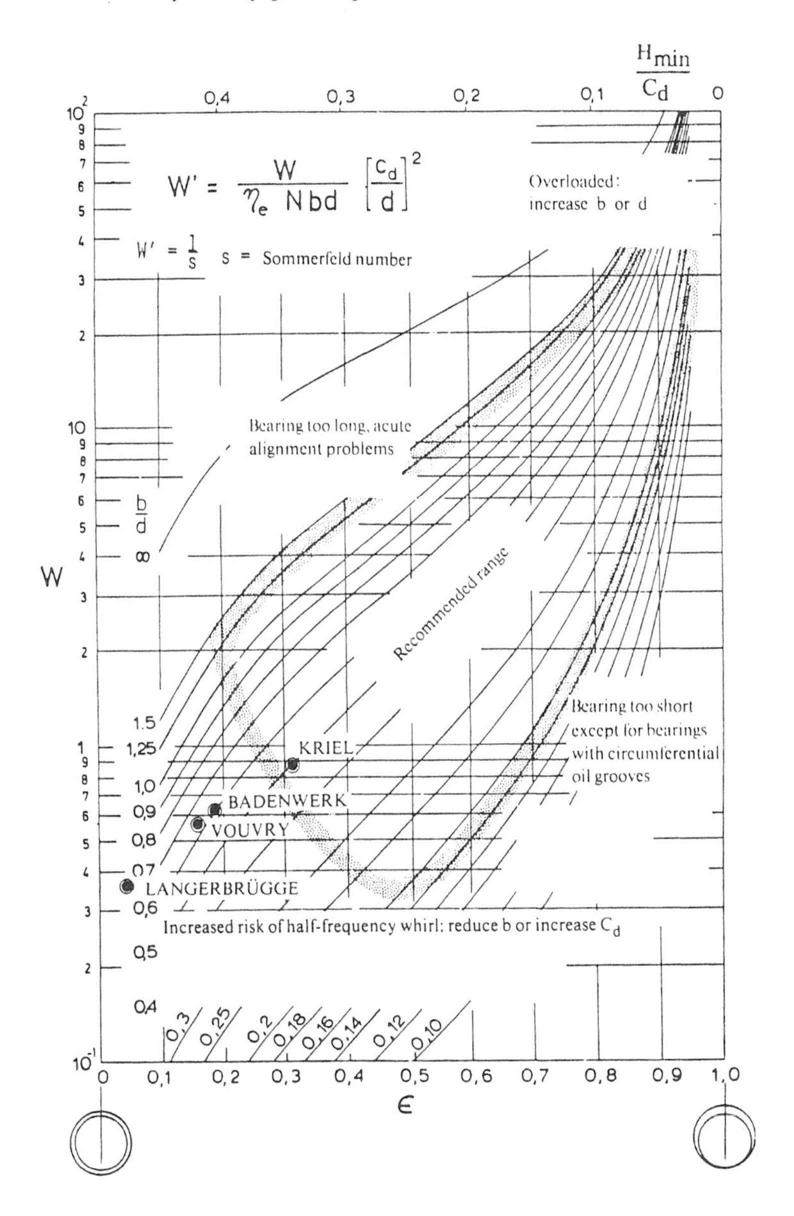

Figure 12.27 Load capacity of bearings [12.19]

from starting up to shutting down. The oil chambers in thrust bearings are relatively isolated systems whose lubrication may be one of the following ways, as illustrated in Figure 12.28.

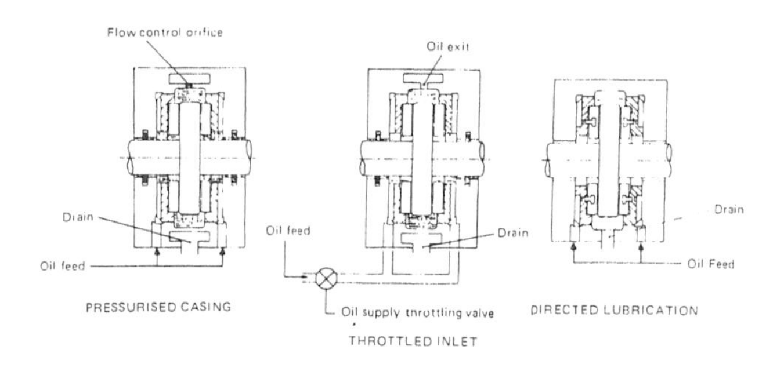

Figure 12.28 Diagrammatical layout of three lubrication systems

1. Pressurized casing – a fixed or adjustable orifice is located at the outlet of the oil chamber so as to maintain the required pressure level and circulation of lubrication oil.

2. Throttle inlet – the circulation of oil is controlled by an adjustable orifice in the oil inlet pipe. Oil is thrown outward by the thrust collar and collects at the bottom of the casing for return to the system reservoir.

3. Directed circulation – cooled oil is sprayed directly on the surface of the thrust collar by oil jets placed between bearing pads and forces out the hot oil.

A typical layout of directed lubrication is shown in Figure 12.29 where oil jet holes are located in hold–down screws that are arranged in a conical array. The oil jets are directed at the oil inlet edge and oil outlet edge, respectively of two adjacent pads, where they scour the hot oil from one pad and inject cold oil into the next.

Figure 12.30 shows a comparison of test results at different specific pressures with temperature rise and power absorbed as functions of oil quantity for thrust bearings equipped with the above lubrication systems at a shaft speed of $12,000 min^{-1}$. It is seen from the diagram, under the same pressure

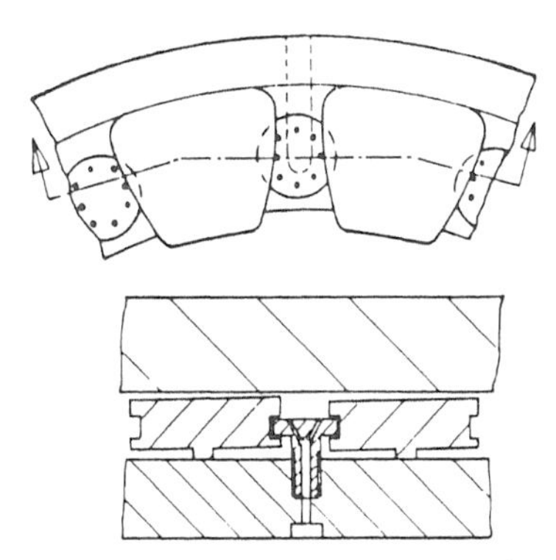

Figure 12.29 Directed system using a conical array of oil jets

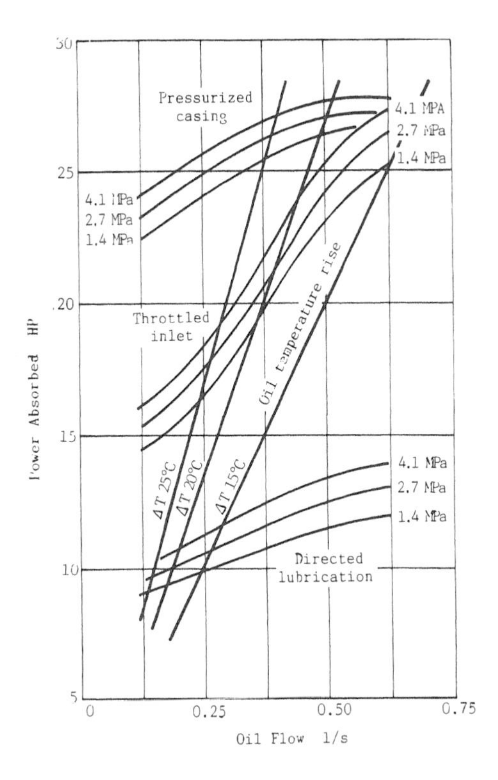

Figure 12.30 Relation between oil flow and power absorbed [12.20]

and oil flow, the temperature rise for the directed lubrication scheme is the lowest.

Other test results from the same source indicate that for the directed lubrication the maximum temperature of the bearing pad is the lowest and the oil pad minimum film thickness is the highest, for the same oil outlet temperature. The above facts all point to the advantages of directed lubrication which effectively raises the load capacity of the bearing and its operating reliability.

12.5.3 Balancing devices

In centrifugal pumps, the axial thrust is proportional to the discharge pressure and may reach very high values in high head multi–stage pumps.

Balancing holes or stub vanes on the back shroud of the impeller have been effective in single stage pumps. But when the pump inlet pressure (system pressure) exceeds that of the discharge pressure, the axial thrust will reverse in direction and render the above devices useless. This reverse thrust must be taken up by bearings. Single stage pumps working in high pressure systems generally have specially designed bearing assemblies and often use double–action thrust bearings of the Kingsbury type. The pump with canned motor which has the pressure balanced within the pump itself is used in special cases. Multi–stage pumps with opposed impellers (Figure 12.21) have the axial thrust balanced hydraulically but are usually equipped with small double–action bearings to take up the residual axial thrust.

The thrust balancing devices for vertically split type multi–stage pumps have evolved through the following stages, as shown in Figure 12.31.

1. Balancing disc – too little axial clearance, often have operating troubles, especially in frequent starting and stopping.

2. Balancing disc (or combination of balancing piston and disc) plus double–action thrust bearing with flexible support – increased axial clearance and capable of frequent starting and stopping, but the adjustment of axial clearance between the balancing disc and the thrust bearing is difficult.

3. Balancing piston plus double–action thrust bearing with rigid support – often requires installation of load transducers under the thrust bearing segment carrier (Figure 12.32).

It is interesting to note that the balancing piston is the earliest in design in multi–stage pumps, it experienced the evolution through the balancing disc and now returns to the balancing piston. But the current design is more advanced with the addition of double–action thrust bearings and load

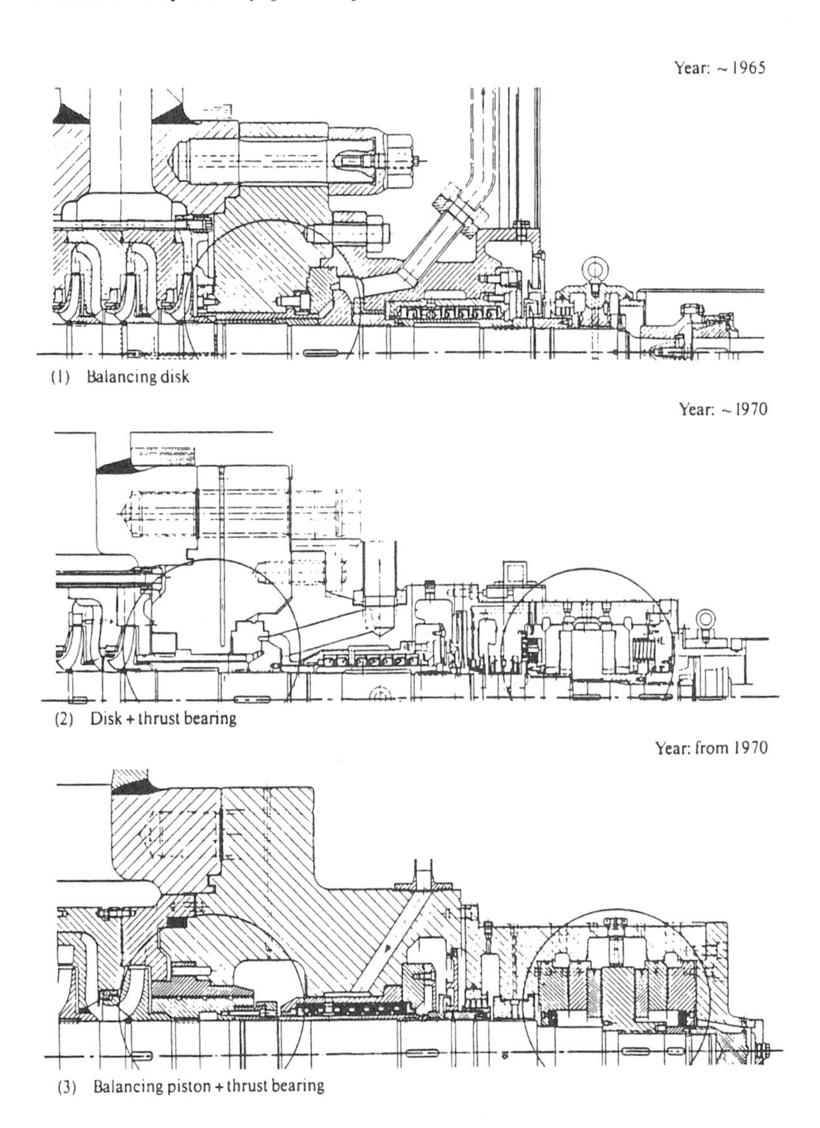

Figure 12.31 Development of thrust balancing devices
(Drawing courtesy Sulzer)

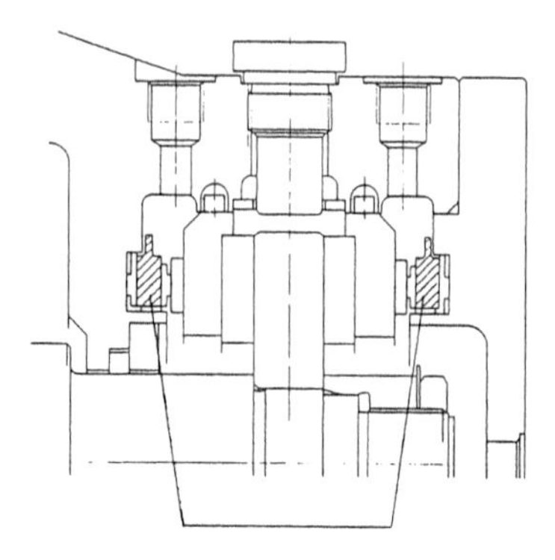

Figure 12.32 Annular load transducers fitted in the axial thrust bearing (Drawing courtesy Sulzer)

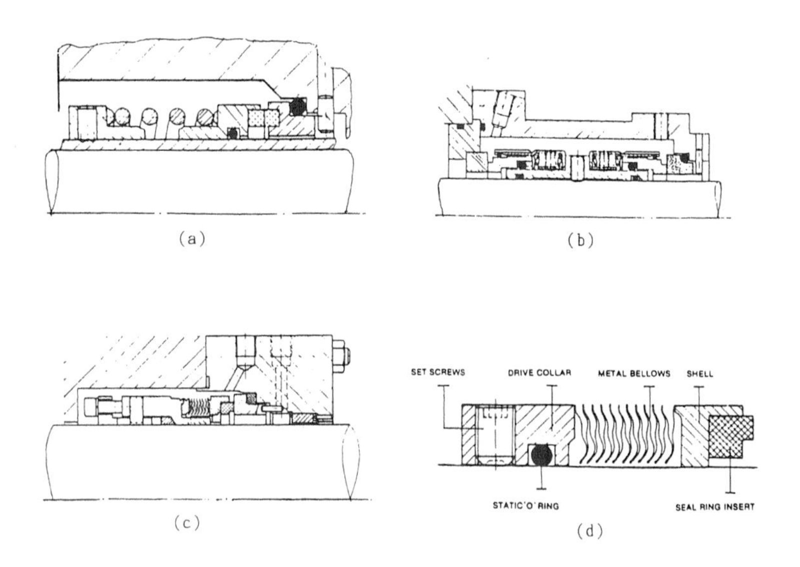

Figure 12.33 Typical construction of mechanical seals (Drawing courtesy Shanghai Pump Works)

transducers to assure a higher degree of reliability. Similar reversals in design concepts also exist in other components of multi-stage centrifugal pumps, each time the design is recycled to a higher stage of perfection.

12.5.4 Shaft seals

Stuffing boxes with packing glands are the earliest and most common kind of shaft seal in centrifugal pump history. The packing gland actually came into being at the same time as the pump because of its simplicity in construction, low cost and ease of maintenance. Very good results may be obtained if the stuffing box and the shaft sleeve are properly cooled, well protected from wear and the correct packing material used. Internal cooling of the shaft sleeve is important because it has effect on a greater length of the contact surfaces.

The mechanical seal is becoming widely used in recent years. It not only replaces the soft-type packing gland in ordinary pumps but also occupies a share in high performance pumps (high PV value, high temperature and corrosive environment, etc.).

Figure 12.33 shows typical constructions of mechanical seals, where (a) is the single spring type, (b) the multi-spring type, (c) the metal bellows type and (d) enlargement of the metal bellows. The metal bellows is successful in replacing the springs as well as the sliding flexible seal ring to give consistent and effective contact. The mechanical seal is able to work in fluids of high temperature or in corrosive and particle entraining environments. The allowable temperature range is -75° up to 225^oC; and the operating pressure range is from vacuum to $2MPa$ for single metal bellows and a maximum of $7MPa$ for double bellows.

Auxiliary circuits for mechanical seals are essential in providing long term safe operation. Different auxiliary circuits are to be used for different operation conditions. Figure 12.34 shows examples of circuits specified in American Petroleum Institute specifications. Figure 12.35 shows a recommended auxiliary system for mechanical seals in a boiler feed pump. Self circulating cooling water systems utilizing the labyrinth pump in the mechanical seal will require coolers, cyclone separators or even magnetic filters.

In case of very high system pressure (greater than $10MPa$), the specific pressure on the sealing surfaces of the mechanical seal exceeds that of the allowable normal operating pressure for contacting materials, then the controlled-leakage seal will have to be used.

The working principle of this seal is that a very small clearance (less than $0.01mm$) between the contact surfaces is established by hydrostatic or hydrodynamic action of liquid so the seal can operate smoothly while keeping the leakage rate within permissible limits. This design is commonly used in main coolant pumps for nuclear power stations. To prevent radioactive

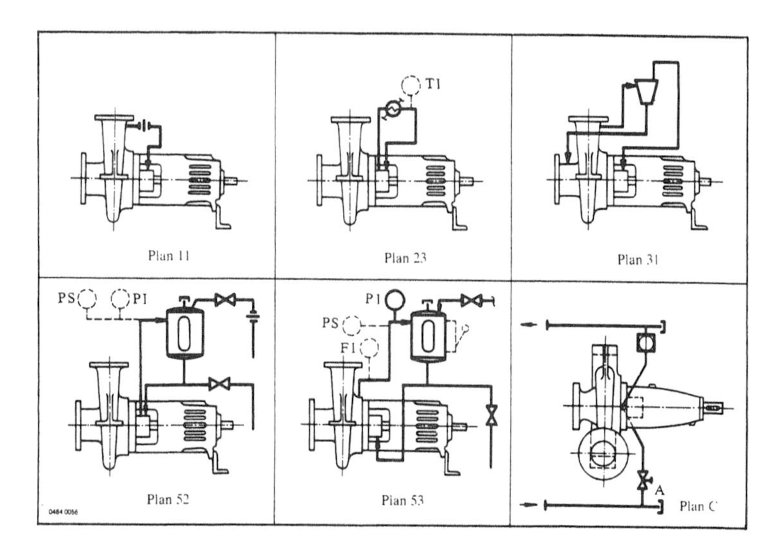

Figure 12.34 Pipe work of auxiliary circuit for mechanical seals (API610)

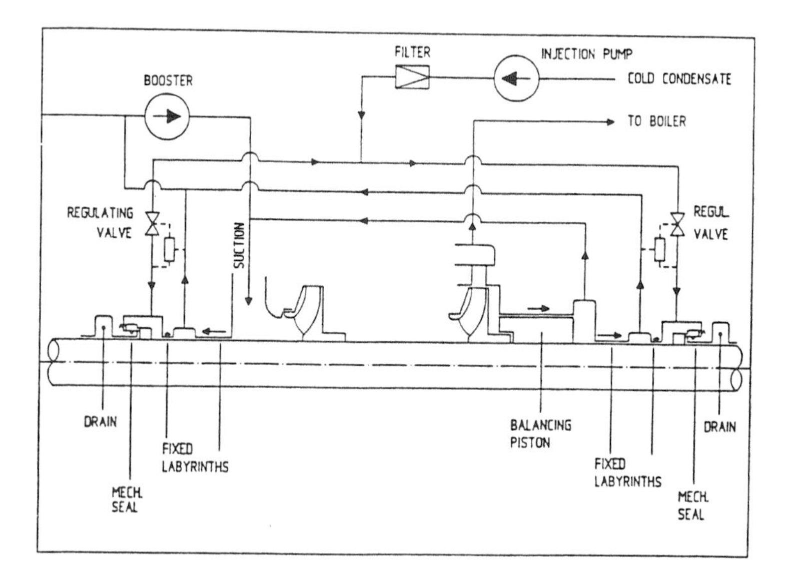

Figure 12.35 Possible shaft sealing arrangement of boiler feed pumps for increased feedwater temperature and suction pressure [12.21]

the leakage rate within permissible limits. This design is commonly used in main coolant pumps for nuclear power stations. To prevent radioactive substances from leaking out, normally the shaft seal is made up of three sets of mechanical seals arranged in series. The controlled leakage fluid is directed to a chemical–volume treatment system and electrolyte–free water is injected to the sealing surfaces of the last stage to bring the leakage of radioactive fluid down to zero. Actually any one of the seals in series is able to sustain alone the total pressure of the system.

Controlled–leakage seals may be classified according to their working principles as:

1. Hydrostatic type – the sealing surface of the seals are made in shapes of small–hole throttles, steps or conical surfaces and are kept from contact with each other under the action of the system pressure. Figure 12.36(a) shows an example of seal construction of the conical surface type.

2. Hydrodynamic type – curved or straight grooves are formed on the sealing surfaces to provide hydrodynamic effects that keep the sealing surfaces from contacting. Figure 12.36(b) shows the application of this type.

General recommendations for shaft seal selection are given in Figure 12.37.

12.5.5 Fluid seals and rotordynamics

The working of the non–contact seals between rotating parts and stationary parts has effect on the efficiency and life of a single stage pump, for multi–stage pumps it will affect the rotordynamics of the pump as well.

Figure 12.38 shows the internal fluid leakage passage of a multi–stage centrifugal pump. Various types of seals for the purpose of reducing leakage (Figure 12.39) may be fitted on wear rings, interstage seal sleeves and balancing devices. Figures 12.41 and 12.42 show the difference in critical speed of a pump rotating in air and in water. When running in water, the clearances at the impeller wearing ring, inter–stage seals and balancing devices all have the effect of water lubricated bearings. Due to the rigidity and damping of these bearings, the first critical speed of the rotor system is increased. The amount of increase depends on the geometry of the seals. Small grooves machined on the seal surfaces for increasing fluid resistance, as (b) and (f) in Figure 12.39, have the effect of reducing the Lomakin–effect. Improved design have shallow grooves, as in Figure 12.40.

Stepped seals as shown in (c, d, e) of Figure 12.39 are commonly used in pumps with high stage heads, and are able to provide a high coefficient of friction and good stability of the rotor system.

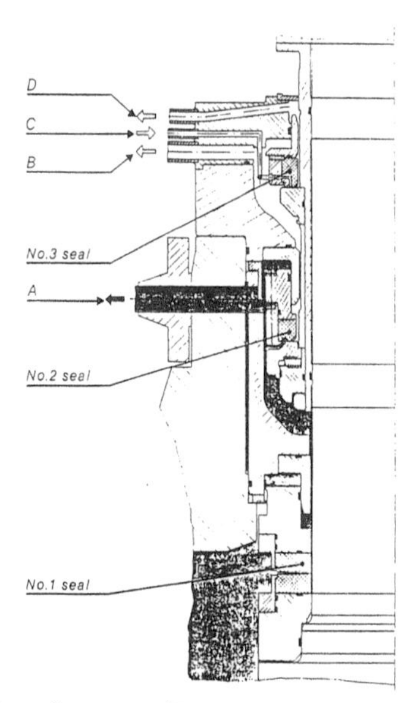

(a) Conical surface type (Drawing courtesy Jeumont–Schneider)

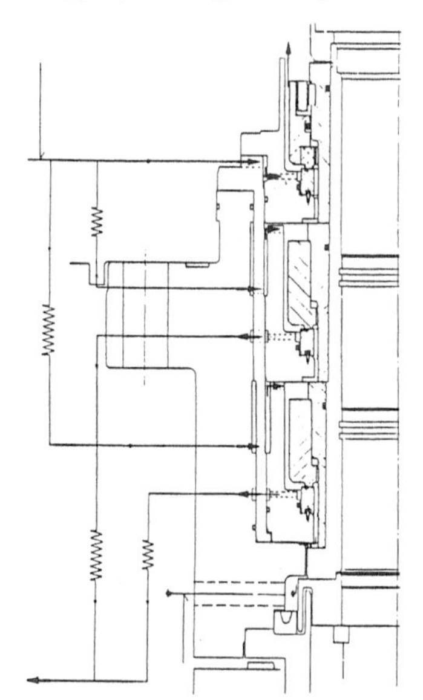

(b) Hydrodynamic type (Drawing coutesy Burgmann)

Figure 12.36 Controlled leakage seals

Type	Section	U_D (m/s)		Disadvantages
Mechanical seal	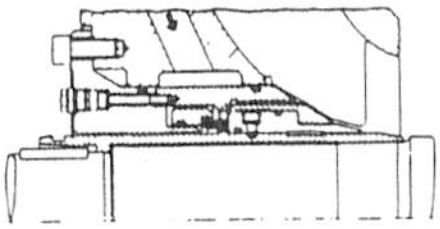	$\leqslant 60$	Almost no leakage	Separate cooling system, seal damaged if it fails Sensitive to: material choice – possible deformation of sliding surfaces – vibration
Floating rings	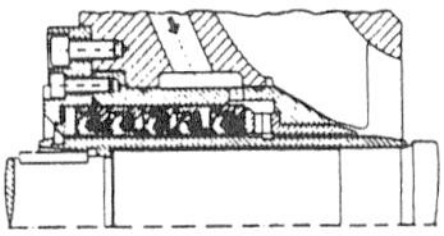	$\leqslant 60$	Little leakage	$Q_{leak} > Q_{mech.\ seal\ leak}$ Sensitive to: vibration – material mating – failure of sealing fluid – evaporation and thence damage
Fixed throttle bush	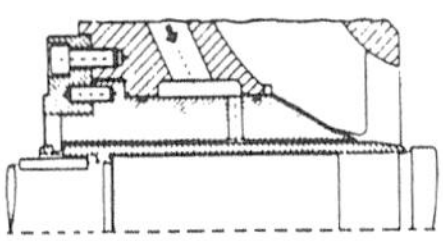	$\leqslant 80$	Simple design, insensitive to vibration, insensitive to evaporation.	$Q_{leak} > Q_{fl.\ ring\ leak}$
Packed stuffing box	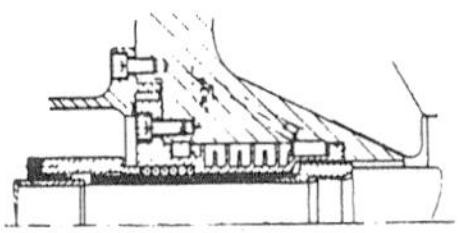	$\leqslant 30$ with cooled shaft having protection sleeve $\leqslant 20$ uncooled	Simple design	Sensitive to: packing method – material mating – surface quality of shaft protection sleeve – vibration – deformation

Figure 12.37 Comparison of different types of shaft seals [12.19]

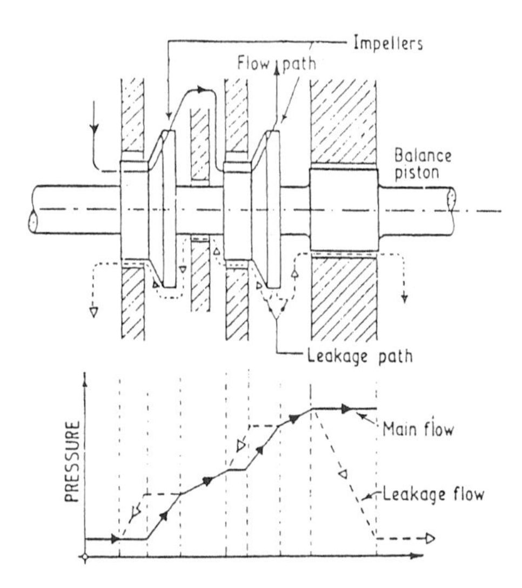

Figure 12.38 Leakage passage through a pump

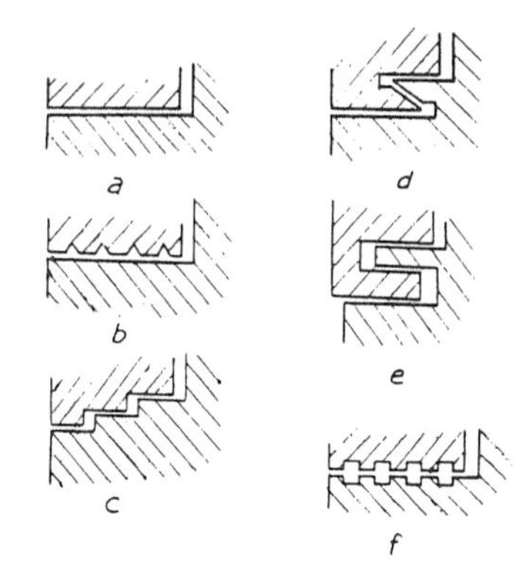

Figure 12.39 Typical clearance seals

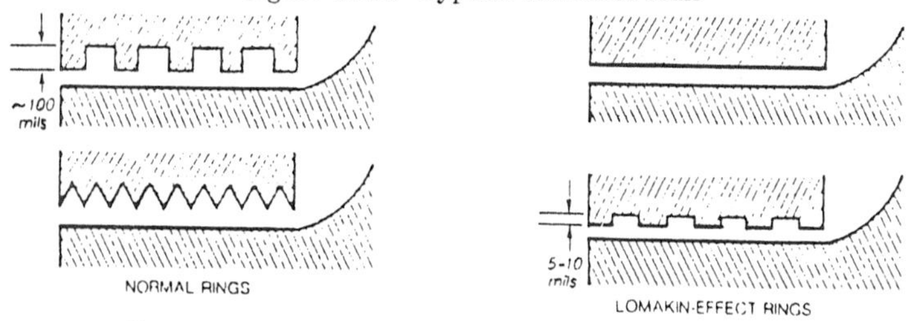

Figure 12.40 Sealing rings working by Lomakin–effect

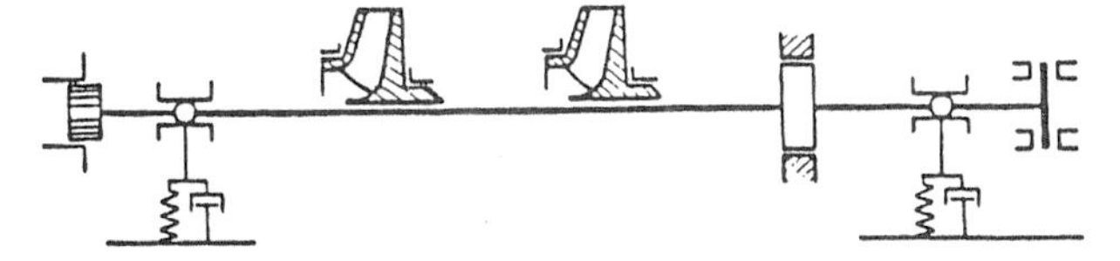

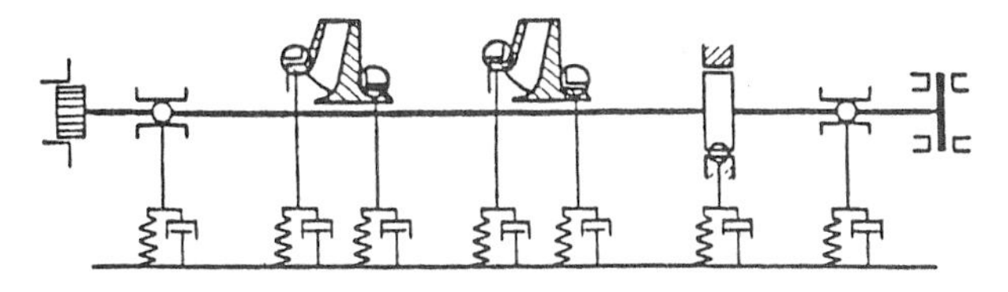

Figure 12.41 Schematic arrangement of pump system

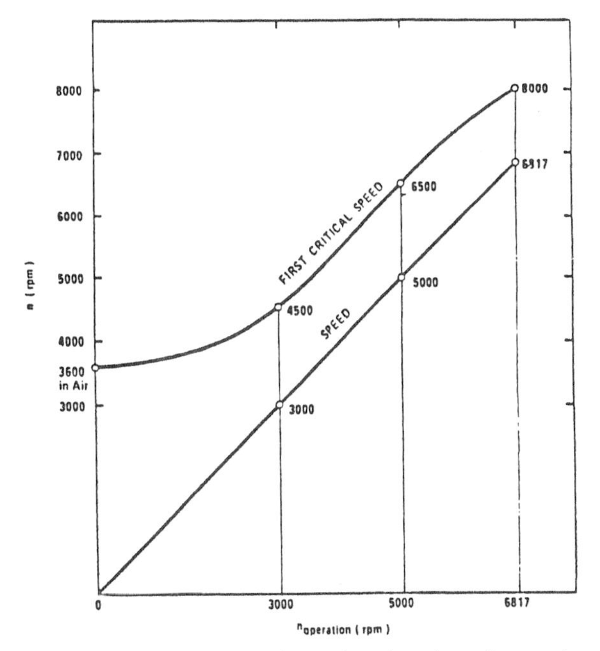

Figure 12.42 First critical speed as function of operating speed rotor [12.22]

critical speed is lowered to the natural frequency of the rotor system. Therefore, it is important to set a limit on the amount of clearance enlargement in the design stage.

It has been proved by tests that the swirl produced by the revolving impeller upstream of the clearance seal will reduce the damping effect of the seal, especially for the case of balancing pistons with long sealing lengths [12.21]. A break (machined slots) before the inlet of the seal will greatly restore the damping effect of the seal, and allows the use of a larger clearance as much as twice the normal value. Figure 12.43(a) shows the construction of a swirl break at the balancing piston, similar breaks may be used at the casing wearing ring of each stage of impeller. Figure 12.43(b) shows effect of the swirl break on damping capacity of the seal by indication of the relative tangential force opposite to the direction of rotation. The research results described above have already been incorporated in recent designs.

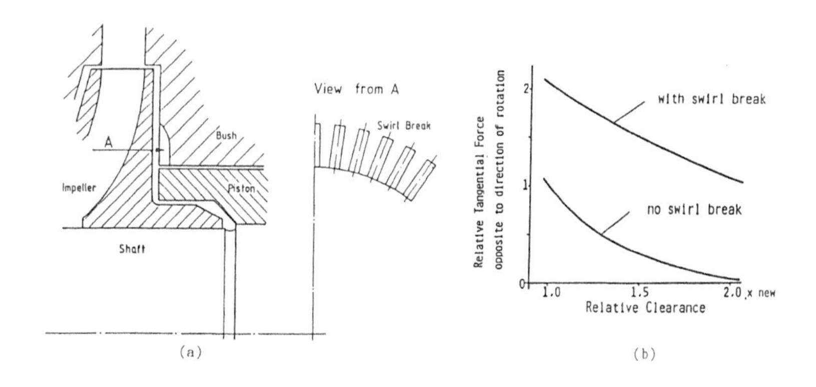

(a) execution of a swirl break at the balancing piston
(b) relative tangential force representing the damping capacity of the swirl break

Figure 12.43 Swirl breaks for seals

12.6 High Performance, High Speed Pumps

12.6.1 High performance centrifugal pumps

Discharge head of a centrifugal pump may be greatly increased by driving it at higher speed with the following advantages:

- Smaller number of stages, smaller impeller diameter and shorter span between bearings that result in greater stiffness of rotor.
- Enabling selection of optimal specific speed for higher efficiency, or by use of variable speed drive to obtain desired performance.
- Making the pump smaller in size, lighter in weight and lower in cost.

Figure 12.44 shows the very apparent difference in size of multi-stage barrel type pumps designed with different speeds for the same specification (discharge head 2,400m).

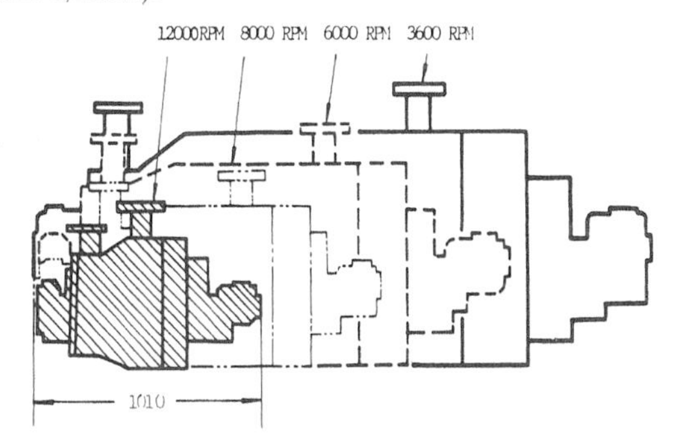

Figure 12.44 Comparison of size of pumps with different design speeds

High speed pumps are first used in the small flow and high head range. Barske developed a non-standard high speed centrifugal pump with open impellers which found application in the region normally reserved for reciprocating pumps, as shown in Figure 12.45. This type of pump is sometimes known as the partial-emission pump or tangential-emission pump. The impeller is of open construction and has no wear rings, the clearance between the impeller and casing is relatively large.

This allows the pump to be capable of running dry or handle fluids containing foreign substance. The pump casing which is concentric with the impeller discharges into a conical diffuser, whose size determines the slope of the head-discharge curve and the break point of the head curve (Figure 12.46).

Centrifugal pumps of conventional construction driven at high speed are now widely used in various industries. Figure 12.47 shows a single stage feed pump in the system of a 762MW boiling water reactor generating unit with $Q = 1,080m^3s^{-1}$, $H = 843m$, $n = 5,850min^{-1}$ and $P = 3,080kW$.

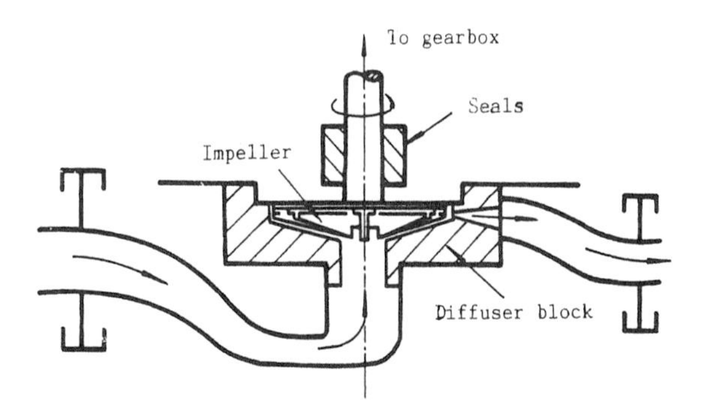

Figure 12.45 Typical open-impeller high speed pump [12.25]

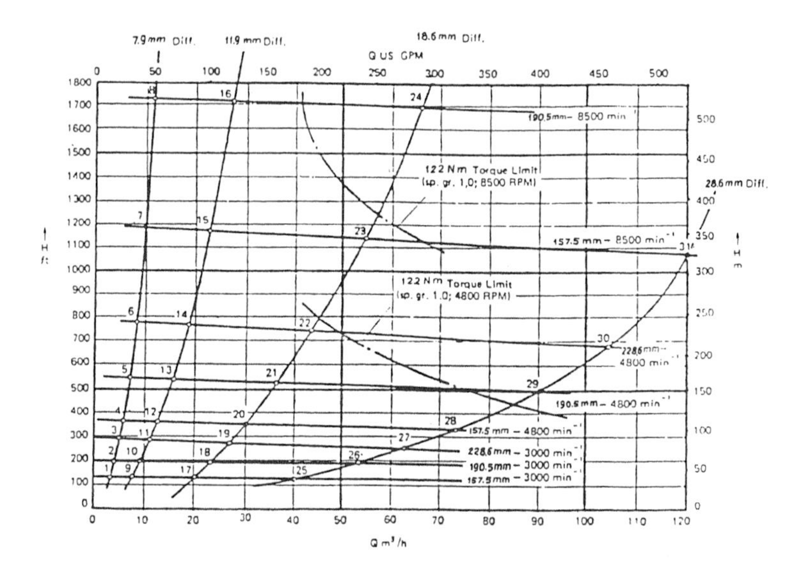

Figure 12.46 Typical characteristic curves of partial-emission pump [12.24]

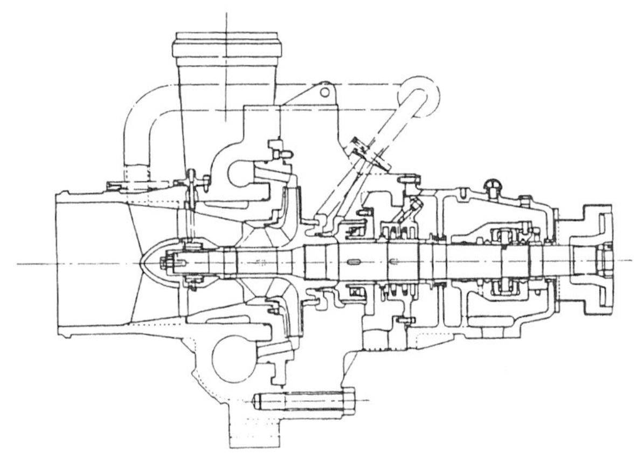

Figure 12.47 Boiling water reactor feed pump
(Drawing courtesy Weir Pumps)

Single stage end suction high speed pumps of similar construction are used to transport carbonate salt solution and hot oil. The barrel type multi-stage double casing pump shown in Figure 12.48 is used for water jet bedrock-crushing in undersea operations. Its speed and discharge head are among the foremost in pumps of similar construction, being $Q = 60m^3h^{-1}$, $H = 5740m$, $n = 20,000min^{-1}$ and $P = 2,400kW$, number of stages 8. The impellers are mounted in two groups symmetrically back-to-back on the shaft and are supplied by two separate suction inlets one on each end of the casing. Studies in rotordynamics and vibration problems as well as designs of pressure sustaining components and their self-sealing joints are very important for pumps of this category.

12.6.2 Suction inducers

Operation of high speed pumps under low *NPSH* conditions led to the development and application of the suction inducer. Figure 12.47 shows the pump fitted with an inducer at the inlet of the centrifugal impeller.

Suction inducers, like the partial-emission pumps, are first used in the fuel system of rocket engines. The inducer is basically an axial flow impeller with lightly loaded blades. The efficiency of the inducer is lower than ordinary axial-flow impellers due to its high surface ratio which is the precondition of lower impeller loading. But the inducer together with the matching centrifugal impeller is able to increase the overall efficiency somewhat, especially in the high flow rate region.

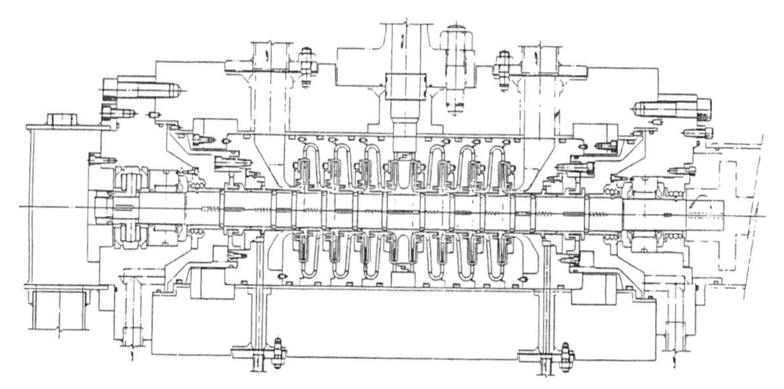

Figure 12.48 Barrel type multistage pump for water jet bed–rock crushing (Drawing courtesy Mitsubishi)

Figure 12.49 shows the basic types of inducers and Figure 12.50 shows the elements of a typical inducer design. In order to increase the head coefficient of inducers, twisted blades, sometimes called variable pitch blades, are used.

References [12.7] and [12.26] give detailed explanation of design criterion and methods for inducers. The Brumfield criterion is used to obtain the optimum flow coefficient ϕ_{opt}

From cavitation factor

$$K = \frac{2\phi opt^2}{1 - 2\phi_{opt}^2} \tag{12.24}$$

it follows that

$$\phi_{opt} = \sqrt{\frac{K}{2(1+K)}} \tag{12.25}$$

The corresponding suction specific speed (the numerically highest) is given by

$$C_{max} = \frac{500}{(1+K)^{1/4} \cdot \sqrt{K}} \tag{12.26}$$

In the above, K is an empirical value which is determined from previous experience. The optimal inducer diameter is calculated from:

$$D_{opt} = 0.115 \left[\frac{Q}{(1-\gamma^2) n \phi_{opt}} \right]^{1/3} \tag{12.27}$$

The various design parameters and variables as given in Table 12.4 are very useful in designs.

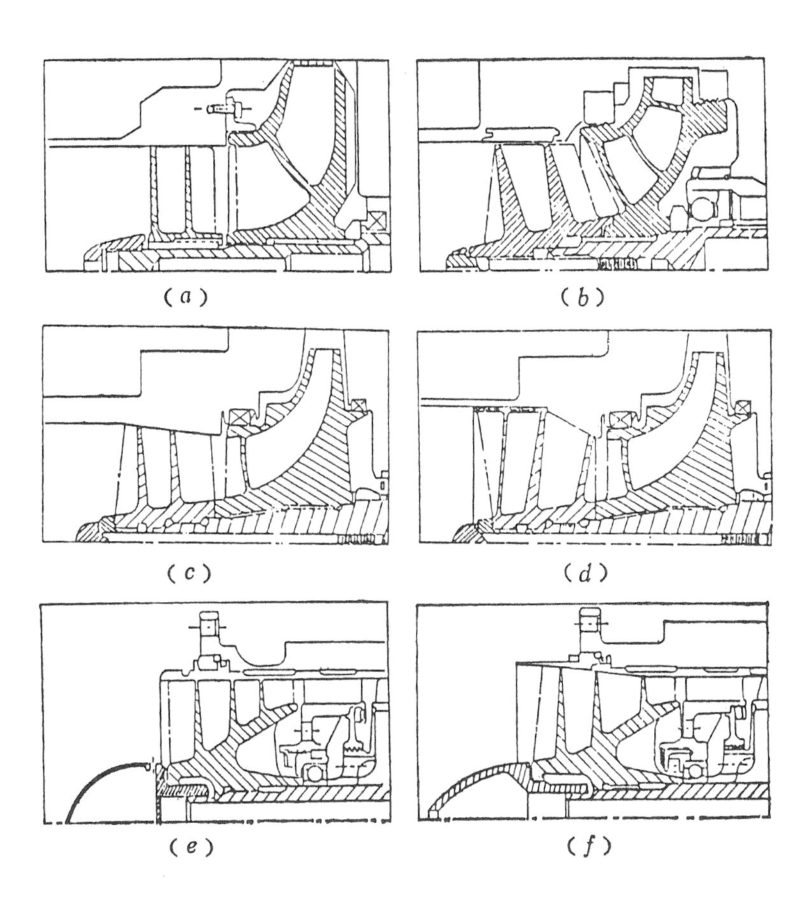

(a) – low–head inducer with cylindrical blade tip and hub
(b) – low–head inducer with cylindrical blade tip and conical hub
(c) – low–head inducer with conical blade tip and hub
(d) – low–head inducer with shroud ring
(e) – high–head inducer with cylindrical blade tip and conical hub
(f) – high–head inducer with conical blade tip and hub

Figure 12.49 Basic types of inducers

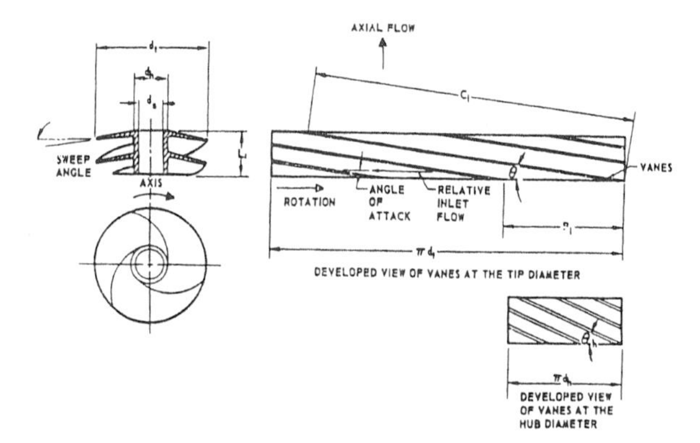

Figure 12.50 Elements of typical inducer design

Table 12.4 Cavitating inducer design parameters and variables [12.26]

PARAMETER OR VARIABLE	TYPICAL DESIGN VALUES	DESIGN REQUIREMENT
Specific speed $(n_s)_{ind}$	6000 to 12 0000	Head – capacity characteristics
Suction specific speed $(n_{ss})_{ind}$	20 000 to 50 000	Suction characteristics
Head coefficient ψ_{ind}	0.06 to 0.15	Head rise
Inlet flow coefficienct Φ_{ind}	0.06 to 0.20	Cavitation performance
Inlet vane angle θ	3° to 16° (measured from plane normal to axis)	Flow coefficient, angle of attack
Angle of attack α	3° to 8°	Performance, flow coefficient vane loading
Diameter ratio r_d	0.2 to 0.5	Performance, shaft critical speed
Vane solidity S_v	1.5 to 3.0 at the tip	Desired flow area
Number of vanes s	3 to 5	Desired solidity
Hub contour	Cylindrical to 15° taper	Compatibility with main impeller and shaft geometry
Tip contour	Cylindrical to 15° taper	Compatibility with main impeller and shaft geometry
Vane loading	Leading edge loading, channel lead	Performance
Leading edge	Swept forward, radial, swept back	Vane stress, performance
Sweep angle	Normal to shaft to 15° forward	Vane stress
Vane thickness	0.07 to 0.30 chord length C_1	Vane stress
Tip clearance (between inducer outside diameter and casing)	0.5 to 1 percent of inducer outside diameter	Shaft axial and radial deflections
Length to tip diameter ratio (L_t/d_t)	0.3 to 0.6	Head – capacity characteristics

Under certain conditions, where the required *NPSH* of the centrifugal impellers are high, inducers must have high discharge heads but with low required *NPSH* themselves. Two–stage tandem inducers may then be used with the first stage of low loading (low ψ value) and the second stage of high loading (high ψ value). Because the ψ values of the two stages are different, the outside diameters and/or the hubs must be made in conical shape.

In some pumps of low flow rate, a transformed inducer design is used to replace the two–stage arrangement for simplification. This is done by adding the same number of short blades with the same blade angle at the outlet of the rather long and slender inducer. Such designs are generally used with partial–emission pumps.

Designs with conical–contour tip and hub are useful in improving stability of the inducer at low flow rate by effectively reducing the reverse flow at inducer inlet. Problems of stability continue to be subjects of research in future designs.

12.7 Large Capacity Low Head Pumps

Volutes and suction bends made in concrete are becoming more popular in applications of vertical pumps of low head and large capacity. The practice was first started in irrigation and drainage duties but later has found use also in the power industry. Figure 12.51 shows the range of application for cooling water pumps with concrete volutes in power stations as compared with other types of pumps. Figure 12.52 shows a typical construction of a mixed–flow pump with concrete volute and suction bend.

Guide vanes used in conjunction with concrete volutes have pronounced influence on the performance stability of pumps [12.27]. As shown by model tests in Figure 12.53, the shorter guide vanes are superior to the longer ones. Guide vanes for this type of pump are primarily structural members that have little function in diffusing the flow, hence short and sturdy vanes, fewer in number are often used. When the impeller blades are adjustable, the number of guide vanes and their installed angle must conform to the best relations determined by model tests to ensure higher efficiency and stable operation of the pump at all blade angles.

Figure 12.54 shows a pump with pressure oil–actuated adjustable blades and concrete volute and suction bend with the following performance ratings: $D = 6,000mm$, $Q = 100m^3s^{-1}$, $H = 7m$, $n = 75min^{-1}$ and $P = 7,000kW$. A number of these pumps manufactured by the Shanghai Pump Works have been in satisfactory service in the South–North Water Diversion Project for a few years now.

The concrete volutes and suction bends sometimes are lined with thin steel plating in order to reduce hydraulic losses, these linings can also serve

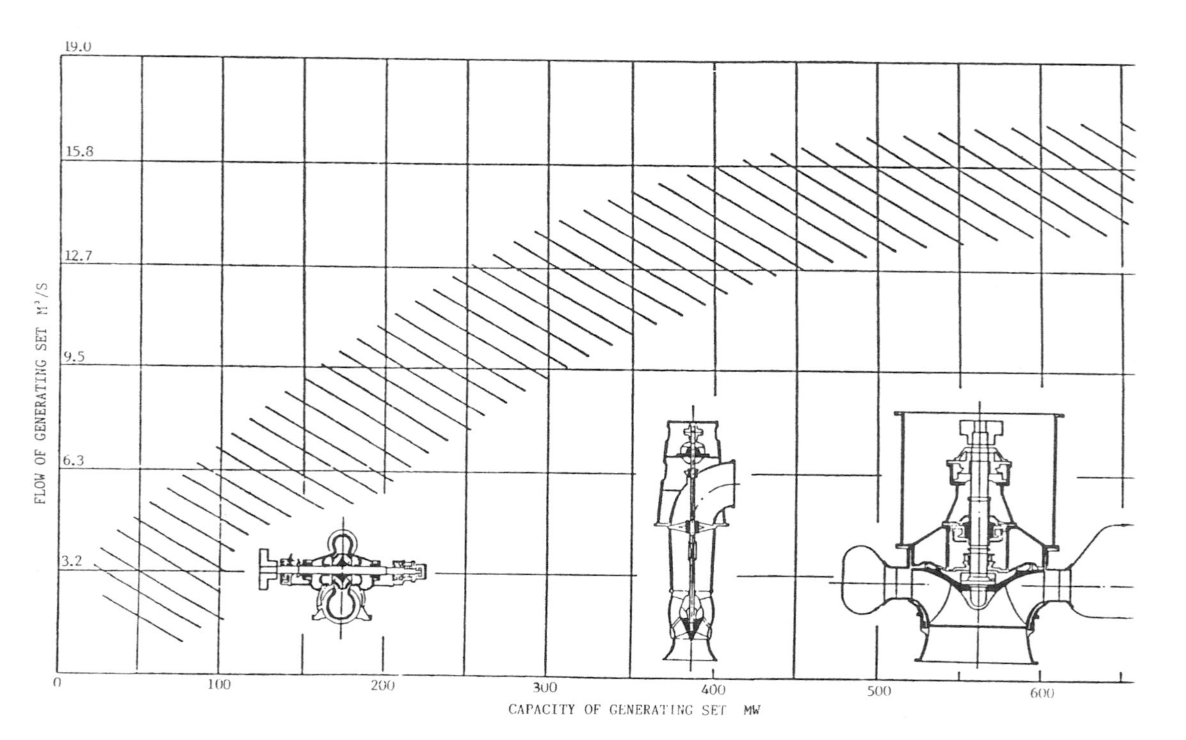

Figure 12.51 Different types of power station cooling water pumps

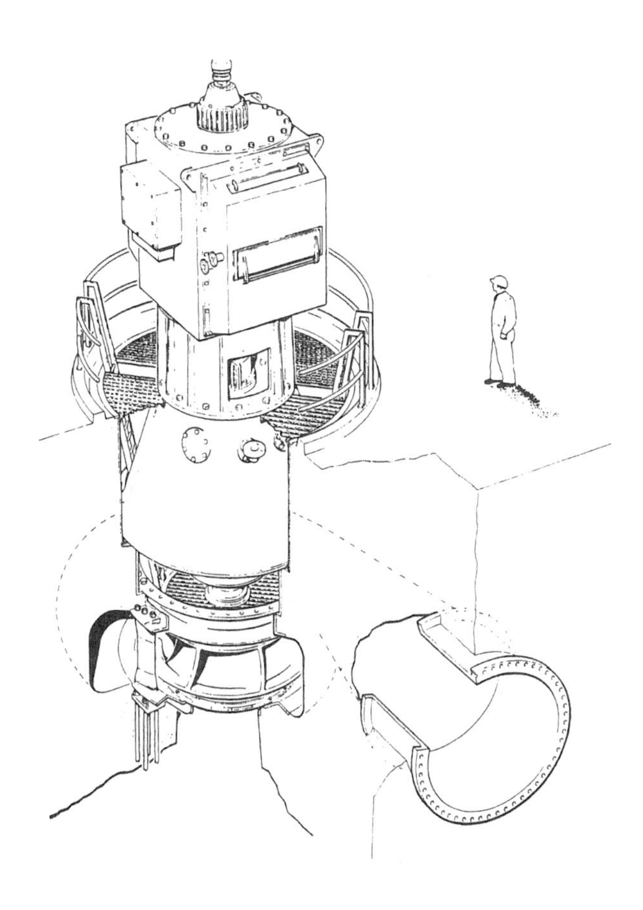

Figure 12.52 Typical construction of concrete volute mixed-flow pump (Drawing courtesy Weir Pumps)

in place of wooden moulds for pouring of concrete.

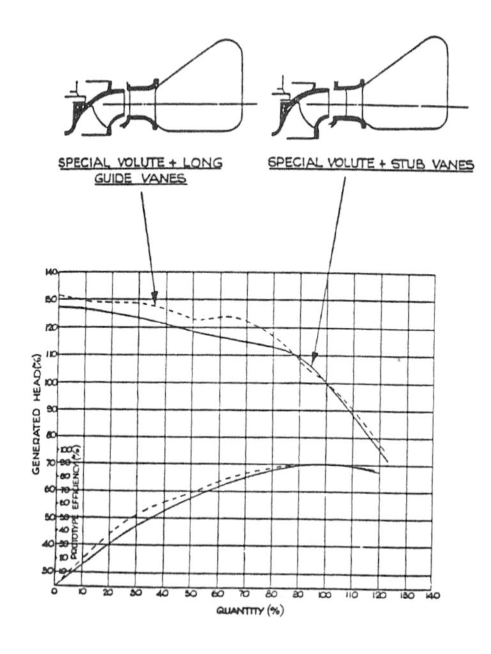

Figure 12.53 Comparison of performance with different guide vane geometry [12.27]

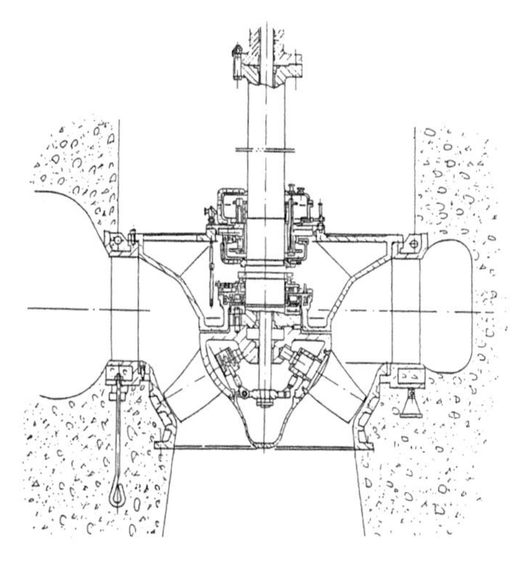

Figure 12.54 Vertical adjustable blade pump with concrete volute and suction bend (Drawing courtesy Shanghai Pump Works)

References

12.1 Stepanoff, A. J. *Centrifugal and Axial-flow Pumps*, 2nd Ed. John Wiley & Sons, New York - London, 1975.

12.2 Stepanoff, A. J. *Pumps and Blowers, Two-phase Flow*, John Wiley & Sons, New York - London, 1965.

12.3 Yedidiah, S. 'Avoiding cavitation in Centrifugal Pumps', *Machine Design*, Vol. 52, No. 5 (1980).

12.4 Yedidiah, S. 'Some Observations Relating to Suction Performance of Inducers and Pumps', ASME Paper 71-WA/FE-25.

12.5 Pfleiderer, C. *Die Kreiselpumpen fur Flussigkeiten und Gase*, Vierte Auflage, Springer Verlag, 1955.

12.6 Oshima, M. 'Cavitation Inception and NPSH of Pumps', *Plumbing Technology*, Vol. 20, No. 2 (1978) (in Japanese)

12.7 Jakobsen, J. K. 'Liquid Rocket Engine Turbopump Inducers', NASA 1971.

12.8 Schoneberger, W. 'Investigation on Cavitation in Radial Impeller of Centrifugal Pumps', *Konstruction*, Vol. 21, No. 7 (1969).

12.9 Florjancic, D. 'Net Positive Suction Head for Feed Pumps', Published by Sulzer Brothers Ltd, Winterthur, 1985.

12.10 Schroder, E. Chart plotted by C. Pfleiderer, Zeitschrift VDI, Vol. 82, Nr .9 (1938), p. 263.

12.11 Makay, E., Barret, J. A. 'Changes in Hydraulic Component Geometries Greatly Increased Power Plant Availability and Reduced Maintenance Costs: Case Histories', First International Pump Symposium, Texas A & M University, 1984.

12.12 Makay, E. 'Feed Pump Suction is Performance Key', *Power*, Vol. 125, No. 9 (1981).

12.13 Xu Hong-yuan, Yao Zhi-min. 'Reverse Flow at Pump Impeller Suction at Small Discharge Rates', *Pump Technology*, 1987, No. 3, 1-5 (in Chinese).

12.14 Weldon, R. 'Boiler Feed Pump Design for Maximum Availability'. I. Mech. E. 'Advanced Class Boiler Feed Pumps', Proceedings 1969-70, Vol. 184. Part 3N.

12.15 Anderson, H. H. *Centrifugal Pumps*, Third edition, Trade & Technical Press Ltd, 1980.

12.16 Anderson, H. H. 'The Area Ratio System', *World Pumps*, June 1984.

12.17 Ogasawara, Y. 'Development of Super–high Pressure Boiler Feed Pump for Ultra–supercritical Thermal Power Plant', *EBARA Engineering Review*, No. 138, 1987 (in Japanese).

12.18 Stewart, W. M. 'Power Station Pumps for Reliability and Availability', Electric Power & Water Expo/China 1984, published by Weir Pumps Ltd.

12.19 Florjancic, D., Eichhorn, G., Frei, A. 'Fifty Years Development of Sulzer Boiler Feed Pumps', published by Sulzer Brothers Ltd, Winterthur, 1984.

12.20 New, N. H. 'Experimental Comparison of Flooded, Directed and Inlet Orifice Type of Lubrication for a Tilting Pad Thrust Bearing', ASME Lubrication Symposium, 1973.

12.21 Florjancic, D., Frei, A., Simon, A. 'Boiler Feed Pumps for Supercritical Power Plants – State of Art', First International Conference on Improved Coal–fired Power Plants.

12.22 Florjancic, D. 'Pumps in Feedwater Circuits of Thermal and Nuclear Power Stations', published by Sulzer Brothers Ltd, Winterthur, 1984.

12.23 Florjancic, D., Bolleter, U., Simon, A. 'Boiler Feed Pumps for Modern Fossil Fired Power Plants', published by Sulzer Brothers Ltd, Winterthur,1987.

12.24 Janes, P. J. A. 'High–head, High–speed Centrifugal Process Pumps', Conference on Process Pumps, I.Mech.E., London 1973.

12.25 Barske, U. M. 'Development of Some Unconventional Centrifugal Pumps', Proc. I.Mech.E. 174 (1960), 437–461.

12.26 Huzel, D. K., Huang, D. H. 'Design of Liquid Propellant Rocket Engines', NASA Ap–125, 1967.

12.27 Weldon, R., Dyson, F. 'Circulating Water Pumps, Proceedings 2nd BPMA Technical Conference, Glasgow, 1971.

Chapter 13

Construction of Axial–Flow, Mixed–Flow and Tubular Pumps

Masuji Matsumura

13.1 General Features of Low Head Pumps

Axial–flow, mixed–flow and tubular pumps are the essential pump types used for water lifting under relatively low heads. Generally speaking, axial–flow pumps are used for head ranges from several metres to less than $10m$ and are applied mainly in the medium to large capacity range. Mixed–flow pumps are used in somewhat higher head ranges and mostly in the small/medium capacities. Both types may be constructed in vertical or horizontal designs and occasionally in the inclined arrangement (S–type), but the majority of these are vertical pumps. There are specific reasons for choosing one type or the other on grounds of preference to hydraulic performance and/or mechanical construction.

The tubular pump is an axial–flow pump set horizontally. Its main feature is that the water flowing through the pump has a straight–through passage with no inlet or discharge bends, consequently it has a higher hydraulic efficiency. The driving motor is either mounted vertically above the pump with transmission through bevel gearing or enclosed in a shell with water flowing around it as is know as a bulb–type pump in large installations.

13.1.1 Variable pitch blade and fixed blade impellers

An example of a fixed blade mixed–flow impeller is shown in Figure 13.1. The impeller may consist of four to eight blades (the larger number for lower specific speed pump) with three–dimensional curvature, fixed to the conical surface of the hub as a one–piece casting. This figure shows the semi–open type of impeller which is used to handle sewage water that contains fibres or solid particles. In general, with decreasing specific speed a closed type impeller with a shroud is used. The ratio of passage breadth to the outlet diameter is higher than with a centrifugal impeller. In the process of manufacture, a blade of stainless steel is apt to deform from the original shape during heat treatment. Therefore, special consideration should be given in controlling the geometric shape of the curved blades in production of one–piece impellers.

Contrary to the fixed blade impeller, it is easy to control the accuracy of blade shapes for a built–up or a variable pitch type impeller. There are two typical methods of adjusting the blade angle. One is shown in Figure 13.2 where the axial movement of a blade operating rod passing through the hollow pump shaft is converted into a rotating movement of the blade stem distributed with the same pitch on the conical surface of the impeller hub. The other method is a rotary type of servomotor which enables the blade arms to rotate around the blade operating rod, as with the Deriaz type water turbine as shown in Figure 3.17. All sliding surfaces between the blades and the hub are sealed with O–rings or other packing material and the inside of the hub is filled with lubricating oil to reduce friction at the blade stem bushings and sliding joints of the link mechanism.

In the case of axial–flow impellers, greater accuracy in forming the blade shape and in setting the blade angle is necessary than with mixed–flow impellers, so it is logical to make the blades separately to form a built–up impeller (with blades fixed by bolts) even if no adjustment is required while the pump is running. This construction will enable resetting the blade angle after the pump has been tested and will facilitate replacement of blades. Figure 13.3 shows an external view and a section of a variable–pitch axial–flow impeller which may have two to six blades and has a linkage mechanism to convert the axial movement of blade operating rod into rotation of blade stems as with mixed–flow impellers.

13.1.2 Hydraulic and mechanical loading on variable–pitch blades

The loading on variable–pitch blades consists of hydraulic and centrifugal forces. To predict the hydraulic force it is necessary to calculate the pressure distribution on the blade surface. Among the several numerical calculation methods available for three dimensional inviscid flow in the impeller, the quasi–three dimensional flow analysis developed by Senoo and Nakase [13.1] is an effective method. This is an iterative procedure between the meridional flow (Figure 13.4b) and the blade–to–blade flow (Figure 13.4c), the latter

Figure 13.1 Fixed blade mixed-flow impeller

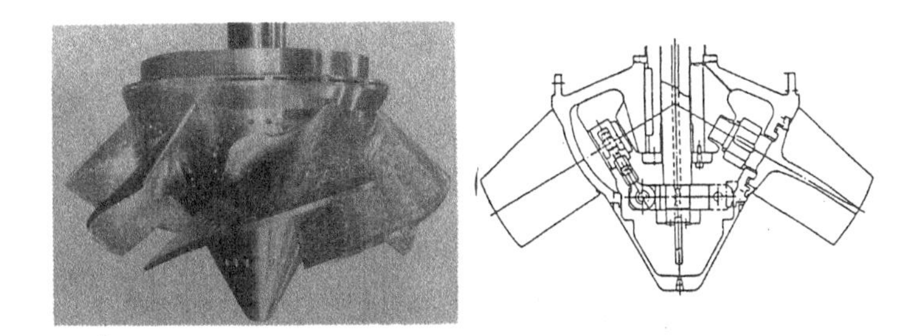

Figure 13.2 Variable-pitch blade mixed-flow impeller

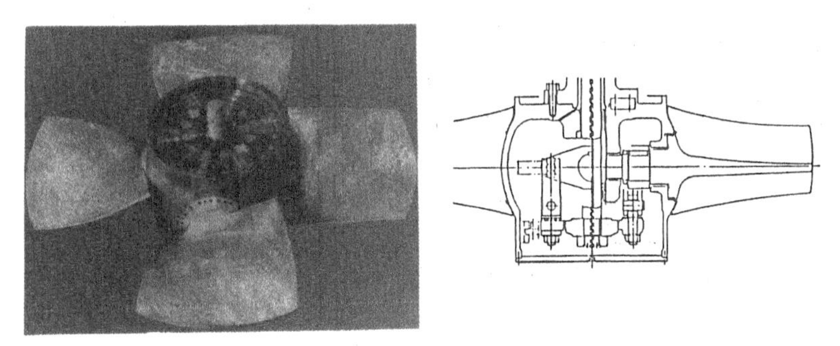

Figure 13.3 Variable-pitch blade axial-flow impeller

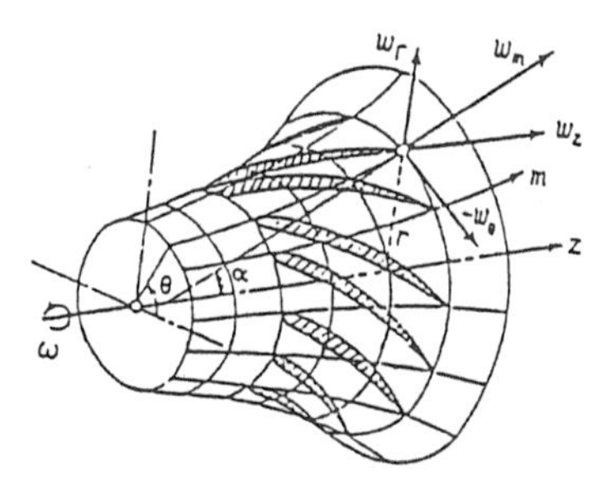

(a) Rotating stream surface [13.1]

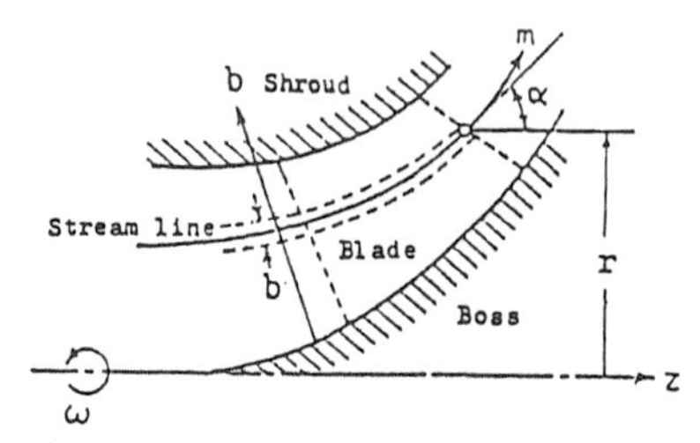

(b) Meridional stream lines [13.1]

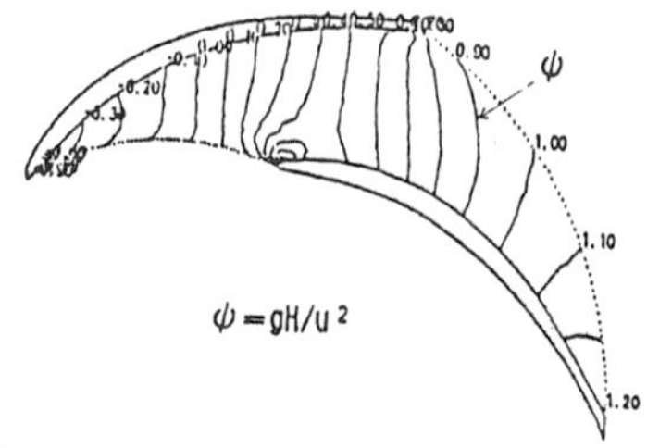

(c) Pressure distribution between blades

Figure 13.4 Three–dimensional flow analysis of mixed–flow pump impeller

being the developed plane from the rotating stream surface as shown in Figure 13.4a. The hydraulic loading on the blade surface is based on the static pressure distribution of the blade-to-blade flow and calculation by small triangular meshes laid out on the blade surface.

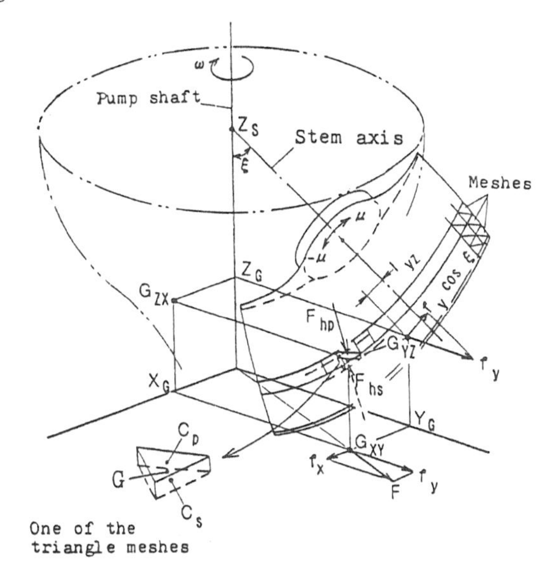

Figure 13.5 Forces acting on a blade element [13.2]

In the triangular meshes (Figure 13.5) F_{hp} is the hydraulic force obtained from the average surface pressure pP on the three apexes and its components can be written as [13.2]

$$\begin{aligned} \mathrm{f}_{hx} &= \overline{p}S_{yz} \\ \mathrm{f}_{hy} &= \overline{p}S_{zx} \\ \mathrm{f}_{hz} &= \overline{p}S_{xy} \end{aligned} \tag{13.1}$$

where S_{yz}, S_{zx} and S_{xy} are the projected areas of the triangle on the planes yz, zx and xy respectively. Then the hydraulic moment loading on the blade stem axis based on the triangular element can be written as

$$\Delta M_h = \mathrm{f}_{hx} \cdot l_{yzp} - (\mathrm{f}_{hy} \cdot \cos\xi + \mathrm{f}_{hz} \cdot \sin\xi) \cdot x_h \tag{13.2}$$

where l_{yzp} is the perpendicular length of point c_p to the blade stem axis z_s

on the plane yz

$$l_{yzp} = -Y_c \cos\xi + (Z_s - Z_c)\sin\xi$$

Centrifugal force component loading on the triangular mass $\rho\Delta V$ (where ρ is the density of the blade, and ΔV is the volume of the triangular element) at the centre of gravity G_c can be written as

$$\begin{aligned} \mathrm{f}_{cx} &= \rho\Delta V\omega^2 X_G \\ \mathrm{f}_{cy} &= \rho\Delta V\omega^2 Y_G \\ \mathrm{f}_{cz} &= 0 \end{aligned} \tag{13.3}$$

and the centrifugal moment on the blade stem axis Z_s is

$$\Delta M_c = -\mathrm{f}_{cx} l_{yzc} + \mathrm{f}_{cy}\cos\xi X_G \tag{13.4}$$

where $l_{yzc} = -Y_G\cos\xi + (Z_s - Z_G)\sin\xi$ (from Figure 13.5).

The total hydraulic moment, the difference between that on the pressure and suction surfaces $M_{hp} - M_{hs}$ and the centrifugal moment M_c are obtained from the sum of each partial moment on individual triangular meshes divided along and perpendicular to the streamlines. Figure 13.6(b) shows the blade torque curves of a mixed–flow impeller at several blade angles and a comparison between experiment and numerical solution obtained from the above calculation method at the design blade angle. Numerical solution was not given in the low flow region, thatis below the stall point. As can be seen from the figure the discrepancy between the two curves has an increasing tendency toward the lower flow rates.

Figure 13.7(a) shows the effect of pump speed at constant blade angle $\beta = 100\%$ (design angle). In this figure the hydraulic moment μ_h and the centrifugal moment μ_c are nearly equal but act in opposite directions so the total blade torque μ_0 is near a balanced condition. Figure 13.7(b) shows the effect of blade angle on the hydraulic and centrifugal moment at the best efficiency points according to each blade angle. In this case the change of hydraulic moment is greater than that of the centrifugal moment so the total torque resembles the hydraulic moment curve.

Figure 13.8 shows the comparison of blade torque curves of theoretical hydraulic moment computed by the boundary element method [13.3] with experimental results for an axial–flow impeller. These calculated curves (excluding the centrifugal moment) show more disagreement as the flow decreases to the stall point due to the enlargement of separation zone on the suction surface.

13.1.3 Stress and vibration mode of variable pitch blades

With the rapid progress of calculation procedures using the three dimensional finite element method, the prediction of stresses for complicated shaped

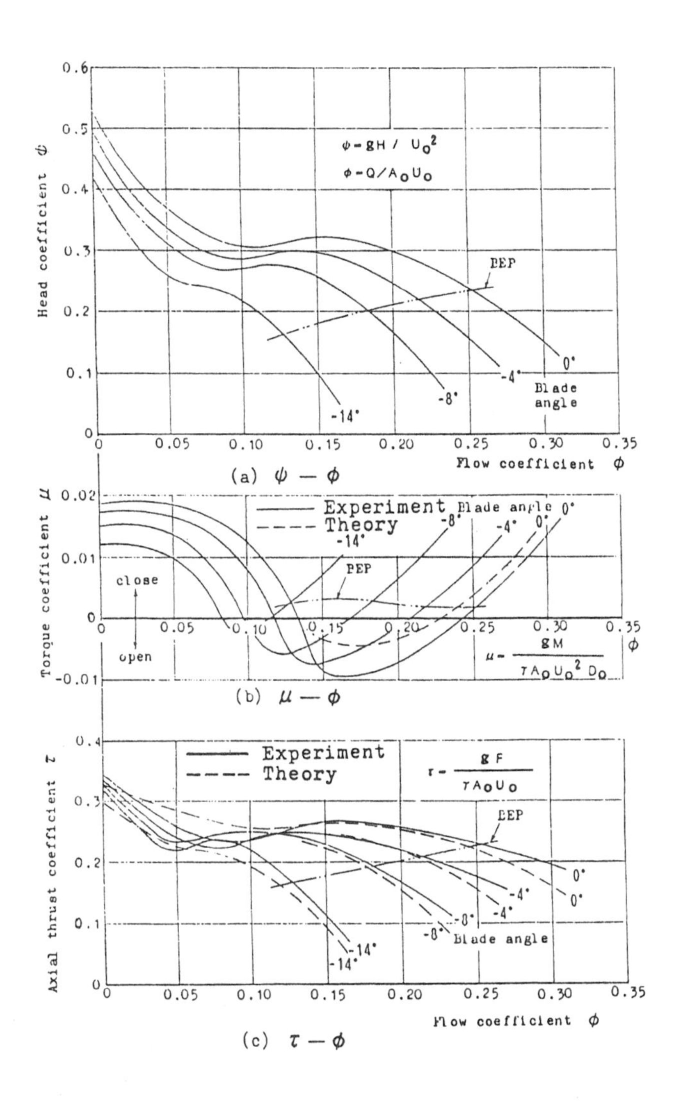

Figure 13.6 Capacity versus head, blade torque and axial thrust of mixed–flow impeller, $N_s = 930$ $(m \cdot m^3 min^{-1})$

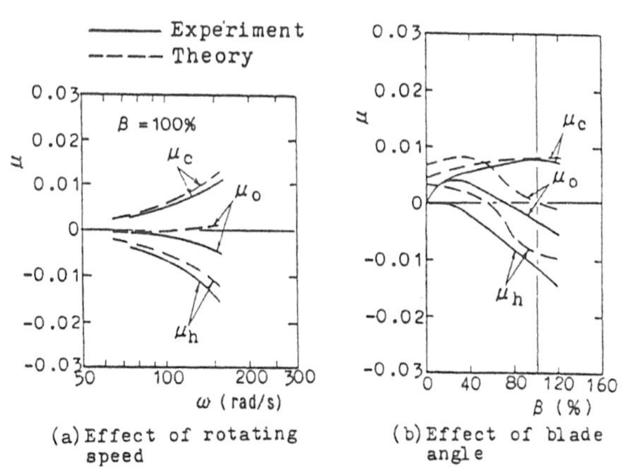

(a) Effect of rotating speed

(b) Effect of blade angle

Figure 13.7 Effect of rotating speed and blade angle on the blade torque [13.2]

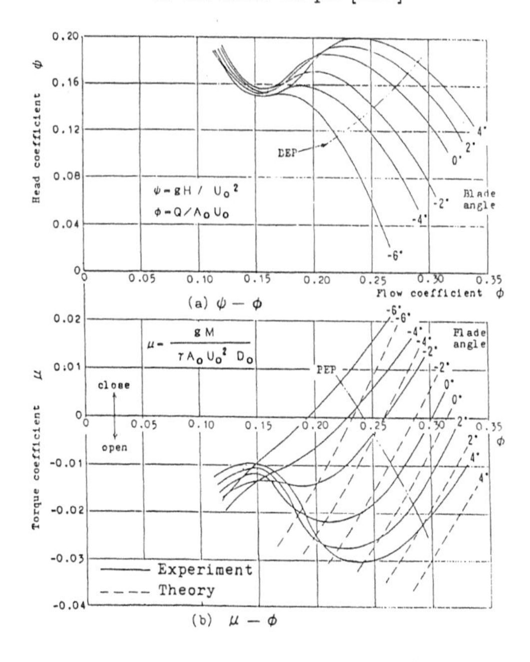

Figure 13.8 Capacity versus head and blade torque of axial–flow impeller $N_s = 1300\ (m \cdot m^3 min^{-1})$ [13.3]

components of hydraulic machinery has been widely performed with the aim of improvement of reliability. In case of the variable–pitch blade the stress concentration is likely to appear at the root of the blade stem because of its overhanging structure. Figure 13.9 shows the stress distribution on a variable–pitch blade with the hydraulic and centrifugal forces described in 13.1.2. for the best efficiency point. Experimental data is obtained from the reading of strain gauges attached to the rotating blade surfaces.

The stress analysis and experimental data are in fairly good agreement except near the root of the blade stem (section X–X) because of the effect of a fillet at the root section. If the mesh at the fillet region is divided into finer elements the accuracy of prediction would be improved. The similarity relations of stress between model and prototype blade is as follows [13.4]

$$\frac{\sigma_m}{\sigma_p} = \frac{1}{S^2} \cdot \frac{\rho_m}{\rho_p} \left(\frac{n_m}{n_p} \right)^2 \tag{13.5}$$

where σ – stress
S – scale ratio
ρ – density of material
n – pump speed

and subscripts m – model
p – prototype.

Hydraulically excited vibrations of the blade due to pressure fluctuation resulting from flow separation, wake, vortices and cavitation can lead to material damage under certain circumstances when the frequencies of these forces approach the natural frequency of the blade. The blade under vibration will exhibit different modal lines at each natural frequency. By means of a frequency generator controlled with an oscillation generator the natural frequency and vibration mode are determined when the Lissajous figure appearing on the oscillograph reaches its full extent and the phase displacement between the exciter and the proximeter becomes 90°.

The analysis of vibration amplitude and vibration mode at the first and second natural frequencies using finite element methods such as SAP IV and NASTRAN in air are shown in Figure 13.10 compared with the experimental results. The vibration frequency of the blade in water is reduced by effect of the added mass of water and this added mass depends on the shape of the blade, blade pitch and the vibration mode. The amount of reduction in natural frequency due to the added mass $(f_a - f_w)/f_a$ (subscripts: a for air, w for water) for the case of a Kaplan turbine of runner diameter $0.72m$ and blade mean chord length $0.34m$ is reported to be in the range of 31% to 37% [13.5].

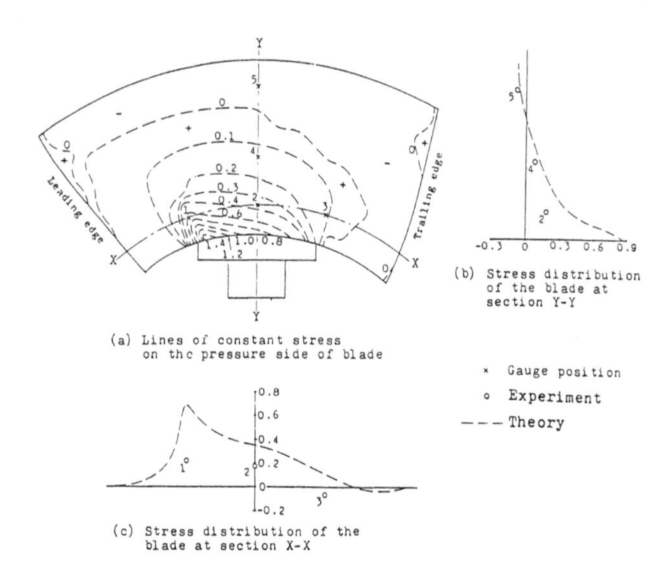

Figure 13.9 Stress analysis of mixed–flow pump blade at best efficiency point

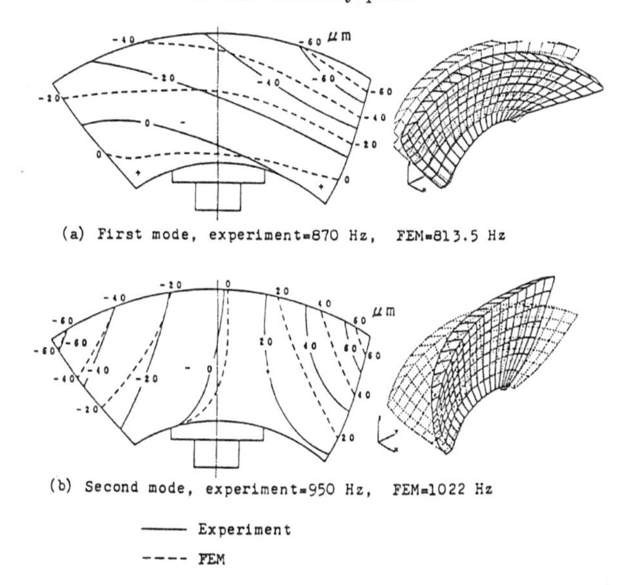

Figure 13.10 Comparison between theoretical and measured values of natural frequency of blade

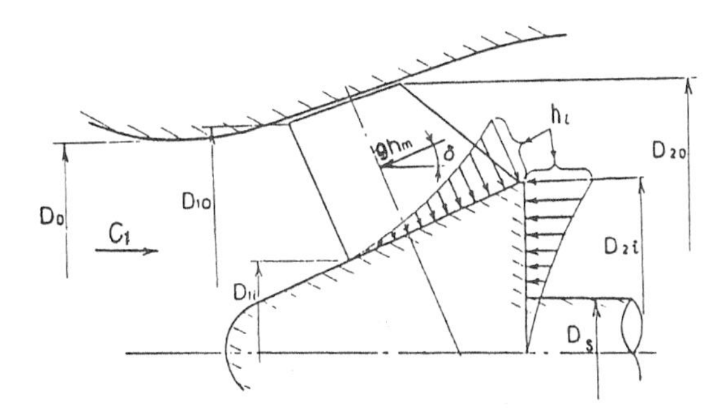

Figure 13.11 Forces acting on mixed–flow pump impeller [13.7]

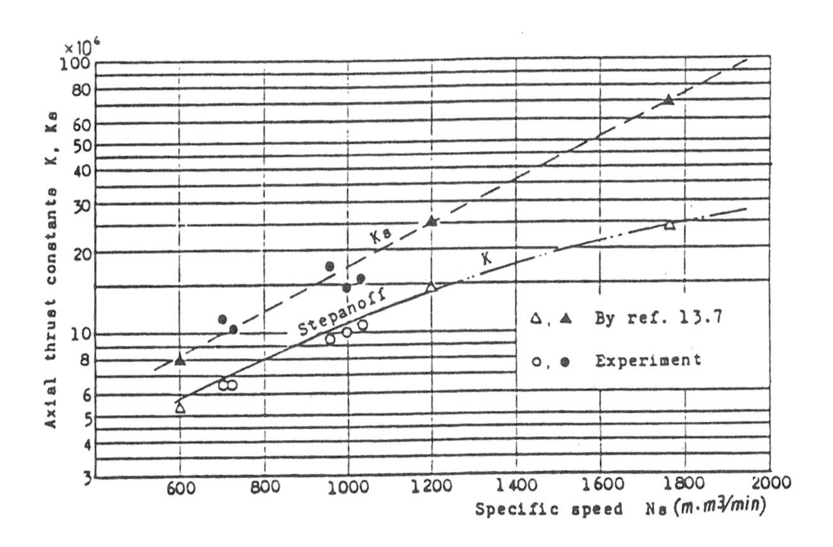

Figure 13.12 Axial thrust constants versus specific speed

13.1.4 Axial thrust

Formula related to calculation of axial thrust on mixed–flow impellers have been proposed by many authors. Among these, Stepanoff gave the simplest and most convenient formula to use [13.6]

$$T = K\rho g H \frac{\pi}{4} D_{10}^2 \tag{13.6}$$

where

- ρ – density of liquid
- H – normal pump total head
- D_{10} – impeller inlet diameter (Figure 13.11)
- K – axial thrust constant varying with specific speed (Figure 13.12)

The thrust constant K given by Stepanoff is in good agreement with experimental results near the best efficiency point but around the shut–off point it gives values too high, a different constant K_s is recommended to be used for the shut–off point where H is defined as normal pump head, as in equation 13.6. The following formula proposed by Oshima [13.7] is relatively simple and its results agree fairly well with the experimental data over a wide flow range (Figure 13.6)

$$\begin{aligned} T = {} & \rho g h_m \cos\delta \frac{S_1 + S_2}{2} + \rho g \left[\frac{\pi}{4} h_1 (D_{2i}^2 - D_s^2) - \pi \frac{\omega^2}{16^2 g}(D_{2i}^2 - D_s^2)^2\right] \\ & - \rho g \frac{1}{8}\pi h_1 (D_{2i}^2 - D_s^2) + \rho v_1^2 \left[S_1\left(\cos\delta - \frac{4S_1}{\pi D_0^2}\right) - \frac{\pi}{4} D_0^2 \left\{1 - \left(\frac{4S_1}{\pi D_0^2}\right)\right\}\right] \end{aligned} \tag{13.7}$$

where h is the static pressure head in the impeller, defined by

$$h = H\left(1 - \frac{gH}{2\eta_h \cdot u_2^2}\right) + \eta_h \cdot \frac{C_1^2 - C_{m2}^2}{2g} \tag{13.8}$$

where
- η_h – hydraulic efficiency
- u – peripheral velocity
- C – absolute velocity
- C_m – meridian velocity
- δ – meridional flow angle to the rotating axis
- S – area normal to the flow passage
- D – diameter
- ω – angular velocity
- g – acceleration of gravity

and subscripts
- 1 – impeller inlet
- 2 – impeller outlet
- m – outlet mean diameter
- i – outlet hub diameter

13.1.5 Blade pitch control mechanism

Typical mechanism that controls the blade operating rod in vertical motion may be of either hydraulic or mechanical in principle. Figure 13.13 shows an example of hydraulic control mechanism in which both the four-way servo valve distributing the pressure oil and the oil supply head are mounted on the top of the hollow driving shaft. Hydraulic pressure is changed by switching the stroke of the four-way valve according to the rotating angle of the control motor. The servomotor which is situated in the middle portion of the pump shaft, moves upward and downward in proportion to the swing angle of the control motor. The return motion mechanism continuously compensates the stroke of the four-way valve to the neutral position where the hydraulic pressure is cut off in all directions.

A certain type of hydraulic control mechanism with oil supply may be mounted above the oil lubricated bearings with the return motion mechanism extracted from the side of the servomotor [13.8]. This arrangement makes it easy to convert fixed blade pumps to the variable-pitch type without major change to the driving shaft of the prime mover.

Figure 13.14 shows a mechanical control mechanism that slides the thrust collar upward and downward by the rotating worm wheel and is operated by an electric actuator. This device can also be used in converting fixed blade pumps to the variable-pitch type.

13.2 Low Head Pump Installation

13.2.1 Typical examples of low head pump installation

A general view of a drainage pumping station is shown in Figure 13.15. Water is drawn into the suction sump through the auto screen (for removing debris from the water) and is lifted to the surge tank by vertical shaft pumps driven by diesel engines through bevel-type reduction gears. Water then flows by way of the delivery conduit up the gradient. This plan shows the pumping system for handling relatively large quantities of water.

For a small capacity installation such as a sewage pumping station, a typical example is shown in Figure 13.16 where the vertical mixed-flow pump is set in the underground pump chamber, the impeller level is below the low water mark of the suction pool. Pumps for this purpose have fewer numbers of impeller blades and a volute casing specially designed for handling sewage water. In general the motor and the delivery valve are installed on the upper floor in order to prevent electric trouble due to high humidity.

Pump installation features such as installation space, cavitation damage, hydraulic losses of water passages, operating and maintenance requirements

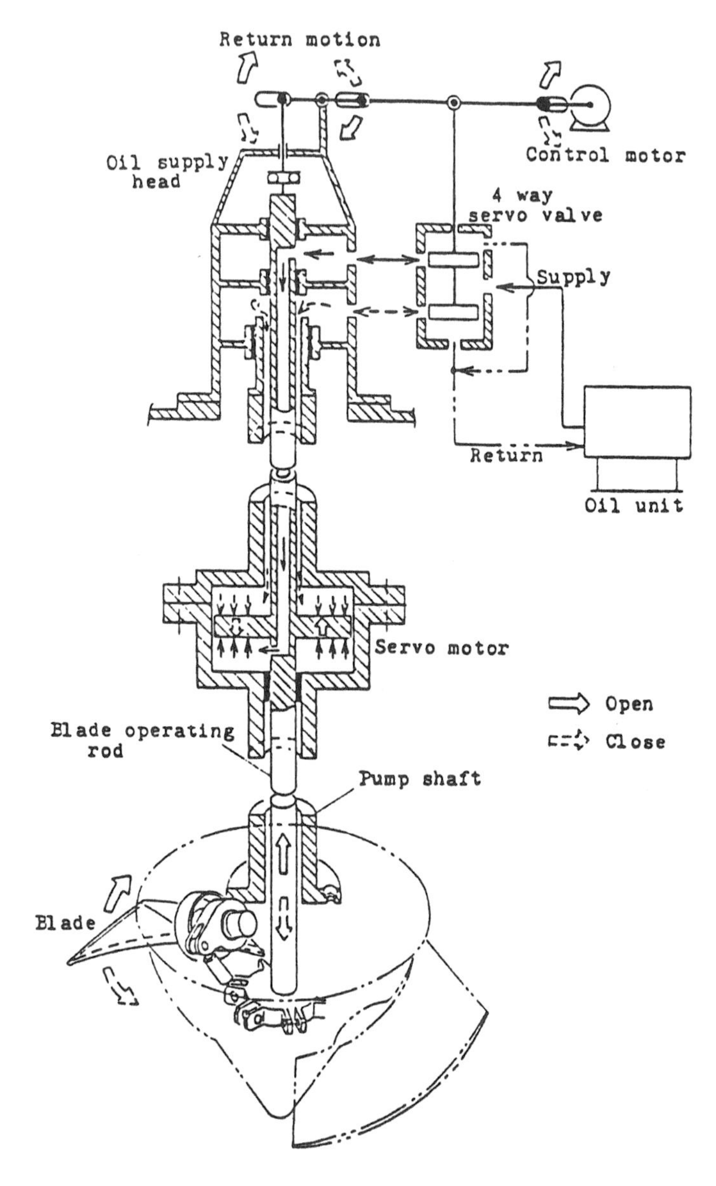

Figure 13.13 Blade pitch hydraulic control mechanism

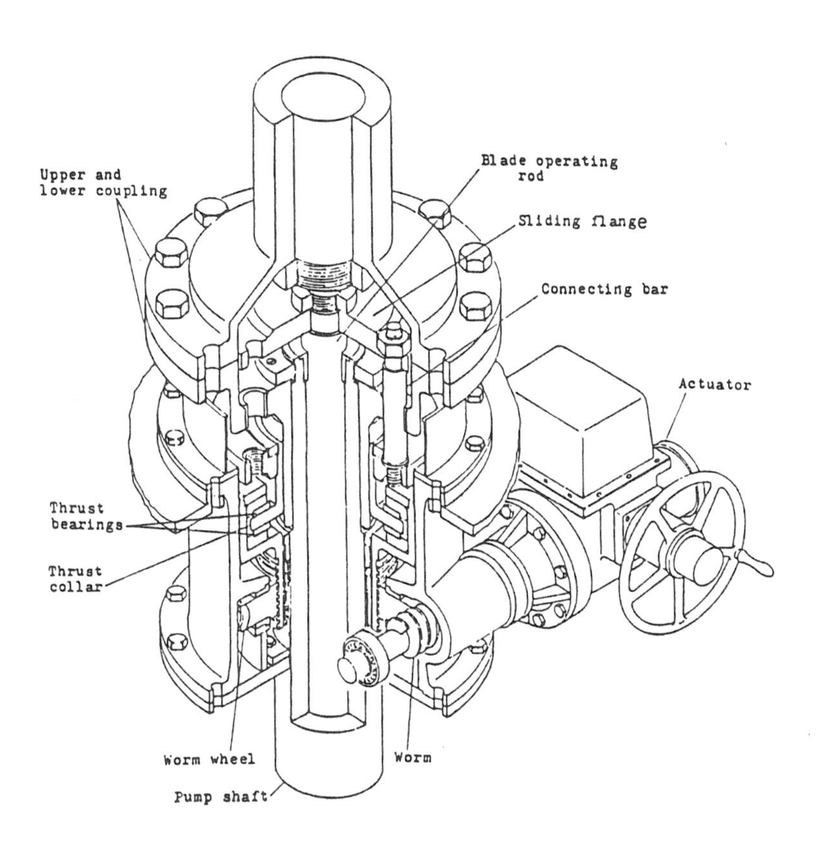

Figure 13.14 Blade pitch mechanical control mechanism

are summarized for pumps of horizontal, vertical and inclined types in Table 13.1.

Table 13.1 Comparison of installation features classified according to pump arrangement

Items	Horizontal shaft	Vertical shaft	Inclined shaft
1. Installation Space	Requires large space	Requires small space	Requires smaller space than horizontal shaft pump.
2. Cavitation	Comparatively lower pump speed must be designed to prevent cavitation. Pump speed must be limited within lower range, so that the pump size will become larger than vertical shaft pump.	Comparatively higher pump speed may be designed, and the pump size will become smaller than horizontal shaft pump.	More advantageous than horizontal shaft pump.
3. Pipe Friction Loss	Large		Small
4. Maintenance, Inspection and Dismantling	Dismantling may be facilitated by lifting the upper half of the casing, without removing the motor.	In most cases, the motor shall be removed prior to dismantling of the pump.	Easier than vertical shaft pump.
5. Priming	Required	Not required. Easy start-up operation	Not required Easy start-up operation
6. Corrosion Resistance	Advantageous	More corrosion will occur in wet pit type pump.	More advantageous than vertical shaft pump.
7. Typical Example	Fig. 13.19	Fig. 13.15.~13.18. Fig. 13.21., 13.22.	

For large capacity ($Q > 5m^3s^{-1}$) and low head ($H < 8m$), pumps with water passages wholly made of concrete are often used for economical reasons (reduction of total construction cost), several examples of which are shown in Table 13.2. In Figures 13.17 to 13.19 typical examples of pump installations with various designs of suction and delivery passages are shown. Figure 13.20 shows the section through a horizontal shaft tubular pump that is suitable for large–quantity low–head water drainage duty.

13.2.2 Recent trends of low head pump installation

The recent trends in construction of pump installations are in the direction of simplifying the auxiliary equipment system, improving quick start–up operation and saving installation space and man–hours for maintenance.

Figure 13.21 shows a vertical pump installation having bearings that are able to start without pre–lubrication and in–line coolers for cooling the water required by the diesel engine and reduction gearing. The pumping station can thus eliminate the cooling and lubricating water reservoir and its pumping system which are representative of the practice in the past.

In order to provide quick starting of pumps for cases where the water level may rise very fast as in the drainage system in an over–populated city, it is necessary to have the pumps running in dry condition before the water

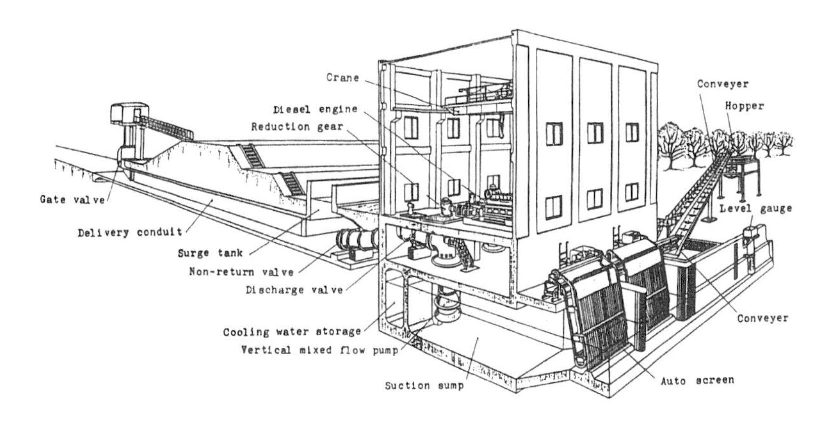

Figure 13.15 General view of a typical low head pumping station

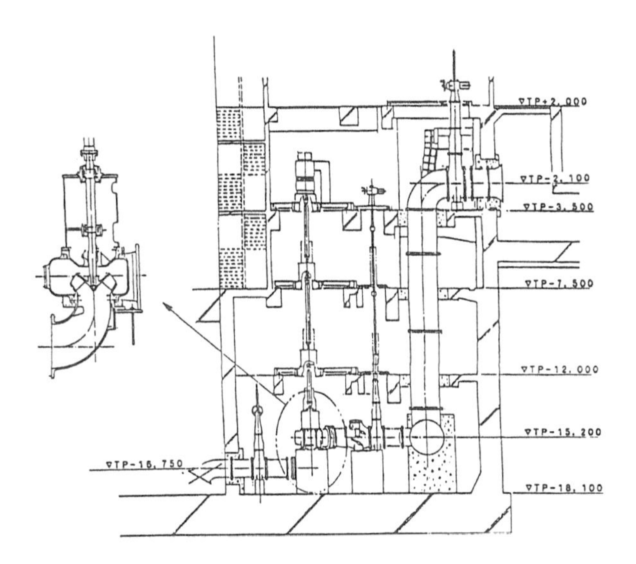

Figure 13.16 Sewage pumping installation for small size mixed–flow pump with volute casing

Table 13.2 Typical water passage types for large capacity low head pumps [13.9]

Sketches	Suction	Delivery	Application		Features	Examples
			Quantity(m^3/s)	Head(m)		
	Bell mouth with cone	Single spiral	≤15	≥4	1. Smaller height dimension 2. Suitable for large capacity pumps 3. Requires large installation space 4. Requires complicated civil works	
		Double spiral	>15	≥4		
	Bell mouth with cone	Umbrella	>20	<6	The same as above and adequate to the axial flow pumps.	Fig. 13.17
	Elbow	Siphone	>20	<8	1. Provides siphonic operation 2. Requires no gate valve: but siphone breaker 3. Minimize the water passage loss and provides high total efficiency 4. Larger height dimension 5. Requires complicated civil works	Fig. 13.18
	Bell mouth with cone	Elbow	≤15	≥4	1. Requires simple civil works 2. Easy installation 3. Maintenance is rather difficult	
	Tubular		> 5	<5	1. Requires very easy civil works 2. Requires very small installation space 3. Maintenance is rather difficult	Fig. 13.19

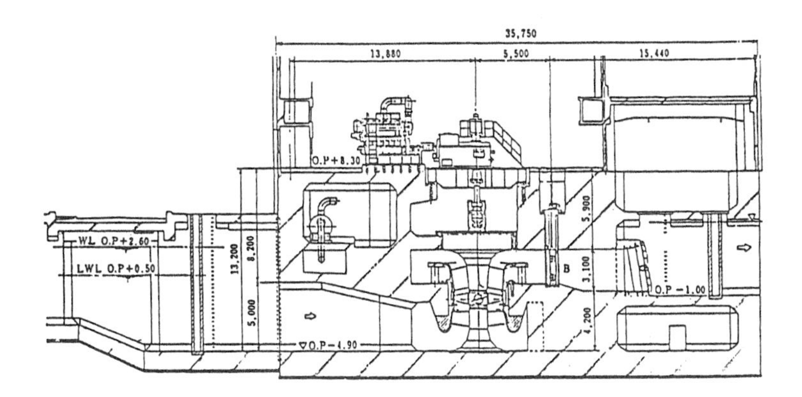

Figure 13.17 Bell–mouth umbrella type concrete water passage pump installation

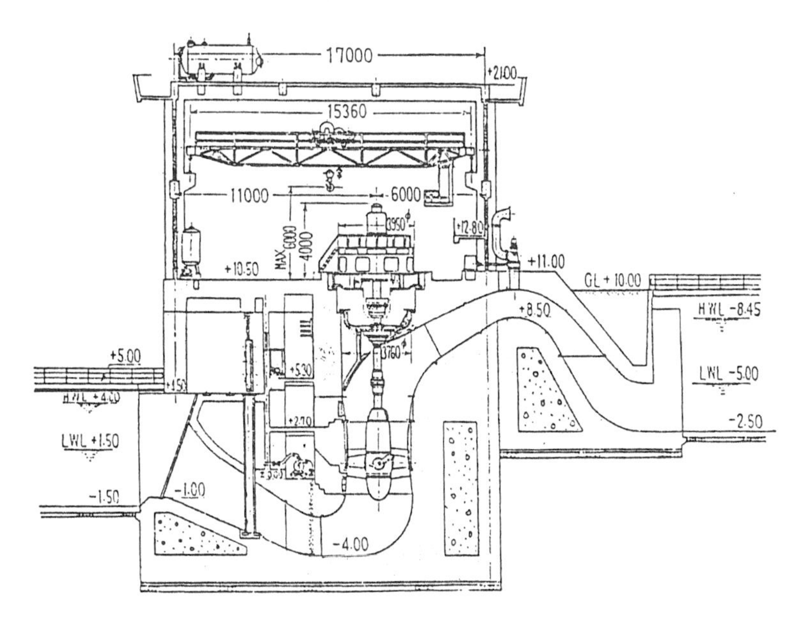

Figure 13.18 Siphon type concrete water passage pump installation (Drawing courtesy Ebara Co)

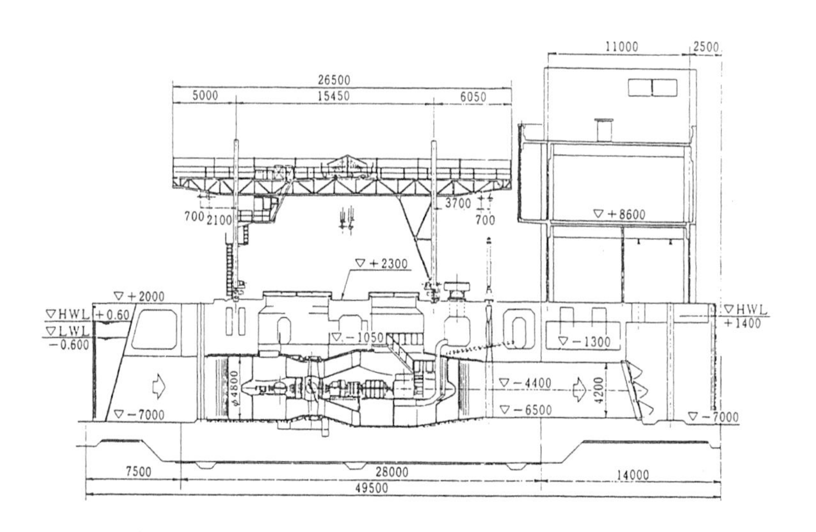

Figure 13.19 Tubular pump installation
(Drawing courtesy Ebara Co)

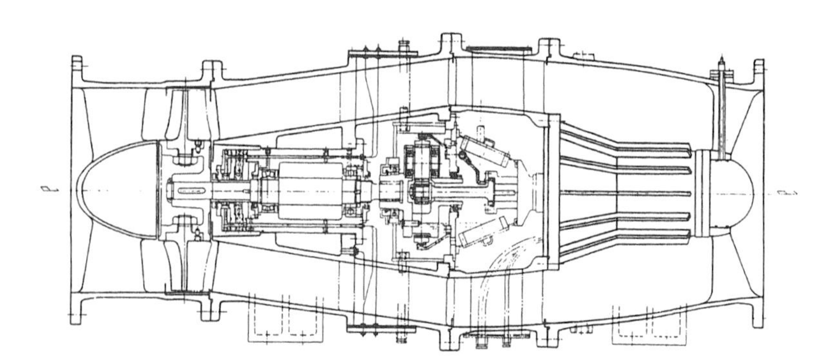

Figure 13.20 Section of tubular pump, $D = 800mm$

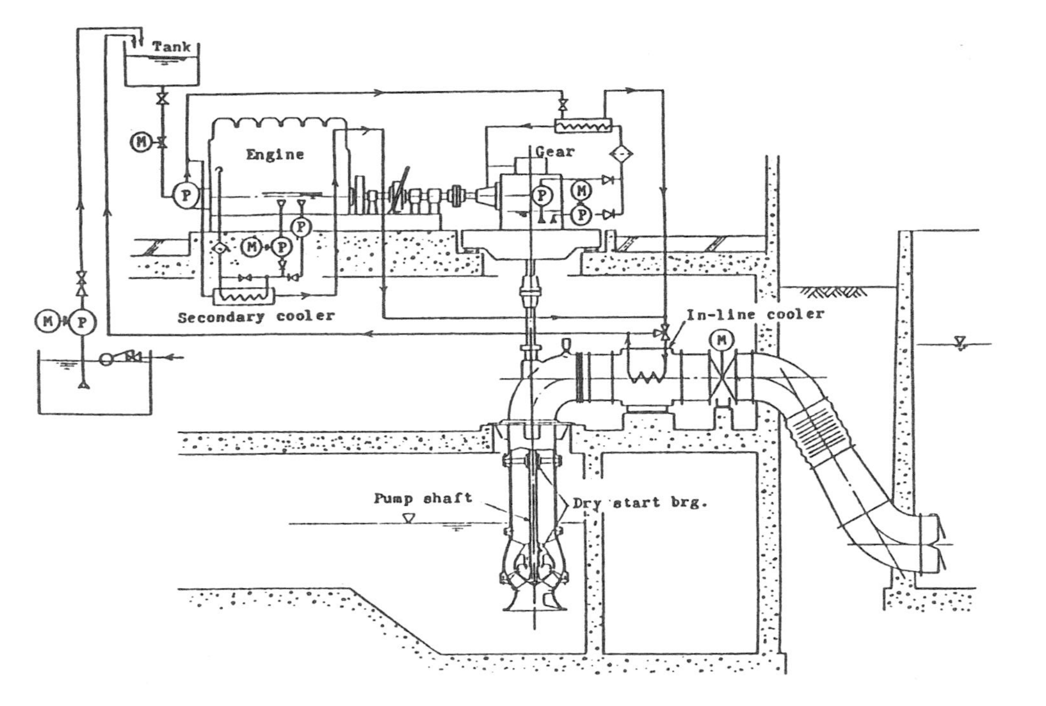

Figure 13.21 Low head pump installation with simplified lubrication and cooling systems

reaches normal suction level. Problems arising from this operation and their counter measures are listed in Table 13.3.

Table 13.3 Transient operating phenomena of vertical pump for increasing suction water level to the impeller

Water Level	Problems	Counter measures
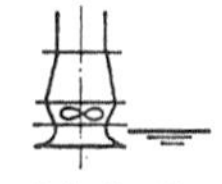 Racing Operation	Galling of the casing liner	Increase of clearance between the impeller and the casing liner
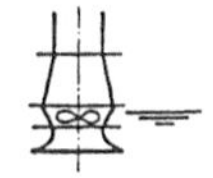 Churning Operation	The impeller and the shaft are subject to shock stress during the transition period of racing and churning operation.	The strength of the impeller and the shaft shall be improved.
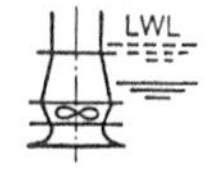 LWL Operation under LWL	Vibration and noise caused by suction of the air.	Flow control according to the water level (By controlling the pump speed or by controlling the blade angle)

For effective use of land space in the metropolis and its surrounding areas, low head pumps of the submersible motor–driven type, shown in Figure 13.22, have seen wide application. This trend is likely to propagate with the successful development of large capacity submersible motors. Monitoring of operation for a system of pumping stations in regard to trouble diagnosis and maintenance can be done with the aid of a computer system.

13.3 Intake Structures for Low Head Pumps

13.3.1 Suction sump design

When deciding on the suction sump design the following factors must be taken into consideration:

1. Air–entraining vortex formation (Figure 13.23) and the submerged vortex formation from the terminal end on a side wall or a bottom surface (Figure 13.23) must be avoided.

2. The suction approach flow must be uniform and must not contain whirls.

3. The intake structure should be planned in an economic way.

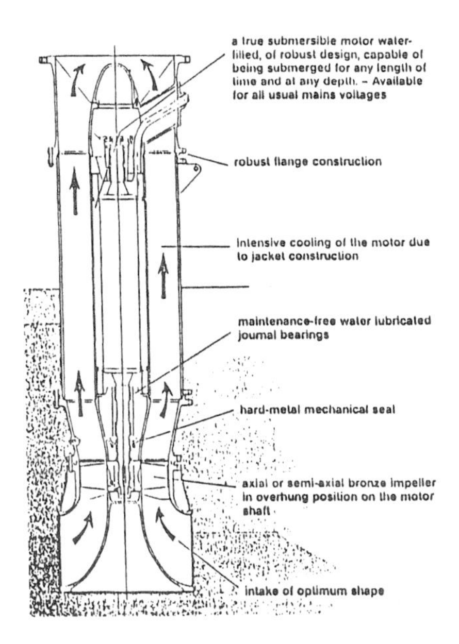

Figure 13.22 Large capacity submersible motor pump (Drawing courtesy Dresser–Pleuger GmbH)

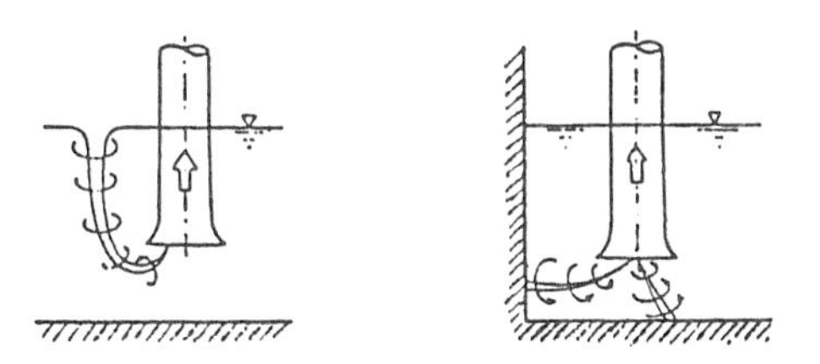

(a) Air–entraining vortex (b) Submerged vortex

Figure 13.23 Typical vortex formation in suction sump

If the suction approach were not uniform and the sump design not suitable, noise and vibration caused by the formation of air entraining vortices and submerged vortices may bring about a drop in pump performance. In the case of a pump with a horizontal shaft the effect of a suction approach to pump performance is relatively small because of the usually longer distance from the suction sump. But for a vertical pump with submerged impeller in the suction sump it is easy for the detrimental effects described above to occur.

To guarantee satisfactory pump performance, the following items must be considered in practice [13.10]:

1. In case ofa multiple pump installation the flow in the suction sump must be divided to provide as even a flow as possible. If necessary, partitions between pumps may be put in the sump.

2. The pump arrangement, the location of bell mouth and the shape of suction sump must be designed to avoid formation of whirls.

3. The average velocity in the vicinity of the pump shall be less than $0.3ms^{-1}$, that at the upstream (about eight times the bell mouth diameter from the pump centre) shall be less than $0.5ms^{-1}$. If the average velocities are to be higher than these values, it is advisable to perform model tests for the suction sump.

4. The suction sump shall not have sudden enlargement or shapes causing sharp turning of the flow.

5. To prevent vortex formation adequate baffles (examples are given in Figure 13.24) or a partition wall shall be provided.

6. In the case of a closed suction sump, that is with no free water surface, the sump design must be so as not to cause air pockets from forming on the upper surface.

7. Examples of desirable and undesirable suction sump designs are given in Figure 13.25. Principal dimensions relating the pump to suction pipe and suction sump should conform to the recommended values given in Figures 13.26 and 13.27.

13.3.2 Suction sump model test

In case of a special shape of suction sump design or if a high approach velocity is to be used, a model test is recommended and should observe the following requirements [13.10]:

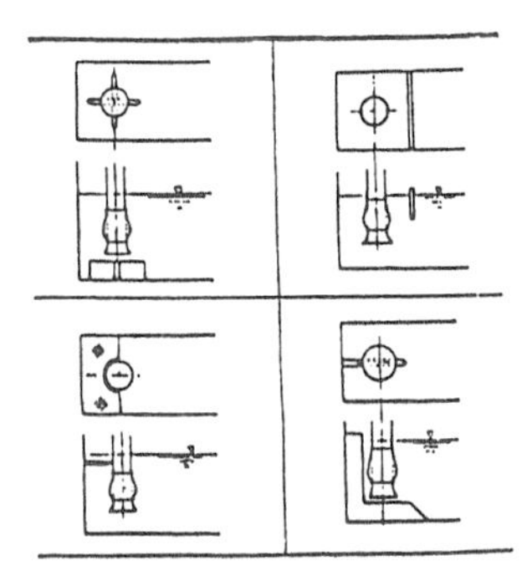

Figure 13.24 Baffles for prevention of vortex formation [13.10]

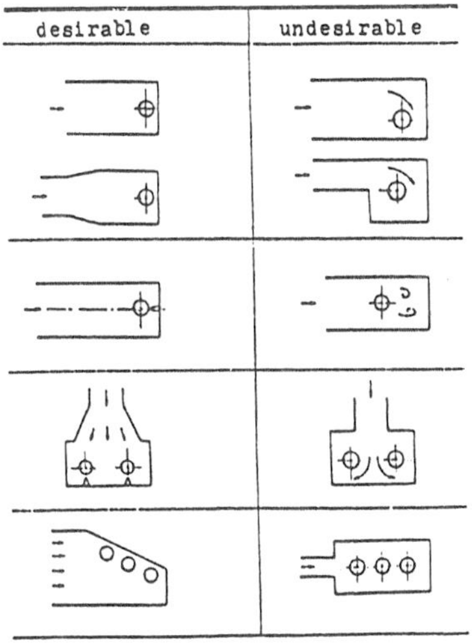

desirable

undesirable

Figure 13.25 Desirable and undesirable suction sump designs [13.10]

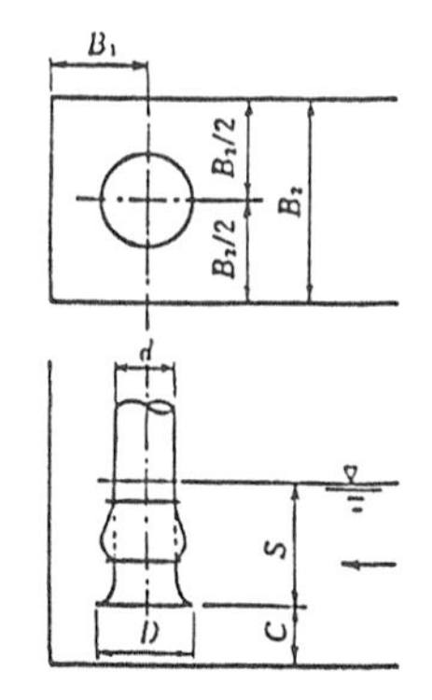

Items	Recommended value
C min	0.5 ~ 0.75 D
S min	1.3 ~ 1.7 D
B_1 max	0.8 ~ 1.0 D
B_2	2.0 ~ 2.5 D

Figure 13.26 Basic sump design and recommended values [13.10]

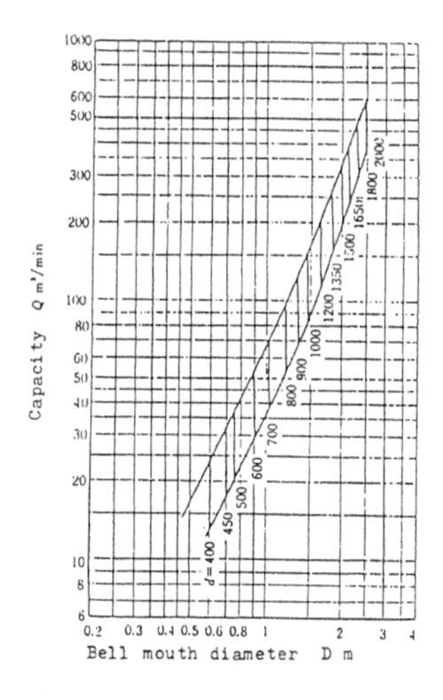

Figure 13.27 Recommended bell mouth diameter [13.10]
(Figures in chart refers to pump diameter)

1. Bell mouth diameter of model shall be greater than $100mm$.

2. In case of an investigation of the total flow in the sump or the formation of both air entraining and submerged vortex, a suction pipe may be used instead of a model pump.

3. The following formula may be used to relate quantity of flow between model and prototype

$$\frac{Q_m}{Q_p} = S^n \tag{13.9}$$

where	Q	– quantity of flow
	S	– scale ratio
subscripts	m	– model
	p	– prototype

the exponent n may take the following values

$n = 2$ – condition of identical velocity, that is $C_m = C_p$, this condition must be followed when investigating submerged vortex formation

$n = 2.2$ – for investigation of air entraining vortex formation

$n = 2.5$ – condition of identical Froude number $F_r = C^2/gD$, applicable to investigation of total flow condition such as whirl flow in the sump.

13.3.3 Example of suction sump problem

Performance of low head vertical pumps is apt to be affected by whirl flow around the suction bell mouth. In Figure 13.28 pump B in the centre of the sump has uniform approach flow while at pump A the direction of flow is the same as the pump impeller so the Euler head is reduced. At pump C the whirl flow is in the reverse direction therefore increases the Euler head.

The performance by field test of pumps A, B and C show slight differences among all three. The ratio of head difference of pump A to C and A to B is proportional to the difference of power consumed (Figure 13.29).

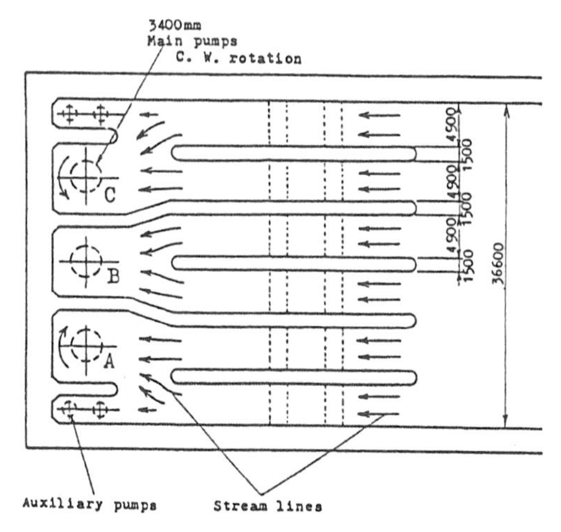

Figure 13.28 Non–uniform flow intake structure for sea water cooling pumps at best efficiency point condition

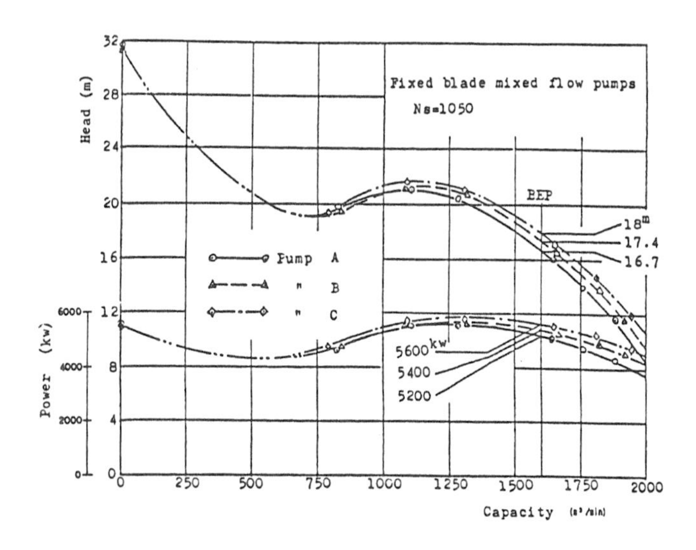

Figure 13.29 Pump characteristic curves affected by whirl flow in the suction sump

13.4 Large–capacity Low–head Pumps with Concrete Volute

R. Canavelis

13.4.1 Scope of application

The first concrete volute pump in the world was designed and installed in France in 1917. Since that time, this type of pump has been more and more widely used in several countries in the world, owing to its numerous advantages.

Concrete volute pipes are particularly adapted for large flows (from 5 to 30 m^3s^{-1}) and low head duties (from 5 to 30m). Most usual applications are: cooling water circuits in power plants and various industrial processes, dry docks and shipyards; port basin water level upholding; irrigation; drainage; flood control; raw water and effluent handling. Special construction techniques allow the extension of the field of application of this type of pump up to heads much higher than 30m and to flows much lower than 5 m^3s^{-1}.

13.4.2 Description of pump

Concrete volute pumps differ from conventional metal casing pump design in that the suction duct and the volute are constructed in concrete as part of the pumping station structure. They are single stage vertical pumps of the *pull-out* type, that is with removable internal parts. Main components are shown in Figure 13.30.

(i) Concrete parts

The suction duct is made of concrete and included in the civil works of the pumping plant. Different shapes may be proposed in order to meet the following requirements (Figure 13.31):

- a suction duct inlet section adjusted to the trashrack structure;
- a suction duct with resistance as low as possible in order to give the highest overall efficiency;
- a quasi–uniform velocity distribution at impeller inlet;
- the bottom level of the suction duct as high as possible to reduce the cost of excavation works.

The shape of the concrete volute is designed to give:

- a hydraulic efficiency as high as possible;

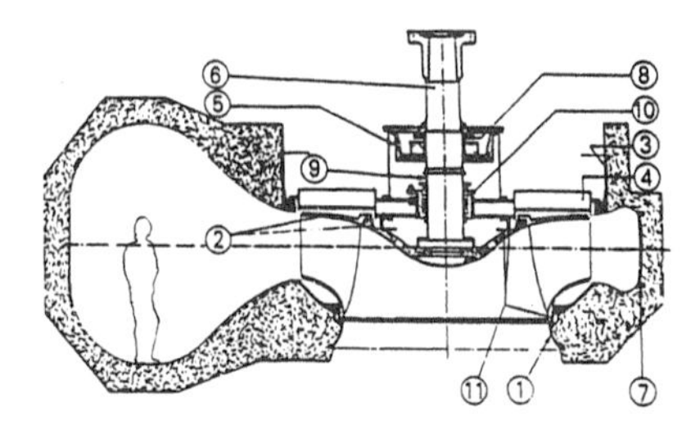

1	– grouted suction part	2	– impeller
3	– grouted upper part	4	– casing cover
5	– bearing	6	– shaft
7	– cut water	8	– journal bearing
9	– shaft sleeve	10	– stuffing box
11	– sealing ring		

Figure 13.30 Cross section of centrifugal pump with concrete volute

- a formwork as simple as possible;
- fabrication of the reinforcement and pouring procedure of the concrete as easy as possible.

Figure 13.32 gives examples of various types of volute shapes. The double curvature volute may lead to a slightly higher efficiency and allows an easier connection with a circular delivery piping. The volute with polygonal sections requires a simpler reinforcement and cheaper framework. It allows an easy connection with a square or rectangular discharge culvert. Purely centrifugal impellers are generally fitted to volutes with a symmetry plane. Mixed–flow impellers are better suited to volutes with a flat bottom.

(ii) The embedded metal parts

An embedded suction ring is used for fixing the suction wearing ring while the upper ring is put in for fixing the pump casing cover. The cut–water of the volute is made longer with the higher pump head in order to avoid high flow velocities at its connection with the concrete surface of the volute.

(iii) The removable parts

The wearing rings on the suction side are fixed in the embedded part. They

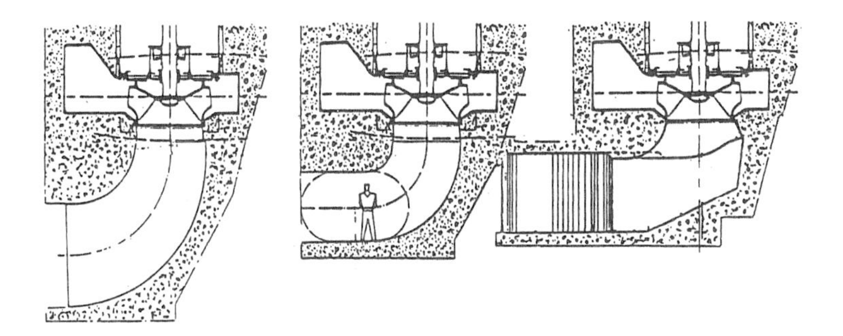

Figure 13.31 Various shapes of suction duct

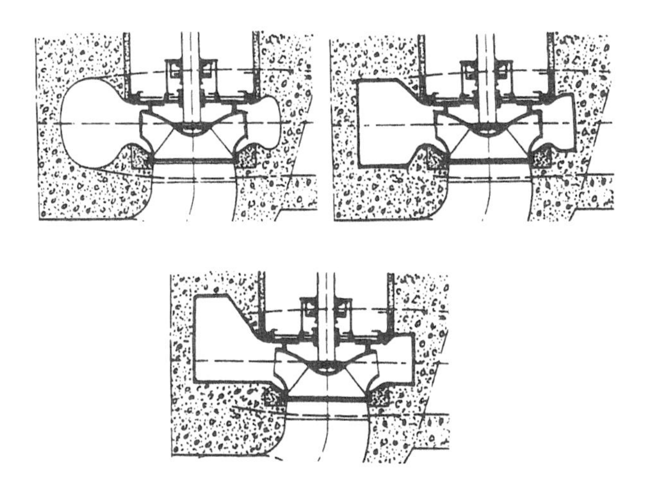

Figure 13.32 Various shapes of concrete volute

can be easily inspected during overhauls. On the upper side, wearing rings, if any, are fixed on the casing cover.

The impeller may be of the centrifugal or mixed–flow types according to the required specific speed. The centrifugal design is generally used for n_q between 70 and 100 while the mixed–flow design for n_q between 100 and 150. The impeller is flanged to the shaft or hooped on it by a hydraulic assembling equipment. Exchangeable sleeves are fitted at the lower part of the shaft under the stuffing box cover to keep it from contact with the pumped liquid.

Impeller outlet diameter and rotational speed ranges are given in Figure 13.33.

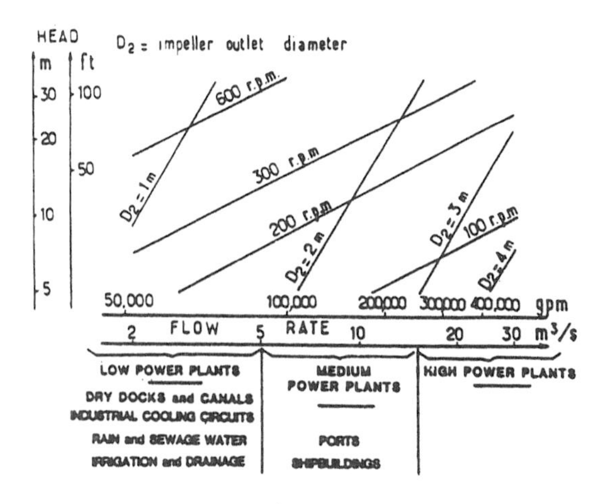

Figure 13.33 Usual operating range of concrete volute pump

Different bearing systems may be provided, with one or two guide bearings. One of the most reliable designs includes only one oil bath journal bearing located above the stuffing box cover. This design assures no wear while providing easy inspection and simplified maintenance. The thrust bearing is a conventional one of the spherical roller or tilting pad type. According to the arrangement of the total assembly, it may be located on the pump cover, on the gear box or in the motor.

In most cases the shaft is sealed by means of soft packing, lantern ring and gland. The stuffing box is connected to cooling water which can be directly delivered from a tap on the discharge duct. When a significant amount of solid particles is present in the cooling water a filter system must be provided.

13.4.3 Materials for concrete volute pumps

The material selection depends on several factors such as flow velocity at impeller blade tip, available *NPSH*, and water quality. Table 13.4 gives examples of currently used materials for different applications.

Table 13.4 Materials for concrete volute pumps

Parts	Clean clear water (1)	Clear water with sand (2)	Pure sea water (3)	Polluted sea water (4)
grouted parts and casing cover	. grey cast iron ASTM A 48 . spheroïdal graphite cast iron ASTM A536	. same as (1) + antiabrasion coating	. low alloyed grey cast iron (z alloys < 3%) + anticorrosion coating . copper aluminium (ASTM B148C-95 800) (if no SH2 present) . austenitic or austeno-ferritic stainless steel (ASTM CF-8M,CF-3M) ι resist cast iron (ASTM.A436, A439/A571)	. low alloyed grey cast iron (z alloys < 3%) + anticorrosion coating . copper aluminium (ASTM B148C-95 800) (if no SH2 present) . austenitic or austeno-ferritic stainless steel (ASTM CF-8M,CF-3M, CD-4MCU)
impeller	copper aluminium ASTM-B148-C95800	martensitic stainless steel (ASTM-CA-6NM)	austenitic or austeno-ferritic stainless steel ASTM CF3M CD - 4 MCU copper aluminium ASTM-B148-C95800	austenitic or austeno-ferritic stainless steel ASTM CD -4MCU,CN-7M
shaft	carbon steel		martensitic carbon steel (AISI 420)	same as (3)
shaft sleeves and wear rings	copper aluminium ASTM B148 C95500	martensitic stainless steel + heat treatment	copper aluminium $(Al\% < 8.2 + \frac{Ni}{2})$	austeno ferritic stainless steel ASTM CD4 MCU

13.4.4 Construction and erection

The concrete reinforcement and the formwork are the responsibility of the civil contractor, according to the civil guide drawings delivered by the pump manufacturer.

The formwork may be locally built by the civil contractor or supplied by the pump manufacturer. In the first case, they are composed of wooden forms and are removed after the concrete has sufficiently hardened (48 hours). In the second case, they may be of prefabricated components embedded in the concrete. The concrete must be of a sound quality. In most cases, no special surface treatment is required, even with sea water or significant sand or silt contents. It must be noted that a quite satisfactory surface roughness can be obtained with conventional wooden formwork.

The grouted parts have to be inserted and adjusted in alignment with each other. This is done either by adjusting the rotor (impeller and shaft) with regard to the wearing rings or with the help of a special positioning tool.

The space around the grouted parts are filled with non–shrinking mortar. All the removable internals may be mounted as soon as the mortar has hardened.

References

13.1 Senoo, Y., Nakase, Y. 'An Analysis of Flow through a Mixed-flow Impeller', Trans. ASME, J. Eng. Power, 1972, 1, 43–50.

13.2 Isogami, A., Ido, A. 'Force Requirement for Blade Actuation in Variable Pitch Mixed-flow Pumps', Proceedings 12th Symposium on Turbomachinery, October. 1982, 7–12 (in Japanese).

13.3 Liu Da-kai, Li Fu-sun. 'Boundary Element Numerical Method and Experimental Research on Blade Torque Performance', Proceedings 2nd China-Japan Joint Conference on Fluid Machinery, Xian, 1987, 23–31.

13.4 Niikura, H. 'Stress Analysis of Turbine and Pump-Turbine by Finite Element Method', *Turbo-machinery*, 1975, Vol. 3, No. 4, 16–24 (in Japanese).

13.5 Osterwalder, J., Sosino, M. 'Investigations Concerning Natural Vibrations and Flow Induced Vibrations on Kaplan Turbine Runner Blades', 5th Conference Fluid Machinery, Budapest, 1975, Vol., 749–760.

13.6 Stepanoff, A. J. *Centrifugal and Axial-flow Pumps*, 2nd edn, John Wiley & Sons, 1957.

13.7 Oshima, M., Endo, H. 'Axial Thrust of Mixed-flow Pumps', *Turbo Machinery*, 1977, Vol. 5, No. 2, 25–30 (in Japanese).

13.8 Troskolanski, T. Lazarkiewicz, S. *Kreiselpumpen*, Birkhause Verlag, Basel und Stuttgart, 1975.

13.9 Komatsu, T., Kuwabara, N., Tsuchiya, M. 'Recent Trends of Large Size Drainage Pumping Plants', *Hitachi Hyoron*, 1977, Vol. 59, No. 4, 81–86 (in Japanese).

13.10 Japan Society of Mechanical Engineers, 'Handbook of Mechanical Engineering', 1984, Vol. B5, 88–89 (in Japanese).

Chapter 14

Material Selection for Pump Construction

Xu Yi–hao

14.1 General Remarks

Pumps are rotodynamic machines used to pump various kinds of fluids. Their service conditions and environments are more severe than most other types of machinery and dictate very stringent selection of materials for pump construction. The following are essential to persons engaged in pump manufacture:

1. A good command of the fundamental principles of materials and corrosion besides general knowledge of mechanical design; thorough investigation of service conditions at site; analysis of causes of previous failures of pump parts. A continual accumulation of practical experience is of utmost importance to the pump engineer.

2. Detailed assessment of each installation and consideration of relevant factors in the course of material selection. In chemical process facilities, not only the properties of fluid such as viscosity, density, volatility, chemical stability and particle entrainment but also the constituents of the solution must be investigated. In many instances, the minor constituents play a more important role than the major ones. For example, there is an increasing tendency of utilizing waste hydrochloric acid in the chemical industry. The corrosive action of waste hydrochloric acid is actually stronger than pure hydrochloric acid or crude hydrochloric acid because iron and copper are added in the flow process, pitting corrosion may appear on stainless steels.

3. Documented reference on the use of various kinds of material, especially experiences of actual applications, are extremely valuable. All

available information sources should be checked before making a selection. Experiments must be resorted to when suitable reference cannot be found. In corrosion tests, site testing in the actual plant or the pilot plant is preferred. Experimental data is not always reliable but may be used for indication of the trend.

4. Other important factors to be considered in material selection are service conditions (continuous or intermittent operation, level of importance of duty), best operating speed (high or low speed), life expectancy together with consideration on machineability, cost and availability of material. The most suitable and economical material for the predetermined purpose is the best selection.

14.2 Environment and Service Condition

Ninety–eight per cent of all material failure in pumps are due to improper choice of material and are not results of material defects. Improper selection of material mainly result from insufficient consideration of the environment and service condition, and the lack of knowledge of corrosion effects.

14.2.1 Service condition of installation

This includes the required operating head and maximum head, physical–chemical properties of the fluid, composition (constituents, concentration, contents and nature of impurities, suspended solids, extent of aeration), the head–capacity range and duration of operation under these conditions, duration and condition of standstill (dry state, wet state or alternating), frequency of start up and shut down, delivery or recirculation, requirement of life and overhaul interval, setting level of impeller or actual submergence.

14.2.2 Properties of fluids

Liquids handled by centrifugal pumps fall into several categories.

(i) Water

Water is the most widely used liquid and has relatively little corrosive effect.

Clear water – fresh water such as rain water, river water, tap water, distiled water and chemically–treated water that contain entrained hydrophobic solid particles of less than $2.5kgf/m^3$.

Sewage – defined as used water. Though the majority part of the sewage is clear water, waste material of various kinds are present, including complex organic and inorganic compounds.

Sea water – the contents of dissolved salts, four fifths of which is sodium chloride, vary greatly between different sea regions. The higher the temperature of the sea water and the concentration of salts, the higher will be the corrosive effect. The pH–value of the sea water depends on the carbon dioxide content and may be as low as four, but the hydrogen sulphide content may vary from trace amount to several ppm. The presence of oxygen in sea water is the necessary condition for metallic corrosion (corrosion will still progress at oxygen content of $\leq$ 1ppm or less). With the increase of acidity of the solution, corrosion may occur even at the absence of oxygen. The oxygen content of sea water will be increased and its pH value changed when the sea water is polluted.

Mineral water – varies from neutral (pH = 7) to the heavily corrosive strong acidic (pH as low as 1.5) and to strongly alkaline (like in caustic mines). Suspended abrasive particles are always present in all subterranean water sources in active mines.

(ii) Acidic solutions

Inorganic acids – sulphuric, nitric and hydrochloric acids are the most commonly used inorganic acids in the industries. The corrosive effects of pure inorganic acids are fairly well known, but when these acids contain other constituents, assessment and selection of materials become more difficult.

Sulphuric acid – when other constituents are present in the sulphuric acid they may cause the solution to be oxidizing or reducing and produce different corrosive effects on the material. For instance, 80% sulphuric acid saturated with chlorine will be much more corrosive than the pure acid, material selection must then be made on the basis of the amount of wet chlorine gas.

Nitric acid – stainless steel parts are generally used for handling this very strong oxidizing acid but non–metallic materials like plastics should not be used. Materials for pumps to handle fuming nitric acid, mixture of nitric acid and hydrofluoric acid or a mixture of nitric acid and hydrochloric acid, require special investigation. In reality, the environment caused by a mixture of nitric acid and other constituents are always more corrosive.

Hydrochloric acid – the addition of a few parts per million of iron into commercially pure hydrochloric acid will generate iron chloride and transform the reducing solution to oxidizing. These chlorides are sufficient to make materials like NiMo alloys, NiCu alloys and zirconium totally unsuitable.

Phosphoric acid - in different stages of production, there will be varying quantities of sulphuric acid, hydrofluoric acid, silicon fluoride and phosphoric acid. Under certain conditions, water in the solutions has very high chloride content which will turn into strongly corrosive hydrochloric acid and cause serious corrosive problems.

Chlorine - wet chlorine, for example mixture of chlorine steam and moisture in air, is highly corrosive.

Organic acids - less corrosive as compared with inorganic acids, such as acidic acid, formic acid and lactic acid which have particularly corrosive properties.

(iii) Alkaline solutions

Alkaline solutions have relatively little corrosiveness below a certain temperature (93^oC). Exception to these are some bleaching agents, basic brines and other solutions containing chlorine. Temperature of molten sodium is $450 - 560^oC$ and is highly corrosive.

(iv) Salt solution

These are normally neutral and are not very corrosive. However, in the process, the pH–value may shift to slightly acidic and become noticeably corrosive.

(v) Petroleum products

Pumps are used in the petroleum industry mainly in mining, transport of crude and refined products, and various processes in refineries. Crude oil coming from underground is normally non–corrosive but may become strongly corrosive when containing hydrogen sulphide. Fluids handled in pipelines may contain crude oil, gasoline, fuel oil, jet engine fuel, liquefied petroleum gas (LPG) and dehydrated ammonia. LPG has poor lubrication property because of its low density (0.35), care should be taken when selecting material for the wearing parts of pumps. The most difficult task is in selecting material for reactor feed pumps which deliver liquids at $280 - 370^oC$ and contain highly corrosive hydrogen sulphide. Different types of metallic material are necessary to make up these pumps and the designer is faced with the problem of balancing thermal expansion of different materials. Oil slurry pumps must be able to withstand abrasion from solids that are always present in crude oil.

(vi) Chemical products

Aside from the acidic and alkaline salt solutions described above, pumps may have to handle very hot liquids such as heat medium salts at $200 - 500^oC$, molten lead at $370 - 900^oC$, molten sodium at $450 - 560^oC$, and on the other hand to handle cryogenic fluids such as liquid oxygen (-183^oC), liquid hydrogen (-253^oC) and liquid nitrogen (-196^oC). For pump parts under tension or torsion, material that exhibits brittleness at low temperature must not be used. These liquids are easily volatile, pump parts that have relative motion to each other should be checked for galling due to lack of lubrication.

(vii) Slurries

Slurries are solid–liquid mixtures such as mud, residues from mining, cinder from power plants, crystalline turbid solution, sand–silt mixture and highly viscous suspension containing fibres from paper mills. The flow pattern of solid–liquid mixtures varies with grain size (measured by mesh number or diameter in μm) and flow velocity. Erosion will increase with the increase of grain size of solids and also with the increase of concentration (by weight). Erosion rate will rapidly increase when the hardness of solids exceed that of the material surface. Grains with sharp edges will erode more than rounded ones.

14.2.3 Constituents of fluids and properties of solutions

(i) Concentration

The increase of concentration of the solution will effect the corrosive action on materials both in the positive and negative directions. For instance, CrNiMo steel exhibits completely different behaviour when placed in sulphuric acid of over 50% concentration (Figure 14.1). The concentration of a solution is closely related to its composition. Simple descriptions like *concentrated*, *diluted* or *trace* are basically meaningless unless the percentage in weight of each constituent or trace element in the solution is given. For instance, when high silicon iron is placed in a solution containing a few ppm of fluoride it will be catastrophically eroded.

(ii) pH–value

The pH–value indicates the concentration of hydrogen ions (H) or hydroxide ions (OH^-) in a solution. The pH–value of a liquid and the order of ionization tendency of the metal greatly influence the corrosive action. The pH–value of a liquid varies with many factors, such as a change in temperature or whether it is polluted. It should be noted that the pH–value of the liquid in the flow process is adjustable so the solution may be acidic or basic. The condition under which this process of changing from acidic to basic is very important in material selection.

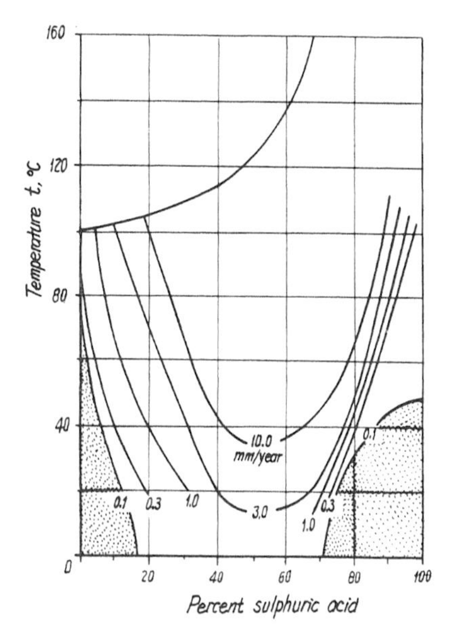

Figure 14.1 Eqi–corrosion lines, for X5CrNiMo 18:10 steel in sulphuric acid [14.4]

(iii) Temperature

The temperature should not be described only as *cold*, *hot*, or *ambient temperature*, rather the maximum and minimum operating temperature in ^oC or ^oF must be specified. A common misconception is that no serious damage will result at slightly higher temperature if the material is corrosion–resistant at a given temperature. Actually, the chemical reaction rate will double for each increase of 10^oC in temperature. The corrosion rate of 18:8 stainless steel in nitric acid at 177^oC is 1000 times that at 121^oC.

(iv) Gas (oxygen) content

The presence of gas (especially oxygen) in a reduction solution generally changes it from reducing to oxidizing and would require a completely different type of material to handle. The presence of other gases such as carbon dioxide or hydrogen sulphide, will also change its corrosion behaviour.

(v) Pollutants

The pollution of a solution as caused by corrosion from chlorides, sulphides and organic substances will generate harmful substances such as chlorine ions, ammonia ions and nitrite ions, potassium permanganate and hydrogen sulphide. The pollutants may be harmful or may not depending on the oxygen content of the water which tends to reduce when polluted.

14.3 Mechanism of Material Damage

14.3.1 Corrosion

Corrosion is described in DIN 50900 as 'the interaction between the material and its environment which will bring about measurable changes and may result in corrosive destruction of the material'. The following forms of corrosion may occur to pumps depending on service conditions
Uniform corrosion

Local corrosion (with no mechanical stress)

1 – couple (galvanic) corrosion
2 – crevice corrosion
3 – pitting corrosion
4 – interchrystalline corrosion
5 – selective corrosion

Local corrosion (with mechanical stress)

1 – stress corrosion
2 – corrosion fatigue
3 – erosion corrosion
4 – cavitation

14.3.2 Uniform corrosion

This is a kind of even corrosion that covers the whole surface of the metal and has a uniform rate of corrosion (Figure 14.2(a)), such as the simple process of rusting. Another example is the use of cast iron or aluminium alloy to handle mineral water of very low pH–value, the metallic surfaces of the pump will rapidly dissolve in the acid and result in serious uniform corrosion that may cause the pump to fail in one week.

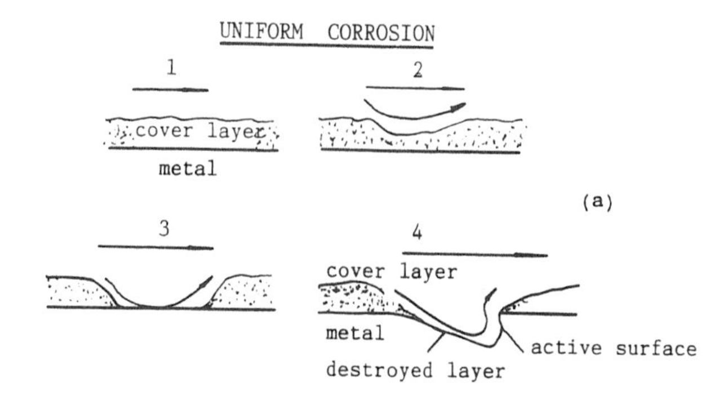

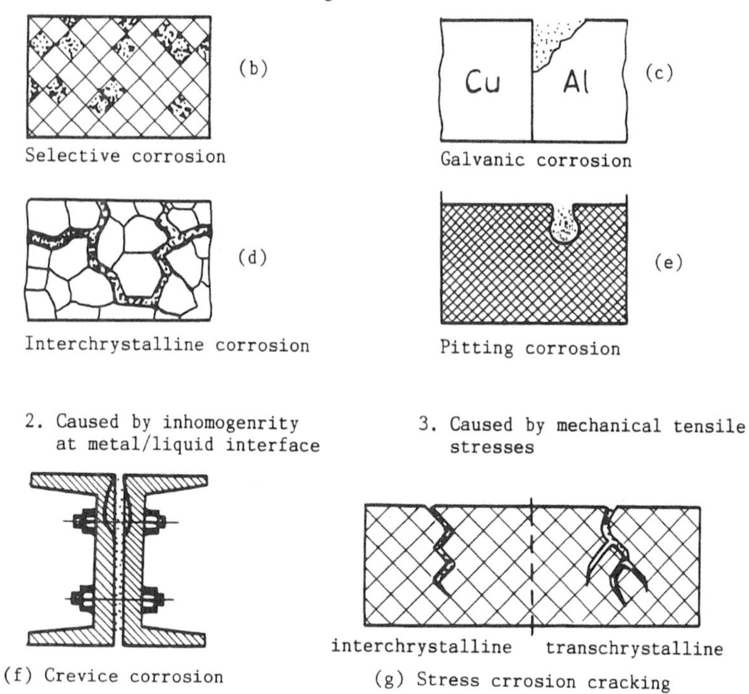

(f) Crevice corrosion

(g) Stress crrosion cracking

Figure 14.2 Classification of corrosion [14.4]

14.3.3 Local corrosion (with no mechanical stress)

Local corrosion may be classified as corrosion with mechanical stress and corrosion without mechanical stress. Pitting corrosion, crevice corrosion, intercrystalline corrosion, electro–couple corrosion and selective corrosion belong to the kind without mechanical stresses. The character of this kind of corrosion is that the corrosion is confined to fixed local positions.

The other kind of corrosion is that where mechanical stresses are created inside the material, such as stress corrosion cracks and fatigue corrosion. In some cases the corrosion is only on the surface of the material, such as erosion corrosion and cavitation. The flow passage surfaces of pumps very seldom have purely chemical corrosion, but because of the high flow velocity and uneven abrasion, corrosion may propagate at a high rate. As compared with other types of equipment, local corrosion such as erosion corrosion and cavitation may be of greater importance or exhibit stronger characteristics.

(i) Galvonic corrosion

It is the result of two different metals in contact or placed very near to each other in a corrosive medium (or electrolyte (Figure 14.2b)). The low potential metal with weaker resistance to corrosion (the less dear) becomes the anode and is corroded while the metal with stronger resistance (the dearer) receives protection. The greater the polarity difference of metals, or the greater apart they are in the potential series, the stronger is the tendency in couple corrosion. Whenever necessary in the case of two different metals in contact, they must be selected with the smallest possible potential difference, otherwise the metal weaker in resistance should be provided with much larger surface area than the stronger metal in order to prevent premature corrosion damage. For instance, in large vertical pumps utilizing stainless steel impeller and cast iron casing, a piece of zinc alloy of sufficient size must be attached to the casing to prevent accelerated corrosion.

(ii) Crevice corrosion

This is a strong local corrosion that appears in narrow gaps and other unexposed areas. The corrosion action is related to the limited amount of stagnant fluid in holes, screw threads, gasket surfaces, crevices and near surface sediments (Figure 14.2(f)). A concentration difference of metal ions or oxygen existing between the stagnant area and the main part of the flow will cause current to flow between the two areas and produce local corrosion in the stagnant region. Ordinary 13Cr or type 304 stainless steel is particularly sensitive to corrosion in mating joints of flanges and gasket contact surfaces when subjected to chloride environment (highly conductive fluid like sea water). High nickel cast iron (Ni–resist) and titanium containing palladium may be used to combat this corrosion. Another remedy is to correctly select gaskets or fill the joints with special coating.

(iii) Pitting corrosion

This highly localized type of corrosion is the most destructive and may cause pitting or hole-through (Figure 14.2(e)). As pointed out in [14.5], pitting corrosion results from local destruction of the passive layer when a small corrosion anode is formed. Due to hydrolysis of the corrosion products, the pH-value in the pit drops considerably, to as low as 2, even with a neutral electrolyte outside the pit. Pitting is triggered above all by chlorides and sulphides, especially with increasing temperature. Pitting occurs if the corrosion potential is more positive than the pitting-dependent breakdown potential. The corrosion potential is strongly influenced by the presence of an oxidant, thus for example, the danger of pitting is far greater in the presence of oxygen than without it. 18:8 (type 304) stainless steel will pit corrode even at low oxygen content while 18:8:3 (type 316) stainless steel will not pit corrode for oxygen content smaller than 30 p.p.b. (parts per billion). When oxygen is considerably polluted, type 316 stainless steel may pit corrode as well. As a counter measure, steels with higher chromium content and the addition of molybdenum are used. For applications in extremely aggressive environments (like brines containing H2S), adequate resistance to pitting can be guaranteed only by using expensive high-alloy materials.

Pitting corrosion is related to crevice corrosion since it is effected by the change in constituency of the stagnant fluid.

Pitting corrosion may start at any rupture of the passive layer and form the first hole without being noticed. If this happens to be in the highly stressed area, it may further develop into propagation of cracks.

(iv) Intercrystalline corrosion

This is a kind of selective corrosion that occurs only at or near crystalline boundaries. It causes crystals to break off and gradually destroys the metallic structure (Figure 14.2(d)). A typical example is the chromium in 18:8 stainless steel (austenitic) which when heated to the range of $427 - 870°C$ will precipitate out as carbide and settle on the boundaries of crystals and form a chromium-depleted zone around the crystals. But during welding this only takes place at the edges of the weld seam (so-called heat effected area). Consequently, crystalline corrosion will quickly develop when the steel is placed in an oxidizing environment (as in nitric acid). If the steel is not adjusted by changing the alloy elements, it may be cured by heating to $1065 - 1093°C$ and quenched rapidly to have the carbides redissolved (known as solid diffusion treatment) and the corrosion resistance restored at the crystalline boundaries. By using low carbon steel (C $< 0.03\%$, called L grade) and ultra–low carbon stainless steel, the tendency toward intercrystalline corrosion may be reduced. Intercrystalline corrosion occurs easily in castings because their crystals are much larger than those of forgings of the same composition. Another important cause for corrosion is the negligence of solid diffusion treatment and allowing castings to cool in the furnace. Other alloy systems may also have intercrystalline corrosion and should be properly heat treated according to technical codes.

(v) Selective corrosion

This corrosion process is the removal of certain elements from the alloy, such as the loss of zinc from brass or the loss of silicon from 3% Si stainless alloy when handling hot sea water. Gray cast iron may show signs of graphitization (graphite lattice filled with iron rust) and will lower the mechanical properties and cause the material to look porous like sponge and become brittle (Figure 14.2(b)). These corrosive activities are internal so they cannot be observed by the naked eyed.

14.3.4 Local corrosion (with mechanical stress)

(i) Stress corrosion

The combined action of static tensile stress (residual stress within the metal or external stress) and corrosion will produce cracks that are normally perpendicular to the direction of stress. The speed of crack propagation is not linear and will increase rapidly as the ultimate strength of the material is reached. Usually cracking across crystals may also happen (Figure 14.2 (g)). The stress corrosion is peculiar in that there is nearly none or very little outward sign of corrosion. Another feature worthy of mention is the different kinds of corrosion resulting under different corrosive environments for the same alloy under the same service conditions. In power plants, especially in nuclear plants, when high pressure pumps deliver high temperature water containing small amounts of chlorides, ordinary 18:8 stainless steel may crack if the chlorides in the water become more concentrated such as in the case of drying from a wet state. Resistance to stress corrosion in this case may be greatly strengthened by controlling the increase of ferrite in austenitic stainless steel structures and by adding chromium. The residual stresses in welded parts must be closely checked.

(ii) Corrosion fatigue

This is a special form of stress corrosion that is essentially started by corrosion and develops fatigue failure under the action of cyclic stresses.

The corrosion fatigue strength should be taken as the limiting factor in material selection under corrosive environment. The steady–state stress and variation of stress due to hydraulic action must not exceed this limit. The allowable stress is determined by considering the mechanical property, corrosion–resistance and the severity of corrosion condition.

Actual stress tests performed on pumps reveal properties different from those obtained in fatigue tests in dry air. The fatigue strength under actual service condition is found to be greatly reduced for the same loading. The thickness of surface chrome overlay on the shaft has marked effect on the fatigue strength, reducing it by 50% for a thickness of $0.15mm$. This is because the chrome overlay is a kind of cathodic coating and will produce harmful residual tensile stresses and surface capillary cracks plus the possibility of hydrogen brittleness.

The quality of castings and forgings is also an important factor that governs the corrosion fatigue strength. If at a place where the tendency of segregation is strong and this also happens to be at an eroded spot caused by various reasons, it may have the effect of stress concentration and finally may lead to shaft rupture. Previously, snapping of shafts of boiler feed pumps were mainly attributed to ordinary fatigue and the damaged parts were replaced by a higher grade steel with better fatigue strength. However, this helped very little since the new steel does not have higher corrosion resistance, rather the propagation of cracks is faster in high strength steels.

The yield strength of the steel may be raised by 9–18% with an increase in corrosion fatigue strength by decreasing the S, N2 and H2 contents in order to reduce segregation in the steel and by other casting techniques. Data from cyclic fatigue stress tests are invaluable in material selection for impellers or rotors. Some rotors work $6 \cdot 10^8$ cycles per year and will exceed $1 \cdot 10^{10}$ cycles in their service life.

(iii) Erosion corrosion

Erosion corrosion is the combined effect by electrolytic corrosion and mechanical abrasion. It often appears in the flow passages of pumps handling fluids at high velocity and is the most common type of damage, becoming more intense when suspended solids are present. In many metals and alloys the corrosion resistance comes from a passive film (mostly oxide films) on the surface. When this film is broken by the mechanical action from high flow velocity (impingement and cavitation, etc.) beyond a certain limit, the balanced state of the metal in the electrochemical process is upset, the metal goes into an active state and corrodes quickly. Under this condition, the rate of erosion corrosion is related to the following factors:

$$h = bV^n \quad (\mu m h^{-1})$$

where

b – factor depending on property of material and coefficient of activity of solution

V – flow velocity (ms^{-1})

n – exponential, normally around 3

This is known as the mechanical–chemical damage of metals. Once the metal is in a state of activation, the recovery of the passive film is slow even when the flow velocity is lowered. Figure 14.3 shows the erosion corrosion behaviour of certain alloy steels in chloride brine containing hydrogen sulphide, and the related flow velocity.

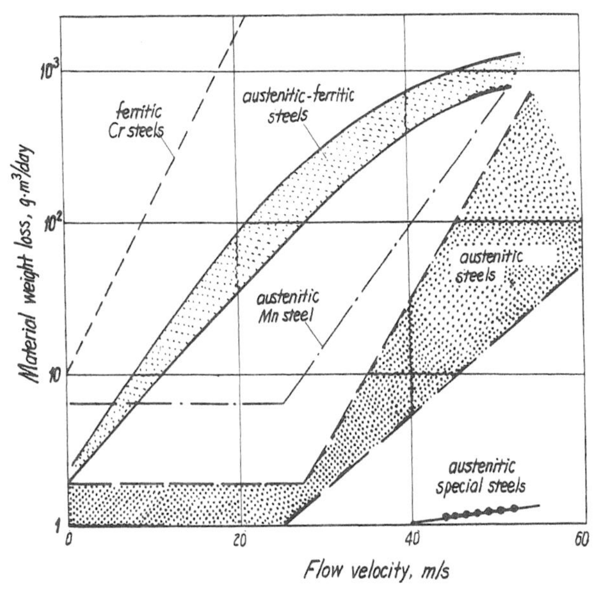

Figure 14.3 Relation pH = 4.5 of corrosion to flow velocity of stainless steel in sour brine, $60°C$, dissolved salt $270{,}000mg$ per litre, 500 ppm H2s, pH = 4.5 [14.5]

Other test results show that in stagnant water (pH = 7.8) 13% chromium steel has a erosion corrosion rate 1,000 times smaller than that of carbon steel.

An interesting characteristic of erosion corrosion is that it has little dependency on the hardness of the material. All metals that have good erosion resistance will usually have also good erosion corrosion resistance. The important thing is that the material must have a strong and dense surface plus a passive film that will not activate in an electrolyte. A typical example is type 316 stainless steel with hardness HB 150 which has a service life six times that of precipitation–hardened type 329 stainless steel (HB 450) when handling high velocity sulphuric acid slurry. But both steels have no corrosion when placed in still fluid. Alloys with too high a carbon content or those that easily precipitate carbide, or with overgrown crystals after heat treatment, or with plastic deformation and uneven internal stress due to various reasons have lower resistance to erosion corrosion. Increased temperature of the fluid tends to accelerate the corrosion. Erosion corrosion is similar to cavitation in the way damage will occur in case of strong vortex turbulence and high velocity impact by the flow. Cavitation in fact may be considered as one form of erosion corrosion.

(iv) Cavitation

The hydraulic design, installation conditions and mode of operation are all important factors in deciding the occurrence of cavitation of a pump. Cavitation is the phenomenon where the local pressure in the underpressure zone caused by pressure drop and high velocity falls below the vapour pressure. This leads to bubble growth on the impeller and sometimes even in the casing. The bubbles implode in the high pressure zone and can inflict damage to the material. The stress caused by cavitation may reach $1{,}400 MPa$ in the most severe case, which far surpasses the ultimate strength of metals.

All too often there is a tendency to regard cavitation damage as a purely mechanical process. This may be true in chemically inert media (like water, sewage, petroleum), but not in aqueous corrosive media. Here the cavitation process consists of a complex interaction of mechanical stressing and chemical reaction. Involved chiefly are [14.5]:

1. Mechanical influences (bubble implosion): destruction of protective films, plastic deformation of the metal surface, decohesion due to material fatigue.

2. Corrosive influences: active anodic metal dissolution, detachment of metal particles, possible formation of local galvanic cells.

The rule that the harder the material the more it resists cavitation no longer holds true under corrosion attack. On the contrary, optimal cavitation resistance is attained by combining hardness with good toughness and corrosion resistance. Some of these requirements are incompatible, consequently, materials must be chosen on the basis of an acceptable compromise.

If corrosion pits already exist in the cavitation region, the rate of propagation of cavitation erosion will greatly increase. In places where there are inclusions, defects, microcracks or residual stresses, serious cavitation of short duration is able to produce high enough instantaneous stress to destroy the material and will aggravate in subsequent operation and bring about ultimate failure of the machine.

14.3.5 Abrasion wear

Hardness is traditionally considered as the gauge to measure wear–resistance of materials. However, unless the hardness of metal exceeds that of the abrasive solid, little can be gained by increasing hardness of the metal. Figure 14.4 gives hardness values of common minerals and metals.

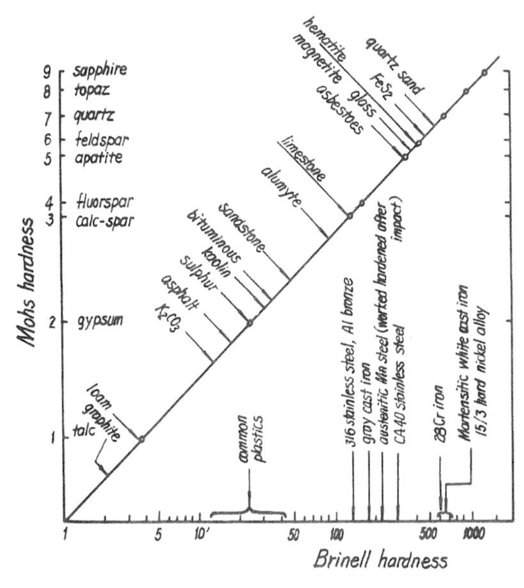

Figure 14.4 Hardness values of common minerals and metals [14.1]

The effective abrasion–resistance of a metal is generally judged by its rating on the Mohs or Knopp hardness scales. But hardness is not a truly reliable criterion when used to gauge a martensitic steel where the most important factor is micro–structure of the material. The amount of abrasive wear depends on the geometry and the flow velocity of the pump, the quantity, size and type of solids entrained in the fluid. Abrasion and wear appear in three ways:

1. Abrasive wear by impact – the result of hammering action on the surface by coarse grains under the force of high velocity. Large chips

may be torn from the attacked surface.

2. Abrasive wear by grinding – crushing action by solids sandwiched between two wearing surfaces.

3. Abrasive wear by erosion – due to impinging action on the surface by particles in free motion (sometimes parallel to the surface) at high or low speed.

The abrasive wear increases with increasing of particle size and concentration (by weight) in the fluid and also with the presence of sharp–edged particles as against rounded ones. Repeated impact on the surface by the sharp–edged particles will create high stress in the metal which in turn will cause plastic deformation and eventual breakage. Surface erosion is accelerated where cavitation is likely to happen as a result of turbulence, vortex and separation of flow. The joint action of abrasion and cavitation will produce series of grooves on metals resembling scales of fish. Material with homogeneous metallurgical structure, high hardness and good surface finish are stronger in resistance to erosion. Major factors to be considered in pump material selection are the speed of moving particles, their kinetic energy and the angle of impact. Metals with high elastic limit are used to resist direct impact by particles whereas metals with high hardness are used to fend impingement at small angles (flow being nearly parallel to surface). A simple rule to estimate the abrasion rate is that it varies with flow velocity to the (2.2–3)th power.

14.3.6 Mechanism of galling and gall resistance

Moving and stationary parts in pumps separated by minute clearances (as in seals and balancing drums) are subject to serious damage with the increasing rotational speeds of modern high–head centrifugal pumps. The parts in relative motion may have a rubbing speed up to $100 ms^{-1}$. Depending on the property of the material and condition of contact, damage may take the form of breaking off, local fusion, flaring, grooving, tearing, cracks and even seizing (fusion of advanced degree).

The resistance to galling by metal is closely related to the combination of mat[illegible]. By proper selection of materials extensive repairs or even outage may be prevented. The usual combinations are 17Cr–2Ni stainless steel paired with 10Sn–10Pb bronze or 37Ni–10Sn–8Pb bronze. Actual tests on combination of stainless steel and nonferrous metals indicate there is an increase in surface damage with speed and boundary pressure. However, the combination of 17Cr–2Ni stainless steel with precipitation – hardened stainless steel (15Cr–5.5Ni–1.5Mo–1.5Cu) or 13Cr–0.2C chrome steel has a decreasing tendency of damage with increase of speed. Other factors that may contribute to the resistance to galling are: improved surface finish, higher absolute hardness of material and a difference of hardness of at least 50 HB between the paired materials. Actual tests reveal that hardened specimens with overlay such as Stellite welded or tungsten carbide sprayed, are definitely inferior under high speed conditions where the overlay may separate from the base material or develop cracks. But this combination works quite well under low speed conditions so it may be suitable for use as water-lubricated bearings for large vertical pumps delivering unclean water. Property of fluid and its temperature, suction condition and thermal expansion of the paired materials are other important factors to be considered.

14.4 Selection of Engineering Materials for Pumps

The reliability, economics and life of a pump depend to a very large extent on the proper choice of material used in the construction of the pump, this is sometimes more important than a good structural design. Factors to be considered, in order of importance, are corrosion-resistance, machineability, mechanical strength (may be the major factor in some applications), cost, availability and outward appearance.

The criterion of corrosion-resistance is not as clear cut as the yield point of materials or their conductivity. It depends not only on the environment medium but also is influenced by many factors as shown in Figure 14.5.

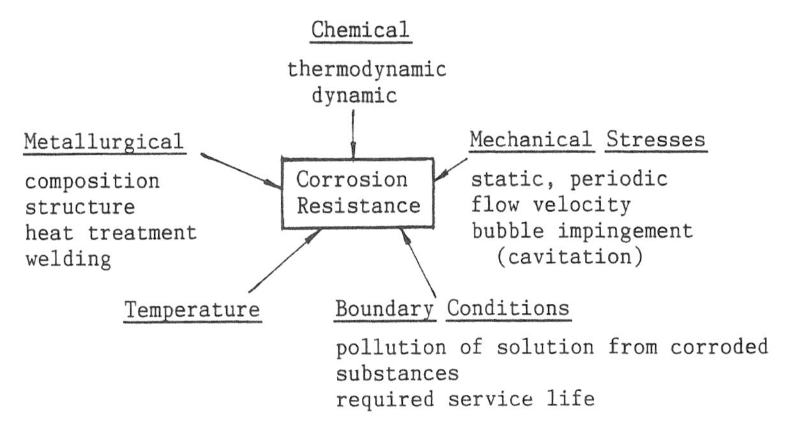

Figure 14.5 Factors that have influence on corrosion resistance [14.4]

14.4.1 Essential structural materials and their classification

Table 14.1 lists the principal materials used in pump construction for handling various liquids.

14.4.2 Examples of pump material selection

Table 14.2 lists the combination of materials for use in refinery process pumps. There are many types of material that satisfy the requirements, the most commonly used ones are given in the table. Care must be exercised in selecting material for rotating parts of pumps as cast iron, bronze and Ni–Resist have low mechanical properties. Table 14.3 lists the materials for pump construction according to service conditions. Table 14.4 gives specifications of materials according to ASTM standards. These two tables may be used as cross references. Table 14.5 lists materials recommended for pumps used in mining applications.

Special care must be given to the adaptability and economics of the materials used. A bad selection may be the using of materials suitable for 10 year's life in a mine intended for only 5 years of service, while the best materials should be used for a mine with 25 years of service. Most metals are unstable in acid solutions, therefore, various stainless steels and alloys listed in Table 14.1 must be first considered. But these alloys likewise have limitations depending on the type of acids handled, concentration and operating temperature, in addition to the fact that they are usually very expensive. Wherever applicable, materials as plastic, porcelain or glass may be considered as substitutes.

Table 14.1 Essential structural materials for pump construction

Type	Major Composition	Remarks
Cast iron	gray cast iron	Used in handling clean water or non−corrosive sewage.
	low Ni−Cr alloyed cast iron	Improved denseness of structure with addition of Ni and Cr, allowing use in parts with greater thickness, longer service life.
	spherodized graphite cast iron	Superior to gray cast iron in compression and impact resistance, also cavitation and corrosion resistance.
Bronze	Al, Fe bronze Sb, Sn, Pb bronze	Superior to gray cast iron in wear−, erosion− and cavitation−resistance.
Carbon steel		Suitable for welding and repair, used in relatively corrosion free environment, physical properties superior to gray cast iron.
Martensitic Cr steel	4−6 Cr	Used to handle low−corrosion liquids like deaerated boiler feed water.
	11.5−13 Cr	Excellent corrosion–resistant to clean water and weak acid solution.
Ferritic Cr steel	17 Cr	Used in handling oxydizing solution, e. g. high concentration nitric acid.
	28 Cr	The higher the Cr content the higher the corrosion resistance. Good resistance to alkaline mixture and solutions. Good abrasion–resistance.
Austenitic Cr Ni steel	18:8 Cr Ni (type 304)	Most widely used stainless steel, erosion−resistance is evidently superior to Cr steel in most cases.
	18:8:3 Cr Ni Mo (type 316) *	Similar to type 304 but with addition of Mo, improved pitting corrosion resistance, especially against chlorides.
	29:20:3.25:2.25 Cr Ni Cu Mo (Durimet 20)	Able to handle all liquids (except the most corrosive ones) including hydrochloric acid and hot sulphuric acid of medium concentration.
Duplex steel (ferrite + austenite)	25:5:3:2 Cr Ni Cu Mo (type CD4M Cu) **	Very good strength and fatigue resistance, also resistant to inter−crystalline corrosion in welding state. Excellent anti−corrosion and anti − erosion properties under different corrosive environments, especially in chloride solutions containing crystalline salts. Insensi−tive to stress corrosion cracks caused by chlorides and corrosion by H_2S. Suitable for use in water injection system of offshore oil fields.
Austenitic cast iron	Type I 13−17 Ni 5− 7 Cu (Ni−Resist)	A type of high Ni cast iron. Good corrosion−resistance to most liquids, widely used in chemical industries for transport of salt slurry.
	Type II18−22 Ni (Ni−Resist)	Does not contain Cu, widely used in transport of strong alkaline liquids. Type I and II both have high coefficients of expansion, care must be taken in treating high temperature applications.

Table 14.1 (cont)

Nickel base alloys	Ni – base alloy with Cu (Monel)	Excellent corrosion – resistance to many non – oxydizing liquids, especially sea water and other chloride fluids. Able to transport hydrochloric acid and hydrofloric acid of high temperature and concentration. Not suitable for nitric acid and most acidic mineral water. Not resistant to sultfuric gases. Strongly resistant to caustic soda.
	Ni – base alloy with Cr (Inconel)	Superior to pure nickel in oxydizing environment but not resistant to hydrochloric acid.
	Ni – base alloy with Mo (Hastelloy)	Mainly used in transport of high temperature high – concentration hydrochloric acid that cannot be handled by pure nickel and Monel. Suitable for reducing corrosive media.
Precipitation hardened stainless steel ***	Addition of small amounts of Cu, Al, Mo	There are some 20 types of materials, most typical are 17–4PH and PH55A. Strength and hardness higher than 18:8. Weldability same as austenitic steels. No intercrystalline corrosion as in 18:8 stainless steel.
Austenitic manganese steel	12 – 14 Mu high Mn steel	Highly resilient, HB 220. Will cold – harden after impact. Low residual stress in casting, can be machined and welded with ordinary methods. Widely used in dredge pumps and heavy duty sand pumps to handle abrasive liquids.
Martensitic white cast iron	4 Ni, 2 Cr (Ni – hard) 15 Cr 3 Mo alloy	Hardness as cast is in range HB 550 – 650, may reach 730 after hest treatment under special conditions. Used in cases of serious erosion and abrasive wear. Impact strength somewhat low.
High silicon cast iron	14.35 Si alphaphase diffusion – hardened white cast iron	Used in many heavily erosive applications, especially in handling hot sulphuric acid with suspended solids. Very high hardness, difficult to machine, easy to crack.

NOTE: * – The sensitivity of stainless steel to crevice corrosion (pitting corrosion) decreases with the increase of Cr and Mo contents, i. e. the increase of stability of the passive film. The order of decrease is 13 Cr < 17 Cr < 18:8:3 CrNiMo steel. To have sufficient resistance to crevice erosion in sea water, the following must be satisfied: (%Cr + 3.3% Mo) > 30.

** – The earliest duplex steel is the type CD4M Cu (25Cr 5Ni 3Cu 2Mo). Many grades have now been developed. The ratio between ferrite and austenite is varied, the latter varying from 25 – 50%. This ratio is mainly controlled by the Cr – Ni constituents. The latest generation of steels use Mn to replace part of the Ni because Mn enhances the solubility on nitrogen and is helpful in resistance to local corrosion.

*** – The precipitation – hardened stainless steel is produced by adding small amounts of Cu, Al and Mo into the Fe – Ni – Cr triplex system. Carbides, phosphates and intermetallic compounds are separated by heat treatment to form precipitation hardening.

Table 14.2 Refinery process pump materials

Trim part	Steel case, cast iron trim	Steel case, bronze trim	Steel case, Ni–Resist trim	Steel case, 11–13 chrome trim	11–13 chrome case and trim	18–8 S. S. case and trim	316 S. S. case and trim
Cover	Cast steel	Cast steel	Cast steel	Cast steel	11–13 chrome	18–8 S. S.	316 S. S.
Case stud	ASTM–A193 GR B7	ASTM–A193 GR B7	ASTM–A193 GR B7	ASTM–A193 GR B7	ASTM–A193 GR B7	ASTM–A193 GR B7	ASTM–A193 GR B7
Case nut	ASTM–A194 GR 2H	ASTM–A194 GR 2H	ASTM–A194 GR 2H	ASTM–A194 GR 2H	ASTM–A194 GR 2H	ASTM–A194 GR 2H	ASTM–A194 GR 2H
Shaft	SAE 4140 HT	SAE 4140 HT	SAE 4140 HT	SAE 4140 HT	SAE 4140 HT	18–8 S. S.	316 S. S.
Impeller	Cast iron	Bronze	Ni–Resist	Cast steel	11–13 chrome	18–8 S. S.	316 S. S.
Impeller wear ring	Cast iron	Bronze	Ni–Resist	11–13 chrome hardened	11–13 chrome hardened	18–8 S. S. hard–faced	316 S. S. hard–faced
Case wear ring	Case iron	Bronze	Ni–Resist	11–13 chrome hardened	11–13 chrome hardened	18–8 S. S. hard–faced	316 S. S. hard–faced
Shaft sleeve packed pump	11–13 chrome hardened	11–13 chrome	11–13 chrome	1020 hard–faced	11–13 chrome hardened	18–8 S. S. hard–faced	316 S. S. hard–faced
Shaft sleeve mechanical seal	11–13 chrome	11–13 chrome	11–13 chrome	11–13 chrome	11–13 chrome	18–8 S. S.	316 S. S.
Gland	Steel	Steel	Steel	Steel	Steel	18–8 S. S.	316 S. S.
Gland stud	ASTM–A193 GR B7	ASTM–A193 GR B7	ASTM–A193 GR B7	ASTM–A193 GR B7	ASTM–A193 GR B7	ASTM–A193 GR B8	ASTM–A193 GR B8M
Lantern ring	Cast iron	Cast rion	Cast iron	Cast iron	Cast iron	18–8 S. S.	316 S. S.
Throut bushing	Cast iron	Bronze	Ni–Resist	11–13 chrome	11–13 chorme	18–8 S. S.	316 S. S.
Throttle bushing	Cast iron	Bronze	Ni–Resist	11–13 chrome hardened	11–13 chrome hardened	18–8 S. S. hard–faced	316 S. S. hard–faced
Gasket, sleeve	18–8 S. S. annealed	18–8 S. S. annealed	18–8 S. S. annealed	18–8 S. S. annealed	18–8 S. S. annealed	18–8 S. S. annealed	316 S. S. annealed
Gasket, case	18–8 S. S. with asbestos	18 8 S. S. with asbestos	18–8 S. S. with asbestos	18–8 S. S. with asbestos	18–8 S. S. with asbestos	18–8 S. S. with asbestos	316 S. S. with asbestos
Impeller nut	Steel	Steel	Steel	Steel	Steel	18–8 S. S.	316 S. S.
Bearing shield	Bronze	Bronze	Bronze	Bronze	Bronze	Bronze	Bronze
Oil rings	Brass	Brass	Brass	Brass	Brass	Brass	Brass
Bearings bracket	Cast iron	Cast iron	Cast iron	Cast iron	Cast iron	Cast iron	Cast iron
Heat exchanger assembly	Steel	Steel	Steel	Steel	Steel	Steel	Steel
Coupling guard	Fab steel	Fab steel	Fab steel	Fab steel	Fab steel	Fab steel	Fab steel
Base plate	Fab steel	Fab steel	Fab steel	Fab steel	Fab steel	Fab steel	Fab steel

Table 14.3 Engineering standard: Materials for centrifugal pumps (Shell Oil Company)

Class	Products Transported		Material			
			Casing	Impeller	Casing wearing ring and inter-mediate bushing	Impeller wearing ring and inter-mediate bushing
1	Brine	under suction	ASTM B143	ASTM A296 CF GR 8M	ASTM B143	AISI 316
		under submergence	CI 1A	ASTM B143 CI 1A	CI 1A	Monel K HB250-300
2	Clean water, boiler feed, all alkaline fluids, gasoline within and out of design scope and non-or weak-corrosive heavy petroleum products out of design scope	under 10kg/cm²	ASTM A278 No. 30	ASTM A48	AISI 420	AISI 420
		over 10kg/cm²	ASTM A216 GR WCB	No. 40	HB225-275	HB325-375
3	All petroleum products within design scope and lighter than gasoline products outside of design scope, non- or weak-corrosive under all pressure and temperature lower than 100 ℃		ASTM A216 GR WCB	ASTM A48 No. 40	AISI 420 HB225-275	AISI 420 HB325-375
4	Non- or weak-corrosive products under all pressures and temp. over 100 ℃		ASTM A216 GR WCB	ASTM A48 No. 40	AISI 420 HB225-275	AISI 420 HB325-375
5	Hot oil products containing sulphides under all presures and temp. over 300 ℃		ASTM A351 CA. GR 15	ASTM A296 CA. GR 15	AISI 420 HB225-275	AISI 420 HB325-375
6	Water solution of oil products containing inorganic acids (H_2SO_4, H_2S, SO_2, etc.) under all pressures and temperatures		ASTM A351 CF. GR 8M	Ni-Resist	Ni-Resist	Ni-Resist
7	Straight-run distillates (acid value over 0.5) containing naphthenate under all pressures and temp. 240-400 ℃		ASTM A351 CF. GR 8M	ASTM A296 CF. GR 8M	AISI 316	AISI 316 plus Colmonoy 6

Table 14.4 Material specification

Material	Castings	Forgings	Bars	Studs	Nuts
Cast iorn	A–48				
Ni–Resist	A–436, Types 1 and 2				
Bronze	B–143 Alloy 1A B–143 Alloy 2B B–144 Alloy 3B B–145 Alloy 4A		B–139 Alloy 510	B–124 Alloy 655	
Carbon steel	A–216 GR WCB	A–266, Class 1 A–266, Class 2	A–108 GR 1018 A–575 GR 1020		A–108 GR 1018
Alloy steel (SAE 4140)			A–434 Class BC or BD	A–193 GR B–7	A–194 GR 2H
11–13 chrome steel	A–296 GR CA–15	A–182 GR F–6 A–336 CL F–6	A–276 Type 410 416, or 420	A–194 GR B6	A–194 GR 6
18–8 stainless steel	A–296 GR CF–8	A–182 GR F–304 A–366 CL F–8	A–276 Type 304	A–193 GR B8	A–194 GR 8
316 stainless steel	A–296 GR CF–8M	A–182 GR F–316 A–366 CL F–8M	A–276 Type 316	A–193 GR B8M	A–194 GR 8

* All entries are ASTM numbers.

Table 14.5 Materials for construction of mine pump

	Neutral water		Acidic water		Alkaline water	
	Moderate head	High head	Moderate head	High head	Moderate head	High head
Casing	cast iron	ductile iron cast steel	316 stainless steel/Alloy 20	17–4PH PH55A	cast iron	ductile iron cast steel
Impeller	28% Cr	28% Cr	PH55A/17–4PH CD4–MCu	PH55A/17–4PH CD4–MCu	28% Cr	28% Cr
Wearing ring	28% Cr	28% Cr	PH55A/17–4PH CD4–MCu	PH55A/17–4PH CD4–MCu	28% Cr	28% Cr
Shaft sleeve	28% Cr or 303 stainless steel ceramic–coated	28% Cr or 303 stainless steel ceramic coated	316 stainless steel or Alloy 20, ceramic–coated, PH55A	316 stainless steel or Alloy 20, ceramic–coated, PH55A	28% Cr or 303 stainless steel ceramic–coated	28% Cr or 303 stainless steel ceramic–coated
Shaft	carbon steel	high tensile strength alloy steel	316 stainless steel/Alloy 20	17–4PH PH55A	carbon steel	high tensile strength alloy steel

References

14.1 Karassik, I. J. (editor). *Pump Handbook*, second edition, McGraw–Hill Book Company, 1986.

14.2 Fontana, M. G. 'Selection of Structural Materials for Pumps', Chemical Engineering Proceedings, Ohio University, 1970 No. 5.

14.3 Bocking, A. et al. 'Investigation of Certain Materials for High Speed Centrifugal Pumps', Proceedings of BPMA Technical Conference.

14.4 'Centrifugal Pump Handbook', 3rd edition, Sulzer Brothers Limited, Winterthur, Switzerland, 1987.

14.5 Weber, J. 'Materials for Seawater Pumps and Related Systems' Sulzer Pump Division.

Chapter 15

Manufacturing Processes of Pumps

Hisashi Morita

15.1 General

This chapter concerns the several phases of the manufacturing of pumps, namely, casting, machining, assembling and quality control. Small and medium size pumps of standardized design are produced in large batches, consequently their manufacturing is of a quantity where production line technology is utilized to certain degrees.

Pumps are tested for performance in the factory which may be considered as a last check of the quality of production. For information concerning testing of pumps, the reader is referred to Volume *Testing of Hydraulic Machinery* of this Book Series.

15.2 Casting

Castings are used in great extent for making components of pumps, such as impellers, casings, bearing housings and base plates. Casting is the most suitable means to form the complicated shape of the hydraulic parts of pumps.

15.2.1 Casting processes

Casting is the manufacturing process of pouring molten metal into a mould. The mould requires a pattern to form the cavity in the mould identical to the object to be produced. This cavity is usually formed between the main mould and cores.

The molten metal and sand of moulds change physically and chemically in the process of pouring and solidification. Hence the casting process should be carried out according to rules derived from theory and experience. A typical casting process of pump impeller is shown in Figure 15.1.

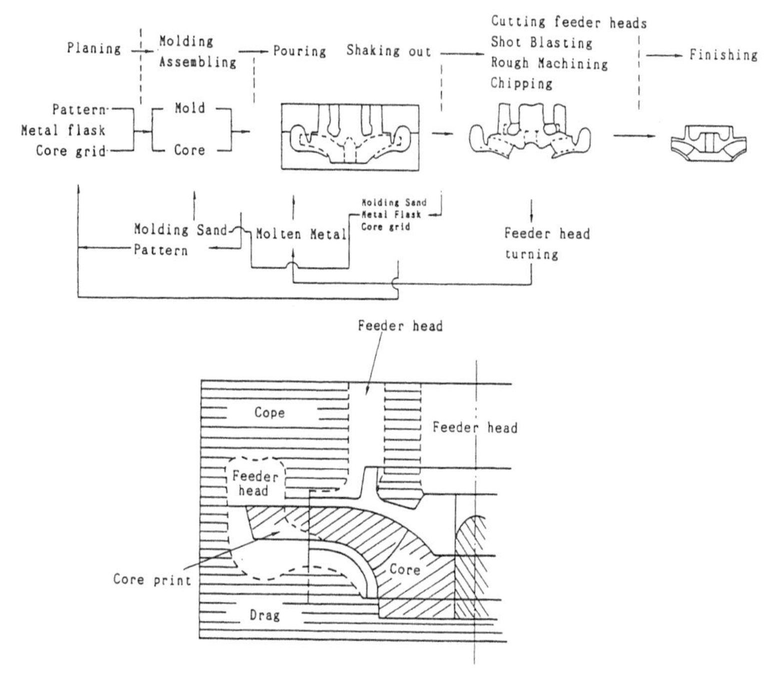

Figure 15.1 Typical casting process of pump impeller

(i) Pattern

Figures 15.2 and 15.3 show patterns for a single–suction impeller and the casing of a double–suction pump. The material making up a pattern may be wood, metal or resin. Wood is most widely used for making patterns because it is inexpensive, easy to work and light in weight. Metal patterns, mostly of aluminium, are mainly used for production in large quantities.

(ii) Casting methods

Although sand casting is the general and basic process, many special sand casting methods are in practice today:

Figure 15.2 Pattern for a single–suction impeller

Figure 15.3 Patterns for a double–suction pump casing

Sand casting
Special sand casting
- Shell moulding
- Hot box process – cold box process
- Oil sand moulding
- CO_2 process
- Self–curing moulding
- Fluid sand mixture process
- Lost wax process
- Full mould process

Sand casting– the mould is made of sand, and molten metal is poured directly into it, so the moulding sand plays a very important role in the casting process and must meet the following requirements:

- The sand must form moulds easily.
- The sand must have sufficient strength to withstand the stress of transportation and the impact by the molten metal.
- The sand must be heat resistant, it must not melt or be damaged by the molten metal.
- The moulding sand must be permeable to allow entrapped air or generated gas to escape.
- The moulding sand must enable a smooth surface to be formed for a good casting.
- The possibility of reuse of the sand, market availability and price.

Sweeping moulds are used for parts which have symmetrical cross sections along the centre lines. The mould is formed by sweeping a board which has the shape of one half of the cross section.

Shell moulding– the mould is made by the mixture of sand and phenolic resin. It is called shell moulding because the mould is thin and is used mainly in mass production.

CO_2 process– the mould is made from silicon with sodium silicate added. After moulding, CO_2 gas is injected into the mould to harden the mould. This process is applied to bronze casting and sweeping moulds.

Self–curing moulding:– the mould hardens at room temperature without special drying or injection of gas. A special binder or additive is used to help harden the mould. The productivity of this process is high in the case of small quantity production.

Lost wax process:– the pattern is made of wax or resin and is encased in sand. Upon heating, the wax or resin is melted and drained out leaving the interior for the pouring of molten metal. Castings of very small sections and of intricate shape can be produced by this process, so it is known as precision casting.

15.2.2 Design considerations of casting

The complicated shape of pump parts may be the cause for defects in castings, careful consideration must therefore be given to design details. The following are typical rules in designing castings.

(i) Thickness of casting

When the thickness of a casting is not uniform, the time required for solidification and the rate of cooling are varied for all parts of the casting. In this condition, molten metal may not be supplied sufficiently and solidification will cause shrinkage, cavity or porosity. The shape of the casting must therefore maintain a correct proportion of thickness as shown in Figure 15.4.

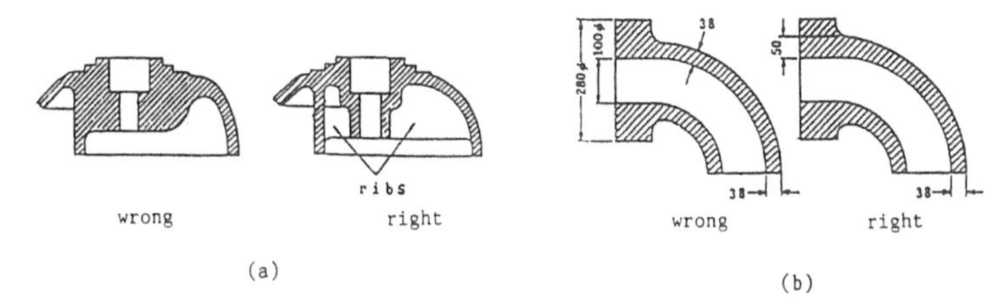

Figure 15.4 Correct and incorrect design of a casting [15.1]

It is desirable from a functional point of view to have the same thickness in each part of the casting. But when the casting is much longer than its width, the centre of the thickness solidifies last and centre line shrinkage may occur. Where the mould is thin, it is difficult to get sufficient molten metal to all parts, so it is preferable to thicken the feeder head side of the wall, as shown by the examples in Figure 15.5.

The minimum thickness is another important consideration in designing the casting. The minimum thickness is decided by the fluidity of the molten metal which is determined not only by the material but also by the condition of pouring or the kind of mould used. Pump casings are usually designed

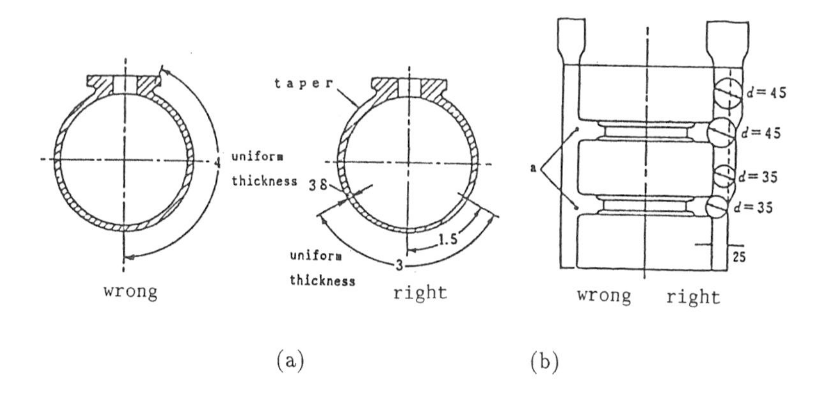

Figure 15.5 Correct and incorrect design of pump parts [15.2]

by satisfying the minimum thickness rule. In case of scale down or small pumps, the minimum thickness of casting must be checked with great care.

(ii) Core

The use of a core enables complex internal surfaces of casting to be formed that cannot be made by any other manufacturing means. Core making is labour–intensive and is difficult to maintain the dimensions accurately. Cavities in the casting may be caused by trapped gas generated by the core. Also cracks in the casting may be caused by resistance of the core when the casting shrinks at the cooling stage.

The core will become overheated and sand burning will occur if the thickness of the casting is much greater than that of the core. If the core is thin and long, it may not be strong enough to withstand the bending stress. Particular attention should be paid to these features of the core in designing volute pump casings or return passages of multi–stage pumps.

15.3 Machining

The machining procedures for pump parts are basically the same as for other rotating machinery. Essential procedures are:

- Marking–off of the castings.
- Machining the mating and sealing surfaces.

- Finishing the hydraulic passages.

15.3.1 Marking–off of castings

The hydraulic passages of pumps are usually formed by casting. Marking–off of these cast parts is important in maintaining dimension control. When the axial position is marked off on an impeller, one side of the cast hydraulic passage (flow channel) should be taken as the base face of machining. If this is not followed, the wall thickness of both sides of the shroud may become unequal, or the axial position of the hydraulic passage may fall out of line during assembly.

In the case of a split volute casing, the machined faces of the upper and lower casings should be marked off in order to obtain adequate dimensions for the volute and flange thickness. These faces must also be marked off in the horizontal direction so as not to cause the misalignment of the upper and lower halves. It will also help to locate the proper axial position of the rotor in relation to the volute.

15.3.2 Machining requirements

Since a pump is a rotating machine which contains liquid that should be sealed within, the following items are important:

(i) Concentricity of diameters and bores

Diameters and bores of rotating parts must be kept concentric, otherwise vibration due to unbalance will be generated. When the rotating parts are machined on the lathe, it is preferable that the diameters and bores are machined all in one setting without removal between operations. The shafts for high speed pumps are especially inspected for concentricity of all diameters.

(ii) Finishing of sealing faces

It is important not to have any leakage from the sealing components. If the sealing face is not well finished or not flat, or if the gasket does not function properly, liquid inside the pump will leak through the gaps across these sealing faces. Needless to say, if any sealing surface is damaged in the manufacturing process, the component is rejection at quality control.

(iii) Finishing of hydraulic passages

Friction is a major source of hydraulic losses of pumps. High flow velocity and/or rough surfaces will cause friction loss to increase. For high performance pumps the flow passages must have a smooth finish which can only be attained by using hand tools. The profile of impeller blades of such pumps are finished by hand tools and the dimensions checked by a gauge or a pass.

15.3.3 Machining of a horizontally split casing

To illustrate the process of machining parts of pumps, the sequence of machining of a horizontally split casing is described in Table 15.1. The flange faces of the upper and lower casing halves are machined separately first and the holes for connecting bolts drilled. The bores where the casing rings and seals will be installed are then machined with the two halves joint. They are again separated and machined for other details. The casing halves are finally checked for alignment of the flow passages of the volute by assembling again.

15.3.4 Automatic machines

Automatic machines are used for machining of various pump parts. Special purpose machines are used in mass production. Numerically controlled machines are used in other cases, such as impellers machined by numerically controlled lathes and casings being machined by numerically controlled milling machines and numerically controlled boring machines, or machining centres which combine these and other operations.

15.4 Assembling and Balancing

15.4.1 Assembling

The basic procedures such as preparation and cleaning of parts, fitting of joining parts, fastening of bolts and nuts are almost the same as for other rotating machines. The following are only particular features of pump assemblies.

(i) Rotating parts

The impeller is installed on the shaft when the following points are observed:

- Cleaning of the mating faces of impeller and shaft for prevention of seizure (sometimes anti–seize compounds may be used).
- Sense of rotation of the impeller.
- Axial location of the impeller.

Other rotating parts such as sleeves are then installed on the shaft. With high speed pumps, it is necessary to check the runout of the rotor and

Table 15.1 Procedures of machining of pump casing

NO	action	face or location to be machined	machine tool (example)	Machined face	
				lower casing	upper casing
1	milling	flange where upper&lower casings are jointed and pump feet	milling machine (plano-miller)	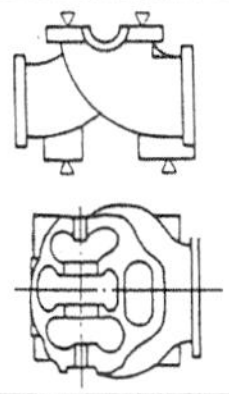	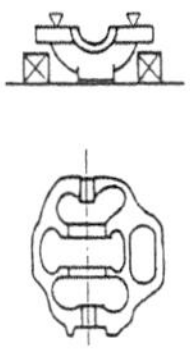
2	drilling & tapping	bolt holes and threads on the above flange	drilling machine (radial drilling machine)	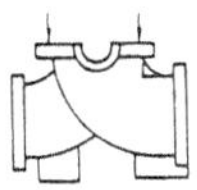	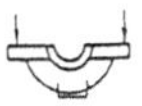
3	joint both casing & boring	bore of casing where casing ring & seals are installed and suction & discharge flange	boring machine (horizontal boring machine)	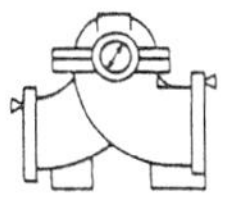	
4	drilling	bolt holes of suction & discharge flanges and feet other holes & threads	drilling machine (radial drilling machine)	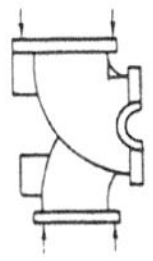	
5	trimming	discrepancy between upper & lower volute casing	hand grinder	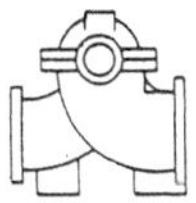	
6	hydraulic test	———————			

∇, ↓ : machined face

balancing is always required (as described below).

(ii) Casing

The rotor is installed with the casing rings placed in the lower casing half. The upper casing half is then placed on the lower casing guided by dowel pins and the flange bolts tightened. At this time, it is important to check the gaskets or O–rings for proper setting.

(iii) Bearings

After both end bearings and bearing housings are installed, it is essential to adjust the clearance between the impeller and casing rings to be almost identical in four directions, top, bottom, left and right. This is done by moving both bearing housings with adjusting screws. Dowel pins are then fitted to fix the location of the rotor. Serious seizure or vibration may occur if the rotor is not properly aligned.

(iv) Seals

The sealing devices are installed after the rotor is set. There are many types of sealing devices used in pumps, such as packing glands, hydrostatic seals and various kinds of mechanical seals, depending on the working liquid, sealing pressure, pump speed and dimension. The assembly procedure for each type of seal is different, detailed instructions should be provided for the assembling work.

(v) Alignment of coupling

When the coupling between the pump and the driver is connected, it must be aligned within the limits of parallel and runout. The alignment is adjusted by moving the pump or the driver and by inserting shims between the equipment and the base plate. The couplings now in use have some flexibility, but the two halves must be aligned properly in the factory or at site as a common rule.

(vi) Others

Anti–seizure compounds may be used for fitting parts and on bolts and nuts. It is necessary to fasten the rotating parts to the required torque and use set screws to lock the nuts of rotating parts.

15.4.2 Balancing

Any unbalance of the rotating parts may cause vibration that will hamper the operation of the pump. When the rotational speed is high, it is particularly necessary to remove whatever unbalance is left in the machine. The unbalance of pump rotors may result from various sources, such as:

- Out of symmetry: certain areas of cast parts, such as impellers, may be left unmachined or is short of dimension.
- Offset: accumulation of tolerances in machining, assembling and fitting that cause the centre of gravity of the rotating parts to be offset.

(i) Static balancing

A simple disc is attached to a shaft and the shaft is allowed to rotate on knife edges, as shown in Figure 15.6(a). It will rotate and rest with the heavy side pointing down if the disc has unbalance. The amount of unbalance can be measured statically, hence the process is called static balancing. Most impellers of pumps are inspected for unbalance in this manner with correction by grinding off some metal or drilling blind holes in the area where it is heavy.

(ii) Dynamic balancing

If the mass of the rotor is distributed along the shaft length such as in multi–stage pumps, even if it is balanced statically, an unbalance force may still act on the shaft and bearings when it is rotating. Such a shaft is not truly balanced and must be corrected by dynamic balancing. A simplified example is given in Figure 15.6(b) where both unbalances are equal and 180° apart. This system is statically balanced but when the shaft rotates, each disc produces a centrifugal force that is different in direction.

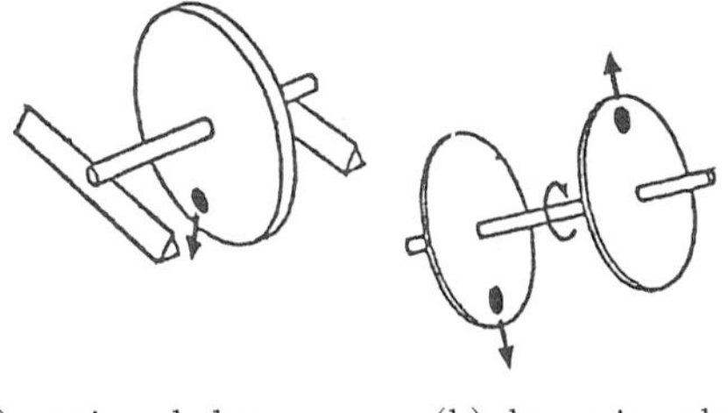

(a) static unbalance (b) dynamic unbalance

Figure 15.6 Schematic of the principle of unbalance [15.3]

Correction for this unbalance is done by grinding or drilling using a balancing machine. The behaviour of the rotor in static unbalance and dynamic unbalance are illustrated in Figure 15.7.

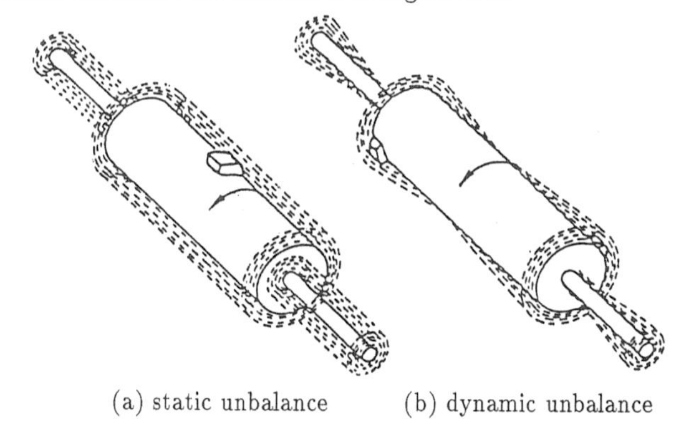

Figure 15.7 Behaviour of rotor under unbalance [15.4]

15.5 Quality Control

The term quality control was first used in 'Economic Control of Quality of Manufactured Products' written by W. E. Shewher in 1931. Quality control means all action taken to prevent manufacturing of disqualified products whilst inspection refers to judgment of whether the products are qualified or not.

15.5.1 Quality of design

Quality design is setting a goal in quality which the products should achieve. The quality of design should be decided by the requirements of the customer. The most basic problem in quality control is how to balance between quality and cost. As the quality of design increases, the cost will increase markedly as shown in Figure 15.8.

Minimum requirements of quality should also be considered. These requirements have a technical and an economical aspect. If the quality of the products decreases too much, malfunction or trouble may occur after the products are delivered, or on the other hand, the cost may increase because of added work load in manufacturing.

An example is the thickness of the casting for pressure sustaining parts. Thickness is decided by a given formula and allowable stress may differ with

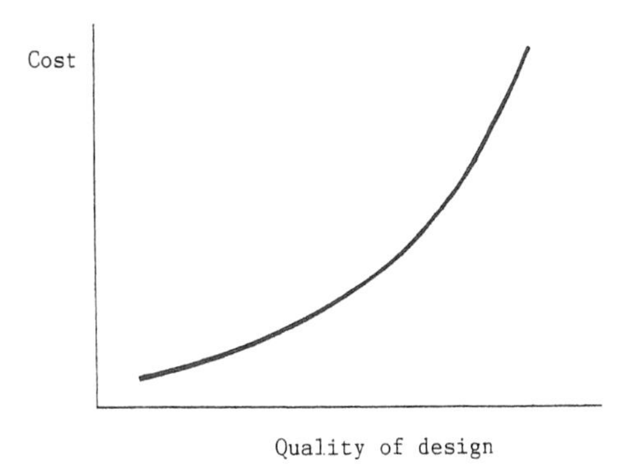

Figure 15.8 Cost versus quality design of pumps

the level of quality control of the material. Higher allowable stress values can be used only when the soundness of the interior of materials is certified by radiographic testing and the quality of the surface is proved by liquid penetrant testing or magnetic particle testing. Under this condition, the walls of the casting can be made thinner. Conversely, these non–destructive examinations can be waived if the casting wall is designed to be thicker. The level of quality is decided by the minimum requirement and economy.

15.5.2 Quality control in manufacturing

Quality is created in the manufacturing stage. The various manufacturing processes all play important roles in quality control. The specifications and the operation standards which reflect the quality of design should be decided before the manufacturing stage.

The operation standards decide technical levels of workers, materials, parts, machines, jigs and fixtures, operation, instruments and measurements. Regulation of operation standards concerning workers, machines, materials, parts, instruments and working procedures help to prevent wide fluctuation of quality levels.

Manufacturing should only start after the specifications and the operation standards are determined. It is necessary that manufacturing is carried out in conformity with the operation standards. Process control sheets may be used to prove these standards are achieved.

Fluctuation in the quality level may occur if strict adherence to the operation standards is not enforced. It is important to find out quickly the fluctuations and to remove the causes.

References

15.1 A. S. M. 'Casting Design Handbook', 1962.

15.2 Chijiwa, K. *Casting Process, Manufacturing Engineering*, Seibundo Shikosha, Japan, 1967 (in Japanese).

15.3 Thomson, W. T. *Mechanical Vibrations*, Prentice–Hall, 1953.

15.4 Collacott, R. A. *Vibration Monitoring and Diagnosis*, George Godwin, 1979.

15.5 Beck, W. W., Karassik, I. J. 'Pump Testing', *Pump Handbook*, 2 edn. Edition, McGraw Hill, 1986.

15.6 Hydraulic Institute. 'Standards for Centrifugal, Rotary and Reciprocating Pumps', U. S. A.

Chapter 16

Pump Drives

A. G. Salisbury

16.1 Pump Speed and Speed Variation

A pump is a machine for converting rotary or reciprocating power into fluid power. It therefore always requires a suitably rated prime mover. The principal selection criteria for matching the driver to the pump are speed and power but there are many other factors to be considered. This chapter introduces the fundamental principles underlying proper driver selection and operation.

Pump speed is determined by considering the pump size, its efficiency and cavitation performance.

Since for rotodynamic machines the total head generated at any specified flowrate is a function of impeller tip speed squared:

$$H = f(U_2^2)$$

and from

$$U_2 = KD_2n$$

it can be seen that the higher the rotational speed the smaller the pump. This should mean that the pump is lighter, cheaper to make and would occupy less space in the pumping installation, and also the driver would be smaller. For positive displacement pumps the fundamental relationships between flowrate, head (or pressure) and speed are different but the same conclusion is true that a faster acting pump is a smaller, cheaper pump.

For most pumps there is a relationship between speed and efficiency, that is the speed may be optimized in order to obtain maximum pump efficiency. Maximum efficiency infers that the prime mover power rating is minimized and also the energy costs of pumping are minimized.

The selection of pump type number Ω (or specific speed n_s) for a given duty has been explained in previous chapters. From the expression for the type number

$$\Omega = \frac{2\pi n}{60} \cdot \frac{\sqrt{Q}}{(gH)^{0.75}}$$

it can be seen the rotational speed n is directly proportional to Ω. For rotodynamic pumps there is a relationship between the net positive suction head available $NPSH_a$ and the pump speed n

$$\Omega_s = \frac{2\pi n}{60} \cdot \frac{\sqrt{Q}}{(gNPSH_a)^{0.75}}$$

For pumps of a particular type operating at design point, Ω_s has specific optimum values (for instance, 3.25 for certain single–entry overhung impeller pumps), hence the maximum acceptable running speed for a specified flowrate Q and given $NPSH$ can be determined by transposing the above equation. Recommended maximum operating speeds based on this type of consideration are available from a number of sources, for example the Hydraulic Institute Standards, USA.

An impression may be formed from the foregoing discussion that pump speed can be determined independently of driver considerations, although normally the decision is made partly based on the driver specification. Since many pumps are driven by alternating current induction motors (for large pumps sometimes by synchronous motors), the pump speed is limited by the supply frequency and the number of poles in the motor. With $50Hz$ alternating current power supply most pumps would run at 730, 980, 1480 or 2980 min^{-1} .

The speed of most prime movers is load dependent so unless there is an independent speed control, for example a governor on an engine or steam turbine, the speed of the pump will vary with pump load. It is important to allow for this variation in pump system interaction calculations especially since most standard performance data is corrected to constant speed. For a.c. induction motors, slip is almost directly proportional to load so that if full load and no load speeds are known for the motor an accurate estimate of pump speed at any other load can be made.

Having established the maximum acceptable drive speed, the variation of speed must be considered. The reasons for selecting a variable speed system will be discussed later but the effect of speed change on pump performance has other important effects during transient conditions, such as the case of pump starting.

From the pump affinity laws, the following rules relating pump performance to speed

Flowrate	$Q = kn$			
Total head	$H = kn^2$	also	$NPSH_r$	$= kn^2$
Power input	$P = kn^3$			

These rules relate similar operating points at different speeds hence any point on a pump characteristic at speed n_1 can be translated to an equivalent point on the characteristic at speed n_2. The relationships hold true for most practical speed changes but extra care should be taken when manipulating power and $NPSH$. Power is dependent on efficiency which does not remain constant as speed changes. As the Reynolds number of the flow will reduce with a decrease in velocity (speed) the flow losses tend to increase and lead to a reduction in efficiency. Figure 16.1 gives an estimate of the change of efficiency with speed based on estimates with the Moody formula and should be taken into account when calculating power input at varied speed. The variation of $NPSH$ given is approximate in case of speed changes for similar reasons.

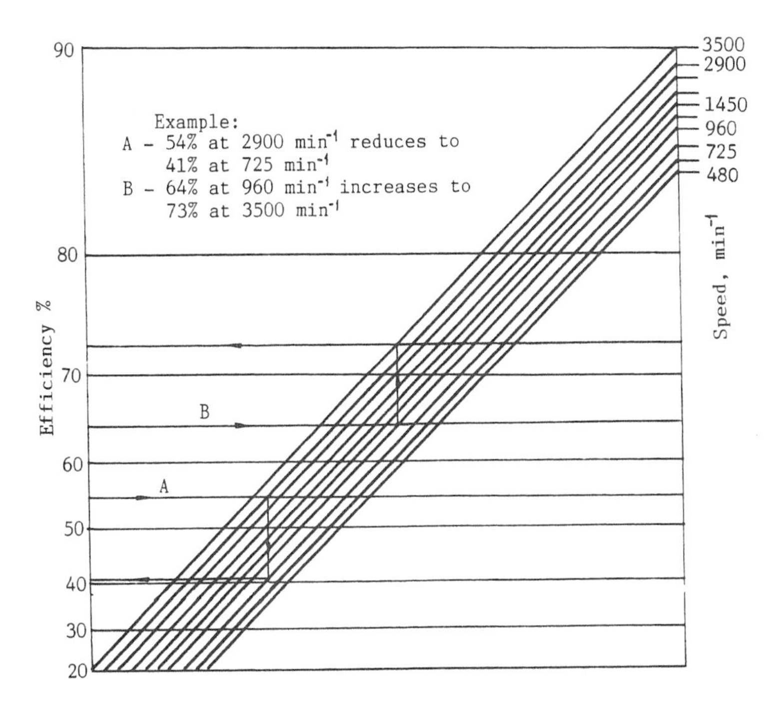

Figure 16.1 Approximate change of efficiency with change in speed for a given pump

16.2 Pump Power Input

The power for any pump duty can be calculated if the efficiency is known. This in turn requires a knowledge of the pump characteristics, especially where the pump operates over a range of flowrate or at different speeds such as a variable speed pump set.

Various factors may influence the actual power absorbed, either initially or during operation. Viscosity has a significant effect and corrections based on *normal* water are available for use to adjust the performance when handling viscous fluid, as shown by the chart in Figure 16.2. Input power varies directly with liquid density which in turn is temperature dependent so full account must be taken of temperature variations. Furthermore, in most cases the dynamic viscosity is also temperature dependent.

Other factors to be considered are variations in pump and system characteristics, caused by inaccuracies in estimation or test, and changes of operating conditions, for example operating range. Low specific speed pumps generally have a power characteristic of positive slope, that is, power increases with flowrate whereas high specific speed axial–flow pumps have a falling power curve (Figure 16.3). This variation in power must be considered when choosing the prime mover and usually margins must be added to pump duty input power.

Ideally a pump with a non–overloading power curve would be chosen but this is not usually feasible, particularly for relatively high heads or low heads. Non–overloading means that there is a maximum power absorbed at or near the design flowrate (Figure 16.3, curve 5) so that when the motor is rated, usually with a margin of 10 to 15% over design pump input power, it will not be overloaded at any other flowrate. The pump designer can control to some extent the degree to which the pump power varies from design power by choosing suitable design parameters but once a pump is built the shape of its power curve is fixed.

16.3 Pump Drivers

It is possible to use any form of rotary output prime mover for industrial pumps. The most common choices are:

- Electric motors
  - a.c. motors, induction and synchronous
  - d.c. motors, variable speed
- Combustion engines
  - petrol (gasoline) engines
  - diesel engines
  - gas engines, gas turbines

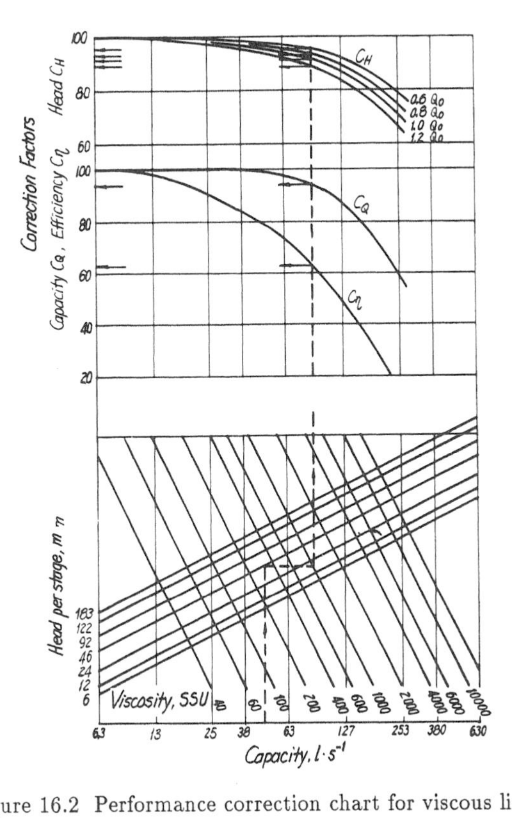

Figure 16.2 Performance correction chart for viscous liquids

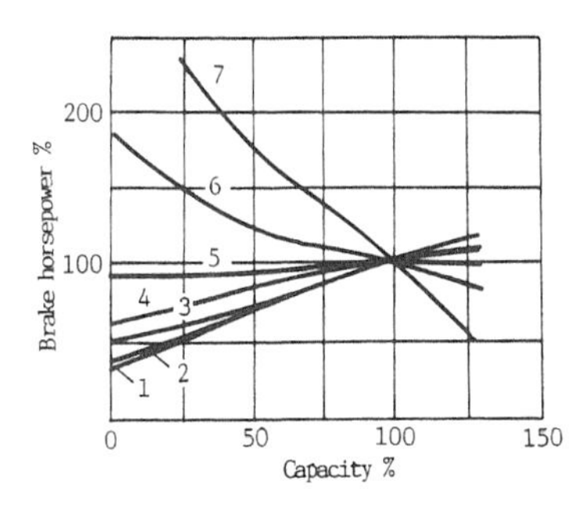

Figure 16.3 Brake horsepower curves for pumps of different specific speed

- Turbines – steam, gas (that is power recovery)

- Hydraulic motors

The primary factors in the selection of a pump drive are:

- Available forms of energy
- Cost of energy
- Capital cost of the driver and associated systems
- Security of drive

The a.c. electric motor is by far the most widely used, due to its availability, standardization and low cost (as a result of larger scale of production than any other industrial drives).

Electric power is widely available in industrial and developing countries over a range of voltages (with suitable local transformers) and powers. In countries where $50Hz$ alternating current is used, the maximum synchronous speed is $3000min^{-1}$ and in countries using $60Hz$ it is $3600min^{-1}$.

Factors affecting the choice of type of electric motors are:

- Environment – ambient temperature, altitude, humidity, hazard conditions, local surroundings,
- Application – type of pump, nature of load, rated duty, type of shaft power transmission,
- Electric power supply conditions – stability, starting restrictions, fault levels,
- Maintenance – accessibility, frequency of planned shut–downs, skill of personnel,
- Cost – relative importance of capital cost versus running costs.

Direct current motors are not in common use now and are utilized only where variable speed or special characteristics are required. They tend to be relatively large in size and expensive compared with a.c. machines.

Internal combustion engines are used where electrical power is unavailable or unreliable, or for portable sets. Diesel or petrol (gasoline) engine–driven pump sets are often used for emergency applications, for example fire fighting pumps. Diesel engines tend to be larger, heavier and slower running but have the advantage of wide fuel availability at advantageous cost.

Gas turbines are available for a wide range of powers but are comparatively expensive for pump drives. A gear box is required in most cases because of the inherent high speed of gas turbines, and the fuel consumption is often uneconomically high. They have the advantage of being small and light for a given power and are reasonably tolerant of a poor operating environment. The main applications to date have been where fuel is readily available, such as oil or gas pipeline pumping overland or on oil platforms.

Steam turbines offer speed up to 12,000 min^{-1} and unlimited power for pumping applications but generation of steam in sufficient quantity at modern turbine pressures is becoming rare except in processes which use steam for other purposes, for example power, refinery and chemical industries. The economic generation of steam often requires boiler pressures and temperatures much higher than required by the process. By utilizing the excess steam energy to drive a pump, expensive and wasteful throttling of control valves can be avoided. The value of energy saved may well offset the additional cost of the steam turbine, particularly for continuous processes. An additional advantage is that a steam turbine can be used in hazardous (high explosion/fire risk) areas. But a gear box is often required for pump drives because of its relatively high speed as with gas turbines.

A pump is of course a reversed turbine and vice versa and so there is now an increasing interest in using pumps as low cost drives for other pumps in areas where there is a supply of pressurized liquid. This may seem peculiar but the example of using high–flow, low–head supply water to drive a high–head, low–flow pumping duty proves to be a case of sound engineering.

Hydraulic motors have the advantages of small size for a given power and the ability to work in hazardous areas. They are not in common use and are relatively inefficient due to the need to generate and transmit the hydraulic power initially. Nevertheless, for some submersible applications, such as cavern storage pumping, they have been successful.

The major advantage of engine and turbine drivers over the common induction motor has in the past been the relative ease with which variable speed and therefore control of pumping rate could be achieved.

16.4 Drive Systems

All of the prime movers described above can be used in conjunction with combinations of the following drive line components:

Rigid couplings
Flexible couplings – pin, gear, flexible member
Belt drives
Gearboxes
Clutches
Fluid couplings
Magnetic couplings

Rigid couplings are normally used to reduce shaft lengths where accurate alignment can be established.

Flexible couplings are used to connect two independent shafts when alignment is more difficult to achieve and maintain. Flexible couplings should tolerate some degree of parallel, axial and angular misalignment but operate best (for example with least transmitted force and least vibration) when the initial misalignment is minimized. Gear type couplings need to be lubricated; all types require regular checks and/or maintenance.

Belt drives are not in common use nowadays but are useful for matching pump/motor speed, periodic speed change and remote drives. Speed changes are achieved by changing pulley sizes. Belts impose radial bearing loads due to belt tensioning so some care is necessary, and occasionally a separate jack shaft is required.

Gearboxes are many and varied. They are commonly used to reduce a high drive output speed to suit a low speed pump (for example a steam or gas turbine driven pump) or alternatively increase the pump speed where the driver speed is limited (for example 2,980 min^{-1} electric motor driving a small high speed pump). Their other main use is in right angle drive systems, such as engine driven vertical shaft fire pumps. The efficiency of oil lubricated gearboxes is normally high, about 97–98%, but they are relatively expensive and require additional space and maintenance. An example of the successful integration of a high speed output step–up gearbox in a competitive pumpset is the Sundyne pump. Ingersoll Rand has recently introduced a pump incorporating similar technology.

Clutches are normally used when either the prime mover is unable to start with the pump load coupled or to allow pump disconnect in drive trains. Occasionally clutches are used for emergency pump isolation. For pumps having dual drives, for example electric motor and power recovery turbine, an overrunning clutch is used so that the electric motor can drive the pump alone without the drag of the de–energized turbine.

Constant filling fluid couplings are often used for reasons similar to those applicable to clutches and to give a soft start facility. Variable filling fluid couplings are used when speed control is required.

Magnetic couplings serve in similar circumstances. They may be of the permanent magnet variety, either synchronous or induction, or separately electrically energized to give speed control. The permanent magnet type is often used for glandless applications as an alternative to the wet or canned electric motor.

16.5 Pump Starting

When considering the starting of a pumpset, the following must be answered:

- Will the driver overcome static friction?
- Will the driver produce enough torque to accelerate the set to full speed and how long will it take?
- What will be the effect on the driver and starting power supply?

Static friction is variable and depends on the construction of the pump and its condition. Factors such as gland friction, which is extremely variable, are difficult to estimate accurately. It is therefore good sense to ensure adequate margin between driver zero speed (or break–away) torque and the estimated friction torque of the pump at zero speed.

The relationship between motor torque and pump torque at any speed has then to be considered. The excess driver torque is the torque which will accelerate the pump and the driver itself plus any intermediate drive system components, such as couplings and gearboxes. This relation is usually expressed as

$$T = I\alpha$$

where T is the excess motor torque, I the total pumpset rotor inertia, and α the instantaneous acceleration rate.

The quantity α is the instantaneous acceleration rate since the torque T is normally speed–dependent as is shown in the typical speed–torque curve of Figure 16.4. The torque available from the driver is usually obtained by test (data may be provided by the driver manufacturer). Pump torque is calculated by matching the pump characteristic to the system characteristic for the starting sequence and calculating the pump power at each intersection for speeds between zero and full speed. Torque is then calculated for each speed. Figures 16.4 and 16.5 show typical starting characteristics, respectively, for radial–and axial–flow pumps. If there is excess driver torque available over the whole run of the speed range then the pumpset will accelerate.

In order to calculate the run–up time it is necessary to calculate acceleration and speed attained for small increments of time, adjusting the torque for subsequent time increments according to the speed attained at the end of the last and finally adding all the time increments.

It will be observed that the system characteristic during starting is very important and can, in some applications, be controlled. Generally low specific speed pumps are best started against a closed valve (Figure 16.5) and high specific speed pumps should always be started with the system open (Figure 16.6). Very long discharge pipes which are empty on start–up are undesirable for low specific speed pumps, and long discharge pipes which are full are undesirable for high specific speed pumps. This type of problem is generally insignificant for pumps with non–overloading power characteristics.

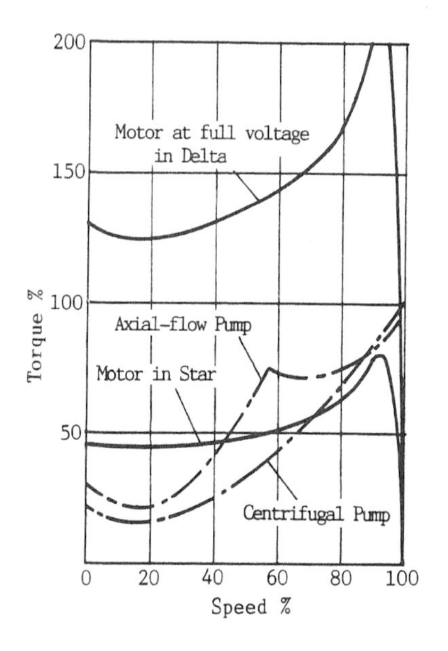

Figure 16.4 A percentage speed/torque curve for pumps related to induction motor characteristics

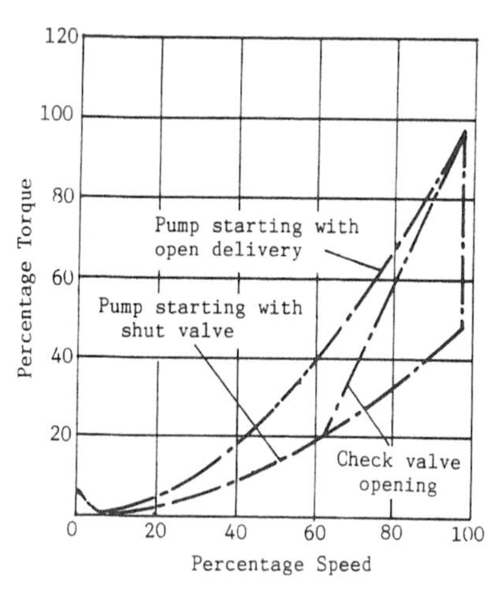

Figure 16.5 Torque/speed curves for a radial–flow pump

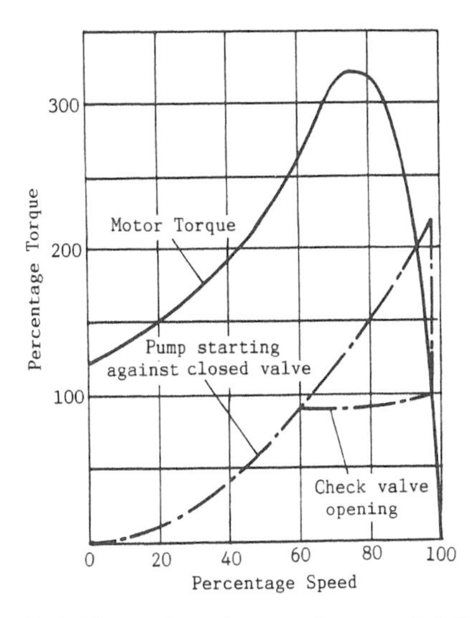

Figure 16.6 Torque/speed curves for an axial–flow pump

The ability to start a pumpset depends on the driver characteristics. In the case of starting a squirrel cage induction machine, the current taken at constant voltage during start–up depends on the torque characteristic. In general starting currents are much higher than (three to eight times) normal full load current. This level of current is usually unacceptable from a supply point of view as it would lead to large voltage drops on the supply system and cause external grid problems. It would also lead to a voltage drop at the motor terminals resulting in reduced torque availability from the motor. For these reasons the starting currents drawn are restricted and it is necessary to find a motor starting system which not only gives acceptable pumpset run–up but also acceptable starting currents. Some systems in common use are the DOL, star–delta, auto transformer, Korndorffer and inverter–fed soft starter systems.

Whichever system is used it will add to the cost of the installation, a sophisticated starter can cost almost as much as the basic squirrel cage motor.

16.6 Variable Speed Drives

16.6.1 Slip couplings

In these couplings there is no positive connection between input and output shafts so the torque on the output shaft cannot exceed that of the input shaft. The engagement between the shafts is fluid or magnetic, and if there is a drop in speed (slip) between the two, power is generated and dissipated in the linkage medium, so there is a loss of power and efficiency between input and output.

Slip couplings can be automated fairly readily, the magnetic type perhaps more so than the hydraulic by having a more rapid response. However, in both this respect and efficiency they are no better, and even marginally less good, than slip–controlled electric motors. Moreover, if substantial power is dissipated in the coupling, the need for artificial cooling may present problems or even be inadmissible, and the magnetic type may be of limited availability for vertical drives.

Advantages of slip couplings:

- Provide unlimited and stepless speed variation.
- Introduce resilience and lessen the transmission of sudden shocks and violent fluctuations in load between driving and driven machines.
- Enable heavy inertia loads to be run up to speed gradually while allowing normal motor run–up conditions.
- Make speed variation available where electrical methods would entail enclosure flameproofing difficulties.
- Make independently variable speeds available on the same line of shaft.

16.6.2 Variable speed motors

Direct current motors give free choice of speed within practical design limits, and are inherently amenable to speed variation. Alternating current motors have fundamental speeds fixed by the frequency of the supply and the detail of the motor winding.

(i) Pole–change motors

Depending on the nature of variation required, pole–change motors may be stepless over a range only at certain set speeds.

If only two set speeds are required and both are simple divisions of the maximum synchronous speed, they can be obtained with a single motor having separate windings for each speed, with corresponding starter provision including change-over rotor connections for slip ring motors which must have two rotor windings to correspond with stator pole changes. Squirrel cage motors which do not involve rotor pole change windings can be made pole amplitude modulated for three or four speeds if all are prime factors of the fundamental speed. If however they are not all as such, all except the maximum speed will have to be obtained by other means.

(ii) Speed variation using a.c. slip ring motors

This is the common method used for small and medium power drives where running cost savings would not justify any slip power recovery scheme and particularly where small speed ratios are involved and stepless variation or unduly numerous and close speed steps are not required. Such cases frequently arise in pumping schemes, for example, the need may be for only one single alternative duty, such as for solo instead of parallel operation, or a single pump performing different alternative services, and the relationship of the two duties is such that they cannot be met with a pole-change two-speed arrangement. A slip ring motor is used with starting equipment having a single continuously rated step of rotor resistance which can remain in circuit or be cut out as required. Such an arrangement is readily amenable to automatic control.

16.6.3 Variable frequency invertors

The invertor, driving a standard a.c. cage motor, is a relatively new introduction to the field which has become more attractive as the cost of its complicated electronics has reduced.

This scheme operates by accepting the normal a.c. power supply, rectifying it to d.c., and from d.c. producing a variable frequency a.c. output to drive a standard a.c. induction motor. As the invertor output frequency is varied, the speed of the motor varies in proportion.

It is becoming increasingly clear that for pump service the current-fed type of invertor circuit has significant advantages. It operates by controlling the current rather than the output voltage (Figure 16.7). It employs components which are less critical and the resulting system is more tolerant of misuse. It incorporates a simple thyristor bridge which is self-commutating, so that only standard converter-grade thyristors are required. Its output is short-circuit proof, and is largely unaffected by dips in the mains supply voltage. Regenerative braking is inherent in the circuit, and since standard thyristors are used, control is possible for higher voltage motors.

The invertor is quite suitable for converting existing fixed speed installations to variable speed, and for providing variable speed drives in hazardous

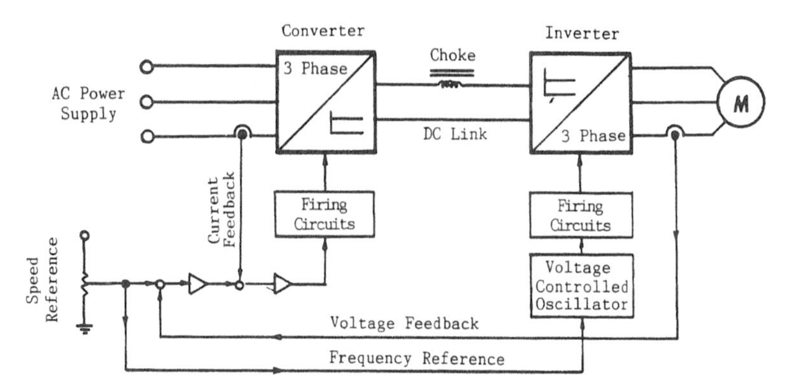

Figure 16.7 Block diagram of current fed invertor

(flame–proof) environments or for submersible pumps. For new installations the high efficiency of the invertor makes it a strong contender while its ability to provide an emergency fixed speed alternative, by running the a.c. motor direct off the mains, gives an added performance security not easily provided by other drives.

The output frequency can of course exceed the input frequency so higher *synchronous* speeds are attainable, although this should not be attempted using standard motors. Potentially a soft–start, high–speed drive is available with this system. Drawbacks of the variable–frequency invertor system are the added cost and space requirement (for air–cooling).

Appendix – Useful Data

The following tables give useful electrical data for selecting and operating electrical pumping plants.

Table 16.1 Approximate full load speed (min^{-1}) of squirrel cage motors

kW	No of poles	Supply frequency, Hz 50	60
0.75 to 2.2	2	2820	3380
	4	1405	1685
	6	930	1115
3 to 7.5	2	2855	3425
	4	1420	1700
	6	945	1130
11 to 22	2	2910	3490
	4	1445	1730
	6	960	1145
	8	720	860
30 to 75	2	2935	3520
	4	1475	1770
	6	975	1170
	8	735	880

Table 16.2 Starting – Squirrel cage motors

Method of starting	Starting torque (approx) % Full load torque	Starting current (approx) % Full load current
Direct	100% - 200%	350% - 700%
Star delta	33% - 66%	120% - 230%
Series parallel	25% - 50%	90% - 175%
Auto transformer	25% - 85%	90% - 300%

The above figures apply to Squirrel Cage motors of normal design and other types are available namely:

High Torque Squirrel Cage machines will give approximately twice the above starting torques with unrestricted currents.

Low Current Squirrel Cage machines restrict the current but give a lower starting torque than the High Torque machines. These two types can now be used in many cases where Slipring machines would have been necessary in the past.

Slipring machines. All Slipring machines must be started by means of a rotor resistance starter. A starting torque of full load torque is obtainable with a starting current of approximately 1.25 full load current, this usually being sanctioned by Supply Authorities for any size of motor.

Table 16.3 Approximate kW absorbed to pump clean water

l/s	1	2	3	5	10	20	30	50	100	200
Average Efficiency %	34	44	50	58	66	72	75	78	81	83
Total head m										
5	0,15	0.23	0.30	0.43	0.75	1.47	1.96	3.15	6.05	11.8
7	0.20	0.31	0.41	0.59	1.04	1.91	2.75	4.40	8.48	16.6
10	0.29	0.45	0.59	0.85	1.49	2.73	3.92	6.30	12.1	23.6
20	0.58	0.89	1.18	1.69	2.97	5.45	7.84	12.6	24.2	47.3
30	0.87	1.34	1.77	2.54	4.46	8.18	11.8	18.9	36.3	71
50	1.44	2.23	2.94	4.23	7.43	13.6	19.6	31.5	60.5	118
70	2.02	3.12	4.12	5.92	10.4	19.1	27.5	44.0	84.8	166
100	2.88	4.46	5.88	8.45	14.9	27.3	39.2	63.0	121	236
200	5.77	8.92	11.8	16.9	29.7	54.5	78.4	126	242	473
300	8.65	13.4	17.7	25.4	44.6	81.8	118	189	363	710
500	14.4	22.3	29.4	42.3	74.3	136	196	315	605	1180

The above average efficiencies are for one duty only with the best pump selected for that duty. These figures can be improved by using high efficiency pumps designed for the actual duty but allow reasonable variations obtainable from having a large range of sizes to give a good selection.

Table 16.4 Average efficiencies and power factors of electric motors

kW	Efficiency %			Power Factor			Full load amps on 3ph 415 v
	Full load	.75 load	.5 load	Full load	.75 load	.5 load	
0.75	74	73	69	0.72	0.65	0.53	2.0
1.5	79	78.5	76	0.83	0.78	0.69	3.2
3	82.5	82	80.5	0.85	0.80	0.73	6.0
5.5	84.5	84.5	83.5	0.87	0.87	0.75	10.5
7.5	85.5	85.5	84.5	0.87	0.83	0.76	14
11	87	87	85.5	0.88	0.84	0.77	20
18.5	88.5	88.5	87	0.89	0.85	0.79	33
30	90	89.5	88	0.89	0.86	0.80	52
45	91	90.5	89	0.89	0.86	0.80	77
75	92	91.5	90	0.90	0.87	0.81	126

Appendix I

Nomenclature of Symbols

1. Introduction

This document is a unified nomenclature, accepted by all authors, to ensure coherence and credibility for the hydraulic machinery book series.

We have tried as far as possible to conform to IEC and ISO, but have accepted some differences which seemed necessary: the main differences are pointed out in the nomenclature.

It is of course evident that any attempt to unify a system of symbols or nomenclature in any field of study will meet with strong reactions due to the habits already established in particular domains. We are conscious of this, but feel that the system based on the principles exposed here has one major advantage: it comes from a coherent internal logic, which proves satisfactory after 20 years of use within the Hydraulic Machines and Fluid Dynamics Institute of the Swiss Federal Institute of Technology in Lausanne, and also by several Swiss manufacturers.

2. Symbols

2.1 Guidelines

Symbols are Latin and Greek letters, both capital and small, and Arabic and Roman numerals.

The system's internal logic lies in the distinct meaning given to each of these forms of symbols.

Exceptions are accepted only when supported by widespread use.

The lists of symbols given below are not exhaustive. Additional symbols may be defined for the description of particular topics, but this must be done in agreement with the system's internal logic.

Different variables may be represented by the same letter. This situation is acceptable when other solutions would be too far-fetched, and the variables are not likely to be found within the same equation.

We suggest "high pressure side" and "low pressure side" to replace the confusing upstream/downstream reference, which is always a cause of misunderstanding, specially in dealing with reversible pump-turbine units.

2.2 Latin and Greek capital letters

Latin and Greek capital letters stand for dimensional variables in absolute value.
Capital Latin letters are also used as subscripts for elements of a plant (see 2.5.1.).

2.3 Latin and Greek small letters

Latin small letters stand for non-dimensional variables, called "factors". They are more often the ratio of a dimensional variable to a reference value for this variable. The symbol for this factor corresponds, in small letters, to the capital letter assigned to the variable.

Examples

hydraulic specific energy transformed by runner blades to mechanical specific energy	E_t	→	energy efficiency factor	$e_t = E_t/E$
flow velocity	C	→	velocity factor	$c = C/\sqrt{2E}$
radius of the runner blade at hub side of high pressure side edge	R_{1i}	→	radius factor	$r_{1i} = R_{1i}/R_{\bar{1}e}$

Small Latin letters are also used as subscripts for the elements of a machine (see 2.5.2.), for the characteristic points for machine power (see 2.5.3.), for the directions (Velocity components,see 2.5.4) and for the operating conditions of the machine (see 2.5.5). Small Greek letters stand for non-dimensional variables, called "coefficients", characteristic of the working conditions of a machine: specific speed, flow and hydraulic energy coefficients, efficiency, Thoma number.

2.4. Roman and Arabic numerals

Roman numerals are used as subscripts for particular points of a plant (see 2.5.1.).

Arabic numerals are used as subscripts for particular points of a machine (see 2.5.2.).

2.5. Subscripts

2.5.1. Elements and particular points of the plant

Elements of a hydraulic plant have capital Latin letters. For elements located on the low pressure side of the machine, letters are topped with the sign ($\bar{}$). (Figure 1)

A	headwater basin
B	intake trash rack
T	power (headrace) tunnel
S	high pressure side surge tank
V	high pressure side valve
P	high pressure penstock
R	manifold with high pressure side valve
M	machine

$\overline{R}$	manifold with low pressure side valve
$\overline{P}$	low pressure penstock
$\overline{V}$	low pressure side valve
$\overline{S}$	low pressure side surge tank
$\overline{T}$	low pressure (tailrace) tunnel
$\overline{B}$	low pressure side trash rack
$\overline{A}$	tailwater basin

Particular points within a hydraulic plant are denoted by Roman numerals. For points located on the low pressure side of the machine, the numerals are topped with the sign ($\overline{\ }$). (Figure 1)

VII	headwater side of headwater trash rack
VI	machine side of headwater trash rack
	headwater section of head race tunnel
V	machine section of headrace tunnel
	headwater section of high pressure side surge tank
IV	machine section of high pressure side surge tank
	headwater section of high pressure side valve
III	machine section of high pressure side valve
	headwater section of high pressure penstock
II	machine section of high pressure penstock
	headwater section of high pressure manifold
I	machine section of high pressure manifold
	high pressure reference section of machine
$\overline{I}$	low pressure reference section of machine
	machine section of low pressure manifold
$\overline{II}$	tailwater section of low pressure manifold
	machine section of low pressure penstock
$\overline{III}$	tailwater section of low pressure penstock
	machine section of low pressure side valve
$\overline{IV}$	tailwater section of low pressure side valve
	machine section of low pressure side surge tank
$\overline{V}$	tailwater section of low pressure side surge tank
	machine section of tailrace tunnel
$\overline{VI}$	tailwater section of tailrace tunnel
	machine side of tailwater trash rack
$\overline{VII}$	tailwater side of tailwater trash rack

Note: the performance guarantees of the machine are referred to the reference sections I and $\overline{I}$ (1 and 2 for IEC Codes, respectively)

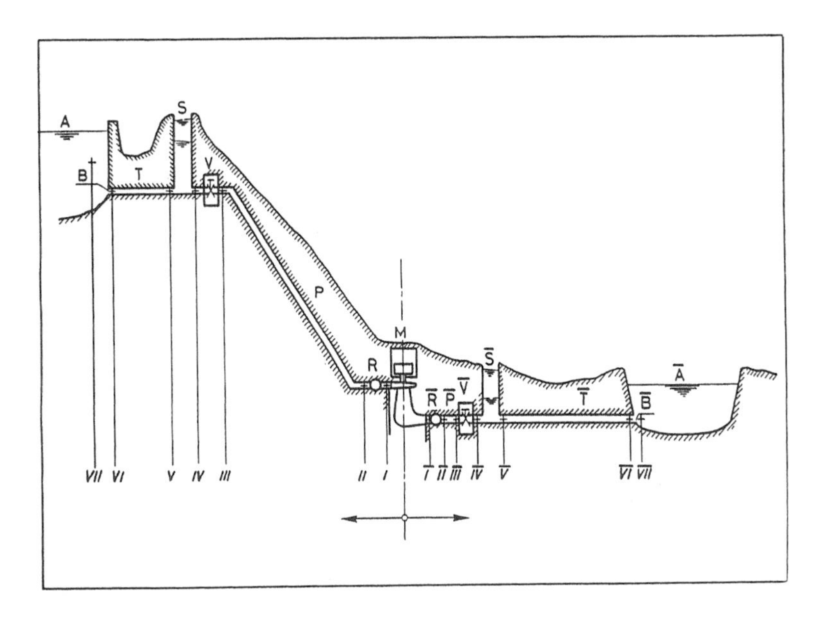

Figure 1 Particular points within a hydraulic plant

2.5.2. Elements and particular points of the machine

Elements of a machine are denoted by small Latin letters.

machine	Subscript	turbine	pump
		Terms	
	c	spiral case	casing
		nozzle	
high pressure side components	v	stay vanes	diffusor
	g	wicket gate	wicket gate
runner	b	runner blades	impeller blades
	i	internal, at the hub	
	e	external, at the band	
low pressure side components	d	draft tube	suction pipe

Arabic numerals denote particular points within the machine. For points located on the low pressure side of the runner blades, the numerals are topped with the sign (‾). (Figure 2)

	turbine		pump
6	spiral case high pressure (inlet) section	6	casing high pressure (outlet) section
5	leading edge of stay vanes	5	trailing edge of diffusor vanes
4	trailing edge of stay vanes	4	leading edge of diffusor vanes
3	leading edge of wicket gate	3	trailing edge of wicket gate
2	trailing edge of wicket gate	2	leading edge of wicket gate
1	high pressure (inlet or leading) edge of runner blades	1	high pressure (outlet or trailing) edge of impeller blades
$\bar{1}$	low pressure (outlet or trailing) edge of runner blades	$\bar{1}$	low pressure (inlet or leading) edge of impeller blades
$\bar{2}$	inlet edge of draft tube fins or stabilizing device	$\bar{2}$	outlet edge of suction pipe fins or stabilizing device
$\bar{3}$	outlet edge of draft tube fins	$\bar{3}$	inlet edge of suction pipe fins
$\bar{4}$	inlet section of draft tube elbow	$\bar{4}$	outlet section of suction pipe elbow
$\bar{5}$	outlet section of draft tube elbow	$\bar{5}$	inlet section of suction pipe elbow
$\bar{6}$	suction (outlet) section of draft tube	$\bar{6}$	suction (inlet) section of suction pipe

2.5.3. Characteristic points for machine power

Small Latin letters denote characteristic points for machine power (Fig.2).

Subscript	Definition and Term
h	hydraulic power turbine : P_h hydraulic power available for producing mechanical power pump : P_h hydraulic power imparted to the water
t	transformed by runner / impeller blades turbine: P_t hydraulic power converted to mechanical power pump: P_t mechanical power converted to hydraulic power
m	at runner (impeller) coupling flange P_m mechanical power of runner
no letter	at machine coupling flange P mechanical power of the machine

Note : IEC Codes do not take into consideration P_t,for sake of simplicity

2.5.4. Directions (see Figure 3)

Subscript	Term
a	axial
m	meridional
r	radial
u	peripheral

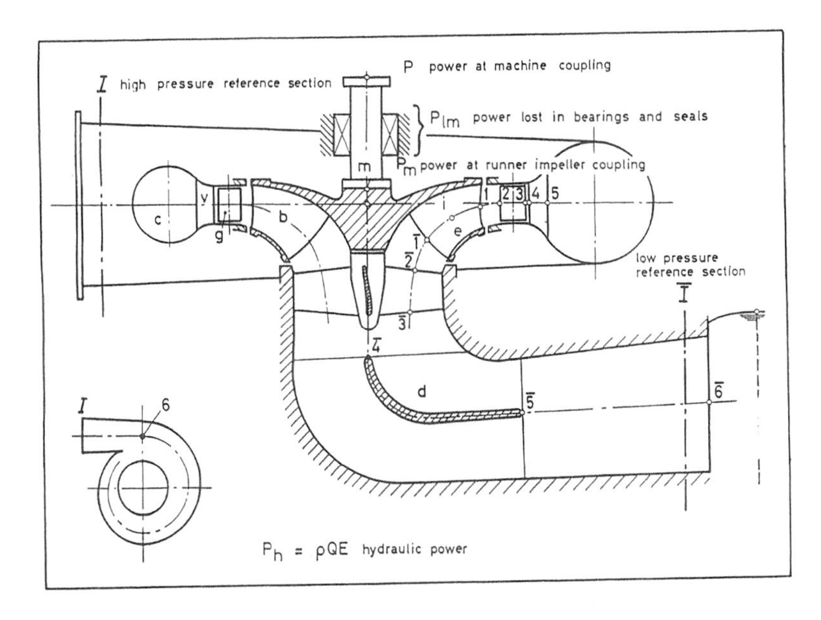

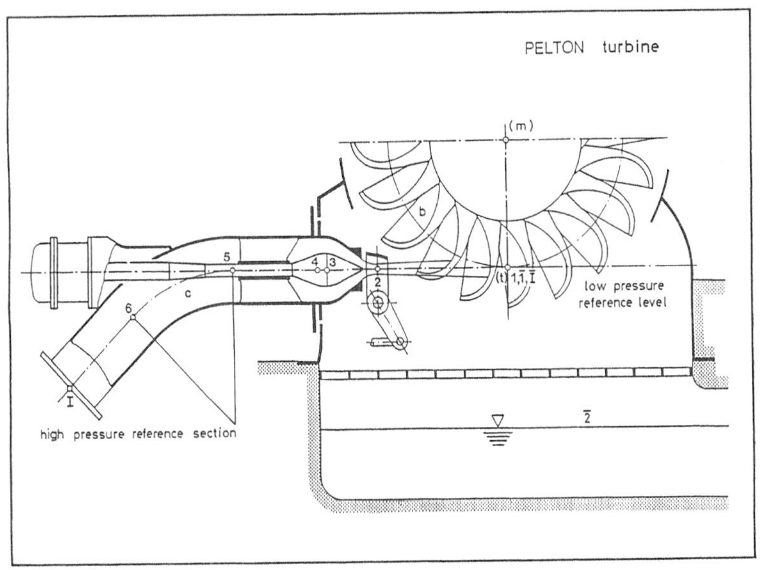

Figure 2 Characteristic points for machine power

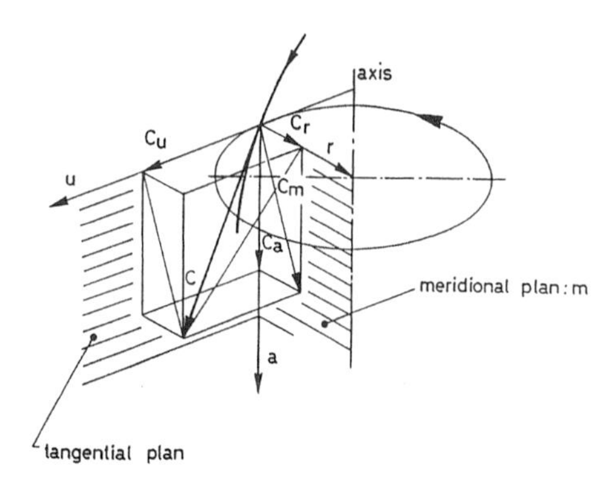

Figure 3 Flow direction

2.5.5. Operating conditions of a machine

Subscript	Term
v	turbine's zero load discharge
^	machine's best efficiency
sp	machine's specified discharge, specific hydraulic energy,power...
max	maximum of specified variable
min	minimum of specified variable
r	turbine's runaway speed

2.5.6. Other subscripts

Subscript	Term	Example
a	atmospheric	p_a atmospheric pressure
c	cavitation	ψ_c cavitation energy coefficient
d	drag	c_d drag coefficient
f	friction	F_f friction force, drag
k	kinetic	E_k kinetic hydraulic specific energy
ℓ	discharge loss	Q_ℓ discharge loss
p	potential	E_p potential hydraulic specific energy (position and pressure)
	pressure	F_p pressure force
r	losses	h_r specific energy loss of turbine
		q_r specific volumic loss of turbine
ref	reference	E_{ref} reference hydraulic energy
s	shear	F_s shear force
va	vapour	p_{va} vapour separation pressure
z	elevation	E_z position hydraulic specific energy

3. Symbol lists

3.1 Alphabetic list of capital Latin letters

Symbol	Term	Definition	Units
A	cross-section		m^2
	opening (of guide vane ,valve or needle)		m
B	guide vane height,bucket width (Pelton)		m
C	absolute velocity		m/s
$\tilde{C}$	compliance	$\tilde{C} = \partial V/\partial gH$	m.s
C_a	axial component of absolute velocity		m/s
C_m	meridional component of absolute velocity		m/s
C_r	radial component of absolute velocity		m/s
C_u	peripheral component of absolute velocity		m/s
D	diameter		m
E	specific (massic) hydraulic energy of machine :		
	delivered to turbine	$E = g_I H_I - g_{\bar{I}} H_{\bar{I}}$	J/kg
	delivered by pump	$E = g_I H_I - g_{\bar{I}} H_{\bar{I}}$	J/kg
E	elasticity (YOUNG) modulus		N/m^2
E_b	specific hydraulic energy at runner / impeller blades		
	turbine: specific hydraulic energy delivered to runner blades	$E_b = E_t + E_{rb}$	J/kg
	pump: specific hydraulic energy delivered by impeller blades	$E_b = E_t - E_{rb}$	J/kg
E_{kd}	specific hydraulic energy delivered to turbine draft tube	$E_{kd} = \left(C_{\bar{I}}^2 - C_{\bar{I}}^2\right)/2$	J/kg

Symbol	Term	Definition	Units
E_r	lost specific hydraulic energy		
	turbine: from I to $\bar{I}$	$E_r = E - E_t$	J/kg
	pump: from $\bar{I}$ to I	$E_r = E_t - E$	J/kg
E_{rb}	specific hydraulic energy lost within runner / impeller blades		
	turbine	$E_{rb} = E_b - E_t$	J/kg
	pump	$E_{rb} = E_t - E_b$	J/kg
E_{rd}	specific hydraulic energy lost within turbine draft tube	$E_{rd} = gH_{\bar{I}} - gH_{\bar{I}}$	J/kg
E_t	transformed specific energy		
	turbine: specific hydraulic energy converted to specific mechanical energy	$E_t = E - E_r$	J/kg
	pump: specific mechanical energy converted to specific hydraulic energy	$E_t = E + E_r$	J/kg
E_{td}	specific hydraulic energy transformed (kinetic → potential) by turbine draft tube	$E_{td} = E_{kd} - E_{rd}$	J/kg
E_s	suction specific hydraulic energy		J/kg
F	force		N
F_d	drag		N
F_l	lift		N
F_p	pressure force		N
F_f	friction force, resistance force		N
Fr	FROUDE number	$Fr = C / \sqrt{gL}$	-
gH	specific hydraulic energy		
	in a section	$gH = g.Z + \frac{p}{\rho} + \frac{C^2}{2}$	J/Kg
gH_k	kinetic specific hydraulic energy in a section	$gH_k = C^2/2$	J/kg

Symbol	Term	Definition	Units
gH_p	potential specific hydraulic energy in a section	$gH_p = g.Z+p/\rho$	J/kg
H	head (hydraulic energy per unit weight)		m
	height		m
H_a	representative height of atmospheric pressure		m
H_r	head loss		m
H_s	suction head		m
H_{va}	representative height of vapour pressure		m
J	clearance		m
K	sand roughness		m
L	length		m
	airfoil/blade length		m
	wave length	$L = S/f$	m
$\tilde{L}$	pipe inertance	$\tilde{L} = \int dL/A$	m^{-1}
M	mass		kg
Ma	MACH number	$Ma = C/S$	-
M_f	bending moment		N.m
M_t	twisting torque		N.m
NPSE	net positive suction energy	$NPSE = \frac{\left(\frac{p_a - p_{va}}{\rho}\right) - g.H_s}{E}$	J/Kg
NPSH	net positive suction head	$NPSH=\frac{NPSE}{g_{\bar{I}}}$	m

Symbol	Term	Definition	Units
P	power		
	mechanical power of the machine		
	(at the machine coupling flange)		
	turbine	$P = P_h . \eta$	W
	pump	$P = P_h / \eta$	W
P_h	hydraulic power		
	turbine hydraulic power available for producing mechanical power		
	pump hydraulic power imparted to the water	$P_h = (\rho . Q_l) . E$	W
P_{lm}	mechanical losses within the turbine (seals, bearings,.)	$P_{lm} = P_m - P$	W
	mechanical losses within the pump (seals, bearings.)	$P_{lm} = P - P_m$	W
P_m	mechanical power of the runner (at the runner / impeller coupling flange)		
	turbine	$P_m = P_h . \eta_m$	W
	pump	$P_m = P_h / \eta_m$	W
P_t	Power transformed by runner / impeller blades		
	turbine:		
	hydraulic power converted to mechanical power	$P_t = \rho . Q_t . E_t = P_h . \eta_h$	W
	pump:		
	mechanical power converted to hydraulic power	$P_t = \rho . Q_t . E_t = P_h / \eta_h$	W
P_{h-t}[1]	total hydraulic losses within the turbine (volumic and energetic)	$P_{h-t} = \rho . Q_l . E_r$	W
	total hydraulic losses within the pump (volumic and energetic)	$P_{h-t} = \rho . Q_l . E_r$	W
P_{rm}	mechanical and disc friction losses		
	within the turbine	$P_{rm} = P_t - P$	W
	within the pump	$P_{rm} = P - P_t$	W

(1) The IEC Codes do not take into consideration P_t(see note at page 8). Therefore P_{h-t} becomes P_{h-m}, total hydraulic losses within the machine, including P_{t-m} assumed as internal "hydraulic" and not mechanical losses within the machine, for simplicity.

Symbol	Term	Definition	Units
P_{t-m} (1)	disc friction and labyrinth losses		
	within the turbine	$P_{t-m} = P_t - P_m$	W
	within the pump	$P_{t-m} = P_m - P_t$	W
Q_I	discharge, volume-flow rate	$Q_I = dV/dt$	m^3/s
	delivered to turbine	$Q_I = Q_t + Q_\ell$	m^3/s
	delivered by pump	$Q_I = Q_t - Q_\ell$	m^3/s
Q_ℓ	discharge loss		
	turbine	$Q_\ell = Q_I - Q_t$	m^3/s
	pump	$Q_\ell = Q_t - Q_I$	m^3/s
Q_t	discharge concerned with energy transformation		
	turbine	$Q_t = Q_I - Q_\ell$	m^3/s
	pump	$Q_t = Q_I + Q_\ell$	m^3/s
R	radius		m
	degree of reaction		-
$\widetilde{R}$	resistance	$\widetilde{R} = \partial E/\partial Q$	$m^{-1}s^{-1}$
Re	REYNOLDS number	$Re = C.L/\upsilon$	-
R_h	hydraulic radius	$R_h = 2.A/U$	m
S	sound (wave) propagation speed		m/s
St	STROUHAL number	$St = C.t/L$	-
T	torque		N.m
	external torque (at machine coupling flange)		N.m
	spacing		m
	period	$T = 1/f$	s
T_h	hydraulic torque		N.m
T_m	internal torque (at runner / impeller coupling flange)		N.m

Symbol	Term	Definition	Units
U	wetted perimeter		m
	peripheral velocity	$U = R.\omega$	m/s
V	volume		m^3
W	relative velocity	$\vec{W} = \vec{C} - \vec{U}$	m/s
W_a	relative axial velocity		m/s
W_m	relative meridional velocity		m/s
W_r	relative radial velocity		m/s
W_u	relative peripheral velocity		m/s
X	abscissa		m
Y	ordinate		m
	admittance	$Y = 1/Z$	m.s
Z	elevation		m
	altitude		m
Z	impedance	$Z = dgH/dQ$	$m^{-1}s^{-1}$
Z_{ref}	reference impedance	$Z_{ref} = S/A$	$m^{-1}s^{-1}$

3.2. Alphabetic list of capital Greek letters

Symbol	Term	Definition	Units
Γ	circulation	$\Gamma_{1-2} = \int_1^2 C.dL = \Phi_2 - \Phi_1$	m^2/s
Δ	amplitude		-
Φ	velocity potential	$\vec{W} = \overrightarrow{grad}\, \Phi$	m^2/s
θ	temperature		K or °C
Ψ	stream function		m^2/s

3.3. Alphabetic list of small Latin letters (see 2.3.)

Symbol	Term	Definition	Units
a	cross-section factor	$a = A/R_{\overline{1}e}^2$	-
b	guide vane height factor	$b = B/R_{\overline{1}e}$	-
c	absolute velocity factor	$c = C/\sqrt{2E}$	-
c_m	meridional component of absolute velocity factor	$c_m = C_m/\sqrt{2E}$	
c_p	pressure coefficient	$c_p = \dfrac{\Delta p}{\frac{1}{2}\rho C^2}$	-
	pressure energy factor	$c_p = (p/\rho)/E$	-
c_{pa}	atmospheric pressure energy factor	$c_{pa} = (p_a/\rho)/E$	-
c_{pva}	vapour pressure energy factor	$c_{pva} = (p_{va}/\rho)/E$	-
c_u	peripheral velocity factor	$c_u = C_u/\sqrt{2E}$	-
c_d	drag coefficient		-
c_l	lift coefficient		-
c_r	head loss coefficient, resistance		-
e	specific hydraulic energy factor of machine, delivered to turbine or delivered by pump	$e = E/E = 1$	-
e_b	specific hydraulic energy factor at runner (impeller) blades - turbine: specific hydraulic energy delivered to blades [1]	$e_b = \dfrac{E_b}{E} = e_t + e_{rh}$	-

---

[1] Reference specific energy: for turbine: E
for pump: E_t

Symbol	Term	Definition	Units
	- pump: specific hydraulic energy delivered by blades (1)	$e_b = \frac{E_b}{E_t} = 1\text{-}e_{rb}$	-
e_{kd}	kinetic specific hydraulic energy factor, delivered to turbine draft tube	$e_{kd} = c_{\bar{1}}^{-2} - c_{\bar{1}}^{2}$	-
e_p	specific potential hydraulic energy factor	$e_p = z+p$	-
e_r	specific hydraulic energy loss factor of turbine	$e_r = 1\text{-}e_t$	-
e_{rt}	specific hydraulic energy loss factor in pump	$e_{rt} = E_r / E_t$	-
e_{rb}	specific hydraulic energy loss factor within runner / impeller blades		
	turbine	$e_{rb} = e_b\text{-}e_t$	-
	pump	$e_{rb} = 1\text{-}e_b$	-
e_{rd}	specific energy loss factor within turbine draft tube	$e_{rd} = e_{\bar{1}} - e_{\bar{I}} = e_{\bar{1}\text{-}\bar{I}}$	-
e_{rr}	specific residual velocity energy factor at turbine runner outlet	$e_{rr} = c_{\bar{1}}^{-2}$	-
e_t	specific energy transformation factor:		
	turbine : specific hydraulic energy to specific mechanical energy conversion factor	$e_t = E_t/E = 1\text{-}e_r$	-
	pump : specific mechanical energy to specific hydraulic energy conversion factor	$e_t = E/E_t$	-
e_{td}	specific hydraulic energy transformation factor (kinetic → potential) by turbine draft tube	$e_{td} = e_d - e_{rd}$	-
e_{tdd}	turbine draft tube energy efficiency factor	$e_{tdd} = e_{td}/e_d = 1\text{-}e_{rd}/e_d$	-
k	relative sand roughness	$k = K/L$	-

Symbol	Term	Definition	Units
$q^{(1)}$	discharge factor		
	delivered to turbine	$q = Q_I/Q_I = 1$	-
	delivered by pump	$q = Q_I/Q_t$	
q_ℓ	specific volumic loss factor		
	turbine	$q_\ell = Q_\ell/Q_I = 1\text{-}q_t$	-
	pump	$q_\ell = Q_\ell/Q_t = 1\text{-}q_t$	-
q_t	volumic efficiency		
	of turbine	$q_t = Q_t/Q_I = 1\text{-}q_\ell$	-
	of pump	$q_t = Q_I/Q_t = 1\text{-}q_\ell$	-
r	radius factor	$r = R/R_{\bar{1}e}$	-
u	peripheral velocity factor	$u = U/\sqrt{2E}$	-
w	relative velocity factor	$w = W/\sqrt{2E}$	-
z	position energy factor	$z = g.Z/E$	-
z_b	number of runner (impeller) blades or buckets		-
z_g	number of adjustable guide vanes		-
z_{gf}	number of fixed guide vanes		-
z_v	number of fixed stay vanes in spiral case		-
z_n	number of nozzles (impulse turbine)		-

3.4 Exceptions

Exceptions due to widespread use are listed hereafter.

Symbol	Term	Definition	Units
a	acceleration		m/s^2
f	frequency	$f = 1/T$	$s^{-1} = Hz$
g	acceleration due to gravity		m/s^2

(1) Reference discharge: for turbine: Q_I
for pump: Q_t

Symbol	Term	Definition	Units
g_n	standard gravity acceleration	$g_n = 9.80665$	m/s^2
i	incidence angle		°
n	rotational speed	$n = \omega/2\pi$	s^{-1}
p	pressure		N/m^2
p_a	atmospheric pressure		N/m^2
p_{va}	vapour pressure		N/m^2
s	needle opening (impulse turbine)		m
$\tilde{s}$	complex pulsation		s^{-1}
t	time		s

3.5 Alphabetic list of small Greek letters

Symbol	Term	Definition	Units
ζ	transient energy ratio	$\zeta = \sqrt{\frac{E}{E_{ref}}}$	-
η	efficiency		-
	of turbine	$\eta = P/P_h = \eta_m \cdot \eta_{em} = \eta_h \cdot \eta_t$	-
	of pump	$\eta = P_h/P = \eta_m \cdot \eta_{me} = \eta_h \cdot \eta_t$	-
η_{em}	mechanical efficiency of turbine (seals and bearings losses)	$\eta_{em} = P/P_m$	-
η_{me}	mechanical efficiency of pump (seals and bearings losses)	$\eta_{me} = P_m/P$	-
η_t	total mechanical efficiency		
	of turbine	$\eta_t = P/P_t$	-
	of pump	$\eta_t = P_t/P$	-
η_m	hydraulic efficiency		
	of turbine	$\eta_m = P_m/P_h = \eta_h \cdot \eta_{mt}$	-
	of pump	$\eta_m = P_h/P_m = \eta_{tm} \cdot \eta_h$	-

Symbol	Term	Definition	Units
η_{mt}	internal mechanical (disc friction losses) efficiency of turbine	$\eta_{mt} = P_m/P_t$	-
η_{tm}	internal mechanical (disc friction losses) efficiency of pump	$\eta_{tm} = P_t/P_m$	-
η_h	internal efficiency (1)		
	of turbine	$\eta_h = P_t/P_h = e_t.q_t$	-
	of pump	$\eta_h = P_h/P_t = e_t.q_t$	-
κ	valve opening	$\kappa = (Q/A_{ref})/\sqrt{2E}$	-
λ	power coefficient	$\lambda = \varphi.\psi = c_m/u^3$	-
		$\lambda_{\bar{1}e} = \dfrac{2P_h}{\rho\pi R_{\bar{1}e}^5 \omega^3}$	-
	head loss coefficient	$gH_r = \lambda L/(2DA^2).Q^2$	-
μ	complex form of STROUHAL number	$\mu = \sqrt{\tilde{C}s(\tilde{L}s+\tilde{R})}$	-
	friction coefficient	$\mu = F_r/F_p$	-
υ	specific speed	$\upsilon = \varphi_{\bar{1}e}^{1/2} / \psi_{\bar{1}e}^{3/4} = \dfrac{\omega.(Q_I/\pi)^{1/2}}{(2E)^{3/4}}$	-

Note: The values $n_q = \dfrac{60\, n(Q_I)^{1/2}}{H^{3/4}}$ or $n_s = \dfrac{60\, n\, P^{1/2}}{H^{5/4}}$ can be used as exceptions but there are so many other definitions of the specific speed that we recommend strongly the use of υ.

Symbol	Term	Definition	Units
ρ	ALLIEVI number	$\rho = Z_{ref}Q/2E_{ref}$	-
σ	THOMA number	$\sigma = \dfrac{NPSE}{E}$	-

(1) IEC Codes define $\eta_h = P_m/P_h$ for turbine and $\eta = P_h/P_m$ for pump, considering the disk friction losses and leakage losses as hydraulic losses (see footnote at page 14)

Symbol	Term	Definition	Units
φ	flow coefficient	$\varphi = c_m/u$; $\varphi_{\bar{1}e} = \dfrac{Q_I}{\pi R_{\bar{1}e}^3 \omega}$	-
	isopotential lines		-
ψ	hydraulic energy coefficient	$\psi = 1/u^2$; $\psi_{\bar{1}e} = \dfrac{2E}{\left(R_{\bar{1}e}\omega\right)^2}$	-
	stream lines		-

3.6 Exceptions

Exceptions due to widespread use are listed hereafter.

Symbol	Term	Definition	Units
α	absolute velocity angle		rad
β	relative velocity angle		rad
β_b	runner / impeller blade setting angle	$\beta_b = \beta + i$	rad
γ	guide vane opening (rotating stroke)		rad
ε	lift/drag ratio	$\varepsilon = c_l / c_d$	rad
ζ	dynamic viscosity	$\zeta = \upsilon . \rho$	kg/m/s
υ	kinematic viscosity	$\upsilon = \zeta/\rho$	m^2/s
ρ	density (volumic mass)	$\rho = M/V$	kg/m^3
φ	polar angle		rad
	phase angle		rad
ω	angular speed		rad/s
	pulsation		rad/s

Appendix II

Reference List of Organisations

AMERICAN HYDRO CORPORATION:
130 Derry, Court, York, PA 17402, USA **Product Range:** Turbines: Francis, Kaplan, pump turbine. Impellers. Propeller. **Service:** Engineering, repairs, rehabilitation: analysis, upgrade. **Contact Name:** William Colwill, V P Adv. Tec.
Tel. No. (717) 764-3587
Fax. No. 717-764-0848

BALAJU YANTRA SHALA(P) LIMITED:
P O Box 209, BID Balaju, Katmandu, Nepal **Product Range:** Turbines up to 400 KW. Penstock and accessories. Control and instrumentation equipment. **Service:** Installation, commissioning, survey and design for small hydro plants. **Contact Name:** S L Vaidya, Head of Hydro Power Dept.
Tel. No. 412379
Fax. No. 2429 BYS NP

BERGERON SA:
155 Boulevard Haussmann 75008 Paris, France **Product Range:** Hydraulic engineering. Water pumping systems. Centrifugal, mixed flow and axial flow pumps for large capacities. **Service:** Transient flow studies. Turnkey pumping stations. **Contact Name:** Jean Louis Bloch, Commercial Director
Tel. No. (33) 145619555
Telex. No. F643557 Bergron

BOETTICHER Y NAVARRO, SA:
Ctra. De Andalucia, KM 9 28021 Madrid, Spain **Product Range:** Gates, valves, manual and self-cleaning trashracks, stoplogs, penstocks, bridge cranes, sliding gates. **Service:** Consulting, engineers and design. **Contact Name:** Javier Castellanos Ybarra, Director
Tel. No. 7978200/7979000
Telex. No. 47964 Bynsae

BOMBAS ELECTRICAS SA:
Ctra.
Mieras, s/n P O Box 47, 17820 Banyoles, Girona, Spain **Product Range:** Electric pumps for fluids, centrifugal pumps, submersible pumps, multi-stage pumps, swimming pool pumps. **Service:** Swimming pools, whirl-pool baths, irrigation, household and gardens. **Contact Name:** Josep Planas, Export Manager
Tel. No. 34-72 570662
Telex. No. 57218

BOVING FOURESS PVT. LTD:
Plot. No. 2, Phase II Peenya Industrial Area Bangalore 560 058, India **Product Range:** Small hydro turbines (up to 5000 KW of both reaction and impulse type in either horizontal or veritical shaft configuration). Governors and associated auxiliaries. **Service:** Design, manufacture, supply erection & commissioning of small hydro turbine systems. Refurbishing of small hydro installations. **Contact Name:** D R Bhutani, Plant Director
Tel. No. 385734 Ex. 30
Fax. No. 0812-385176
Telex No. 0845-5086

CHENGDU HYDROELECTRIC INVESTIGATION AND DESIGN INSTITUTE OF MWREP:
Chengdu, Sichuan Province, People's Republic of China 610072 **Service:** The institute comprises five specialised divisions, computer centre and a scientific research department. The hydraulic machinery section engages in designing hydraulic mechanical parts of various hydro-

electric power stations, pressure vessels and scientific researches as well as operating personnel training. **Contact Name:** Hu Dun-Yu, Chief of Electro Mechanical Department
Tel. No. 24023
Telex. No. 60158 CSDI CN

D M W CORPORATION:
28-4 Kamata 5-chome, Ota-Ku, Tokyo 144, Japan **Product Range:**Pumps. Fans, blowers and compressors. Valves. Water treatment system. Underwater dredging robot. Jet cutter. **Contact Name:** T Fujii, Manager of Sales Department
Tel. No. 03 739-9312
Telex. No. 02466391 DGSTOK J

EBARA CORPORATION:
11-1 Haneda-Asahi-Cho Ohta-Ku, Tokyo 144, Japan **Product Range:** Pumps: centrifugal, axial mixed, submersible, motor, self-priming. Turbines: Francis, propeller, Kaplan, Pelton. Pumps. Valves. Filters. **Service:** Engineering and construction for various pumping stations. **Contact Name:** International Sales Div. **Address:** Asahi Bldg, 6-7 Ginza 6-Chome, Chuo-Ku, Tokyo 104, Japan
Tel. No. 03 572 5611 Telex. No. TOEBARA J26976

ELC-ELECTROCONSULT SPA:
20151 Milan - Via Chiabrera, N.8, Italy **Product Range:** Power plants: hydroelectric, thermoelectric, geothermal, nuclear. Transmission and distribution lines. **Service:** Master plans, feasibility studies, contract and tender documents. Construction supervision, project management training, environmental studies. **Contact Name:** G E Casartelli, Business Development Manager
Tel. No. 02-30071
Telex. No. 331103 MILELC I

ELECTROWATT ENGINEERING SERVICES LTD:
Bellerivestrasse 36, PO Box CH-8022 Zurich, Switzerland **Service:** Consulting engineering services for all types of hydraulic machinery, hydro mechanical and power plant equipment, and related fields: planning, design and specification plant and equipment rehabilitation operation and maintenance planning quality control and factory inspection execution supervision and acceptance tests technical assistance **Contact Name:** Helmut Muller, Vice President
Tel. No. 01/385 22 61
Telex. No. 815 115

ENERGOPROJEKT: CONSULTING AND ENGINEERING CO. WATER RESOURCES DEVELOPMENT DEPT.
11070 N Beograd, Lenjinov Bulevar 12, PO Box 20, Yugoslavia **Service:** Investigation, design, consulting and engineering services in waterpower, water economics and infra- structure facilities and systems. **Contact Name:** R Zivojinovic, Deputy Director
Tel. No. 011 144 491
Telex. No. 11181 ENERGO

ENERSA:
Poligono Industrial de Malpica, Calle A, Parcela 20 50016 Zaragoza, Spain **Product Range:** Turbines: Kaplan, Francis, Pelton. Gates and valves. Speed increasers. **Service:** Construction. Installation and starting. **Contact Name:** Xarier Segui Puntas, Director General

Tel. No. 76 57 07 84
Telex. No. 58 163

ERHARD-ARMATUREN:
Postfach 1280, D-7920 Heidenheim, West Germany **Product Range:** Valves: butterfly, gate, flap, tapping, diaphragm, float, check, needle, control, hydrants, non-return, air, cone outlet, flow guards, knife gate. Penstocks. **Service:** Water hammer and cavitation calculations. Valve calibration. **Contact Name:** H Hahnel, Advertising Manager
Tel. No. 7321 320196
Telex. No. 714872

FU CHUN JIANG HYDRAULIC MACHINERY WORKS:
Tonglu, Zhejiang, People's Republic of China 311504 **Product Range:**Various types of hydraulic turbines. Electric generator set. Electro-hydraulic governor. Static thyristor excitation equipment. Gates, cranes and hoists. **Contact Name:** Xu Xiao Hua
Tel. No. 181 Cable. 1381.

GE CANADA INC:
795 First Avenue, Lachine, Quebec H8S 2S8, Canada **Product Range:** Hydraulic turbines. Hydro generators. Exciters. Bus ducts. **Service:** Design, manufacturing, testing laboratory, installation, diagnostic analysis. **Contact Name:** G E Drew, Manager - Marketing
Tel. No. 514 634 3411
Telex No. 05 821673

HANG ZHOU PUMP WORKS:
Qing Tai Men Wai, Hang Zhou City, People's Republic of China **Product Range:** Pumps: submersible, centrifugal. Mechanical seals. **Service:** Various advanced equipment and facilities for manufacture and test. Research institute with more than 70 engineers and technicians. **Contact Name:** Jiang Wenhai
Tel. No. 26192

HARBIN ELECTRICAL MACHINERY WORKS (HEMW):
35 Daqing Rd, Harbin, People's Republic of China 150040 **Product Range:** Hydro generator, governor and oil pressure equipment, turbogenerator, synchronous condenser, large capacity synchronous machine, large and medium capacity AC and DC machine set, automation system and excitation system. **Contact Name:** Wu Xin Run, Deputy Chief Engineer
Tel. No. 52871
Telex. No. 87015 HEMW CN

HIDROELECTRICA ESPANOLA SA:
Heimosilla 3, 28001 Madrid, Spain **Product Range:** Production & distribution of electric energy **Contact Name:** El Secretario General
Tel. No. 34 1 4024020
Telex. No. 23786

HIDROWATT SA:
C/Aragon, No 295, 7a Planta, 08009, Barcelona, Spain
Product Range: Screening equipment. Gates, valves. Control and communication equipment. Electrical equipment. Flow meters. **Service:** Consulting services: small hydro specialists, feasibility studies, turnkey services, project & construction management. **Contact Name:** Joan Fajas, Manager
Tel. No. 93/215 02 09
Telex. No. 50439-E

HYDROART SPA:
Via Stendhal 34, 20144, Milano, Italy **Product Range:** Governors. Pump turbines. Storage pumps. Valves, Water turbines. **Contact Name:** Ing. S. Moroni, Export Sales Manager
Tel. No. 02 479 104
Telex. No. 332281 HY ART I

IMPSA INTERNATIONAL INC:
Manor Oak II - Suite 536, 1910 Cochran Rd, Pittsburgh PA 15220, USA **Product Range:** Gates. Valves. Power house cranes. Turbines: Kaplan, Francis, pit, bulb, tubular. **Contact Name:** Raul Chaluleu, Director of Marketing
Tel. No. 412 344-7003
Telex. No. 710 664 2025

DEPARTMENT OF HYDRAULIC MACHINERY. INSTITUTE OF WATER CONSERVANCY & HYDRO ELECTRIC POWER RESEARCH (IWHR):
Al FuXing Rd, Beijing, 100038, People's Republic of China **Service:** Co-operative research & development on model hydraulic turbine, pump and pump turbines. Providing optimised model of turbine for hydro power project. International acceptance test on model of hydraulic turbine. Consultancy service on hydraulic turbine R & D and laboratory technology. **Contact Name:** Wang Haian, Deputy Head
Tel. No. 86.7078
Telex. No. 22786 ITCES-CN

INSTRUMENTATION LTD A GOVERNMENT OF INDIA ENTERPRISE:
Kanjikode West, Palghat, 678 623, Kerala, India **Product Range:** Butterfly valves, safety relief valves, orifice plates, flow nozzles, process control valves, electric and pneumatic actuators, positioners, accessories and electro magnetic flow meters. **Service:** Complete design, engineering, manufacture supply and consultancy for selection, application and training. **Contact Name:** Shri R.G. Kini, General Manager
Tel. No. 24452
Telex. No. 0852 205 ILP IN

IRRIGATION AND DRAINAGE MACHINERY RESEARCH INSTITUTE OF CAAGM:
No. 1 Beishatan, Deshengmen Wai, Beijing, People's Republic of China 100083 **Product Range:** Pumps: large axial, mixed, centrifugal, deep well, submersible, hand. Equipment for sprinkler and drip irrigation. **Service:** Designs of engineering for drainage or irrigation water, sprinkler irrigation. Consultant and training management, operation and maintenance of pump station. **Contact Name:** Nie Jinhuang, Director
Tel. No. 441331 2366

ISHIKAWAJIMA-HARIMA HEAVY INDUSTRIES CO. LTD:
2-1 Ohtemachi 2-Chome, Chiyoda-Ku, Tokyo, Japan **Product Range:** Pumps: centrifugal, mixed flow, volute, volute type mixed flow and axial flow. **Service:** Pump engineering (including modification), design, manufacture, installation and after sales service. **Contact Name:** M Watanabe, Manager
Tel. No. 03-244-5483
Telex. No. J22232 IHICO

KIRLOSKAR BROTHERS LTD:
Udyog Bhavan, Tilak Road, Pune 411 002, India **Product Range:** Pumps: rotody-

namic, end suction, double suction, vertical turbine, vacuum, axial flow.Valves: sluice, check, gate, butterfly. **Service:**Irrigation, water works,fire protection, sewage handling, process industries, chemical industries, thermal power stations, mining, refineries. **Contact Name:** Mr S C Kirloskar, Managing Director
Tel. No. 58133
Telex. No. 0145-247 KBPN IN

KSB, KLEIN SCHANZI & BECKER, AG:
Joh-Klein-Str. 9, PO Box 225 D-6710 Frankenthal, West Germany **Product Range:** Centrifugal pumps for: power stations, thermal, nuclear water supply, irrigation, drainage, process industry, environmental, engineering, domestic and general industrial engineering Valves: cast iron, steel. **Contact Name:** Peter Hergt, Manager hydraulic research development
Tel. No. 06233-86-2442
Telex. No. 465211-0 KS

KUBOTA LTD:
1-3 Nihonbashi-Muromachi 3-Chome, Chuo-Ku, Tokyo 103, Japan **Product Range:** Pumps: mixed flow, double suction volute, volute type mixed flow, single suction volute, multi-stage single suction, submersible, vaneless. **Service:** Pumping system from design to turnkey completion for city water, industrial water, desalination, chemical, sewage treatment, irrigation. **Contact Name:** Mr Masaru Tsuboi, Manager Pump Export Dept.
Tel. No. 03-245-3456
Telex. No. 222-3922

KVAERNER BRUG A/S:
Kvaerner V.10, PO Box 3610 GB 0135 Oslo 1, Norway **Product Range:** Hydro turbines, valves, governors, gates. **Service:** Repair. Refurbishment and upgrading. Training. **Contact Name:** Knut Pettersen, Sales Manager
Tel. No. 472 666020
Telex. No. 71650 KBN

LABEIM (LABORATORIOS DE EUSAYOS E INVESTIFACING INDES):
C/Westa De Olabeaga 16, 48013 Bilbao, Spain **Service:** Research and development. Model test on turbo- machinery **Contact Name:** Andomi Larreategui, Head of hydraulic machinery section
Tel. No. 34 4 4419300

MECANICA DE LA PENA SA:
Aita Gotzon No. 37, 48610 Urduliz (Vizcaya) Apartado 1.308 - 48080 Bilbao, Spain. **Product Range**: Turbines: Kaplan, bulb, Francis, Pelton, pump turbines. Transfer pumps. Valves: ball, butterfly. Coefferdams, penstock. Small hydraulic plants. **Service:** Design, construction and installation. Maintenance. **Contact Name:** Jesus Urquidi, Director
Tel. No. 4 676.10.11
Fax. No. 4 676.28.81
Telex No. 3301 MELPE-E

MITSUBISHI HEAVY INDUSTRIES LTD:
5-1 Marunouchi 2-Chome, Chiyoda-Ku, Tokyo, Japan **Product Range:**Water turbine. Pump turbine. Compressor. Pump and mechanical drive turbine. **Service:** Power plant, process and industrial plant, water works, sewage irrigation, flood control. **Contact Name:** H Nish-

ioka, Manager
Tel. No. 03-212-3111
Telex. No. J22443 HISHIJU

NEYRPIC:

75 Ave General Mangin BP 75-38041 Grenoble Cedex, France **Product Range:** Large water turbines from 15 to 1000 MW. Small water turbines from 0.1 to 15 MW. Standardised mini- turbines from 0.1 to 5 MW. Spherical and butterfly valves. Gates. Automatic systems and speed governors. **Service:** Rehabilitation and modernisation of power plants. **Contact Name:** M F de Vitry, Chairman M Y Couchet, Deput General Manager
Tel. No. 76 39 30 00
Telex. No. 320750 F

NORTH CHINA INSTITUTE OF WATER CONSERVANCY AND POWER:

Handan, Hebei, People's Republic of China 056021 **Service:** Internal flow analysis of hydraulic machinery. Cavitation mechanism and detection of hydraulic machinery. Hydraulic transient computer simulation of hydraulic machinery. Computer monitoring and control of hydraulic machinery. CAD of hydro-power station. **Contact Name:** Z Y Liu, Professor
Tel. No. 25951

O'HAIR GROUP:

7 Victoria Terrace, Bowen Hills, Brisbane, Queensland, Australia **Product Range:** Pumps: centrifugal, axial, mixed-flow, gear, screw, piston/plunger. Water turbines. Power recovery turbines. Blowers. Turbo-compressors. Sewage and fish hatchery aerators. Mechanical seals. **Service:** Consultants for specifications, design, installation and testing all hydraulic and fluid machinery and associated civil facilities. Can arrange supply of machinery and installation and testing, agencies, joint ventures, import/export finance. Spare and repair & re-designs submersible motor units. CAD software. **Contact Name:** J Brian O'Hair, Director
Tel. No. 61-7-2528001
Fax. No. 61-7-1257
Telex No. AA140472 Attention O'Hair Group

QIAN JIANG PUMP WORKS:

Xiao Shan City, Zhejiang Province, People's Republic of China **Product Range:** ISO 2858 Standard: IB IS type single-stage centrifugal pump. High quality. High efficiency. **Contact Name:**
Tel. No. 22328

RADE KONCAR:

Fallerovo Setaliste 22 41000 Zagreb, Yugoslavia **Product Range:** Generators and transformers. Electrical equipment for power plants. Control and instrumentation for power plants. **Service:** Design, production, erection, contracting of complete electrical equipment for power plants. **Contact Name:** Kozina Josip, Sales Manager
Tel. No. 041 316726
Telex. No. 21-159, 21-104

SHANGHAI PUMP WORKS:

Jiang Chuan Road, Min Hang, Shanghai, 200240 People's Republic of China **Product Range:** Pump: auxiliary for nuclear power station, boiler feed, circulating, condensate, drainage for power station, process, hot water circulation (West Germany KSB licence) for petrochemical enterprises, big variable or invariable

mixed and axial flow for irrigation, sewage for city engineering (USA DRESSER licence); 50 or 60 HZ marine (West Germany KSB licence). ISO standard mechanical seal and welded metal bellow seal (USA Sealol licence). **Contact Name:** Gu Xian
Tel. No. 358191
Telex. No. 33546 SMUDC CN

SHI SHOU PUMP WORKS:
Zhong Shan Road, Shi Shou City, Hubei Province, Peoples's Republic of China **Product Range:** Pumps: impurity, screw, mine drainage multi-stage, boiler feed, single-stage centrifugal clean water. **Service:** Technical services offered: development of erosion-resistant materials, design of special pumps for consumer. **Contact Name:** Sun Dong He
Tel. No. 2391

SHIN NIPPON MACHINERY CO. LTD:
Seio Bldg, 1-28 Shiba 2-Chome, Minato-Ku, Tokyo 105, Japan **Product Range:** Pumps: axial flow, mixed flow, centrifugal, ring section, barrel, liquid ring vacuum, slurry and screw. **Service:** Irrigation, water treatment, water supply, boiler feed water, process pump and chemical pump. **Contact Name:** Mr T Ikeshita, General Manager
Tel. No. 03 454-1412
Telex No. 242 4302 SNZOKIJ

SIGMA KONCERN SE:
Kosmonautu 6, 772 23 Olomouc
Czechoslovakia **Product Range:** Pumps. Valves. Irrigation, water treatment. Waste water purification. **Contact Name:** J Holada, Director
Tel. No. 02 235 77 48
Fax. No. 02 265616
Telex No. 12 12 05 C

SOCIETE HYDROTECHNIQUE DE FRANCE SHF:
199 rue de Grenelle, 75007 Paris, France **Product Range:** Scientific association concerned with the development of knowledge and techniques for the engineering of fluids and water management. **Contact Name:** P Constans
Tel. No. 1 47051337

STORK POMPEN BV:
Lansinkesweg 30, PO Box 55, 7550 AB Hengelo, The Netherlands **Product Range:** Centrifugal pumps for: process and petrochemical duties, irrigation and drainage, drydocks, industrial applications, drinking water supply, power stations. **Service:** Industrial measurements and consultancy on pumps and pumping systems. Facilities for pump tests at works (model) and at site. **Contact Name:** N Van Vuren, Manager Sales Department
Tel. No. 074 404000
Fax. No. 074 425696
Telex No. 44324+ SPH +

SULZER-ESCHER WYSS LTD:
Escher Wyss Platz, 8023 Zurich, Switzerland **Product Range:** Equipment for hydro-electrical power plants. All types and sizes of water turbines, Storage pumps. Governors. Pump turbines. Shut-off valves. Penstock and manifolds **Service:** Maintenance, overhaul, modernisation, modification **Contact Name:** Ch. Habegger, Asst. Vice-President
Tel. No. 1-278-22-11
Telex No. 822 900 11 SECH

TEXMO INDUSTRIES:
MTP Rd, G N Mills, P O Coimbat-

ore 641 029, Tamil Nadu State, India **Product Range:** Pumps: irrigation and agricultural, shallow and deep well, borehole, household water supply and sewage. Sprinkler supply systems. **Contact Name:** Mr C Balaram, Marketing Manager
Tel. No. 33455

THOMPSONS, KELLY & LEWIS LTD:

26 Faigh Street, Mulgrave, Victoria 3170, Australia **Product Range:** Pumps: axial flow, propellor, boiler feed, centrifugal, mixed flow, multi-stage, vertical turbine, vertical inline, concrete volute. **Contact Name:** A Grage, Sales Director
Tel. No. 03 562 0744
Telex No. AA31365

TORISHIMA PUMP MFG. CO. LTD:

1-1-8 Miyata-cho, Takatsuki City, Osaka, Japan **Product Range:** Pumps. Mechanical seals. Cast products **Service:** Installation of pumps and their relative equipments. Engineering of pumping stations and after sales service. **Contact Name:** Mitsuma Kitajima, Manager General Affairs Dept.
Tel. No. 0726 95 0551
Telex No. 5336568 TORIPU J

TOSHIBA CORPORATION:

1-1 Shibaura 1-Chome Minatoku, Tokyo 105, Japan **Product Range:** Hydraulic turbines. Pump turbines. Inlet valves. Governors. **Service:** Engineering design, supply, finance **Contact Name:** Mr Hiroji Morimoto, Senior Manager
Tel. No. 03 457 4828 31
Telex No. J22587

DIVISION OF HYDRAULIC MACHINERY, DEPARTMENT OF HYDRAULIC ENGINEERING, TSINGHAU UNIVERSITY:

Beijing, People's Republic of China **Service:** Internal flow investigation and analysis on hydraulic machinery. Cavitation and flow-induced vibrations. Two-phase flow. Facilities in the hydraulic machinery laboratory include two closed-circuit test stands for turbine and pump, open-flow test stand and a slurry pump test stand. **Contact Name:** Lin RuChang, Professor Dept. of Hydraulic Engineering
Tel. No. 282451-2251
Telex 22617 QHTSC CN Fax. 86-01-2562768

VEVEY ENGINEERING WORKS LTD:

CH 1800 Vevey, Switzerland **Product Range:** Vevey specialises in the field of energy (hydro-power products) high, medium and low head water turbines, e.g. Pelton, Francis, Kaplan and bulbs, reversible pump turbines, single- and multi-stage and isogyre-type. **Contact Name:** Michon, Sales Manager
Tel. No. 4121 9257111
Telex No. 451104 VEYCH

WATER CONSERVANCY AND WATER POWER ENGINEERING RESEARCH INSTITUTION:

Dalian Institute of Technology, Dalian, People's Republic of China 116024 **Service:** Vibration problems of turbo- generator units, turbine, Penstock pumps. Hydropower house, pump house, equipment of hydropower station and pumping stations. **Contact Name:** Dong Yu Xin, Professor

Tel. No. 471511-519
Cable No. 7108

XIN CHANG SPRINKLER IRRIGATION WORKS:
Cheng Guan, Xin Chang, Zhejiang Province, People's Republic of China **Product Range:** Sprinkler irrigation systems. Self priming pumps. Centrifugal pumps. Sprinklers. Low and high temperature oil pumps. **Contact Name:**
Tel. No. 23600-22545

ZHEJIANG RESEARCH INSTITUTE OF MECHANICAL SCIENCE:
122 Laodong Rd, Hang Zhan, Zhejiang Province, People's Republic of China 310002 **Service:** Over 30 year's research experience in pumps, hydraulics and automatic control etc. The test centre is equipped with fully computerised, high precision test stands. **Contact Name:**
Cheng Ji Zhong
Tel. No. 27778

Appendix III

List of Educational Establishments

AUSTRALIA

UNIVERSITY OF ADELAIDE

Department of Civil Engineering, Adelaide, S.A. Dr. A. Simpson. Water column separation

UNIVERSITY OF MELBOURNE

Department of Civil & Argicultural Engineering, Grattan St., Parkville, Victoria 3052. Tel: (03) 344 6789. Dr. H.R. Graze. Waterhammer, Air chambers, Water column separation

UNIVERSITY OF QUEENSLAND

Department of Civil Engineering, St. Lucia, Queensland. Prof. C. Apelt. Waterhammer

UNIVERSITY OF TASMANIA

Department of Civil and Mechanical Engineering, Hobard, Tasmania. Dr. S. Montes. Surge tanks

The following is a supplementary list of Engineering Faculties in Australia:

UNIVERSITY OF ADELAIDE

Dept. of Electrical and Electronic Engineering, Tel: (08) 228 5277: Fax: (08) 224 0464. Dr. Donald W. Griffin

AUSTRALIAN MARITIME COLLEGE

School of Engineering, Tel: (003) 260 757: Fax: (003) 260 717. Mr. John J. Seaton, Head

SOUTH AUSTRALIAN INSTITUTE OF TECHNOLOGY

Faculty of Engineering, Tel: (08) 343 3219: Fax: (08) 349 6939. Prof. K.J. Atkins, Dean

UNIVERSITY OF WESTERN AUSTRALIA

Faculty of Engineering, Tel: (09) 380 3105/3106: Fax: (09) 382 4649. Prof. Alan R. Billings, Dean

BALLARAT COLLEGE OF ADVANCED EDUCATION

Faculty of Engineering, Tel: (053) 339 100: Fax: (053) 339 545. Mr. Derek Woolley, Dean

BENDIGO COLLEGE OF ADVANCED EDUCATION

School of Engineering, Tel: (054) 403 339: Fax: (054) 403 477. Dr. Tim Dasika, Head

CANBERRA COLLEGE OF ADVANCED EDUCATION

Dept. of Electronics and Applied Physics, Tel: (062) 522 515: Fax: (062) 522 999. Dr. Paul Edwards, Head

CAPRICORNIA INSTITUTE OF ADVANCED EDUCATION

Faculty of Engineering, Tel: (079) 360 543: Fax: (079) 361 361. Mr. Frank Schroder, Dean

CHISHOLM INSTITUTE OF TECHNOLOGY

Faculty of Engineering, Tel: (03) 573 2162: Fax: (03) 572 1298. Dr. Brian Jenney, A/g Dean

CURTIN UNIVERSITY OF TECHNOLOGY

Faculty of Engineering, Tel: (09) 350 7093: Fax: (09) 458 4661. Assoc. Prof. Lachlan Millar, Dean

DARLING DOWNS INSTITUTE OF ADVANCED EDUCATION

School of Engineering, Tel: (076) 312 527:

Fax: (076) 301 182. Dr. Tom Ledwidge, Dean

DEFENCE ACADEMY
Dept. of Mechanical Engineering, Tel: (062) 688 274. Dr. Ray Watson

FOOTSCRAY INSTITUTE OF TECHNOLOGY
Faculty of Engineering, Tel: (03) 688 4244: Fax: (03) 689 4069. Mr. Ivan A. Bellizzer, Dean

GIPPSLAND INSTITUTE OF ADVANCED EDUCATION
School of Engineering, Tel: (051) 220 461: Fax: (051) 222 876. Dr. Ken Spriggs, Head

JAMES COOK UNIVERSITY
Faculty of Engineering, Tel: (02) 697 5001. Prof. Noel L. Svensson, Dean

ROYAL MELBOURNE INSTITUTE OF TECHNOLOGY
Faculty of Engineering, Tel: (03) 660 2523: Fax: (03) 663 2764. Dr. Bill Carroll, Dean

UNIVERSITY OF MELBOURNE
Faculty of Engineering, Tel: (03) 344 6619: Fax: (03) 347 1343. Assoc. Prof. Bill W.S. Charters, Dean

MONASH UNIVERSITY
Faculty of Engineering, Tel: (03) 563 3400: Fax: (03) 565 3409. Prof. Peter Le Darvall, Dean

UNIVERSITY OF NEWCASTLE
Faculty of Engineering, Tel: (049) 685 395: Fax: (049) 674 946. Prof. Alan W. Roberts, Dean

UNIVERSITY OF NSW
School of Chemical Engineering & Industrial Chemistry, Prof. Chris J.D. Fell

QUEENSLAND INSTITUTE OF TECHNOLOGY
Faculty of Engineering, Tel: (07) 223 2415: Fax: (07) 229 1510. Dr. John J.B. Corderoy, Dean

UNIVERSITY OF QUEENSLAND
Faculty of Engineering, Tel: (09) 380 3105/3106: Fax: (09) 382 4649. Prof. Alan R. Billings, Dean

SWINBURNE INSTITUTE OF TECHNOLOGY
Faculty of Engineering, Tel: (03) 819 8282: Fax: (03) 819 5454. Dr. Murray M Gillin, Dean

UNIVERSITY OF TECHNOLOGY, SYDNEY
Faculty of Engineering, Tel: (02) 218 9272: Fax: (02) 282 2498. Dr. Ken Faulkes, Dean

UNIVERSITY OF TECHNOLOGY, SYDNEY
Faculty of Engineering, Tel: (02) 218 9272: Fax: (02) 281 2498. Prof. J. Paul Gostelow, Dean

UNIVERSITY OF SYDNEY
Faculty of Engineering, Tel: (02) 692 2329: Fax: (02) 692 2012. Prof. John R. Glastonbury, Dean

TASMANIAN STATE INSTITUTE OF TECHNOLOGY
School of Engineering, Tel: (003) 260 576. Mr. Peter Crewe, Head

UNIVERSITY OF TASMANIA
Faculty of Engineering, Tel: (002) 202 129: Fax: (002) 202 186. Mr. Eric Middleton, Dean

UNIVERSITY OF WOLLONGONG
Faculty of Engineering, Tel: (042) 270 491: Fax: (042) 270 477. Prof. Brian H. Smith, Dean

CANADA

BC HYDRO HEAD OFFICE AND GENERAL ADMINISTRATION
970 Burrard, Vancouver, BC V6Z 1Y3. Tel: (604) 663 2212: Fax: (613) 663 3597. Mr H Lang, P. Eng.

CANADIAN DEPARTMENT OF ENERGY, MINES AND RESOURCES
7th Floor, 580 Booth Street, Ottawa, Ont. K1A 0E4. Tel: (613) 996 6119: Fax: (613) 996 6424. Mr T Tung, P. Eng. Operation Engineer Hydro

CENTRE DE RECHERCHES DE L'HYDRO-QUÉBEC
Varennes, PQué, Boîte Postale 1000. Tel: (514) 652 8090: Fax: 05-2677486. Dr B Dubé, Ing.

2687 Dautrive, Ste Foy, Qué, G1W 2C8. Tel: (418) 651 2189: Fax: (418) 656 6425. Dr H Netsch, Ing

ECOLE POLYTECHNIQUE
Dept. de Mathématiques Appliqués, Université de Montréal, Box 6079, Montréal, P.Q. H3C 3A7. Tel: (514) 340 4639: Fax: 340 4440. Prof. Dr. R Camerero

HYDRO-QUÉBEC
5655 Rue Marseille, Montréal, Qué, H1N 1J4. Tel: (514) 251 7043: Fax: (514) 251 7117. M J M Lévesque, Ing., Ingénieur Specialiste, Expertises Techniques

LABORATOIRE HYDRAULIQUE
Montréal Rd, Ottawa, Ont, K1A 0R6. Tel: (613) 993 6650: Fax: (613) 952 7679. M Thierrý D Faure, Ing

2368 Chemin des Foulons, Sillery, Qué, G1T 1X4. Tel: (418) 656 6394: Fax: (418) 656 6425. M Y Jean, Ing M.Sc., Spécialiste

UNIVERSITÈ LAVAL
Dept de Génie méçanique Québec, Qué G1K 7P4. Tel: (418) 656 5359: Fax: (418) 656 5902. Professeure Dr C Deschaînes, Ing

CHINA

BEIJING INSTITUTUE OF ECONOMIC MANAGEMENT OF WATER POWER ENGINEERING
Postgraduate Division, Xizhimen Wei, Ziahuyuan, Beijing. 100044 Tel: 8414894. Prof. Duan Chang Guo, Prof. Shen Zhenju. Silt erosion; Hydraulic transients of hydraulic machinery; Measurement technology of hydraulic system

BEIJING UNIVERSITY OF ARGRICULTURAL ENGINEERING
Department of Hydraulic Engineering, East Qinghua Road, Beijing, 100083 Tel: 2017622. Prof. Liu Shankun - Director of Teaching and Research Division. Turbodynamics and design of pumps; Transients of turbine; Sprinkler irrigation facility and systems

GEZHOBA INSTITUTUE OF HYDRAULIC ENGINEERING
Water Power Engineering Department, Yichang City, Hubei Province. Tel: 22011-3615. Assoc. Prof. Zeng Dean. Hydraulic machinery. Selection, design

and operation of hydraulic turbines (including auxiliary equipment).

HUAZHONG UNIVERSITY OF SCIENCE AND TECHNOLOGY
Department of Electrical Engineering, Wuhan, Hubei, 430074. Tel: Wuhan 870154: Telex: 40131 HBU. Prof. Tan Yuechan. Turbodynamics: Optimisation design theory; Cavitation, vibration, structure and strength design of hydraulic machinery.

JIANGSU INSTITUTE OF TECHNOLOGY
Department of Power Machinery, Zhenjiang City, Jiangsu Province, 212013. Tel: 34071: Telex: 2894. Prof. Cha Sen. Hydraulic performance of hydraulic machinery; Two-phase flow pump; Erosion resistant material, Sprinkler and drip irrigation system; Pump station and irrigation and drainage engineering.

NORTH CHINA INSTITUTE OF WATER CONSERVANCE AND POWER
Department of Power Engineering, Handan, Hebei, 056021. Tel: 22775. Assoc. Prof. Liu Z.Y. - Instrumentation, Prof. Li Shengcai - Hydraulic Machinery. Internal flow of turbines; Cavitation mechanism and its detection; CAD of hydropower station; Hydraulic transients of turbines; Computer control of power station.

NORTH-EAST COLLEGE OF HYDRAULIC AND ELECTRICAL ENGINEERING
Department of Electrical Engineering, Changchun City, Julin Province. Tel: 55991-95. Mr. Shi Zhensheng - Lecturer. CAD of selection design of hydraulic machinery; Test rig and data logger system

NORTH-WEST UNIVERSITY OF AGRICULTURE
Hydraulic Engineering Department, Yangling City, Shaanxi Province. Professor Yang Songpu. Operation, cavitation and erosion resistance of hydraulic machinery

TSINGHUA UNIVERSITY
Department of Hydraulic Engineering, Beijing, 400084 Tel: 282451: Telex: 22617 QHTSC CN. Prof. Lin Ruchang. Internal flow of hydraulic machinery; Regulation and cavitation; Erosion and two-phase flow of hydraulic machinery; Cavitation and performance test of turbine and pump; Solid-liquid two-phase flow pumps

SHAANXI INSTITUTE OF MECHANICAL ENGINEERING
College of Hydro-electrical Engineering, Xian City, Shaanxi Province, 710048. Tel: 721236; Cable: Xian 8503. Assoc. Prof. Chen Jiamo. Cavitation and regulation system of turbine automation of hydro power station

SICHUAN INSTITUTE OF TECHNOLOGY
Power Engineering Department, Changdu City, Sichuan Province, 611744. Tel: 68271/68287: Cable: 6922. Prof. Dutong - Prof. of Hydraulic Machinery. Cavitation, erosion and corrosion of hydraulic machinery; Development of the turbine and pump; Feasibility study of small pumping storage power station

WUHAN INSTITUTE OF HYDRAULIC AND ELECTRICAL ENGINEERING
Research Institute of Irrigation and Drainage Engineering, Luojia Hill, Wuchang City, Hubei Province, 430072. Tel: Wuhan 812212-280: Telex:

40170 WCTEL CN

YUNNAN INSTITUTE OF TECHNOLOGY

Department of Electrical Engineering, Kunming City, Yunnan Province. Tel: 29031. Assoc. Prof. Huang Fenjie. Impulse turbine; Operation of high head Francis turbine.

EGYPT

UNIVERSITY OF ALEXANDRIA

Faculty of Engineering, Mechanical Dept. Alexandria, Horria Str. Prof. Esam Salem. Fluid mechanics and lubrication technology.

UNIVERSITY OF ASSIAT

Faculty of Engineering, Mechanical Dept. Assiat. Prof. A. Huzzien. Fluid mechanics and pumping machinery

UNIVERSITY OF CAIRO

Dept. of Mech. Power Eng, Giza. Prof. A. Mobarak. Experimental and theoretical analysis of flow in turbomachines

UNIVERSITY OF EIN-SHAMS

Dept. of Mech. Eng, Abbasia, Cairo. Prof. M. El Sebaie. Fluid mechanics and hydraulic machines

FRANCE

ENSEEIHT HYDRAULIC LABORATORY

2 Rue Camichel, 31071 Toulouse. Tel: (61) 588200. Fax: (61) 620976. Telex: 530171. Prof. J Gruat, Prof. C Thirriot, Dr.P Crausse

ENSHMG DOMAINE UNIVERSITY, GRENOBLE

38402 St. Martin d'Heres, Grenoble. Tel: (76) 825000: Fax: (76) 825001: Telex: 980668 Hymegre. Prof. P Bois, Prof. P Trompette

INDIA

INDIAN INSTITUTE OF SCIENCE

Department of Civil Engineering, Bangalore 560 012, India. Tel: (0812) 344411 Ex. 245. Telex: (0845) 8349. Rama Prasad. Research and teaching - hydraulic turbines and pumps

INDIAN INSTITUTE OF TECHNOLOGY, DELHI

Department of Applied Mechanics, Hauz Khas, New Delhi 110 016, India. Tel: 653458: Telex: 31-3687 IIT IN. Prof. V. Seshadri. Under graduate and post graduate testing; Research in analysis of flow through hydraulic machines. Performance tests on hydraulic machinery; Microhydel devices and their development; Flow instrumentation and calibration.

INDIAN INSTITUTE OF TECHNOLOGY

Hydroturbines Laboratory, Madras 600 036, India. Tel: 415342 Ex 274. Telex: 041 21062. Prof. H.C. Radha Krishna, Prof. M. Ravindran, Prof. P.A. Aswathanarayana, Dr. V. Balabaskaran. Flow research through hydroturbomachines; Cavitation, Vibration and Noise; Fully reversible axial flow hydroturbo- machines, bulb turbines, Allied research for hydraulic machines

MOTILAL NEHRU REGIONAL ENGINEERING COLLEGE

Mech. Eng. Department, Allahabad

211004 (U.P.) India. Tel: 53520: Telex: 0540 269 MNEC IN. Dr. Y.V.N. Rao - Principal, Dr. R. Yadav, Professor, Dr. S.K. Agrawal, Reader, Dr. V.K. Nema, Reader. Theoretical and experimental investigation in the following fields: centrifugal pumps, centrifugal and axial flow compressors

ITALY

UNIVERSITY OF BOLOGNA
Dept. of Mechanical Engineering, Viale Risorgimento, 2 - 40136. Tel: 051 582162/Bologna 582163. Prof. C. Bonacini

UNIVERSITY OF CAGLIARI
Dept. of Mechanical Engineering, Piazza D'Armi - 09100 Cagliari. Tel: 070 2900071/72/73. Prof. R. Masala

UNIVERSITY OF FIRENZE
Dept. of Civil Engineering, Via. S. Marta, 3 - 50139 Firenze. Tel: 055-47961. Prof. G. Federici

UNIVERSITY OF FIRENZE
Dept. of Enegetics, Via S. Marta, 3 - 50139 Firenze. Tel: 055-47961. Prof. E. Carnevale

UNIVERSITY OF GENOVA
Dept. of Hydraulics, Via Montallegro, 1 - 16145 Genova. Tel: 010-303416. Prof. F. Siccardi

UNIVERSITY OF GENOVA
Dept. of Mechanical Engineering, Via Montallegro, 1 - 16145 Genova. Tel: 010-303416. Prof. O. Acton, Prof. A. Satta

POLITECHNIC OF MILANO
Dept. of Energetics, Piazza Leonardo da Vinci, 32 - 20133 Milano. Tel: 23991. Prof. E. Macchi, Prof. F. Bassi

UNIVERSITY OF PADOVA
Dept. of Mechanical Engineering, Via Venezia, 1 - 35100 Padova. Tel: 049-8071988. Prof. G. Ventrone, Prof. V. Quaggiotti

UNIVERSITY OF PAVIA
Dept. of Hydraulics, Piazza Leonardo da Vinci - 27100 Pavia. Tel: 0382-31325/0382-21636. Prof. R. Sala

UNIVERSITY OF POTENZA
Dept. of Hydraulics, Via N. Sauro, 85 - 85100 Potenza. Tel: 0971-334611. Prof. B. De Bernardinis

UNIVERSITY OF ROMA
Dept. of Mechanical and Aeronautical Eng. Via Eudossiana, 18 - 00184 Roma. Tel: 06-4687314. Prof. C. Caputo, Prof. U. Pighini

JAPAN

EHIME UNIVERSITY
Dept. of Mechanical Engineering, 3 Bunkyo-cho, Matsuyama-shi 790. Tel: (0899) 24 7111. Prof. K. Ayukawa. Internal flow of turbomachines

UNIVERSITY OF FUKUI
Dept. of Mechanical Engineering, 3-9-1 Bunkyo, Fukui-shi 910. Tel: (0776) 23 0500. Prof. I. Ashino. Numerical analysis of flow in turbomachines

HIROSHIMA INSTITUTE OF TECHNOLOGY
Dept. of Mechanical Engineering, 725 Miyake, Itsukaishi-cho, Saeki-ku, Hiroshima 731-51. Tel: (0829) 21 3121. Prof. H. Murai. Cavitation, min-hydroturbines

UNIVERSITY OF HOKKAIDO
Dept. of Mechanical Engineering, Kita 13 Jo, Nishi 8 Chome, Kita-ku, Sapporo-shi 060. Tel: (011) 711 2111. Prof. M. Kiya. Numerical analysis and measurement of flow in turbomachines

KANAGAWA INSTITUTE OF TECHNOLOGY
Dept. of Mechanical Engineering, 1030 Hagino, Atsugi-shi 243-02. Tel: (0462) 41 1211. Prof. S. Akaike. Mini-hydroturbines, flow analysis of hydraulic turbines

KANAGAWA UNIVERSITY
Dept. of Mechanical Engineering, 3-22 Rokkakubashi, Kanagawa-ku, Yokohama-shi 221. Tel: (045) 481 566. Prof. T. Ida. Cavitation, similarity law of pumps and hydraulic turbines

KANAZAWA UNIVERSITY
Dept. of Mechanical Engineering, 2-40-20 Kotatsuno, Kanazawa-shi 920. Tel: (0762) 61 2101. Prof. S. Miyae. Performance of pumps

KEIO UNIVERSITY
Dept. of Mechanical Engineering, 3-14-1 Hiyoshi-sho, Kohoku-ku Yokohama-shi 223. Tel: (044) 63 1141. Prof. T. Ando. Analysis and measurement of flow in turbomachines

KOBE UNIVERSITY
Dept. of Mechanical Engineering, 1-1 Rokkadai-cho, Nada-ku, Kobe-shi 657. Tel: (078) 881 1212. Prof. T. Iwatsubo. Two-phase flow, rotodynamic problems of turbomachines

KYOTO UNIVERSITY
Dept. of Mechanical Engineering, Yoshida-Honmachi, Sakyo-ku, Kyoto-shi 606. Tel: (075) 751 2111. Prof. H. Akamatsu. Cavitation and two-phase flow phenomena

KYUSHU INSTITUTE OF TECHNOLOGY
Dept. of Mechanical Engineering II, Sensui-cho, Tobata-ku, Kitakyushu-shi 804. Tel: (093) 871 1931. Prof. M. Nishi. Flow measurement by small 5-hole probes, unsteady performances

UNIVERSITY OF KYUSHU
Dept. of Mechanical Engineering, 6-10-1 Kakazaki, Higashi-ku, Fukuoka-shi 813. Tel: (092) 641 1101. Prof. Y. Takamatsu. Numerical analysis and measurement of flow in turbomachines, cavitation

MIE UNIVERSITY
Dept. of Mechanical Engineering, 1515 Kamihama-cho, Tsu-shi, Mie-ken 514. Tel: (0592) 32 1211. Prof. Y. Shimizu. Flow in draft tube, self-priming pumps, jet pumps

MURORAN INSTITUTE OF TECHNOLOGY
Dept. of Mechanical Engineering II, 27-1 Mizumoto-cho, Muroran-shi 050. Tel: (0143) 44 4181. Prof. T. Watanabe. Cavitation, performance of pumps and hydraulic turbines

NAGOYA INSTITUTE OF TECHNOLOGY
Dept. of Mechanical Engineering, Gokiso-cho, Showa-ku, Nagoya-shi 466. Tel: (052) 7322111. Prof. Y. Yamada. Disk friction loss, friction on rotating cones

NAGOYA UNIVERSITY
Dept. of Mechanical Engineering, Furo-cho, Chijusa-ku, Nagoya-shi 464. Tel: (052) 781 5111. Prof. K. Kikuyama. Two-phase flow in pumps, turbulent boundary layer and secondary flow

UNIVERSITY OF OSAKA
Faculty of Eng., Dept of Mechanical Eng. 2-1 Yamodaoka, Suita-shi, Osaka 565. Tel: (06) 877 5111. Prof. Y. Miyaka. Numercial analysis of flow in turbo-machines, turbulence modelling

UNIVERSITY OF OSAKA
Faculty of Eng. Science, Dept. of Mechanical Eng. 1-1 Machikaneyana-cho, Toyonaka-shi 560. Tel: (06) 844 1151. Prof. Y. Tsujimoto. Rotating stall, unsteady phenomena in centrifugal impellers

SOPHIA UNIVERSITY
Dept. of Mechanical Engineering, 7 Kioi-cho, Chiyoda-ku, Tokyo-to 102. Tel: (03) 238 3305. Prof. K. Takahashi. Slurry pumps, unsteady radial and axial thrust of pumps

TOHUKU UNIVERSITY
Dept. of Mechanical Engineering, Aza-Aoba, Aramaki, Sendai-shi 980. Tel: (0222) 22 1800. Prof. H. Daiguji. Numerical analysis of flow in turnbomachines, cavitation

TOHOKU UNIVERSITY
Institute of High Speed Mechanics, 2-1-1 Katahira, Sendai-shi 980. Tel: (0222) 27 6200. Prof. R. Ohba. Cavitation and supercavitation, performance of hydraulic turbines

TOKAI UNIVERSITY
Dept. of Production Engineering, 1117 Kita-Kaname-cho, Hiratsuka-shi 259-12. Tel: (0463) 58 1211. Prof. Y. Nakayama. Flow visualisation of flow in pumps and hydraulic turbines, cavitation

UNIVERSITY OF TOKUSHIMA
Dept. of Mechanical Engineering, 2-1 Mishima-cho, Janjo, Tokushima-shi 770. Tel: (0886) 23 2311. Prof. T. Nakase. Cross-flow turbines, tidal and wave power turbines

TOKYO DENKI UNIVERSITY
Dept. of Mechanical Engineering, 2-2 Nishiki-cho, Kanda, Tokyo-to 101. Tel: (03) 294 1551. Prof. Y. Hosoi. Draft tube surge, performance of pumps and hydraulic turbines

TOKYO METROPOLITAN UNIVERSITY
Dept. of Mechanical Engineering, 2-1-1 Fukazawa, Setagaya-ku, Tokyo-to 158. Tel: (03) 717 0111. Prof. H. Kato. Flow in diffusers, turbulent boundary layer of turbomachines

TOKYO INSTITUTE OF TECHNOLOGY
Dept. of Mechanical Engineering, 2-12-1 Ohokayama, Meguro-ku, Tokyo-to 152. Tel: (03) 726 1111. Prof. R. Yamane. Cavitation, internal flow in turbomachines

SCIENCE UNIVERSITY OF TOKYO
Dept. of Mechanical Engineering, 2641 Higashi-kameyama, Yamazaki, Noda-shi 278. Tel: (0471) 24 1501. Prof. H. Maki. Flow analysis and measurement of radial diffusers, flowmeters

UNIVERSITY OF TOKYO
Dept. of Mechanical Engineering, Hongo 7-3-1, Bunkyo-ku, Tokyo 113. Tel: (03) 812 2111. Prof. H. Ohashi. Cavi-

tation, transient phenomena in conduit, two-phase cascade flow

UNIVERSITY OF TOKYO

Institute of Industrial Science, 7-22-1 Roppongi, Minato-ku, Tokyo-to 153. Tel: (03) 294 1551. Prof. T. Kobayashi. Numerical analysis and flow visualisation of flow in turbomachines

TOYOTA INSTITUTE OF TECHNOLOGY

Dept. of Mechanical Engineering, 2-12 Hisakata, Tenpaku-ku, Nagoya-shi 468. Tel: (052) 802 1111. Prof. S. Murata. Numerical analysis and measurement of flow in impellers

UNIVERSITY OF TSUKUBA

Institute of Engineering Mechanics, 1-1-1 Tennodai, Sakura-mura, Niihari-gun, Ibaraki 305. Tel: (0298) 53 2111. Prof. H. Tahara. Performance of centrifugal pumps, fluid force on whirling impellers

WASEDA UNIVERSITY

Dept. of Mechanical Engineering, 3-4-1 Ohkubo, Shinjuku-ku, Tokyo-to 160. Tel: (03) 209 3211. Prof. K. Yamamoto. Flow instability of pumping system, similarity law of suction reservoir

YOKOHAMA NATIONAL UNIVERSITY

Department of Mechanical Engineering II, 2-31-1 Oooka, Minami-ku, Yokohama-shi 233. Tel: (045) 335 1451. Prof. T. Toyokura. Study of annular cascades, internal flow and axial thrust of pumps

NORWAY

THE NORWEGIAN INSTITUTE OF TECHNOLOGY

Division of Hydro - and Gas - Dynamics, Water Power Laboratory. Tel: 47 7 593856: Fax: 47 7 593854 Alfred Getz eg 4, N7034 Trondheim NTH. Prof. H. Brekke. Numerical analysis and experimental tests of unsteady flow; Numerical analysis and model test of turbomachinery

SOUTH AMERICA

UNIVERSITY OF LA PLATA

Faculty of Engineering, Calle 2/7, No. 200, 1900 - La Plata. Telex: 31151 Rpca. Argentina BULAP AR. Head of Laboratory: Prof. Eng. Fernando Zarate. Hydraulics Laboratory "Guillermo Cèspedes"

UNIVERSIDAD NICIONAL AUTÒNOMA DE MEXICO

Apdo. 770472, Deleg. Coyoacàn 04510, Mexico D.F., Mexico. Head of the Hydromechanical Section: Prof. Eng. A. Palacios. Instituto de Ingenieria

UNIVERSITY OF SAO PAULO

Cidade Universitaria, CXP 11014, 05508, SP, Brasil. Head of the Laboratory: Prof. Eng. Giorgio Brighetti. Centro Tecnològico de Hidràulica

UNIVERSITY OF URUGUAY

Faculty of Engineering, J. Herrera y Reissig 565, Montevideo, Uruguay. Tel: 71 03 61. Head of the Institute: Prof. Dr. Rafael Guarga. Instituto de Mecànica de los Fluìdos e Ingenierìa Ambiental

SPAIN

UNIV. AUTÒNOMA DE BARCELONA

Fac. de Ciencias, Campus Univ. de Bellaterra, Barcelona 08071. Tel: (93) 2038900.

UNIV. DE CÀDIZ
E U Ing. Tècn. Industrial, C/Sacramento, 82, Càdiz 11071. Tel: (956) 224359.

UNIV. POL. DE CANARIAS
E T S Ing. Industriales, Cuarto Pab. Seminario de Tarifa Baja, P G Canaria 35071. Tel: (928) 350004

UNIV. DE CANTABRIA
E T S I Caminos, Canales y Puertos, Avda. de los Castros, s/n, Santander 39071. Tel: (942) 275600.

UNIV. DE CAST - LA MANCHA
E U Ing. Tècn. Industrial, Avda. de Portugal, s/n, Toledo 45071. Tel: (925) 223400.

UNIV. POL. DE CATALUNYA
E T S Ing. Caminos, Canales y Puertos, C/Jordi Girona, 31, Barcelona 08071. Tel: (93) 2048252

UNIV. PONTIF. DE COMILLAS
E T S Ing. Industriales, Alberto Aguilera, 23, Madrid 28015. Tel: (91) 2483600

UNIV. SAND. DE COMPOSTELA
E T S Ing. Industriales, C/La Paz, s/n, Vigo 36071. Tel: (986) 373012

UNIV. DE CÒRDOBA
E U Ing. Tècn. Industrial, Avda. Menendez Pidal, s/n, Còrdoba 14071. Tel: (957) 291555

UNIV. NAC. EDUC. DISTANCIA
E T S Ing. Industriales, Ciudad Universitaria - UNED, Madrid 28071. Tel: (91) 4493600

UNIV. DE EXTREMADURA
E U Pol. de Mèrida, C/Calvario, 2, Mèrida 06071. Tel: (924) 318712

UNIV. DE GRANADA
E U Ing. Tècn. Industrial, Avda. de Madrid, 35, Jaen 23071. Tel: (953) 250448

UNIV. DE LEÒN
E U Ing. Tècn. Industrial, C/Jesùs Rubio, 2, Leòn 24071. Tel: (987) 204052

UNIV. POL. DE MADRID
E T S Ing. Industriales, C/Josè Gutierrez Abascal, 2, Madrid 28071. Tel: (91) 2626200

UNIV. DE MÀLAGA
E U Ing. Tècn. Industrial, Plaza El Ejido, s/n, Màlaga 29071. Tel: (952) 250100

UNIV. DE MURCIA
E U Politècnica, P. Alfonso XIII, 34, Caragena 30071. Tel: (968) 527378

UNIV. DE NAVARRA
E T S Ing. Industriales, Urdaneta, 7, San Sebastian 20006. Tel: (943) 466411

UNIV. DE OVIEDO
E T S Ing. Industriales, Ctra. de Castiello, s/n, Gijòn 33071. Tel: (985) 338680

UNIV. DE SALAMANCA
E U Ing. Tècn. Industrial, Av. Ferna 'ndo Ballesteros, s/n, Bejar 37071. Tel: (923) 402416

UNIV. DE SEVILLA
E T S Ing. Industriales, Avda. Reina Mercedes, s/n, Sevilla 41071. Tel: (954) 611150

UNIV. POL. DE VALENCIA
E T S Ing. Industriales, Camino de Vera, s/n, Valencia 46071. Tel: (96) 3699862

UNIV. DE VALLADOLID

E T S Ing. Industriales, Avda. Santa Teresa, 30, Valladolid 47071. Tel: (983) 353608

UNIV. DEL PAIS VASCO

E U Ing. Tècn. Ind. (CEI), Avda. Bilbao, 29, Eibar 20071. Tel: (943) 718444

UNIV. DE ZARAGOZA

E T S Ing. Industriales, Ciudad Universitaria, Zaragoza 50071. Tel: (976) 350058

SWITZERLAND

SWISS FEDERAL INSTITUTE OF TECHNOLOGY

Hydraulic Machines and Fluid Mechanics Institute, 1015 Lausanne. Tel: 19 06 1988: Fax: 19 06 1988: Telex: 455 806. Prof. P Henry. Numerical analysis of flow in turbomachines, cavitation, boundary layer, friction losses, unsteady behaviour of hydraulic machines

U. K.

UNIVERSITY OF BIRMINGHAM

Chemical Engineering, PO Box 363, Birmingham B15 2TT. Tel: 021-472-1301 Ext. 2105. Dr N Thomas. Hydrodynamics, multi-phase flows, fluid mechanics

UNIVERSITY OF BRADFORD

Civil Engineering, Bradford, West Yorkshire BD7 1DP. Tel: 0274-733466 Ext. 8391. Prof. R A Falconer. Modelling, hydrodynamics, turbulence, water quality, tides, sediment transport, dispersion, circulation

UNIVERSITY OF BRISTOL

Civil Engineering, University Walk, Bristol, Avon BS8 1TR. Tel: 0272-303280. Dr R H J Sellin. Rivers, drag reduction, hydraulic models, flood plains, sewers

CAMBRIDGE UNIVERSITY

Engineering, Trumpington Street, Cambridge CB2 1PZ. Tel: 0223-332632. Dr J F A Sleath. Beaches, coastal engineering, erosion, offshore structures, pipelines, sediment transport

CITY UNIVERSITY

Dept of mechanical engineering, London EC1. Dr P A Lush. Cavitation, fluid machines

CRANFIELD INSTITUTE OF TECHNOLOGY

School of mechanical engineering, Cranfield, Bedford MK43 0AL. Tel: 0234 750111: Fax: 0234 750728. Prof. R Elder. Computer-aided pump studies

UNIVERSITY OF DUNDEE

Civil Engineering, Dundee, Scotland DD1 4HN. Tel: 0382-23181 Ext. 4340. Prof. A E Vardy. Unsteady flow, pressure transients, surge, numerical analysis

UNIVERSITY OF GLASGOW

Civil Engineering, Oakfield Avenue, Glasgow, Scotland G12 8QQ. Tel: 041-8855 Ext. 7210. Dr D A Ervine. Aeration, air entrainment, channels, waterways dams, floods, spillways

IMPERIAL COLLEGE OF SCIENCE AND TECHNOLOGY

Civil Engineering, Imperial College Road, London SW7 2BU. Tel: 01-589-5111 Ext. 4864. Dr. J D Hardwick. Vibration, hydraulic structures, hydroelastic modelling

KING'S COLLEGE LONDON

Civil Engineering, Strand, London WC2R 2LS. Tel: 01-836-5454 Ext. 2723. Mr J H Loveless. Sediment, drainage conduits, hydraulic structures cavitation, coastal hydraulics, air-entrainment

UNIVERSITY OF LIVERPOOL

Civil Engineering, Brownlow Street, PO Box 147, Liverpool L69 3BX. Tel: 051-709-6022 Ext. 2461. Mr T S Hedges. Hydrodynamics, waves, wave-current interaction, nearshore processes, mathematical modelling

UNIVERSITY COLLEGE LONDON

Civil and Municipal Engineering, Gower Street, London WC1E 6BT. Tel: 01-387-7050 Ext. 2709. Dr A J Grass. Boundary layer turbulence, sediment transport, fluid loading on structures and pipelines

LOUGHBOROUGH UNIVERSITY OF TECHNOLOGY

Dept. of Mechanical Engineering, Loughborough, Leics LE11 3TU. Tel: 0509 223206: Telex: 34319: Fax: 0509 232029. Mr R K Turton, Senior Lecturer. Gas/liquid pumping: Pump inducer studies

LOUGHBOROUGH UNIVERSITY OF TECHNOLOGY

WEDC Dept of Civil Engineering, Loughborough, Leics LE11 3TU. Tel: 0509 222390: Telex: 34319. Prof. J A Pickford. Water engineering for developing countries

UNIVERSITY OF NEWCASTLE UPON TYNE

Civil Engineering, Newcastle upon Tyne NE1 7RU. Tel: 091-232-8511 Ext. 2399. Dr C Nalluri. Sediment transport, flood channels, sewers, sea outfalls

UNIVERSITY OF NEWCASTLE UPON TYNE

Mechanical Engineering, Stephenson Building, Claremont Road, Newcastle upon Tyne NE1 7RU. Tel: 091-232-8511. Dr A Anderson. Cavitation, hydroelectric power, penstocks, pumped storage, pumps, turbines, valves, water hammer

UNIVERSITY OF NOTTINGHAM

Civil Engineering, University Park, Nottingham, Nottinghamshire Ng7 2RD. Tel: 0602-566101 Ext. 3537. Dr C J Baker. Sediment transport, sandwaves (fluvial and aeolian), Scour around structures

UNIVERSITY OF NOTTINGHAM

Dept of Mechanical Engineering, University Park, Nottingham. Dr A Lichtarowicz. Cavitation, erosion studies

UNIVERSITY OF OXFORD

Engineering Science, Parks Road, Oxford OX1 3PJ. Tel: 0865-273000. Dr. R E Franklin. Cavitation, inception, noise, nuclei, bubble dynamics, bubbly flows, gas content

UNIVERSITY OF READING

Engineering, Whiteknights Park, Reading, Berkshire RG6 2AY. Tel: 0734-875123 Ext. 7315. Dr J D Burton. Turbine, pump, installation, micro-hydro, inertia flow, alternating flow hydraulics

UNIVERSITY OF SALFORD

Civil Engineering, Salford, Lancashire M5 4WT. Tel: 061-736-5843 Ext. 7116. Dr R Baker. Concrete revetment blocks, spillways, models, sediment transport

UNIVERSITY OF SALFORD

Civil Engineering, Salford, Lancashire M5 4WT. Tel: 061-736-5843 Ext. 7122. Prof.

E M Wilson. Hydrology, hydroelectricity, tidal energy

UNIVERSITY OF SHEFFIELD

Civil and Structural Engineering, Mappin Street, Sheffield S1 3JD. Tel: 0742-768555 Ext. 5418 or 5059. Dr F A Johnson. Dams, failurès, floods, routing, waves, reservoirs control

UNIVERSITY OF SOUTHAMPTON

Dept. of Mechanical Engineering, Southampton, Hants SO9 5NH. Prof. S P Hutton, Prof. M Thew. Fluid metering, fluid machines

UNIVERSITY OF STIRLING

Environmental Science, Stirling, Scotland FK9 4LA. Tel: 0786-73171. Dr R I Ferguson. Sediment transport, gravel, non-uniform flow, braided rivers, field measurement

UNIVERSITY OF STRATHCLYDE

Civil Engineering, 107 Rottenrow, Glasgow, Scotland G4 0HG. Tel: 041-552-4400 Ext. 3168. Prof. G Fleming. Hydrology, erosion, simulation, management, sedimentation, reservoirs, land-use, dredging

THAMES POLYTECHNIC

Civil Engineering, Oakfield Lane, Dartford, Kent DA1 2SZ. Tel: 0322-21328 Ext. 318. Mr A Grant. Sediment transport, urban drainage

UNIVERSITY OF WARWICK

Dept of Engineering, Hydrotransient Simulation Unit, Coventry CV4 7AL. Tel: 0203-523086. Dr A P Boldy. Simulation, transients, turbines, hydroelectric

UWIST

Civil Engineering and Building Technology, Colum Drive, Cardiff CF1 3EU. Tel: 0222-42588 Ext. 2802. Dr P W France. Finite difference, hydraulic structures, flow measurement, weirs

U S A

CALTECH, DIVISION OF ENGINEERING AND APPLIED SCIENCE

Thomas Laboratory, 104-44 Pasadena, California 91125. Tel: (818) 356-4106 M E Dept.: Fax: (818) 568-2719. A J Acosta, C E Brennen. Experimental work on cavitation, rotor dynamic forces, unsteady flow effects on axial and centrifugal hydraulic machines, primarily pumps. Rotor-stator blade interactions, inlet shear flows, shroud boundary layer flow. The experimental test facility is capable of cavitation and dynamic testing at rotor power levels of about 15 kW with rotational speeds up to 7000 rpm. Unsteady force and flow instrumentation.

GEORGIA INSTITUTE OF TECHNOLOGY

School of Civil Engineering, Atlanta, Georgia 30332, U. S. A. Tel: 404-894-2224: Telex: 542507: Fax: 404-894-2224. Professor C Samuel Martin. Research and consultation regarding pump-turbine characteristics and hydraulic transient analysis

UNIVERSITY OF MINNESOTA

St Anthony Falls Hydraulic Laboratory, Mississippi River at 3rd Avenue S E. Tel: (612) 627-4010: Fax: (612) 627-4609. Dr Roger E A Arndt, Charles C S Song, John Gulliver. An independent turbine test stand for turbine acceptance testing. Physical and mathematical modelling works on intake structure, turbine rotor and stator, draft tube, cavitation and bubble dynamics are being conducted. Mathematical model for economic evaluation of small hydropower development.

U S S R

MOSCOW INSTITUTE OF HYDROTECHNICAL ENGINEERING AND LAND RECLAMATION

Moscow 127550, Prianisnikova St, 19. Tel: 216-11-85. Prof. Vadim Phirsovitch Chebaevski. Cavitation in blade pumps; design of reclamation pumping plants

RESEARCH INSTITUTE OF POWER MACHINE-BUILDING OF THE MOSCOW HIGHER TECHNICAL SCHOOL

Moscow 107005, Second Bauman St, 5. Tel: 261-59-89. Vladimir Ivanovitch Petrov, Senior Researcher. Cavitation in blade pumps

WEST GERMANY

TECHNISCHE UNIVERSITÄT BERLIN,

Dept. of Civil Eng. Institut für Wasserbau und Wasserwirtschaft, Strasse des 17. Juni 142-144, 1000 Berlin 12. Prof. Dr Ing. P Franke. Numerical analysis

TECHNISCHE UNIVERSITÄT BERLIN,

Dept. of Mech. Eng. Lehrstuhl für Maschinenkonstruktionen, Hydraulische Maschinen und Anlagen, Steinplatz 1, 1000 Berlin 1. Prof. Dr Ing. E Siekmann. Numerical analysis and measurement of flow in turbomachines

TECHNISCHE UNIVERSITÄT BOCHUM

Lehrstuhl
für Regelsysteme und Steuerungstechnik, Dept. of Mech. Eng. Postfach 10 21 48, 4630 Bochum. Prof. Dr Ing. K H Fasol. Numerical analysis

TECHNISCHE HOCHSCHULE BRAUNSCHWEIG PFLEIDERER

Institut für Strömungsmaschinen, Dept. of Mech. Eng. Langer Kamp 6, 3300 Braunschweig. Prof. Dr Ing. H Petermann, Prof. Dr Ing. G Kosyna. Numerical analysis and measurement of flow in turbomachines

TECHNISCHE HOCHSCHULE DARMSTADT

Institut für Hydraulische Machinen, Dept. of Mech. Eng. Magdalenenstr. 8 - 10, 6100 Darmstadt. Prof. Dr Ing. B Stoffel. Numerical analysis and measurement of flow in turbomachines

TECHNISCHE UNIVERSITÄT HANNOVER

Institut für Strömungsmechanik, Dept. of Civil Eng. Callinstr. 32, 3000 Hannover. Prof. Dr Ing. W Zielke. Numerical analysis

TECHNISCHE UNIVERSITÄT HANNOVER

Institut für Strömungsmaschinen, Dept. of Mech. Eng. Appelstr. 9, 3000 Hannover. Prof. Dr Ing. M Rautenberg. Numerical analysis and measurement of flow in turbomachines

TECHNISCHE UNIVERSITÄT KAISERSLAUTERN,

Dept. of Mech. Eng. Erwin-Schrödinger

strasse, 6700 Kaiserslautern. Prof. Dr Ing. F Eisfeld. Numerical analysis and measurement of flow in turbomachines

TECHNISCHE UNIVERSITÄT KARLSRUHE

Institut für Hydromechanik, Dept. of Civil Eng. Kaiserstr. 12, 7500 Karlsruhe 1. Prof. Dr Ing. E Naudascher, Prof. Dr Ing. H Thielen. Numerical analysis and measurement of flow

TECHNISCHE UNIVERSITÄT KARLSRUHE,

Deptartment of Mechanical Eng. Institut für Strömungslehre und Strömungsmaschinen, Kaiserstr. 12, 7500 Karlsruhe 1. Prof. Dr Ing. K O Felsh, Dr Ing. J Zierep. Numerical analysis and measurement of flow in turbomachines

TECHNISCHE UNIVERSITÄT MÜNCHEN,

Dept. of Civil Eng. Lehrstuhl für Hydraulik und Gewässerkunde, Arcisstr. 21, D8000 München 2. Prof. Dr Ing. F Valentin. Numerical analysis

TECHNISCHE UNIVERSITÄT MÜNCHEN,

Dept. of Mech. Eng. Lehrstuhl und Laboratorium für Hydraulische Maschinen und Anlagen, Arcisstr. 21, D8000 München 2. Tel: (089) 2105 3453. Prof. Dr Ing. Habil Joachim Raabe. Analysis and dynamic measurement of quick response of real flow in diffuser, draft tube, rotor channel and pipe, including waterhammer and two-phase flow

TECHNISCHE UNIVERSITÄT MÜNCHEN,

Dept. of Mech. Eng. Lehrstuhl und Laboratorium für Hydraulische Maschinen und Anlagen, Arcisstr. 21, D8000 München 2. Prof. Dr Ing. R Schilling. Numerical analysis and measurement of flow in turbomachines

UNIVERSITÄT STUTTGART

Institut für Hydraulische Strömungsmaschinen, Dept. of Mech. Eng. Pfaffenwaldring 10, 7000 Stuttgart 80. Prof. Dr Ing. G Lein. Numerical analysis and measurement of flow in turbomachines

UNIVERSITÄT STUTTGART

Institut für Wasserbau, Dept of Civil Eng. Pfaffenwaldring 10, 7000 Stuttgart 80. Prof. Dr Ing. J Giesecke, Dr Ing. H B Horlacher. Numerical analysis and measurement of flow

YUGOSLAVIA

UNIVERSITY OF BELGRADE

Faculty of Mechanical Engineering, 27 Marta 80, 11000 Belgrade. Tel: 329 021. Prof. Stanislav Pejovic. Education, test facilities, turbines, pumps, transients, hydropower and pumping systems, measurements, consultancy.

For Product Safety Concerns and Information please contact our
EU representative GPSR@taylorandfrancis.com Taylor & Francis
Verlag GmbH, Kaufingerstraße 24, 80331 München, Germany